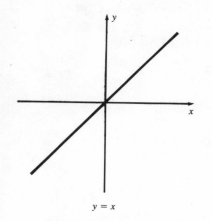

$y = x$

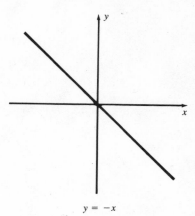

$y = -x$

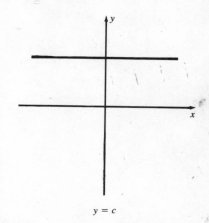

$y = c$

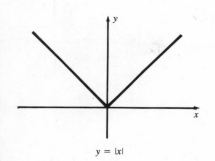

$y = |x|$

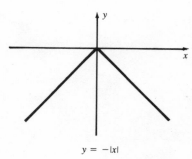

$y = -|x|$

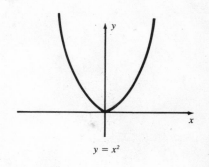

$y = x^2$

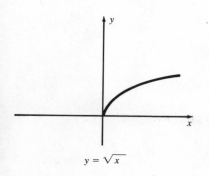

$y = \sqrt{x}$

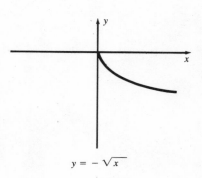

$y = -\sqrt{x}$

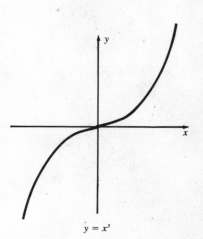

$y = x^3$

PRECALCULUS

FUNCTIONS & GRAPHS

PRECALCULUS
FUNCTIONS & GRAPHS

BERNARD KOLMAN
Drexel University

ARNOLD SHAPIRO
Drexel University

ACADEMIC PRESS, INC.
(Harcourt Brace Jovanovich, Publishers)
Orlando San Diego San Francisco New York London
Toronto Montreal Sydney Tokyo São Paulo

TO THE MEMORY OF LILLIE
B. K.

Cover art: *Galaxy* by Vasarely
courtesy of Vasarely Center,
New York

Academic Press, Inc.
Orlando, Florida 32887

United Kingdom Edition Published by
Academic Press, Inc. (London) Ltd.
24/28 Oval Road, London NW1 7DX

ISBN: 0-12-417894-4
Library of Congress Catalog Card Number: 83-71284

Printed in the United States of America

CONTENTS

PREFACE

The past several decades have seen a growing awareness of the need for comprehension of advanced mathematical concepts as background to many disciplines. As a result, many colleges are requiring an increasing percentage of students to study calculus. Our goal has been to provide a complete and self-contained presentation of the basic mathematical techniques and ideas required for the successful completion of a calculus course.

Our classroom experience has shown that understanding the concept of a function, using function notation, and being able to sketch graphs of functions with ease constitute the most important group of skills that students must master in preparation for the study of calculus. Chapter 2, entitled "Functions and Graphs," forms the cornerstone of this book. Attention is given not only to the function concept and function notation, but also to graphing techniques. We have insisted that the student become familiar with the "standard" graphs, and we have included techniques for quickly sketching a wide variety of functions (see "Aids in Graphing").

We have employed in this book many of the pedagogic devices that instructors have found useful in our other introductory college mathematics texts, and we have added a few new stimuli.

PRESENTATION The style is informal, supportive, "user-friendly." Many algebraic procedures are described with the aid of a "split screen" that displays simultaneously both the steps of an algorithm and a worked-out example.

PROGRESS CHECKS At carefully selected places, problems similar to those worked in the text have been inserted (with answers) to enable the student to test his or her understanding of the material just described.

WARNINGS To help eliminate misconceptions and prevent bad mathematical habits, we have inserted numerous **Warnings** (indicated by the symbol shown in the margin) that point out the incorrect practices most commonly found in homework and exam papers.

FEATURES In each chapter we have inserted one or two features, elements that are independent of the text yet are often related to the mathematical concepts. The features are intended to catch the attention of the student and heighten interest in the material. (We hope the features will provide interesting reading for the instructor as well.)

END-OF-CHAPTER
MATERIAL   Every chapter contains a summary, including

> *Terms and Symbols*   with appropriate page references;
>
> *Key Ideas for Review*   to stress the concepts;
>
> *Review Exercises*   to provide additional practice;
>
> *Progress Tests*   to provide self-evaluation and reinforcement.

ANSWERS   The answers to all **Review Exercises** and **Progress Tests** appear in the back of the book.

SOLUTIONS   Worked-out solutions to selected **Review Exercises** appear in a separate section at the back of the book. The solved problems provide one more level of reassurance to the student using the **Review Exercises** in preparation for the **Progress Tests.**

EXERCISES   Abundant, carefully graded exercises provide practice in the mechanical and conceptual aspects of algebra and trigonometry. Exercises requiring a calculator are indicated by the symbol shown in the margin. Answers to odd-numbered exercises appear at the back of the book. Answers to even-numbered exercises appear in the *Instructor's Manual*. The *Instructor's Manual* and an extensive *Test Bank* are available to the instructor on request.

ACKNOWLEDGMENTS

We thank the following for their review of the manuscript and their helpful comments: Professor Terry Herdman at Virginia Polytechnic Institute and State University, Professor Monty Strauss at Texas Tech University, Professor Patricia Lock at St. Lawrence University, Professor William Blair at Northern Illinois University, and Professor Jack Goldberg at the University of Michigan.

We would also like to thank the following for solving the exercises, proofreading, and providing sketches of the graphs for the answer section: Michael Carchidi, Carol Gray, Stephen Kolman, Isadora Kubert, Nancy Rorabaugh, Todd Rimmer, and Jacqueline Shapiro. A special thanks is due Dwight Spence of Marymount College for his thorough reading of the text and checking of all examples, progress checks, and worked-out solutions.

Finally, our grateful thanks to the staff of Academic Press for their ongoing commitment to excellence.

TO THE STUDENT

If you are like the great majority of students using this book, you are doing so to prepare for a calculus course. Calculus was developed in the seventeenth century by Sir Isaac Newton in England and by Gottfried von Leibniz in Germany. One of the outstanding developments in the history of science, calculus is the mathematics of change. Since everything around us is constantly changing, calculus is used in the solution of many commonplace problems in almost every discipline.

Contrary to what you may have heard, calculus is not a difficult subject, *provided you have prepared yourself* by learning the needed algebra and trigonometry. This book provides you with all the background material necessary for the successful study of calculus. This book was written for you. It gives you every possible chance to succeed—if you use it properly.

We would like to have you think of mathematics as a challenging game—but not as a spectator sport. This wish leads to our primary rule: *Read this textbook with pencil and paper handy*. Every new idea or technique is illustrated by fully worked-out examples. As you read the text, carefully follow the examples and then do the **Progress Checks.** The key to success in a math course is working problems, and the **Progress Checks** are there to provide immediate practice with the material you have just learned.

Your instructor will assign homework from the extensive selection of exercises that follows each section in the book. *Do the assignments regularly, thoroughly, and independently*. By doing lots of problems you will develop the necessary skills in algebra, and your confidence will grow. Since algebraic techniques and concepts build on previous results, you can't afford to skip any of the work.

To help prevent or eliminate improper habits and to help you avoid those errors that we see each semester as we grade papers, we have interspersed **Warnings** throughout the book. The **Warnings** point out common errors and emphasize the proper method.

There is important review material at the end of each chapter. The **Terms and Symbols** should all be familiar by the time you reach them. If your understanding of a term or symbol is hazy, use the page reference to find the place in the text where it is introduced. Go back and read the definition.

It is possible to become so involved with the details of techniques that you lose track of the broader concepts. The list of **Key Ideas for Review** at the end of each chapter will help you focus on the principal ideas.

The **Review Exercises** can be used as part of your preparation for examinations. The section covering each exercise is indicated so that, if needed, you can go back to restudy the material. Answers to all review exercises are in the **Answers** section in the back of the book; solutions to the review exercises whose numbers are in color are in the **Solutions** section, also in the back of the book.

You are then ready to try **Progress Test A.** By checking your answers against the ones in the **Answers** section, you will soon pinpoint your weak spots. Go back for further review and more exercise in those areas. Only then should you proceed to **Progress Test B.**

We believe that the eventual ''payoff'' in studying calculus is an improved ability to tackle practical problems in your field of interest. Since the mathematics presented in this book is widely used in almost all fields and is the basic stepping-stone to success in calculus, the study of this material is well worth your effort.

PRECALCULUS
FUNCTIONS & GRAPHS

CHAPTER 1

THE FOUNDATIONS OF ALGEBRA

Algebra provides us with the basic tools used in all branches of mathematics. A firm grasp of algebraic concepts and techniques is essential for the student who intends to begin a study of calculus, the gateway to advanced mathematics.

It is easy to show that $1 + 2 = 2 + 1$ by using specific items, such as matchsticks or marbles. Algebra enables us to abstract this result and to say that

$$x + y = y + x$$

for all positive integers. The key to this general rule is the introduction of symbols such as x and y to represent a variety of numbers. Symbols are the language of algebra. They allow us to write relationships in a symbolic manner, which facilitates the process of finding solutions.

In general, the numbers denoted by a symbol are restricted to a particular number system, and this is where our study begins. We will also review the fundamentals of polynomials, exponents, and radicals. Much of the work of this chapter is devoted to a study of inequalities, since this topic is not often treated adequately in secondary school mathematics.

1.1
THE REAL NUMBER SYSTEM

SETS

We will need to use the notation and terminology of sets from time to time. A **set** is simply a collection of objects or numbers, which are called the **elements** or **members** of the set. The elements of a set are written within braces so that

$$A = \{4, 5, 6\}$$

tells us that the set A consists of the numbers 4, 5, and 6. If every element of set A is also a member of set B, then A is a **subset** of B. For example, the set of all robins is a subset of the set of all birds.

THE REAL NUMBER SYSTEM

Much of our work in algebra deals with the real number system and we now review the composition of this number system.

The numbers 1, 2, 3, . . ., used for counting, form the set of **natural numbers**. The set of **integers**

$$. . ., -2, -1, 0, 1, 2, . . .$$

consists of the positive integers (the natural numbers), the negative integers, and zero.

The set of **rational numbers** consists of those numbers that can be written as a ratio of two integers p/q where q is not equal to zero. Examples of rational numbers are

$$0 \qquad \frac{2}{3} \qquad -4 \qquad \frac{7}{5} \qquad \frac{-3}{4}$$

We make use of rational numbers when we divide two apples equally among four people, since each person then gets half, or $\frac{1}{2}$, an apple. By writing an integer n in the form $n/1$, we see that every integer is a rational number. The decimal number 1.3 is also a rational number, since $1.3 = \frac{13}{10}$.

We have now seen three fundamental number systems: the natural number system, the system of integers, and the rational number system. Each later system includes the previous system or systems, and each is more complicated than the one before. However, the rational number system is still inadequate for sophisticated uses of mathematics, since there exist numbers that are not rational, that is, numbers that cannot be written as the ratio of two integers. These are called **irrational numbers**. It can be shown that the number a that satisfies $a \cdot a = 2$ is such a number. The number π, which is the ratio of the circumference of a circle to its diameter, is also such a number.

The decimal form of a rational number will either terminate, as

$$\frac{3}{4} = 0.75 \qquad -\frac{4}{5} = -0.8$$

or will form a repeating pattern, as

$$\frac{2}{3} = 0.666 \ldots \qquad \frac{1}{11} = 0.090909 \ldots \qquad \frac{1}{7} = 0.1428571 \ldots$$

Remarkably, the decimal form of an irrational number *never* forms a repeating pattern.

The rational and irrational numbers together form the **real number system**.

THE REAL NUMBER LINE

There is a simple and very useful geometric interpretation of the real number system. Draw a horizontal straight line; pick a point on this line, label it with the number 0, call it the origin, and denote it by O. Designate the side to the right of the origin as the **positive direction** and the side to the left as the **negative direction**.

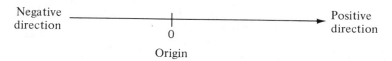

Next, select a unit of length for measuring distance. With each positive real number r we associate the point that is r units to the right of the origin, and with each negative number $-r$ we associate the point that is r units to the left of the origin. Thus, the set of real numbers is identified with all possible points on a straight line. For every point on the line there is a real number and for every real number there is a point on the line. The line is called the **real number line,** and the number associated with a point is called its **coordinate**. We can now show some points on this line.

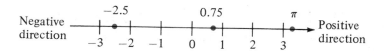

The numbers to the right of zero are called **positive,** the numbers to the left of zero are called **negative**. The positive numbers and zero together are called the **nonnegative** numbers.

We will frequently turn to the real number line to help us picture the results of algebraic computations.

INEQUALITIES

If a and b are real numbers, we can compare their positions on the real number line by using the relations of **less than, greater than, less than or equal to,** and **greater than or equal to,** denoted by the **inequality symbols** $<$, $>$, $\leq$, and $\geq$, respectively. Table 1 describes both algebraic and geometric interpretations of the inequality symbols.

TABLE 1

Algebraic Statement	Equivalent Statement	Geometric Statement
$a > 0$	a is positive	a lies to the right of the origin
$a < 0$	a is negative	a lies to the left of the origin
$a > b$	$a - b$ is positive	a lies to the right of b
$a < b$	$a - b$ is negative	a lies to the left of b
$a \geq b$	$a - b$ is zero or positive	a coincides with b or lies to the right of b
$a \leq b$	$a - b$ is zero or negative	a coincides with b or lies to the left of b

Expressions involving inequality symbols, such as $a < b$ and $a \geq b$, are called **inequalities**. We often combine these expressions so that $a \leq b < c$ means both $a \leq b$ and $b < c$. For example, $-5 \leq x < 2$ is equivalent to $-5 \leq x$ and $x < 2$.

PROGRESS CHECK

Verify that the following inequalities are true by using either the "Equivalent Statement" or "Geometric Statement" of Table 1.

(a) $-1 < 3$ (b) $2 \leq 2$ (c) $-2.7 < -1.2$

(d) $-4 < -2 < 0$ (e) $-\dfrac{7}{2} < \dfrac{7}{2} < 7$

The real numbers satisfy the following useful properties of inequalities.

Properties of Inequalities

Let a, b, and c be real numbers.
(a) One and only one of the following relations holds:

$$a < b, \ a > b, \ a = b \quad \textbf{Trichotomy property}$$

(b) If $a < b$ and $b < c$, then $a < c$. **Transitive property**
(c) If $a < b$, then $a + c < b + c$.
(d) If $a < b$ and $c > 0$, then $ac < bc$. When an inequality is multiplied by a positive number, the sense of the inequality is preserved.
(e) If $a < b$ and $c < 0$, then $ac > bc$. When an inequality is multiplied by a negative number, the sense of the inequality is reversed.

EXAMPLE 1
(a) Since $-2 < 4$ and $4 < 5$, then $-2 < 5$.
(b) Since $-2 < 5$, $-2 + 3 < 5 + 3$, or $1 < 8$.

(c)   Since $3 < 4$, $3 + (-5) < 4 + (-5)$, or $-2 < -1$.
(d)   Since $2 < 5$, $2(3) < 5(3)$, or $6 < 15$.
(e)   Since $-3 < 2$, $(-3)(-2) > 2(-2)$, or $6 > -4$.

ABSOLUTE VALUE

When we are interested in the magnitude of a number a, and don't care about the direction or sign, we use the concept of **absolute value,** which we write as $|a|$. The formal definition of absolute value is stated this way.

$$|a| = \begin{cases} a & \text{if} \quad a \geq 0 \\ -a & \text{if} \quad a < 0 \end{cases}$$

Since distance is independent of direction and is always nonnegative, we can view $|a|$ as the distance from the origin to either point a or point $-a$ on the real number line.

EXAMPLE 2
(a) $|4| = 4$ $|-4| = 4$ $|0| = 0$
(b)   The distance on the real number line between the point labeled 3.4 and the origin is $|3.4| = 3.4$. Similarly, the distance between point -2.3 and the origin is $|-2.3| = 2.3$.

In working with the notation of absolute value, it is important to perform the operations within the bars first. Here are some examples.

EXAMPLE 3
(a) $|2 - 5| = |-3| = 3$ (b) $|3 - 5| - |8 - 6| = |-2| - |2| = 2 - 2 = 0$

The following properties of absolute value follow from the definition.

Properties of Absolute Value	For all real numbers a and b, (i) $|a| \geq 0$ (ii) $|a| = |-a|$ (iii) $|a - b| = |b - a|$

We began by showing a use for absolute value in denoting distance from the origin without regard to direction. We will conclude by demonstrating the use of absolute value to denote the distance between *any* two points a and b on the real number line. In Figure 1, the distance between the points labeled 2 and 5 is 3 units and can be obtained by evaluating either $|5 - 2|$ or $|2 - 5|$. Similarly, the

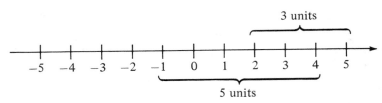

FIGURE 1

distance between the points labeled -1 and 4 is given by either $|4 - (-1)| = 5$ or $|-1 - 4| = 5$. Using the notation $\overline{AB}$ to denote the distance between the points A and B, we provide the following definition.

Distance on the Real Number Line	The **distance** $\overline{AB}$ between points A and B on the real number line, whose coordinates are a and b, respectively, is given by $$\overline{AB} = |b - a|$$

Property (iii) then tells us that $\overline{AB} = |b - a| = |a - b|$. Viewed another way, Property (iii) states that the distance between any two points on the real number line is independent of the direction.

EXAMPLE 4

Let points A, B, and C have coordinates -4, -1, and 3, respectively, on the real number line. Find the following distances.
(a) $\overline{AB}$ (b) $\overline{CB}$ (c) $\overline{OB}$

SOLUTION

Using the definition, we have
(a) $\overline{AB} = |-1 - (-4)| = |-1 + 4| = |3| = 3$
(b) $\overline{CB} = |-1 - 3| = |-4| = 4$
(c) $\overline{OB} = |-1 - 0| = |-1| = 1$

PROGRESS CHECK

The points P, Q, and R on the real number line have coordinates -6, 4, and 6, respectively. Find the following distances.
(a) $\overline{PR}$ (b) $\overline{QP}$ (c) $\overline{PQ}$

ANSWERS

(a) 12   (b) 10   (c) 10

APPENDIX TO 1.1
PROPERTIES OF
THE REAL NUMBERS

With respect to the operations of addition and multiplication, the real number system has properties that are fundamental to algebra. The letters a, b, and c will denote real numbers.

Closure

Property 1. The sum of a and b, denoted by $a + b$, is a real number.
Property 2. The product of a and b, denoted by $a \cdot b$ or ab, is a real number.

We say that the set of real numbers is **closed** with respect to the operations of addition and multiplication, since the sum and product of two real numbers are also real numbers.

Commutative Laws

Property 3. $a + b = b + a$ **Commutative law of addition**
Property 4. $ab = ba$ **Commutative law of multiplication**

That is, we may add or multiply real numbers in any order.

Associative Laws

Property 5. $(a + b) + c = a + (b + c)$ **Associative law of addition**
Property 6. $(ab)c = a(bc)$ **Associative law of multiplication**

That is, when adding or multiplying real numbers we may group them in any order.

Identities

Property 7. There is a unique real number, denoted by 0, such that $a + 0 = 0 + a = a$ for every real number a.
Property 8. There is a unique real number, denoted by 1, such that $a \cdot 1 = 1 \cdot a = a$ for every real number a.

The real number 0 of Property 7 is called the **additive identity;** the real number 1 of Property 8 is called the **multiplicative identity.**

Inverses

Property 9. For every real number a, there is a unique real number, denoted by $-a$, such that

$$a + (-a) = (-a) + a = 0$$

Property 10. For every real number $a \neq 0$, there is a unique real number, denoted by $1/a$, such that

$$a\left(\frac{1}{a}\right) = \left(\frac{1}{a}\right)a = 1$$

The number $-a$ of Property 9 is called the **negative** or **additive inverse** of a. The number $1/a$ of Property 10 is called the **reciprocal** or **multiplicative inverse** of a.

Distributive Laws	*Property 11.* $a(b + c) = ab + ac$ *Property 12.* $(a + b)c = ac + bc$

EQUALITY

When we say that two numbers are **equal,** we mean that they are identical. Thus, when we write

$$a = b$$

(read "a equals b"), we mean that a and b represent the same number. For example, $2 + 5$ and $4 + 3$ are different ways of writing the number 7, so we can write

$$2 + 5 = 4 + 3$$

Equality satisfies four basic properties.

Properties of Equality	Let a, b, and c be elements of a set. **1.** $a = a$ **Reflexive property** **2.** If $a = b$, then $b = a$. **Symmetric property** **3.** If $a = b$ and $b = c$, then $a = c$. **Transitive property** **4.** If $a = b$, then a may be replaced by b in any statement that involves a or b. **Substitution property**

THEOREMS

Using Properties 1–12, the properties of equality, and rules of logic, we can *prove* many other properties of the real numbers.

Theorem 1 If a, b, and c are real numbers, and $a = b$, then   (a) $a + c = b + c$ (b) $ac = bc$

This theorem, which will be used often in working with equations, allows us to add the same number to both sides of an equation and to multiply both sides of an equation by the same number. We will prove Theorem 1a and leave the proof of Theorem 1b as an exercise.

PROOF OF THEOREM 1a	
Statement	**Reason**
$a + c$ is a real number	Closure property
$a + c = a + c$	Reflexive property
$a + c = b + c$	Substitution property with $a = b$

The following theorem is the converse of Theorem 1.

Cancellation Laws	**Theorem 2** Let a, b, and c be real numbers. (a) If $a + c = b + c$, then $a = b$. (b) If $ac = bc$ and $c \neq 0$, then $a = b$.

Part b of Theorem 2 is often called the **cancellation law of multiplication.** We'll prove this theorem to offer another example of the method to be used.

PROOF OF THEOREM 2b	
Statement	**Reason**
ac, bc are real numbers	Closure
$\dfrac{1}{c}$ is a real number	Inverse
$(ac)\left(\dfrac{1}{c}\right) = (bc)\left(\dfrac{1}{c}\right)$	Theorem 1b
$a\left(c \cdot \dfrac{1}{c}\right) = b\left(c \cdot \dfrac{1}{c}\right)$	Associative law
$a \cdot 1 = b \cdot 1$	Multiplicative inverse
$a = b$	Multiplicative identity

We can restate part a of Theorems 1 and 2 in this way: If a, b, and c are real numbers, then $a + c = b + c$ if and only if $a = b$. The connector "if and only if" is used to indicate that both statements are true or both statements are false.

Theorem 3 Let a and b be real numbers.   (a) $a \cdot 0 = 0 \cdot a = 0$ (b) If $ab = 0$, then $a = 0$ or $b = 0$.

The real numbers a and b are said to be **factors** of the product ab. Part b of Theorem 3 says that a product of two real numbers can be zero only if at least one of the factors is zero.

The next theorem gives us the usual rules of signs.

Theorem 4 Let a and b be real numbers. Then
(a) $-(-a) = a$
(b) $(-a)(b) = -(ab) = a(-b)$
(c) $(-1)(a) = -a$
(d) $(-a)(-b) = ab$
(e) $-(a + b) = (-a) + (-b)$

It is important to note that $-a$ is not necessarily a negative number. In fact, Theorem 4a shows that $-(-3) = 3$.

We next introduce the operations of subtraction and division. If a and b are real numbers, the **difference** between a and b, denoted by $a - b$, is defined by

$$a - b = a + (-b)$$

and the operation is called **subtraction**. Thus,

$$6 - 2 = 6 + (-2) = 4 \qquad 2 - 2 = 0 \qquad 0 - 8 = -8$$

It is easy to show that the distributive laws hold for subtraction, that is,

$$a(b - c) = ab - ac$$
$$(a - b)c = ac - bc$$

If a and b are real numbers and $b \neq 0$, then the **quotient** of a and b, denoted by a/b, is defined by

$$\frac{a}{b} = a \cdot \frac{1}{b}$$

and the operation is called **division**. We also write a/b as $a \div b$ and speak of the **fraction** a over b. The numbers a and b are called the **numerator** and **denominator,** respectively, of the fraction a/b. Observe that we have not defined division by zero, since 0 has no reciprocal.

The following theorem summarizes the familiar properties of fractions.

Theorem 5 Let a, b, c, and d be real numbers with $b \neq 0$, $d \neq 0$. Then

Example

(a) $\dfrac{a}{b} = \dfrac{c}{d}$ if and only if $ad = bc$ $\qquad$ $\dfrac{2}{3} = \dfrac{4}{6}$ since $2 \cdot 6 = 3 \cdot 4$

(b) $\dfrac{a}{b} = \dfrac{ad}{bd}$ $\qquad\qquad\qquad\qquad\qquad$ $\dfrac{6}{12} = \dfrac{6 \cdot 3}{12 \cdot 3} = \dfrac{18}{36}$

(c) $\dfrac{a}{b} + \dfrac{c}{b} = \dfrac{a + c}{b}$
  $\dfrac{2}{3} + \dfrac{5}{3} = \dfrac{2 + 5}{3} = \dfrac{7}{3}$

(d) $\dfrac{a}{b} + \dfrac{c}{d} = \dfrac{ad + bc}{bd}$
  $\dfrac{2}{5} + \dfrac{3}{4} = \dfrac{2 \cdot 4 + 5 \cdot 3}{5 \cdot 4} = \dfrac{23}{20}$

(e) $\dfrac{a}{b} \cdot \dfrac{c}{d} = \dfrac{ac}{bd}$
  $\dfrac{2}{3} \cdot \dfrac{4}{5} = \dfrac{2 \cdot 4}{3 \cdot 5} = \dfrac{8}{15}$

(f) $\dfrac{\dfrac{a}{b}}{\dfrac{c}{d}} = \dfrac{a}{b} \cdot \dfrac{d}{c}$ (if $c \neq 0$)
  $\dfrac{\dfrac{2}{3}}{\dfrac{7}{5}} = \dfrac{2}{3} \cdot \dfrac{5}{7} = \dfrac{2 \cdot 5}{3 \cdot 7} = \dfrac{10}{21}$

EXERCISE SET 1.1

In Exercises 1–14 determine whether the given statement is true (T) or false (F).

1. -14 is a natural number.
2. $-\frac{4}{5}$ is a rational number.
3. $\pi/3$ is a rational number.
4. $1.75/18.6$ is an irrational number.
5. -1207 is an integer.
6. 0.75 is an irrational number.
7. $\frac{4}{5}$ is a real number.
8. 3 is a rational number.
9. 2π is a real number.

10. The sum of two rational numbers is always a rational number.
11. The sum of two irrational numbers is always an irrational number.
12. The product of two rational numbers is always a rational number.
13. The product of two irrational numbers is always an irrational number.
14. The difference of two irrational numbers is always an irrational number.

Indicate the following sets of numbers on a real number line.

15. The natural numbers less than 8.
16. The natural numbers greater than 4 and less than 10.

17. The integers that are greater than 2 and less than 7.
18. The integers that are greater than -5 and less than or equal to 1.

Express each statement as an inequality.

19. a is nonnegative.
21. b is greater than or equal to 5.
23. b is less than or equal to -4.

20. -6 is less than -2.
22. a is between 3 and 7.
24. a is between $\frac{1}{2}$ and $\frac{1}{4}$.

State a property of inequalities that justifies each of the following statements.

25. Since $-3 < 1$, then $-1 < 3$. neg. no
27. Since $14 > 9$, then $-14 < -9$. neg no
29. Since $-1 < 6$, then $-3 < 18$. positive no.

26. Since $-5 < -1$ and $-1 < 4$, then $-5 < 4$. Transitive Property
28. Since $5 > 3$, then $5 \neq 3$. Trichotomy property
30. Since $6 > -1$, then 7 is a positive number.

Find the value of each of the following.

31. $|2 - 3|$ 1
32. $|2 - 2|$ 0
33. $|2 - (-2)|$ 4
34. $|2| + |-3|$ 5
35. $\dfrac{|14 - 8|}{|-3|}$ $\frac{6}{3} = 2$
36. $\dfrac{|2 - 12|}{|1 - 6|}$ $\frac{-10}{-5}$ 2
37. $\dfrac{|3| - |2|}{|3| + |2|}$ $\frac{1}{5}$
38. $\dfrac{|3 - 2|}{|3 + 2|}$ $\frac{1}{5}$

The coordinates of points A and B are given in each of the following. Find $\overline{AB}$.

39. 2, 5 $= 3$ 40. $-3, 6$ 9 41. $-3, -1$ 2 42. $-4, \frac{11}{2}$ $9\frac{1}{2}$

43. $-\frac{4}{5}, \frac{4}{5}$ $\frac{3}{3}$ 44. 2, 2 0

In the following, the letters represent real numbers. Identify the property or properties of real numbers that justify each statement.

45. $a + x = x + a$. Commutative law addition 46. $(xy)z = x(yz)$. associative law mulip.

47. $xyz + xy = xy(z + 1)$. Distritive 48. $x + y$ is a real number. closure

49. $(a + b) + 3 = a + (b + 3)$. associative law (+) 50. $5 + (x + y) = (x + y) + 5$. commutative law (+)

51. cx is a real number. closure mulip 52. $(a + 5) + b = (a + b) + 5$. "

53. $uv = vu$. commutative law mulip 54. $x + 0 = x$. additive identity

55. $a(bc) = c(ab)$. associative law mulip & comm. 56. $xy - xy = 0$. additive inverse

57. $5 \cdot \frac{1}{5} = 1$. reciprocal 58. $xy \cdot 1 = xy$. multiplicative identity

Find a counterexample for each of the following statements; that is, find real values for which the statement is false.

59. $a - b = b - a$ $a + b = b + a$ 60. $\dfrac{a}{b} = \dfrac{b}{a}$

61. $a(b + c) = ab + c$ 62. $(a + b)(c + d) = ac + bd$

Indicate the property or properties of equality that justify each statement.

63. If $3x = 5$, then $5 = 3x$. Symmetric 66. If $x + 2y + 3z = r + s$ and $r = x + 1$, then $x + 2y + 3z = x + 1 + s$.

64. If $x + y = 7$ and $y = 5$, then $x + 5 = 7$.

65. If $2y = z$ and $z = x + 2$, then $2y = x + 2$. substitution

In each of the following, a, b, and c are real numbers. Use the properties of the real numbers and the properties of equality to prove each theorem.

67. If $a = b$, then $ac = bc$. (Theorem 1b)

68. If $a = b$ and $c \neq 0$, then $a/c = b/c$.

69. If $a + c = b + c$, then $a = b$. (Theorem 2a)

70. If $ac = bc$ and $c \neq 0$, then $a = b$. (Theorem 2b)

71. $a(b - c) = ab - ac$

72. Prove that the real number 0 does not have a reciprocal. (*Hint:* Assume $b = 1/0$ is the reciprocal of 0. Supply a reason for each of the following steps.

$$1 = 0 \cdot \frac{1}{0}$$
$$= 0 \cdot b$$
$$= 0$$

Since this conclusion is impossible, the original assumption must be false.)

1.2
POLYNOMIALS AND FACTORING

In the introduction to this chapter we said that "symbols are the language of algebra." Unfortunately, we have to introduce many definitions so that statements can distinguish precisely between different types of symbols and their relationships. Fortunately, you have seen these definitions before and this section will serve to refresh your memory.

A **variable** is a symbol to which we can assign values. For example, in Section 1.1 we defined a rational number as one that can be written as p/q, where

p and q are integers (and q is not zero). The symbols p and q are variables, since we can assign values to them. A variable can be restricted to a particular number system (for example, p and q must be integers) or to a subset of a number system (note that q cannot be zero).

If we invest P dollars at an annual interest rate of 6%, then we will earn $0.06P$ dollars interest per year, and we will have $P + 0.06P$ dollars at the end of the year. We call $P + 0.06P$ an **algebraic expression**. Note that an algebraic expression involves **variables** (in this case P), **constants** (such as 0.06), and **algebraic operations** (such as $+$, $-$, $\times$, $\div$). Virtually everything we do in algebra involves algebraic expressions, sometimes as simple as our example and sometimes very involved.

An algebraic expression takes on a **value** when we assign a specific number to each variable in the expression. Thus, the expression

$$\frac{3m + 4n}{m + n}$$

is **evaluated** when $m = 3$ and $n = 2$ by substituting these values for m and n:

$$\frac{3(3) + 4(2)}{3 + 2} = \frac{9 + 8}{5} = \frac{17}{5}$$

We often need to write algebraic expressions in which a variable multiplies itself repeatedly. We use the notation of exponents to indicate such repeated multiplication. Thus,

$$a^1 = a \qquad a^2 = a \cdot a \qquad a'' = \underbrace{a \cdot a \cdot \ \ldots \ \cdot a}_{n \text{ factors}}$$

where n is a natural number and a is a real number. We call a the **base** and n the **exponent** and say that a'' is the nth **power** of a. When $n = 1$, we simply write a rather than a^1.

It is convenient to define a^0 for all real numbers $a \neq 0$ by having $a^0 = 1$. We will provide motivation for this seemingly arbitrary definition in Section 1.3.

WARNING Note the difference between

$$(-3)^2 = (-3)(-3) = 9$$

and

$$-3^2 = -(3 \cdot 3) = -9$$

Later in this chapter we will need an important rule of exponents. Observe that if m and n are natural numbers and a is any real number, then

$$a^m \cdot a^n = \underbrace{a \cdot a \cdot \ldots \cdot a}_{m \text{ factors}} \cdot \underbrace{a \cdot a \cdot \ldots \cdot a}_{n \text{ factors}}$$

Since there are a total of $m + n$ factors on the right side, we conclude that

$$a^m a^n = a^{m+n}$$

EXAMPLE 1
Multiply.
(a) $x^2 \cdot x^3$ (b) $(3x)(4x^4)$

SOLUTION
(a) $x^2 \cdot x^3 = x^{2+3} = x^5$
(b) $(3x)(4x^4) = 3 \cdot 4 \cdot x \cdot x^4 = 12x^{1+4} = 12x^5$

POLYNOMIALS

A polynomial is an algebraic expression of a certain form. Polynomials play an important role in the study of algebra, since many word problems translate into equations or inequalities that involve polynomials. We first study the manipulative and mechanical aspects of polynomials; this will serve as background for dealing with their applications in later chapters.

Let x denote a variable and let n be a nonnegative integer. The expression ax^n, where a is a constant real number, is called a **monomial in x**. A **polynomial in x** is an expression that is a sum of monomials and has the general form

$$P = a_n x^n + a_{n-1} x^{n-1} + \cdots + a_1 x + a_0, \quad a_n \neq 0 \qquad (1)$$

Each of the monomials in Equation (1) is called a **term** of P, and a_0, $a_1, \ldots, a_n$ are constant real numbers that are called the **coefficients** of the terms of P. Note that a polynomial may consist of just one term; that is, a monomial is also considered to be a polynomial.

EXAMPLE 2
(a)   The following expressions are polynomials in x:

$$3x^4 + 2x + 5 \qquad 2x^3 + 5x^2 - 2x + 1 \qquad \frac{3}{2}x^3$$

Notice that we write $2x^3 + 5x^2 + (-2)x + 1$ as $2x^3 + 5x^2 - 2x + 1$.
(b)   The following expressions are not polynomials in x:

$$2x^{1/2} + 5 \qquad 3 - \frac{4}{x} \qquad \frac{2x - 1}{x - 2}$$

Remember that each term of a polynomial in x must be of the form ax^n where a is a real number and n is a nonnegative integer.

The **degree of a monomial in x** is the exponent of x. Thus, the degree of $5x^3$ is 3. A monomial in which the exponent of x is 0 is called a **constant term** and is said to be of **degree zero**. The nonzero coefficient a_n of the term in P with highest degree is called the **leading coefficient** of P and we say that P is a **polynomial of degree n**. A special case is the polynomial all of whose coefficients are zero. Such a polynomial is called the **zero polynomial**, is denoted by 0, and is said to have no degree.

EXAMPLE 3
Given the polynomial

$$P = 2x^4 - 3x^2 + \frac{4}{3}x - 1$$

The terms of P are $\qquad\qquad 2x^4, \quad 0x^3, \quad -3x^2, \quad \frac{4}{3}x, \quad -1.$

The coefficients of the terms are $\qquad 2, \quad 0, \quad -3, \quad \frac{4}{3}, \quad -1.$

The degree of P is 4 and the leading coefficient is 2.

OPERATIONS WITH POLYNOMIALS

If P and Q are polynomials in x, then the terms cx^r in P and dx^r in Q are said to be **like terms;** that is, like terms have the same exponent in x. For example, given

$$P = 4x^2 + 4x - 1$$

and

$$Q = 3x^3 - 2x^2 + 4$$

then the like terms are $0x^3$ and $3x^3$; $4x^2$ and $-2x^2$; $4x$ and $0x$; -1 and 4. We define equality of polynomials in this way.

> Two polynomials are equal if all like terms are identical.

EXAMPLE 4
Find A, B, C, and D if

$$Ax^3 + (A + B)x^2 + Cx + (C - D) = -2x^3 + x + 3$$

SOLUTION
Equating the coefficients of like terms, we have

$$A = -2 \quad A + B = 0 \quad C = 1 \quad C - D = 3$$
$$B = 2 \qquad\qquad\qquad D = -2$$

If P and Q are polynomials in x, the **sum** $P + Q$ is obtained by forming the sums of all pairs of like terms. The sum of cx^r in P and dx^r in Q is $(c + d)x^r$. Similarly, the **difference** $P - Q$ is obtained by forming the differences, $(c - d)x^r$, of like terms.

EXAMPLE 5
(a) Add $2x^3 + 2x^2 - 3$ and $x^3 - x^2 + x + 2$.
(b) Subtract $2x^3 + x^2 - x + 1$ from $3x^3 - 2x^2 + 2x$.

SOLUTION
(a) Adding the coefficients of like terms,

$$(2x^3 + 2x^2 - 3) + (x^3 - x^2 + x + 2) = 3x^3 + x^2 + x - 1$$

(b) Subtracting the coefficients of like terms,

$$(3x^3 - 2x^2 + 2x) - (2x^3 + x^2 - x + 1) = x^3 - 3x^2 + 3x - 1$$

Note that the coefficients of a polynomial are real numbers and that the variable x is assigned real values. It then follows from the definition of addition of polynomials that the commutative and associative laws of addition hold for polynomials. The zero polynomial

$$0 = 0x^n + 0x^{n-1} + \cdots + 0x + 0$$

plays the role of the additive identity. The additive inverse of the polynomial given in Equation (1) is

$$-P = (-a_n)x^n + (-a_{n-1})x^{n-1} + \cdots + (-a_1)x + (-a_0)$$

since $P + (-P) = 0$.

Multiplication of polynomials is based on the rule for exponents developed earlier in this section,

$$a^m a^n = a^{m+n}$$

and on the distributive laws

$$a(b + c) = ab + ac$$
$$(a + b)c = ac + bc$$

EXAMPLE 6
Multiply $(x + 2)(3x^2 - x + 5)$.

SOLUTION
$(x + 2)(3x^2 - x + 5)$

$= x(3x^2 - x + 5) + 2(3x^2 - x + 5)$ Distributive law

$= 3x^3 - x^2 + 5x + 6x^2 - 2x + 10$ Distributive law and $a^m a^n = a^{m+n}$

$= 3x^3 + 5x^2 + 3x + 10$ Adding like terms

PROGRESS CHECK

Multiply.

(a) $(x^2 + 2)(x^2 - 3x + 1)$ (b) $(x^2 - 2xy + y)(2x + y)$

ANSWERS

(a) $x^4 - 3x^3 + 3x^2 - 6x + 2$ (b) $2x^3 - 3x^2y + 2xy - 2xy^2 + y^2$

The multiplication in Example 6 can be carried out in "long form" as follows.

$$
\begin{array}{r}
3x^2 - x + 5 \\
x + 2 \\
\hline
3x^3 - x^2 + 5x \\
6x^2 - 2x + 10 \\
\hline
3x^3 + 5x^2 + 3x + 10
\end{array}
$$

$= x(3x^2 - x + 5)$
$= 2(3x^2 - x + 5)$
$=$ sum of above lines

In Example 6, the product of polynomials of degrees one and two is seen to be a polynomial of degree three. From the multiplication process it is easy to derive the following useful rule.

> The degree of the product of two nonzero polynomials is the sum of the degrees of the polynomials.

Products of the form $(2x + 3)(5x - 2)$ or $(2x + y)(3x - 2y)$ occur often, and we can handle them mentally by the familiar method:

$$
\begin{array}{c}
10x^2 \quad -6 \\
(2x + 3)\ (5x - 2) \qquad = 10x^2 + 11x - 6 \\
15x \\
\dfrac{-4x}{\text{Sum} = 11x}
\end{array}
$$

A number of special products occur frequently and it is worthwhile knowing them.

Special Products

$(a + b)^2 = (a + b)(a + b) = a^2 + 2ab + b^2$

$(a - b)^2 = (a - b)(a - b) = a^2 - 2ab + b^2$

$(a + b)(a - b) = a^2 - b^2$

FACTORING

Now that we can find the product of two polynomials, let's consider the reverse problem: Given a polynomial, can we find factors whose product will yield the given polynomial? This process, known as **factoring,** is one of the basic tools of algebra. In this chapter a polynomial with *integer* coefficients is to be factored as a product of polynomials of lower degree with *integer* coefficients; a polynomial with *rational* coefficients is to be factored as a product of polynomials of lower degree with *rational* coefficients.

The simplest type of factoring involves the removal of a common factor. For example, given the polynomial

$$x^2 + x$$

we see that the factor x is common to both terms and write

$$x^2 + x = x(x + 1)$$

Note that x and $x + 1$ are both polynomials of degree one. We have thus factored a polynomial of second degree with integer coefficients into the product of polynomials of lower degree with integer coefficients.

Before proceeding to other methods of factoring, we offer a useful rule.

Always remove common factors before attempting any other factoring techniques.

To factor a second-degree polynomial

$$ax^2 + bx + c$$

where a, b, and c are integers and $a \neq 0$, we must have

$$ax^2 + bx + c = (rx + u)(sx + v) = (rs)x^2 + (rv + su)x + uv$$

where r, s, u, and v are integers. Equating the coefficients of like terms,

$$rs = a \quad rv + su = b \quad uv = c$$

These three equations give candidates for r, s, u, and v. The final choices from among the candidates are determined by trial and error, which is made easier by using mental multiplication.

EXAMPLE 7
Factor $2x^2 - x - 6$.

SOLUTION
The term $2x^2$ can result only from the factors $2x$ and x, so the factors must be of the form

$$2x^2 - x - 6 = (2x \quad)(x \quad)$$

The constant term, -6, must be the product of factors of opposite signs, so we may write

$$2x^2 - x - 6 = \begin{cases} (2x + \quad)(x - \quad) \\ \qquad \text{or} \\ (2x - \quad)(x + \quad) \end{cases}$$

The integer factors of 6 are

$$1 \cdot 6 \quad 6 \cdot 1 \quad 2 \cdot 3 \quad 3 \cdot 2$$

By trying these we find that

$$2x^2 - x - 6 = (2x + 3)(x - 2)$$

PROGRESS CHECK
Factor.
(a) $3x^2 - 16x + 21$ (b) $2x^2 + 3x - 9$

ANSWERS
(a) $(3x - 7)(x - 3)$ (b) $(2x - 3)(x + 3)$

WARNING   The polynomial $x^2 - 6x$ can be written as

$$x^2 - 6x = x(x - 6)$$

and is then a product of two polynomials of positive degree. Students often fail to consider x to be a "true" factor.

EXAMPLE 8
Factor.
(a) $2x^3 - 12x^2 + 16x$ (b) $2ab + b + 2ac + c$

SOLUTION
(a)   We first remove the common factor $2x$.

$$2x^3 - 12x^2 + 16x = 2x(x^2 - 6x + 8)$$
$$= 2x(x - 2)(x - 4)$$

(b)   It is sometimes possible to discover common factors by first grouping terms. Grouping those terms containing b and those terms containing c,

$$\begin{aligned} 2ab + b + 2ac + c &= (2ab + b) + (2ac + c) \quad \text{Grouping} \\ &= b(2a + 1) + c(2a + 1) \quad \text{Common factors } b,\ c \\ &= (2a + 1)(b + c) \quad\quad\ \text{Common factor } 2a + 1 \end{aligned}$$

PROGRESS CHECK

Factor.

(a) $2x^3 - 2x^2y - 4xy^2$ (b) $2m^3n + m^2 + 2mn^2 + n$

ANSWERS

(a) $2x(x + y)(x - 2y)$ (b) $(2mn + 1)(m^2 + n)$

SPECIAL FACTORS

There are three expressions that occur frequently and deserve special attention. The first is the difference of two squares, the second is the sum of two cubes, and the third is the difference of two cubes.

Special Factors	$a^2 - b^2 = (a + b)(a - b)$
	$a^3 + b^3 = (a + b)(a^2 - ab + b^2)$
	$a^3 - b^3 = (a - b)(a^2 + ab + b^2)$

When using these formulas, be careful with the placement of plus and minus signs.

EXAMPLE 9

Factor.

(a) $4x^2 - 25$ (b) $\dfrac{1}{27}u^3 + 8v^3$

SOLUTION

(a) Since

$$4x^2 - 25 = (2x)^2 - (5)^2$$

we may use the formula for the difference of two squares with $a = 2x$ and $b = 5$. Thus,

$$4x^2 - 25 = (2x + 5)(2x - 5)$$

(b) Note that

$$\frac{1}{27}u^3 + 8v^3 = \left(\frac{1}{3}u\right)^3 + (2v)^3$$

and then use the formula for the sum of two cubes.

$$\frac{1}{27}u^3 + 8v^3 = \left(\frac{u}{3} + 2v\right)\left(\frac{u^2}{9} - \frac{2}{3}uv + 4v^2\right)$$

"NO FUSS" FACTORING FOR SECOND-DEGREE POLYNOMIALS

Factoring involves a certain amount of trial and error, which can become frustrating, especially when the lead coefficient is not 1. You might want to try a rather neat scheme that will greatly reduce the number of candidates.

We'll demonstrate the method for the polynomial

$$4x^2 + 11x + 6 \tag{1}$$

Using the lead coefficient of 4, write the pair of incomplete factors

$$(4x \quad)(4x \quad) \tag{2}$$

Next, multiply the coefficient of x^2 and the constant term in (1) to produce $4 \cdot 6 = 24$. Now find two integers whose product is 24 and whose sum is 11, the coefficient of the middle term of (1). It's clear that 8 and 3 will do nicely, so we write

$$(4x + 8)(4x + 3) \tag{3}$$

Finally, within each parenthesis in (3) discard any common divisor. Thus $(4x + 8)$ reduces to $(x + 2)$ and we write

$$(x + 2)(4x + 3) \tag{4}$$

which is the factorization of $4x^2 + 11x + 6$.

Will the method always work? Yes—if you first remove all common factors in the original polynomial. That is, you must first write

$$6x^2 + 15x + 6 = 3(2x^2 + 5x + 2)$$

Try the method on these second-degree polynomials.

$3x^2 + 10x - 8$

$6x^2 - 13x + 6$

$4x^2 - 15x - 4$

$10x^2 + 11x - 6$

and apply the method to the polynomial $2x^2 + 5x + 2$.

(For a proof that the method works, see M. A. Autrie and J. D. Austin, "A Novel Way to Factor Quadratic Polynomials," *The Mathematics Teacher* 72, no. 2 [1979].)

We'll use the polynomial $2x^2 - x - 6$ of Example 7 to demonstrate the method when some of the coefficients are negative.

Factoring $ax^2 + bx + c$	Example: $2x^2 - x - 6$
Step 1. Use the lead coefficient a to write the incomplete factors $$(ax \quad)(ax \quad)$$	*Step 1.* The lead coefficient is 2, so we write $$(2x \quad)(2x \quad)$$
Step 2. Multiply a and c, the coefficients of x^2 and the constant term.	*Step 2.* $a \cdot c = (2)(-6) = -12$
Step 3. Find integers whose product is $a \cdot c$ and whose sum equals b. Write these integers in the incomplete factors of Step 1.	*Step 3.* Two integers whose product is -12 and whose sum is -1 are 3 and -4. We then write $$(2x + 3)(2x - 4)$$
Step 4. Discard any common factor *within each parenthesis* in Step 3. The result is the desired factorization.	*Step 4.* Reducing $(2x - 4)$ to $(x - 2)$, we have $$2x^2 - x - 6 = (2x + 3)(x - 2)$$

PROGRESS CHECK

Factor.

(a) $16x^2 - 9$ (b) $8s^3 - 27t^3$ (c) $125r^3 + \dfrac{1}{125}s^3$

ANSWERS

(a) $(4x + 3)(4x - 3)$ (b) $(2s - 3t)(4s^2 + 6t + 9t^2)$

(c) $\left(5r + \dfrac{s}{5}\right)\left(25r^2 - rs + \dfrac{s^2}{25}\right)$

Are there polynomials with integer coefficients that cannot be written as products of polynomials of lower degree with integer coefficients? The answer is yes. Examples are the polynomials $x^2 + 1$ and $x^2 + x + 1$. A polynomial is said to be **prime** or **irreducible** if it cannot be written as a product of two polynomials of positive degree. Thus, $x^2 + 1$ is irreducible over the integers.

EXERCISE SET 1.2

1. Evaluate $\frac{2}{3}r + 5$ when $r = 12$.

2. Evaluate $\frac{9}{5}C + 32$ when $C = 37$.

3. If P dollars are invested at a simple interest rate of r percent per year for t years, the amount on hand at the end of t years is $P + Prt$. Suppose $2000 is invested at 8% per year ($r = 0.08$). How much money is on hand after
   (a) one year?   (b) half a year?
   (c) 8 months?

Evalute the given expressions in Exercises 7 and 8.

7. $\dfrac{|a - 2b|}{2a}$ when $a = 1, b = 2$

Carry out the indicated operations in Exercises 9–14.

9. $b^5 \cdot b^2$ 10. $x^3 \cdot x^5$

13. $\left(\dfrac{3}{2}x^3\right)(-2x)$ 14. $\left(-\dfrac{5}{3}x^6\right)\left(-\dfrac{3}{10}x^3\right)$

15. Which of the following expressions are not polynomials?
   (a) $-3x^2 + 2x + 5$ (b) $-3x^2y$
   (c) $-3x^{2/3} + 2xy + 5$ (d) $-2x^{-4} + 2xy^3 + 5$

Indicate the leading coefficient and the degree of each polynomial in Exercises 17–20.

17. $2x^3 + 3x^2 - 5$ 18. $-4x^5 - 8x^2 + x + 3$

4. The perimeter of a rectangle is given by the formula $P = 2(L + W)$, where L is the length and W is the width of the rectangle. Find the perimeter if
   (a) $L = 2$ feet, $W = 3$ feet
   (b) $L = \frac{1}{2}$ meter, $W = \frac{1}{4}$ meter

5. Evaluate $0.02r + 0.314st + 2.25t$ when $r = 2.5$, $s = 3.4$, and $t = 2.81$.

6. Evaluate $10.421x + 0.821y + 2.34xyz$ when $x = 3.21$, $y = 2.42$, and $z = 1.23$.

8. $\dfrac{|x| + |y|}{|x| - |y|}$ when $x = -3, y = 4$

11. $(4y^3)(-5y^6)$ 12. $(-6x^4)(-4x^7)$

16. Which of the following expressions are not polynomials?
   (a) $4x^5 - x^{1/2} + 6$ (b) $\dfrac{2}{5}x^3 + \dfrac{4}{3}x - 2$
   (c) $4x^5y$ (d) $x^{4/3}y + 2x - 3$

19. $\dfrac{3}{5}x^4 + 2x^2 - x - 1$ 20. $-1.5 + 7x^3 + 0.75x^7$

21. An investor buys x shares of G.E. stock at $55 per share, y shares of Exxon stock at $45 per share, and z shares of A.T.&T. stock at $60 per share. What does the polynomial $55x + 45y + 60z$ represent?

22. A field consists of a rectangle and a square arranged as shown in Figure 2.

    What does each of the following polynomials represent?

    (a) $x^2 + xy$ (b) $2x + 2y$ (c) $4x$

    (d) $4x + 2y$

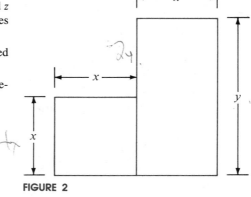

FIGURE 2

Perform the indicated operations in Exercises 23–32.

23. $(2x^2 + 3x + 8) - (5 - 2x + 2x^2)$

24. $(4x^2 + 3x + 2) + (3x^2 - 2x - 5)$

25. $3xy^2z - 4x^2yz + xy + 3 - (2xy^2z + x^2yz - yz + x - 2)$

26. $(x^2 + 3)(2x^2 - x + 2)$

27. $(2y^2 + y)(-2y^3 + y - 3)$

28. $(x^2 + 2x - 1)(2x^2 - 3x + 2)$

29. $(a^2 - 4a + 3)(4a^3 + 2a + 5)$

30. $(2a^2 + ab + b^2)(3a - b^2 + 1)$

31. $(-3a + ab + b^2)(3b^2 + 2b + 2)$

32. $5(2x - 3)^2$

33. An investor buys x shares of IBM stock at $260 per share at Thursday's opening of the stock market. Later in the day, he sells y shares of G&W stock at $13 per share and z shares of Holiday Inn stock at $17 per share. Write a polynomial that expresses the money transactions for the day.

34. An artist takes a rectangular piece of cardboard whose sides are x and y and cuts out a square of side $x/2$ (Figure 3) to obtain a mat for a painting. Write a polynomial giving the area of the mat.

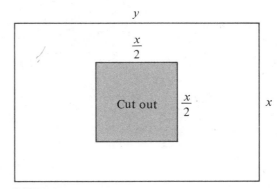

FIGURE 3

In Exercises 45–72, factor completely.

35. $5bc + 25b$

36. $2x^4 + x^2$

37. $-3y^2 - 4y^5$

38. $3abc + 12bc$

39. $3x^2 + 6x^2y - 9x^2z$

40. $9a^3b^3 + 12a^2b - 15ab^2$

41. $x^2 + 4x + 3$

42. $x^2 + 2x - 8$

43. $x^2 - 5x - 14$

44. $x^2 - 8x - 20$

45. $2x^2 - 3x - 2$

46. $2x^2 + 7x + 6$

47. $10x^2 - 13x - 3$

48. $9x^2 + 24x + 16$

49. $6x^2 + 13x + 6$

50. $4x^2 + 20x + 25$

odd

51. $4(x + 1)(y + 2) - 8(y + 2)$

52. $2(x + 1)(x - 1) + 5(x - 1)$

53. $3(x + 2)^2(x - 1) - 4(x + 2)^2(2x + 7)$

54. $4(2x - 1)^2(x + 2)^3(x + 1) - 3(2x - 1)^5(x + 2)^2(x + 3)$

55. $y^2 - \dfrac{1}{9}$

56. $4a^2 - b^2$

57. $16 - 9x^2y^2$

58. $x^{12} - 1$

59. $x^4 - y^4$

60. $a^4 - 16$

61. $x^3 + 27y^3$

62. $8x^3 + 125y^3$

63. $27x^3 - y^3$

64. $64x^3 - 27y^3$

65. $a^3 + 8$

66. $8r^3 - 27$

67. $\dfrac{1}{8}m^3 - 8n^3$

68. $8a^3 - \dfrac{1}{64}b^3$

69. $(x + y)^3 - 8$

70. $27 + (x + y)^3$

71. $8x^6 - 125y^6$

72. $a^6 + 27b^6$

1.3
EXPONENTS, RADICALS, AND COMPLEX NUMBERS

POSITIVE INTEGER EXPONENTS

In Section 1.2 we defined a^n for a real number a and a positive integer n as

$$a^n = \underbrace{a \cdot a \cdot \ldots \cdot a}_{n \text{ factors}}$$

and we showed that if m and n are positive integers then $a^m a^n = a^{m+n}$. The method we used to establish this rule was to write out the factors a^m and a^n and count the total number of occurrences of a. The same method can be used to establish the rest of the following rules when m and n are positive integers.

Rules for Exponents

$$a^m a^n = a^{m+n} \qquad (a^m)^n = a^{mn}$$

$$(ab)^m = a^m b^m \qquad \left(\frac{a}{b}\right)^m = \frac{a^m}{b^m}, \ b \neq 0$$

If $a \neq 0$,

$$\frac{a^m}{a^n} = \begin{cases} a^{m-n} & \text{when } m > n \\ \dfrac{1}{a^{n-m}} & \text{when } n > m \\ 1 & \text{when } m = n \end{cases}$$

A rigorous proof of the rules listed above requires the method of mathematical induction (see Chapter 11).

EXAMPLE 1
Simplify the following.

(a) $(4a^2b^3)(2a^3b)$

(b) $(2x^2y)^4$

(c) $\dfrac{x^{2n+1}}{x^n}$

SOLUTION

(a) $(4a^2b^3)(2a^3b) = 4 \cdot 2 \cdot a^2a^3b^3b = 8a^5b^4$

(b) $(2x^2y)^4 = 2^4(x^2)^4y^4 = 16x^8y^4$

(c) $\dfrac{x^{2n+1}}{x^n} = x^{2n+1-n} = x^{n+1}$

ZERO AND NEGATIVE EXPONENTS

We now repeat the definition of a^0 and introduce the definition of a^{-m}, $a \neq 0$ and m a positive integer.

$$a^0 = 1 \qquad a^{-m} = \frac{1}{a^m}$$

For example,

$$3^0 = 1 \qquad 2^{-3} = \frac{1}{8} \qquad \left(\frac{1}{4}\right)^{-1} = 4$$

These definitions for zero and negative integer exponents were chosen because they are compatible with the rules for positive integer exponents. For instance, if $a \neq 0$, we would expect the rule $a^m a^n = a^{m+n}$ to produce

$$a^m a^0 = a^{m+0} = a^m$$

Using the definition $a^0 = 1$, we have

$$a^m a^0 = a^m \cdot 1 = a^m$$

which demonstrates that this rule holds with $a^0 = 1$. In a similar way, you can verify that *all* of the rules for exponents hold for our definition of zero and negative exponents.

EXAMPLE 2

Simplify and write the answers using only positive exponents.

(a) $4(xy)^0$ (b) $(x^2y^{-3})^{-5}$ (c) $\dfrac{9a^2b^{-3}}{3a^{-2}b^5}$

SOLUTION

(a) $4(xy)^0 = 4(1) = 4$

(b) $(x^2y^{-3})^{-5} = (x^2)^{-5}(y^{-3})^{-5} = x^{-10}y^{15} = \dfrac{y^{15}}{x^{10}}$

(c) $\dfrac{9a^2b^{-3}}{3a^{-2}b^5} = \dfrac{9}{3}\dfrac{a^2}{a^{-2}}\dfrac{b^{-3}}{b^5} = 3a^4b^{-8} = \dfrac{3a^4}{b^8}$

RATIONAL EXPONENTS AND RADICALS

If n is a natural number, we define a root by saying

---
a is an **nth root** of b if $a^n = b$

When $n = 2$, we say that a is a **square root** of b, and when $n = 3$, we say that a is a **cube root** of b. Thus, 5 and -5 are square roots of 25 since $(5)^2 = (-5)^2 = 25$; -2 is a cube root of -8 since $(-2)^3 = -8$. More generally, if $b > 0$ and a is a square root of b, then $-a$ is also a square root of b. If $b < 0$, there is no real number a such that $a^2 = b$, since the square of a real number is always nonnegative. The cases are summarized in Table 2.

TABLE 2

b	n	Number of nth roots of b such that $b = a^n$	Form of nth roots	b	Examples
> 0	Even	2	$a, -a$	4	Square roots are 2, -2
< 0	Even	None	None	-1	No square roots
> 0	Odd	1	$a > 0$	8	Cube root is 2
< 0	Odd	1	$a < 0$	-8	Cube root is -2
0	All	1	0	0	Square root is 0

We would like to define rational exponents in a manner that will be consistent with the rules for integer exponents. If the rule $(a^m)^n = a^{mn}$ is to hold, then we must have

$$(b^{1/n})^n = b^{n/n} = b$$

But a is an nth root of b if $a^n = b$. Then for every natural number n, we say that

---
$b^{1/n}$ is an nth root of b

If n is even and b is positive, Table 2 indicates that there are two numbers, a and $-a$, that are nth roots of b. To avoid ambiguity we always *choose* the positive number to be the nth root and call it the **principal nth root** of b. Thus, $b^{1/n}$ denotes the principal nth root of b.

EXAMPLE 3
Evaluate.

(a) $144^{1/2}$ (b) $(-8)^{1/3}$ (c) $(-25)^{1/2}$ (d) $-\left(\dfrac{1}{16}\right)^{1/4}$

WHEN IS A PROOF NOT A PROOF?

Books of mathematical puzzles love to include "proofs" that lead to false or contradictory results. Of course, there is always an incorrect step hidden somewhere in the proof. The error may be subtle, but a good grounding in the fundamentals of mathematics will enable you to catch it.

Examine the following "proof."

$$1 = 1^{1/2} \tag{1}$$
$$= [(-1)^2]^{1/2} \tag{2}$$
$$= (-1)^{2/2} \tag{3}$$
$$= (-1)^1 \tag{4}$$
$$= -1 \tag{5}$$

The result is obviously contradictory: we can't have $1 = -1$. Yet each step seems to be legitimate. Did you spot the flaw? The rule

$$(b^m)^{1/n} = b^{m/n}$$

used in going from (2) to (3) doesn't apply when n is even and b is negative. Any time the rules of algebra are abused the results are unpredictable!

SOLUTION

(a) $144^{1/2} = 12$

(b) $(-8)^{1/3} = -2$

(c) $(-25)^{1/2}$ is not a real number

(d) $-\left(\dfrac{1}{16}\right)^{1/4} = -\dfrac{1}{2}$

If m is an integer, n a natural number, and b a real number, we define $b^{m/n}$ by

$$b^{m/n} = (b^{1/n})^m = (b^m)^{1/n}$$

where b must be positive when n is even. With this definition, all the rules of exponents continue to hold when the exponents are rational numbers.

EXAMPLE 4

Simplify.

(a) $(-8)^{4/3}$ (b) $x^{1/2} \cdot x^{3/4}$ (c) $(x^{3/4})^2$ (d) $(3x^{2/3}y^{-5/3})^3$

SOLUTION

(a) $(-8)^{4/3} = [(-8)^{1/3}]^4 = (-2)^4 = 16$

(b) $x^{1/2} \cdot x^{3/4} = x^{1/2+3/4} = x^{5/4}$

(c) $(x^{3/4})^2 = x^{(3/4)(2)} = x^{3/2}$

(d) $(3x^{2/3}y^{-5/3})^3 = 3^3 \cdot x^{(2/3)(3)}y^{(-5/3)(3)} = 27x^2y^{-5} = \dfrac{27x^2}{y^5}$

3 $1/3$

PROGRESS CHECK

Simplify.

(a) $27^{4/3}$ (b) $(a^{1/2}b^{-2})^{-2}$ (c) $\left(\dfrac{x^{1/3}y^{2/3}}{z^{5/6}}\right)^{12}$

ANSWERS

(a) 81 (b) $\dfrac{b^4}{a}$ (c) $\dfrac{x^4 y^8}{z^{10}}$

The symbol $\sqrt{b}$ is an alternative way of writing $b^{1/2}$; that is, $\sqrt{b}$ denotes the nonnegative square root of b. The symbol $\sqrt{}$ is called a **radical sign** and $\sqrt{b}$ is called the **principal square root** of b. Thus,

$$\sqrt{25} = 5 \qquad \sqrt{0} = 0 \qquad \sqrt{-25} \text{ is undefined}$$

In general, the symbol $\sqrt[n]{b}$ is an alternative way of writing $b^{1/n}$, the principal nth root of b. Of course, we must apply the same restrictions to $\sqrt[n]{b}$ that we established for $b^{1/n}$. In summary,

$$\sqrt[n]{b} = b^{1/n} = a \quad \text{where } a^n = b$$

with these restrictions:

- if n is even and $b < 0$, $\sqrt[n]{b}$ is not a real number;
- if n is even and $b \geq 0$, $\sqrt[n]{b}$ is the *nonnegative* number a satisfying $a^n = b$.

WARNING Many students are accustomed to writing $\sqrt{4} = \pm 2$. This is incorrect, since the symbol $\sqrt{}$ indicates the *principal* square root, which is nonnegative. Get in the habit of writing $\sqrt{4} = 2$. If you want to indicate *all* square roots of 4, write $\pm\sqrt{4} = \pm 2$.

In short, $\sqrt[n]{b}$ is the **radical form** of $b^{1/n}$. We can switch back and forth from one form to the other. For instance,

$$\sqrt[3]{7} = 7^{1/3} \qquad (11)^{1/5} = \sqrt[5]{11}$$

Finally, we treat the radical form of $b^{m/n}$ where m is an integer and n is a natural number as follows.

$$b^{m/n} = (b^m)^{1/n} = \sqrt[n]{b^m}$$

and

$$b^{m/n} = (b^{1/n})^m = (\sqrt[n]{b})^m$$

EXAMPLE 5

Change from radical form to rational exponent form or vice versa.

(a) $(2x)^{-3/2}$ (b) $\dfrac{1}{\sqrt[7]{y^4}}$ (c) $(-3a)^{3/7}$ (d) $\sqrt{x^2 + y^2}$

SOLUTION

(a) $(2x)^{-3/2} = \dfrac{1}{(2x)^{3/2}} = \dfrac{1}{\sqrt{8x^3}}$ (b) $\dfrac{1}{\sqrt[7]{y^4}} = \dfrac{1}{y^{4/7}} = y^{-4/7}$

(c) $(-3a)^{3/7} = \sqrt[7]{-27a^3}$ (d) $\sqrt{x^2 + y^2} = (x^2 + y^2)^{1/2}$

PROGRESS CHECK

Change from radical form to rational exponent form or vice versa.

(a) $\sqrt[4]{2rs^3}$ (b) $(x + y)^{5/2}$

(c) $y^{-5/4}$ (d) $\dfrac{1}{\sqrt[4]{m^5}}$

ANSWERS

(a) $(2r)^{1/4}s^{3/4}$ (b) $\sqrt{(x + y)^5}$

(c) $\dfrac{1}{\sqrt[4]{y^5}}$ (d) $m^{-5/4}$

WARNING Note that

$$\sqrt{16} + \sqrt{9} \neq \sqrt{25}$$

and, in general,

$$\sqrt{a} + \sqrt{b} \neq \sqrt{a + b}$$

Since radicals are just another way of writing exponents, the properties of radicals can be derived from the properties of exponents.

Properties of Radicals

If n is a natural number, a and b are real numbers, and all radicals denote real numbers, then

(1) $\sqrt[n]{b^m} = (b^m)^{1/n} = (b^{1/n})^m = (\sqrt[n]{b})^m$

(2) $\sqrt[n]{a} \cdot \sqrt[n]{b} = a^{1/n} \cdot b^{1/n} = (ab)^{1/n} = \sqrt[n]{ab}$

(3) $\dfrac{\sqrt[n]{a}}{\sqrt[n]{b}} = \dfrac{a^{1/n}}{b^{1/n}} = \left(\dfrac{a}{b}\right)^{1/n} = \sqrt[n]{\dfrac{a}{b}},\ b \neq 0$

(4) $\sqrt[n]{a^n} = \begin{cases} a & \text{if } n \text{ is odd} \\ |a| & \text{if } n \text{ is even} \end{cases}$

Here are some examples using these properties.

EXAMPLE 6
Simplify.

(a) $\sqrt{18}$ (b) $\sqrt[3]{-54}$ (c) $2\sqrt[3]{8x^3y}$ (d) $\sqrt{x^6}$

SOLUTION

(a) $\sqrt{18} = \sqrt{9 \cdot 2} = \sqrt{9}\sqrt{2} = 3\sqrt{2}$

(b) $\sqrt[3]{-54} = \sqrt[3]{(-27)(2)} = \sqrt[3]{-27}\sqrt[3]{2} = -3\sqrt[3]{2}$

(c) $2\sqrt[3]{8x^3y} = 2\sqrt[3]{8}\sqrt[3]{x^3}\sqrt[3]{y} = 2(2)(x)\sqrt[3]{y} = 4x\sqrt[3]{y}$

(d) $\sqrt{x^6} = \sqrt{x^2} \cdot \sqrt{x^2} \cdot \sqrt{x^2} = |x| \cdot |x| \cdot |x| = |x|^3$

WARNING The properties of radicals state that

$$\sqrt{x^2} = |x|$$

It is a common error to write $\sqrt{x^2} = x$, but this leads to the conclusion that $\sqrt{(-6)^2} = -6$. Since the symbol $\sqrt{}$ represents the principal or nonnegative square root of a number, the result cannot be negative. It is therefore essential to write $\sqrt{x^2} = |x|$ (and, in fact, $\sqrt[n]{x^n} = |x|$ whenever n is even) unless we know that $x \geq 0$, in which case we can write $\sqrt{x^2} = x$.

RATIONALIZING THE DENOMINATOR

It is customary to write fractions so that the denominator is free of radicals. This is done by multiplying the fraction by a properly chosen form of unity, and the process is called **rationalizing the denominator.** For example, to rationalize the denominator of $1/\sqrt{3}$, we write

$$\frac{1}{\sqrt{3}} = \frac{1}{\sqrt{3}} \cdot \frac{\sqrt{3}}{\sqrt{3}} = \frac{\sqrt{3}}{\sqrt{3^2}} = \frac{\sqrt{3}}{3}$$

In this connection, a useful formula is

$$(\sqrt{m} + \sqrt{n})(\sqrt{m} - \sqrt{n}) = m - n$$

which we will apply in the following example.

EXAMPLE 7
Rationalize the denominator. Assume all radicals denote real numbers.

(a) $\sqrt{\dfrac{x}{y}}$ (b) $\dfrac{4}{\sqrt{5} - \sqrt{2}}$ (c) $\dfrac{5}{\sqrt{x} + 2}$ (d) $\dfrac{5}{\sqrt{x + 2}}$

SOLUTION

(a) $\sqrt{\dfrac{x}{y}} = \dfrac{\sqrt{x}}{\sqrt{y}} = \dfrac{\sqrt{x}}{\sqrt{y}} \cdot \dfrac{\sqrt{y}}{\sqrt{y}} = \dfrac{\sqrt{xy}}{\sqrt{y^2}} = \dfrac{\sqrt{xy}}{|y|}$

(b) $\dfrac{4}{\sqrt{5} - \sqrt{2}} = \dfrac{4}{\sqrt{5} - \sqrt{2}} \cdot \dfrac{\sqrt{5} + \sqrt{2}}{\sqrt{5} + \sqrt{2}}$

$= \dfrac{4(\sqrt{5} + \sqrt{2})}{5 - 2} = \dfrac{4}{3} (\sqrt{5} + \sqrt{2})$

(c) $\dfrac{5}{\sqrt{x} + 2} = \dfrac{5}{\sqrt{x} + 2} \cdot \dfrac{\sqrt{x} - 2}{\sqrt{x} - 2} = \dfrac{5(\sqrt{x} - 2)}{x - 4}$

(d) $\dfrac{5}{\sqrt{x} + 2} = \dfrac{5}{\sqrt{x} + 2} \cdot \dfrac{\sqrt{x} + 2}{\sqrt{x} + 2} = \dfrac{5\sqrt{x} + 2}{x + 2}$

PROGRESS CHECK

Rationalize the denominator. Assume all radicals denote real numbers.

(a) $\dfrac{-9xy^3}{\sqrt{3xy}}$ (b) $\dfrac{-6}{\sqrt{2} + \sqrt{6}}$ (c) $\dfrac{4}{\sqrt{x} - \sqrt{y}}$

ANSWERS

(a) $-3y^2\sqrt{3xy}$ (b) $\dfrac{3}{2} (\sqrt{2} - \sqrt{6})$ (c) $\dfrac{4(\sqrt{x} + \sqrt{y})}{x - y}$

COMPLEX NUMBERS

One of the central problems in algebra is that of finding solutions to a given polynomial equation. This problem will be discussed in later chapters of this book. However, observe at this point that there is no real number that satisfies a simple polynomial equation such as

$$x^2 = -4$$

since the square of a real number is always nonnegative.

To resolve this problem, mathematicians created a new number system built upon an "imaginary unit" i defined by $i = \sqrt{-1}$. This number i has the property that when we square both sides of the equation we have $i^2 = -1$, a result that cannot be obtained with real numbers. By definition,

$$i = \sqrt{-1}$$
$$i^2 = -1$$

We also assume that i behaves according to all the algebraic laws we have already developed (with the exception of the rules for inequalities for real numbers). This allows us to simplify higher powers of i. Thus,

$$i^3 = i^2 \cdot i = (-1)i = -i$$
$$i^4 = i^2 \cdot i^2 = (-1)(-1) = 1$$

Now it's easy to simplify i^n when n is any natural number. Since $i^4 = 1$, we simply seek the highest multiple of 4 that is less than or equal to n. For example,

$$i^5 = i^4 \cdot i = (1) \cdot i = i$$
$$i^{27} = i^{24} \cdot i^3 = (i^4)^6 \cdot i^3 = (1)^6 \cdot i^3 = i^3 = -i$$

EXAMPLE 8
Simplify.
(a) i^{51} (b) $-i^{74}$

SOLUTION
(a) $i^{51} = i^{48} \cdot i^3 = (i^4)^{12} \cdot i^3 = (1)^{12} \cdot i^3 = i^3 = -i$
(b) $-i^{74} = -i^{72} \cdot i^2 = -(i^4)^{18} \cdot i^2 = -(1)^{18} \cdot i^2 = -(1)(-1) = 1$

It is easy also to write square roots of negative numbers in terms of i. For example,

$$\sqrt{-25} = i\sqrt{25} = 5i$$

and, in general, we define

$$\boxed{\sqrt{-a} = i\sqrt{a} \text{ for } a > 0}$$

Any number of the form bi, where b is a real number, is called an **imaginary number.**

WARNING $\sqrt{-4}\sqrt{-9} \neq \sqrt{36}$

The rule $\sqrt{a} \cdot \sqrt{b} = \sqrt{ab}$ holds only when $a \geq 0$ and $b \geq 0$. Instead, write

$$\sqrt{-4}\sqrt{-9} = 2i \cdot 3i = 6i^2 = -6$$

Having created imaginary numbers, we next combine real and imaginary numbers. We say that $a + bi$, where a and b are real numbers, is a **complex number.** The number a is called the **real part** of $a + bi$ and b is called the **imaginary part.** The following are examples of complex numbers.

$$3 + 2i \qquad 2 - i \qquad -2i \qquad \frac{4}{5} + \frac{1}{5}i$$

Note that every real number a can be written as a complex number by choosing $b = 0$. Thus,

$$\boxed{a = a + 0i}$$

We see therefore that the real number system is a subset of the complex number system. Once again we have established a number system that incorporates all of the previous number systems and is itself more complicated than the earlier systems.

EXAMPLE 9
Write as a complex number.

(a) $-\dfrac{1}{2}$ (b) $\sqrt{-9}$ (c) $-1 - \sqrt{-4}$

SOLUTION

(a) $-\dfrac{1}{2} = -\dfrac{1}{2} + 0i$

(b) $\sqrt{-9} = i\sqrt{9} = 3i = 0 + 3i$

(c) $-1 - \sqrt{-4} = -1 - i\sqrt{4} = -1 - 2i$

We next seek to define operations with complex numbers in such a way that the rules for real numbers and the imaginary unit i continue to hold. We begin with equality and say that two complex numbers are **equal** if their real parts are equal and their imaginary parts are equal; that is,

$$a + bi = c + di \quad \text{if} \quad a = c \text{ and } b = d$$

EXAMPLE 10
Solve the equation $x + 3i = 6 - yi$ for x and for y.

SOLUTION
Equating the real parts, we have $x = 6$; equating the imaginary parts, $3 = -y$ or $y = -3$.

Complex numbers are added and subtracted by adding or subtracting the real parts and by adding or subtracting the imaginary parts. That is,

$$(a + bi) + (c + di) = (a + c) + (b + d)i$$
$$(a + bi) - (c + di) = (a - c) + (b - d)i$$

Note that the sum or difference of two complex numbers is again a complex number. Thus, the complex numbers are closed under addition. It is easy to verify that the associative and commutative laws of addition also hold under this definition.

EXAMPLE 11
Perform the indicated operations.
(a) $(7 - 2i) + (4 - 3i)$ (b) $14 - (3 - 8i)$

SOLUTION
(a) $(7 - 2i) + (4 - 3i) = (7 + 4) + (-2 - 3)i = 11 - 5i$
(b) $14 - (3 - 8i) = (14 - 3) + 8i = 11 + 8i$

PROGRESS CHECK
Perform the indicated operations.
(a) $(-9 + 3i) + (6 - 2i)$ (b) $7i - (3 + 9i)$

ANSWERS
(a) $-3 + i$ (b) $-3 - 2i$

We now define multiplication of complex numbers in a manner that permits the commutative, associative, and distributive laws to hold, along with the definition $i^2 = -1$. We must have

$$(a + bi)(c + di) = a(c + di) + bi(c + di)$$
$$= ac + adi + bci + bdi^2$$
$$= ac + (ad + bc)i + bd(-1)$$
$$= (ac - bd) + (ad + bc)i$$

The rule for multiplication is

$$(a + bi)(c + di) = (ac - bd) + (ad + bc)i$$

This result is significant since it demonstrates that the product of two complex numbers is again a complex number. It need not be memorized; simply use the distributive law to form all the products and the substitution $i^2 = -1$ to simplify.

EXAMPLE 12
Find the product of $(2 - 3i)$ and $(7 + 5i)$.

SOLUTION
$$(2 - 3i)(7 + 5i) = 2(7 + 5i) - 3i(7 + 5i)$$
$$= 14 + 10i - 21i - 15i^2$$
$$= 14 - 11i - 15(-1)$$
$$= 29 - 11i$$

PROGRESS CHECK
Find the product.
(a) $(-3 - i)(4 - 2i)$ (b) $(-4 - 2i)(2 - 3i)$

ANSWERS

(a) $-14 + 2i$ (b) $-14 + 8i$

EXERCISE SET 1.3

Simplify, using the rules for exponents. Write the answers using only positive exponents.

1. $(y^4)^{2n}$

2. $\dfrac{(-4)^6}{(-4)^{10}}$

3. $-\left(\dfrac{x}{y}\right)^3$

4. $-3r^3r^3$

5. $\dfrac{(r^2)^4}{(r^4)^2}$

6. $[(3b + 1)^5]^5$

7. $\left(\dfrac{3}{2}x^2y^3\right)^n$

8. $\dfrac{(-2a^2b)^4}{(-3ab^2)^3}$

9. $2^0 + 3^{-1}$

10. $(xy)^0 - 2^{-1}$

11. $\dfrac{3}{(2x^2 + 1)^0}$

12. $(-3)^{-3}$

13. $\dfrac{1}{3^{-4}}$

14. x^{-5}

15. $\dfrac{3a^5b^{-2}}{9a^{-4}b^2}$

16. $\left(\dfrac{x^3}{x^{-2}}\right)^2$

17. $\left(\dfrac{2a^2b^{-4}}{a^{-3}c^{-3}}\right)^2$

18. $\dfrac{2x^{-3}y^2}{x^{-3}y^{-3}}$

19. $(a - 2b^2)^{-1}$

20. $\left(\dfrac{y^{-2}}{y^{-3}}\right)^{-1}$

21. $\dfrac{a^{-1} + b^{-1}}{a^{-1} - b^{-1}}$

22. $\left(\dfrac{a}{b}\right)^{-1} + \left(\dfrac{b}{a}\right)^{-1}$

23. Show that $\left(\dfrac{a}{b}\right)^{-n} = \left(\dfrac{b}{a}\right)^n$

Evaluate each expression in Exercises 24–26.

24. $[(1.20)^2]^{-1}$

25. $[(-3.67)^2]^{-1}$

26. $\left[\dfrac{(7.65)^{-1}}{7.65^2}\right]^2$

Simplify and write answers using only positive exponents.

27. $16^{3/4}$

28. $(-125)^{-1/3}$

29. $\left(\dfrac{x^{3/2}}{x^{2/3}}\right)^{1/6}$

30. $\dfrac{125^{4/3}}{125^{2/3}}$

31. $(x^{1/3}y^2)^6$

32. $(x^6y^4)^{-1/2}$

Write each of the following in radical form.

33. $\left(\dfrac{1}{4}\right)^{2/5}$

34. $x^{2/3}$

35. $a^{3/4}$

36. $(-8x^2)^{2/5}$

Write each of the following in exponent form.

37. $\sqrt[4]{8^3}$

38. $\sqrt[5]{3^2}$

39. $\dfrac{1}{\sqrt[5]{(-8)^2}}$

40. $\dfrac{1}{\sqrt[3]{x^7}}$

Evaluate.

41. $\sqrt[4]{-81}$

42. $\sqrt[3]{\dfrac{1}{27}}$

43. $\sqrt{(-5)^2}$

44. $\sqrt{\left(-\dfrac{1}{3}\right)^2}$

45. $\sqrt{\left(\dfrac{5}{4}\right)^2}$

46. $(14.43)^{3/2}$

In Exercises 47 and 48 provide real values for x and y to demonstrate the result.

47. $\sqrt{x^2 + y^2} \neq x + y$

48. $\sqrt{x + y} \neq \sqrt{x} + \sqrt{y}$

Simplify each of the following and write the answer in simplified form. Every variable represents a positive real number.

49. $\sqrt{48}$
50. $\sqrt{200}$
51. $\sqrt[3]{54}$
52. $\sqrt{x^8}$

53. $\sqrt[3]{y^7}$
54. $\sqrt[4]{b^{14}}$
55. $\sqrt[4]{96x^{10}}$
56. $\sqrt{x^5y^4}$

57. $\sqrt{\dfrac{1}{5}}$
58. $\dfrac{4}{3\sqrt{11}}$
59. $\dfrac{1}{\sqrt{3y}}$
60. $\sqrt{\dfrac{2}{y}}$

Rationalize the denominator.

61. $\dfrac{3}{\sqrt{2}+3}$
62. $\dfrac{-3}{\sqrt{7}-9}$
63. $\dfrac{-3}{3\sqrt{a}+1}$
64. $\dfrac{4}{2-\sqrt{2y}}$

65. $\dfrac{\sqrt{2}+1}{\sqrt{2}-1}$
66. $\dfrac{\sqrt{5}+\sqrt{3}}{\sqrt{5}-\sqrt{3}}$
67. $\dfrac{\sqrt{6}+\sqrt{2}}{\sqrt{3}-\sqrt{2}}$

68. Prove that $|ab| = |a|\,|b|$. (*Hint:* Begin with $|ab| = \sqrt{(ab)^2}$.)

Simplify.

69. i^{60}
70. i^{27}
71. i^{-84}
72. $-i^{39}$

Write as a complex number in the form $a + bi$.

73. $-\dfrac{1}{2}$
74. -0.3
75. $\sqrt{-25}$
76. $-\sqrt{-5}$

77. $3 - \sqrt{-49}$
78. $-\dfrac{3}{2} - \sqrt{-72}$
79. $0.3 - \sqrt{-98}$
80. $-0.5 + \sqrt{-32}$

Solve for x and for y.

81. $(3x - 1) + (y + 5)i = 1 - 3i$
82. $\left(\dfrac{1}{2}x + 2\right) + (3y - 2)i = 4 - 7i$

83. $(2y + 1) - (2x - 1)i = -8 + 3i$
84. $(y - 2) + (5x - 3)i = 5$

Compute the answer and write it in the form $a + bi$.

85. $2i + (3 - i)$
86. $-3i + (2 - 5i)$
87. $2 + 3i + (3 - 2i)$
88. $(3 - 2i) - \left(2 + \dfrac{1}{2}i\right)$

89. $i\left(-\dfrac{1}{2} + i\right)$
90. $\dfrac{i}{2}\left(\dfrac{4 - i}{2}\right)$
91. $(2 - i)(2 + i)$
92. $(5 + i)(2 - 3i)$

In Exercises 93–96 evaluate the polynomial $x^2 - 2x + 5$ for the given complex value of x.

93. $1 + 2i$
94. $2 - i$
95. $1 - i$
96. $1 - 2i$

97. Prove that the commutative law of addition holds for the set of complex numbers.

98. Prove that the commutative law of multiplication holds for the set of complex numbers.

99. Prove that $0 + 0i$ is the additive identity and $1 + 0i$ is the multiplicative identity for the set of complex numbers.

100. Prove that $-a - bi$ is the additive inverse of the complex number $a + bi$.

101. Prove the distributive property for the set of complex numbers.

1.4
LINEAR EQUATIONS AND LINEAR INEQUALITIES IN ONE UNKNOWN

Expressions of the form

$$x - 2 = 0 \qquad x^2 - 9 = 0 \qquad 3(2x - 5) = 3$$

$$2x + 5 = \sqrt{x - 7} \qquad \frac{1}{2x + 3} = 5 \qquad x^3 - 3x^2 = 32$$

are examples of equations in the unknown x. An **equation** states that two algebraic expressions are equal. We refer to these expressions as the **left-hand** and **right-hand sides** of the equation.

Our task is to find values of the unknown for which the equation holds true. These values are called **solutions** or **roots** of the equation, and the set of all solutions is called the **solution set.** For example, 2 is a solution of the equation $3x - 1 = 5$ since $3(2) - 1 = 5$. However, -2 is *not* a solution since $3(-2) - 1 \neq 5$.

The solutions of an equation depend upon the number system we are using. For example, the equation $2x = 5$ has no integer solutions but does have a solution among the rational numbers, namely $\frac{5}{2}$. Similarly, the equation $x^2 = -4$ has no solutions among the real numbers but does have solutions if we consider complex numbers, namely $2i$ and $-2i$. The solution sets of these two equations are $\{\frac{5}{2}\}$ and $\{2i, -2i\}$, respectively.

We say that an equation is an **identity** if it is true for every real number for which both sides of the equation are defined. For example,

$$x^2 - 1 = (x + 1)(x - 1)$$

is an identity since it is true for all real numbers; that is, every real number is a solution of the equation. The equation

$$x - 5 = 3$$

is a false statement for all values of x except 8. If, as in that equation, there are real-number values of x for which the sides of the equation, although both defined, are unequal, the equation is called a **conditional equation.**

When we say that we want to "solve an equation," we mean that we want to find *all* solutions or roots. If we can replace an equation by another, simpler equation that has the same solutions, we will have an approach to solving equations. Equations having the same solutions are called **equivalent equations.** For example, $3x - 1 = 5$ and $3x = 6$ are equivalent equations since it can be shown that $\{2\}$ is the solution set of both equations.

There are two important rules that allow us to replace an equation with an equivalent equation.

Equivalent Equations	The solutions of a given equation are not affected by the following operations. (1) Addition or subtraction of the same number or expression on both sides of the equation. (2) Multiplication or division of both sides of the equation by a number other than 0.

EXAMPLE 1
Solve $3x + 4 = 13$.

SOLUTION
We apply the preceding rules to this equation. The strategy is to isolate x, so we *subtract 4 from both sides* of the equation.

$$3x + 4 - 4 = 13 - 4$$
$$3x = 9$$

Dividing both sides by 3, we obtain the solution

$$x = 3$$

To be technically accurate, the *solution* of the equation in Example 1 is 3, while $x = 3$ is an equation that is *equivalent* to the original equation. Now that this distinction is understood, we will join in the common usage that says that the equation $3x + 4 = 13$ "has the solution $x = 3$."

LINEAR EQUATIONS

An equation in which the variable appears only in the first degree is called a **first-degree equation in one unknown,** or more simply, a **linear equation.** The general form of such an equation is

$$ax + b = 0$$

where a and b are real numbers and $a \neq 0$. Solving this equation for x produces the following result.

Roots of a Linear Equation	The linear equation $ax + b = 0$, $a \neq 0$, has exactly one solution: $-b/a$.

LINEAR INEQUALITIES

Much of the terminology of equations carries over to inequalities. A **solution of an inequality** is a value of the unknown that satisfies the inequality, and the **solution set** is composed of all solutions. The properties of inequalities listed in Section 1.1 enable us to use the same procedures in solving inequalities as those used in solving equations *with one exception.*

Multiplication or division of an inequality by a negative number reverses the sense of the inequality.

We will concentrate for now on solving a **linear inequality,** that is, an inequality in which the unknown appears only in the first degree.

EXAMPLE 2
Solve the inequality.

$$2x + 11 \geq 5x - 1$$

SOLUTION
We perform addition and subtraction to collect terms in x just as we did for equations.

$$2x + 11 \geq 5x - 1$$
$$2x \geq 5x - 12$$
$$-3x \geq -12$$

We now divide both sides of the inequality by -3, a negative number, and therefore reverse the sense of the inequality.

$$\frac{-3x}{-3} \leq \frac{-12}{-3}$$
$$x \leq 4$$

PROGRESS CHECK
Solve the inequality $3x - 2 \geq 5x + 4$.

ANSWER
$x \leq -3$

There are three methods commonly used to describe subsets of the real numbers: graphs on a real number line, interval notation, and set-builder notation. Since there will be occasions when we want to use each of these schemes, this is a convenient time to introduce them and to apply them to inequalities.

The **graph of an inequality** is the set of all points satisfying the inequality. The graph of the inequality $a \leq x < b$ is shown in Figure 4. The portion of the real number line that is in color is the solution set of the inequality. The circle at point a has been filled in to indicate that a is also a solution of the inequality; the circle at point b has been left open to indicate that b is not a member of the solution set.

a b
FIGURE 4

An **interval** is a set of numbers on the real number line that forms a line segment, a half-line, or the entire real number line. The subset shown in Figure 4 would be written in **interval notation** as $[a, b)$, where a and b are the **endpoints** of the interval. A bracket, [or], indicates that the endpoint is included, while a parenthesis, (or), indicates that the endpoint is not included. The interval $[a, b]$

is called a **closed interval** because both endpoints are included. The interval (a, b) is called an **open interval** because neither endpoint is included. Finally, the intervals $[a, b)$ and $(a, b]$ are called **half-open intervals.**

The set of all real numbers satisfying a given property P is written as

$$\{x \mid x \text{ satisfies property } P\}$$

which is read as "the set of all x such that x satisfies property P." This form, called **set-builder notation,** provides a third means of designating subsets of the real number line. Thus, the interval $[a, b)$ shown in Figure 4 is written as

$$\{x \mid a \le x < b\}$$

which indicates that x must satisfy the inequalities $x \ge a$ and $x < b$.

EXAMPLE 3

Graph each of the given intervals on a real number line and indicate the same subset of the real number line in set-builder notation.

(a) $(-3, 2]$ (b) $(1, 4)$ (c) $[-4, -1]$

SOLUTION

(a) $\qquad\qquad\qquad\qquad\qquad\qquad\qquad\qquad \{x \mid -3 < x \le 2\}$

(b) $\qquad\qquad\qquad\qquad\qquad\qquad\qquad\qquad \{x \mid 1 < x < 4\}$

(c) $\qquad\qquad\qquad\qquad\qquad\qquad\qquad\qquad \{x \mid -4 \le x \le -1\}$

To describe the inequalities $x > 2$ and $x \le 3$ in interval notation, we need to introduce the symbols ∞ and $-\infty$ (read "infinity" and "minus infinity," respectively). The inequalities $x > 2$ and $x \le 3$ are then written as $(2, \infty)$ and $(-\infty, 3]$, respectively, in interval notation and would be graphed on a real number line as shown in Figure 5. Note that ∞ and $-\infty$ are symbols (not numbers) indicating that the intervals extend indefinitely. An interval using one of these symbols is called an **infinite interval.** The interval $(-\infty, \infty)$ designates the entire real number line. Square brackets must never be used around ∞ and $-\infty$, since they are not real numbers.

FIGURE 5

EXAMPLE 4

Graph each inequality and write the solution set in interval notation.

(a) $x \le -2$ (b) $x \ge -1$ (c) $x < 3$

SOLUTION

(a) $(-\infty, -2]$

(b) $[-1, \infty)$

(c) $(-\infty, 3)$

EXAMPLE 5
Solve the inequality.

$$\frac{x}{2} - 9 < \frac{1 - 2x}{3}$$

Graph the solution set and write the solution set in both interval notation and set-builder notation.

SOLUTION
To clear the inequality of fractions, we multiply both sides by the least common denominator (L.C.D.) of all fractions, which is 6.

$$3x - 54 < 2(1 - 2x)$$
$$3x - 54 < 2 - 4x$$
$$7x < 56$$
$$x < 8$$

We may write the solution set as $\{x | x < 8\}$ or as the infinite interval $(-\infty, 8)$. The graph of the solution set is shown in Figure 6.

FIGURE 6

EXAMPLE 6
Solve the inequality.

$$\frac{2(x + 1)}{3} < \frac{2x}{3} - \frac{1}{6}$$

SOLUTION
The L.C.D. of all fractions is 6. Multiplying both sides of the inequality by 6, we obtain

$$4(x + 1) < 4x - 1$$
$$4x + 4 < 4x - 1$$
$$4 < -1$$

Our procedure has led to a contradiction, indicating that there is no solution to the inequality.

PROGRESS CHECK

Solve, and write the answers in interval notation.

(a) $\dfrac{3x - 1}{4} + 1 > 2 + \dfrac{x}{3}$ (b) $\dfrac{2x - 3}{2} \geq x + \dfrac{2}{5}$

ANSWERS

(a) $(3, \infty)$ (b) no solution

We can solve double inequalities such as

$$1 < 3x - 2 \leq 7$$

by operating on both inequalities at the same time.

$$3 < 3x \leq 9 \quad \text{Add } +2 \text{ to each member.}$$
$$1 < x \leq 3 \quad \text{Divide each member by } 3.$$

The solution set is the half-open interval $(1, 3]$.

EXAMPLE 7

Solve the inequality $-3 \leq 1 - 2x < 6$ and write the answer in interval notation.

SOLUTION

Operating on both inequalities we have

$$-4 \leq -2x < 5 \quad \text{Add } (-1) \text{ to each member.}$$
$$2 \geq x > -\dfrac{5}{2} \quad \text{Divide each member by } -2.$$

The solution set is the half-open interval $\left(-\dfrac{5}{2}, 2\right]$.

PROGRESS CHECK

Solve the inequality $-5 < 2 - 3x < -1$ and write the answer in interval notation.

ANSWER

$\left(1, \dfrac{7}{3}\right)$

EXAMPLE 8

A taxpayer may choose to pay a 20% tax on the gross income or a 25% tax on the gross income less $4000. Above what income level should the taxpayer elect to pay at the 20% rate?

SOLUTION

If we let $x =$ gross income, then the choice available to the taxpayer is
(a) pay at the 20% rate on the gross income, that is, pay $0.20x$, or
(b) pay at the 25% rate on the gross income less \$4000, that is, pay $0.25(x - 4000)$.
To determine when (a) produces a lower tax than (b), we must solve

$$0.20x \le 0.25(x - 4000)$$
$$0.20x \le 0.25x - 1000$$
$$-0.05x \le -1000$$
$$x \ge \frac{1000}{0.05} = 20,000$$

The taxpayer should choose to pay at the 20% rate if the gross income is \$20,000 or more.

PROGRESS CHECK

A customer is offered the following choice of telephone services: unlimited local calls at a fixed \$20 monthly charge, or a base rate of \$8 per month plus 6¢ per message unit. At what level of usage does it cost less to choose the unlimited service?

ANSWER

When the anticipated use exceeds 200 message units.

EXERCISE SET 1.4

Solve the given linear equation and check your answer in the following exercises.

1. $-2x + 6 = -5x - 4$

2. $6x + 4 = -3x - 5$

3. $2(3b + 1) = 3b - 4$

4. $-3(2x + 1) = -8x + 1$

5. $4(x - 1) = 2(x + 3)$

6. $-3(x - 2) = 2(x + 4)$

7. $2(x + 4) - 1 = 0$

8. $3a + 2 - 2(a - 1) = 3(2a + 3)$

9. $-4(2x + 1) - (x - 2) = -11$

10. $3(a + 2) - 2(a - 3) = 0$

Solve for x.

11. $kx + 8 = 5x$

12. $8 - 2kx = -3x$

13. $2 - k + 5(x - 1) = 3$

14. $3(2 + 3k) + 4(x - 2) = 5$

Indicate whether the equation is an identity (I) or a conditional equation (C).

15. $x^2 + x - 2 = (x + 2)(x - 1)$

16. $(x - 2)^2 = x^2 - 4x + 2$

17. $2x + 1 = 3x - 1$

18. $3x - 5 = 4x - x - 2 - 3$

Solve the inequality and graph the result.

19. $x + 4 < 8$

20. $x + 5 < 4$

21. $x + 3 < -3$

22. $x - 2 \le 5$

23. $x - 3 \ge 2$

24. $x + 5 \ge -1$

25. $2 < a + 3$

26. $-5 > b - 3$

27. $2y < -1$

28. $3x < 6$

29. $2x \geq 0$

30. $-\dfrac{1}{2}y \geq 4$

31. $2r + 5 < 9$

32. $3x - 2 > 4$

33. $3x - 1 \geq 2$

34. $\dfrac{-1}{2x + 3} > 0$

35. $\dfrac{4}{5 - 3x} < 0$

36. $\dfrac{3}{3x - 1} > 0$

Solve the given inequality in Exercises 37–60 and write the solution set in interval notation.

37. $4x + 3 \leq 11$

38. $\dfrac{1}{2}y - 2 \leq 2$

39. $\dfrac{3}{2}x + 1 \geq 4$

40. $-5x + 2 > -8$

41. $4(2x + 1) < 16$

42. $3(3r - 4) \geq 15$

43. $2(x - 3) < 3(x + 2)$

44. $4(x - 3) \geq 3(x - 2)$

45. $3(2a - 1) > 4(2a - 3)$

46. $2(3x - 1) + 4 < 3(x + 2) - 8$

47. $\dfrac{2}{3}(x + 1) + \dfrac{5}{6} \geq \dfrac{1}{2}(2x - 1) + 4$

48. $\dfrac{1}{4}(3x + 2) - 1 \leq -\dfrac{1}{2}(x - 3) + \dfrac{3}{4}$

49. $\dfrac{x - 1}{3} + \dfrac{1}{5} < \dfrac{x + 2}{5} - \dfrac{1}{3}$

50. $\dfrac{x}{5} - \dfrac{1 - x}{2} > \dfrac{x}{2} - 3$

51. $3(x + 1) + 6 \geq 2(2x - 1) + 4$

52. $4(3x + 2) - 1 \leq -2(x - 3) + 15$

53. $-2 < 4x \leq 5$

54. $3 \leq 6x < 12$

55. $-4 \leq 2x + 2 \leq -2$

56. $5 \leq 3x - 1 \leq 11$

57. $3 \leq 1 - 2x < 7$

58. $5 < 2 - 3x \leq 11$

59. $-8 < 2 - 5x \leq 7$

60. $-10 < 5 - 2x < -5$

61. A student has grades of 42 and 70 in the first two tests of the semester. If an average of 70 is required to obtain a C grade, what is the minimum score the student must achieve on the third exam to obtain a C?

62. A compact car can be rented from firm A for $160 per week with no charge for mileage, or from firm B for $100 per week plus 20 cents for each mile driven. If the car is driven m miles, for what values of m does it cost less to rent from firm A?

63. An appliance salesperson is paid $30 per day plus $25 for each appliance sold. How many appliances must be sold for the salesperson's income to exceed $130 per day?

64. A pension trust invests $6000 in a bond that pays 5% simple interest per year. Additional funds are to be invested in a more speculative bond paying 9% sim-ple interest per year, so that the return on the total investment will be at least 6%. What is the minimum amount that must be invested in the more speculative bond?

65. A book publisher spends $19,000 on editorial expenses and $6 per book for manufacturing and sales expenses in the course of publishing a psychology textbook. If the book sells for $12.50, how many copies must be sold to show a profit?

66. If the area of a right triangle is not to exceed 80 square inches and the base is 10 inches, what values may be assigned to the altitude h?

67. A total of 70 meters of fencing material is available with which to enclose a rectangular area. If the width of the rectangle is 15 meters, what values can be assigned to the length L?

1.5
ABSOLUTE VALUE
IN EQUATIONS
AND INEQUALITIES

In Section 1.1 we discussed the use of absolute value notation to indicate distance and provided this formal definition.

$$|x| = \begin{cases} x & \text{if } x \geq 0 \\ -x & \text{if } x < 0 \end{cases}$$

The following example illustrates the application of this definition to the solution of equations involving absolute value.

EXAMPLE 1
Solve the equation $|2x - 7| = 11$.

SOLUTION
We apply the definition of absolute value to the two cases.

Case 1. $2x - 7 \geq 0$ *Case 2.* $2x - 7 < 0$

With the first part of the definition,   With the second part of the definition,

$$|2x - 7| = 2x - 7 = 11$$ $$|2x - 7| = -(2x - 7) = 11$$

$$2x = 18$$ $$-2x + 7 = 11$$

$$x = 9$$ $$x = -2$$

PROGRESS CHECK
Solve each equation and check the solution(s).

(a) $|x + 8| = 9$ (b) $|3x - 4| = 7$

ANSWERS

(a) $1, -17$ (b) . $\dfrac{11}{3}, -1$

When used in inequalities, absolute value notation plays an important and frequently used role in higher mathematics. To solve inequalities involving absolute value, we recall that $|x|$ is the distance between the origin and the point on the real number line corresponding to x. For $a > 0$, the solution set of the inequality $|x| < a$ is then seen to consist of all real numbers whose distance from the origin is less than a, that is, all real numbers in the open interval $(-a, a)$ shown in Figure 7. Similarly, if $|x| > a > 0$, the solution set consists of all real

FIGURE 7

numbers whose distance from the origin is greater than a, that is, all points in the infinite intervals $(-\infty, -a)$ and (a, ∞) as shown in Figure 8. Of course, $|x| \leq a$ and $|x| \geq a$ would include the endpoints a and $-a$, and the circles would be filled in.

FIGURE 8

EXAMPLE 2
Solve $|2x - 5| \leq 7$, graph the solution set, and write the solution set in interval notation.

SOLUTION
We must solve the equivalent double inequality

$$-7 \leq 2x - 5 \leq 7$$
$$-2 \leq 2x \leq 12 \qquad \text{Add } +5 \text{ to each member.}$$
$$-1 \leq x \leq 6 \qquad \text{Divide each member by 2.}$$

The graph of the solution set is then

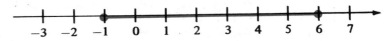

Thus, the solution set is the closed interval $[-1, 6]$.

PROGRESS CHECK
Solve each inequality, graph the solution set, and write the solution set in interval notation.
(a) $|x| < 3$ (b) $|3x - 1| \leq 8$ (c) $|x| < -2$

ANSWERS

(a) $(-3, 3)$

(b) $\left[-\dfrac{7}{3}, 3\right]$

(c) No solution. Since $|x|$ is always nonnegative, $|x|$ cannot be less than -2.

EXAMPLE 3
Solve the inequality $|2x - 6| > 4$, write the solution set in interval notation, and graph the solution.

SOLUTION
We must solve the equivalent inequalities

$$2x - 6 > 4 \qquad\qquad 2x - 6 < -4$$
$$2x > 10 \qquad\qquad\quad 2x < 2$$
$$x > 5 \qquad\qquad\quad\;\; x < 1$$

The solution set consists of the real numbers in the infinite intervals $(-\infty, 1)$ and $(5, \infty)$. The graph of the solution set is then

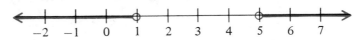

PROGRESS CHECK

Solve each inequality, write the solution set in interval notation, and graph the solution.

(a) $|5x - 6| > 9$ (b) $|2x - 2| \geq 8$

ANSWERS

(a) $\left(-\infty, -\dfrac{3}{5}\right), (3, \infty)$

(b) $(-\infty, -3], [5, \infty)$

WARNING Students sometimes write

$$1 > x > 5$$

This is a misuse of the inequality notation since it states that x is simultaneously less than 1 *and* greater than 5, which is impossible. What is usually intended is the pair of infinite intervals $(-\infty, 1)$ and $(5, \infty)$, and the inequalities that must be written are

$$x < 1 \qquad x > 5$$

EXERCISE SET 1.5

Solve and check.

1. $|x + 2| = 3$
2. $|r - 5| = \dfrac{1}{2}$
3. $|2x - 4| = 2$
4. $|5y + 1| = 11$

5. $|-3x + 1| = 5$
6. $|2t + 2| = 0$
7. $3|-4x - 3| = 27$
8. $\dfrac{1}{|x|} = 5$

9. $\dfrac{1}{|s - 1|} = \dfrac{1}{3}$

Solve and graph the solution set.

10. $|x + 3| < 5$
11. $|x + 1| > 3$
12. $|3x + 6| \leq 12$
13. $|4x - 1| > 3$

14. $|3x + 2| \geq -1$
15. $\left|\dfrac{1}{3} - x\right| < \dfrac{2}{3}$

Solve and write the solution set using interval notation.

16. $|x - 2| \leq 4$
17. $|x - 3| \geq 4$
18. $|2x + 1| < 5$
19. $\dfrac{|2x - 1|}{4} < 2$

20. $\dfrac{|3x + 2|}{2} \leq 4$
21. $\dfrac{|2x + 1|}{3} < 0$
22. $\left|\dfrac{4}{3x - 2}\right| < 1$
23. $\left|\dfrac{5 - x}{3}\right| > 4$

24. $\left|\dfrac{2x + 1}{3}\right| \leq 5$

In Exercises 25 and 26, x and y are real numbers.

25. Prove that $\left|\dfrac{x}{y}\right| = \dfrac{|x|}{|y|}$. (*Hint:* Treat as four cases.)

26. Prove that $|x|^2 = x^2$.

27. A machine that packages 100 vitamin pills per bottle can make an error of 2 pills per bottle. If x is the number of pills in a bottle, write an inequality, using absolute value, that indicates a maximum error of 2 pills per bottle. Solve the inequality.

28. The weekly income of a worker in a manufacturing plant differs from $300 by no more than $50. If x is the weekly income, write an inequality, using absolute value, that expresses this relationship. Solve the inequality.

1.6
SECOND-DEGREE INEQUALITIES

To solve a second-degree inequality, such as

$$x^2 - 2x > 15$$

2 unknowns

we rewrite the inequality in the form

$$x^2 - 2x - 15 > 0$$

or, after factoring,

$$(x + 3)(x - 5) > 0$$

With the right-hand side equal to 0, this inequality requires that the product of the two factors, which represent real numbers, be positive. That means that both factors must have the same sign. We must therefore analyze the *signs* of $(x + 3)$ and $(x - 5)$.

In any situation like this we are interested in knowing all values of x for which the general expression $(ax + b)$ will be positive and those values for which it will be negative. Since $ax + b = 0$ when $x = -b/a$, we see that

why not quadratic equation

The linear factor $ax + b$ equals 0 at the **critical value** $x = -b/a$ and has opposite signs to the left and right of the critical value on a number line.

A practical means for solving such problems as the current example is illustrated in Figure 9. Since the critical values occur where $x + 3 = 0$ and $x - 5 = 0$, the values -3 and $+5$ are displayed on a real number line. The rows above the real number line display the *signs* of the factors $x + 3$ and $x - 5$ for all real values of x. The row below the real number line displays the *signs* of the product $(x + 3)(x - 5)$. The product is positive when the factors have the same sign, is negative when the factors are of opposite sign, and is zero when either factor is zero. The row below the real number line shows the solution set of the inequality $(x + 3)(x - 5) > 0$ to be

$$\{x \mid x < -3 \text{ or } x > 5\}$$

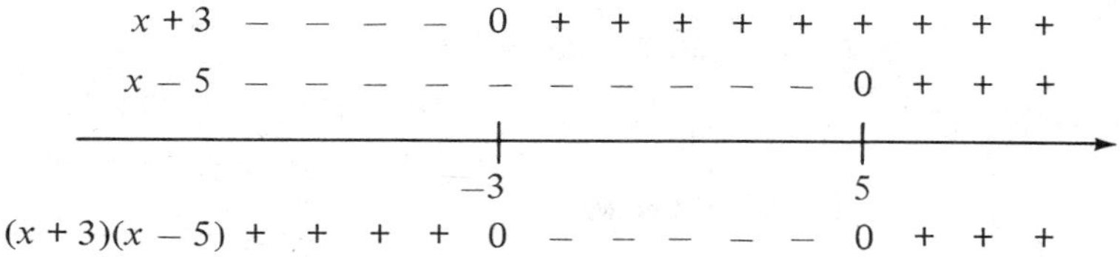

FIGURE 9

which consists of the real numbers in the open intervals $(-\infty, -3)$ and $(5, \infty)$. The solution set is shown in Figure 10.

FIGURE 10

EXAMPLE 1

Solve the inequality $x^2 \leq -3x + 4$ and graph the solution set on a real number line.

SOLUTION

We rewrite the inequality and factor.

$$x^2 + 3x - 4 \leq 0$$
$$(x - 1)(x + 4) \leq 0$$

We seek values of x for which the factors $(x - 1)$ and $(x + 4)$ have opposite signs or are zero. The critical values occur where $x - 1 = 0$ and where $x + 4 = 0$, that is, at $+1$ and -4. Figure 11 gives an analysis of the signs of the factors $x - 1$

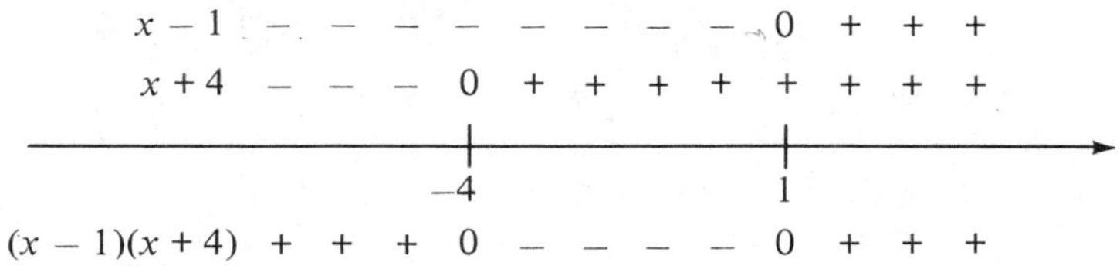

FIGURE 11

and $x + 4$ as well as the signs of their product, $(x - 1)(x + 4)$. We see that the solution set consists of all real numbers

$$\{x \mid -4 \leq x \leq 1\}$$

which is the closed interval $[-4, 1]$, shown in Figure 12.

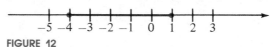

FIGURE 12

PROGRESS CHECK

Solve the inequality $2x^2 \geq 5x + 3$ and graph the solution set on a real number line.

ANSWERS

$$\left\{x \mid x \leq -\frac{1}{2} \text{ or } x \geq 3\right\}$$

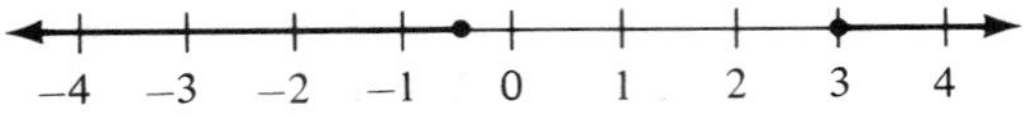

Although

$$\frac{ax + b}{cx + d} < 0$$

is not a second-degree inequality, the solution of this inequality is the same as the solution of the inequality

$$(ax + b)(cx + d) < 0$$

since both inequalities require that the two expressions composing them have different signs.

EXAMPLE 2

Solve the inequality $\dfrac{y + 1}{2 - y} \leq 0$.

SOLUTION

Figure 13 gives an analysis of the signs of $y + 1$ and $2 - y$. The critical values occur where $y + 1 = 0$ and where $2 - y = 0$, that is, at -1 and $+2$. The bottom row shows the signs of the quotient $(y + 1)/(2 - y)$, from which we see that the solution set is $\{y \mid y \leq -1 \text{ or } y > 2\}$ or all real numbers in the intervals $(-\infty, -1]$, $(2, \infty)$. Note that $y = 2$ would result in division by 0 and must be excluded.

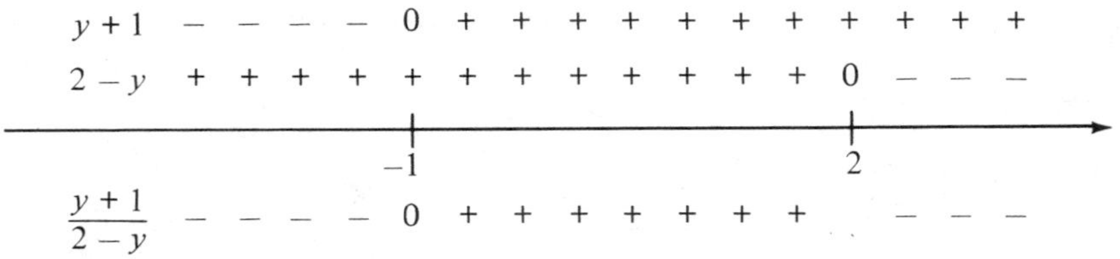

FIGURE 13

PROGRESS CHECK

Solve the inequality $\dfrac{2x - 3}{1 - 2x} \geq 0$.

ANSWERS

$$\left\{x\mid\frac{1}{2}<x\le\frac{3}{2}\right\}\quad\text{or}\quad\left(\frac{1}{2},\frac{3}{2}\right]$$

EXAMPLE 3

Solve the inequality $(x-2)(2x+5)(3-x)<0$.

SOLUTION

Although this is a third-degree inequality, the same approach will work. Figure 14 gives an analysis of the signs of $x-2$, $2x+5$, and $3-x$. The product of

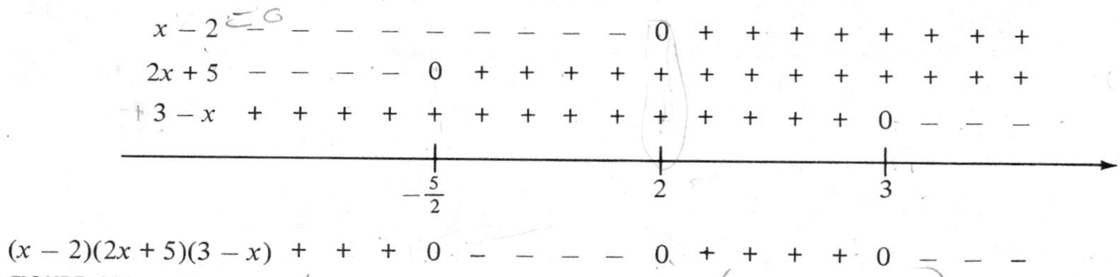

FIGURE 14

three factors is negative when there are an odd number of negative factors. The solution set is then

$$\left\{x\mid-\frac{5}{2}<x<2 \text{ or } x>3\right\}\quad\text{or}\quad\left(\frac{5}{2},2\right),\ (3,\infty)$$

PROGRESS CHECK

Solve the inequality $(2y-9)(6-y)(y+5)\ge0$.

ANSWERS

$$\left\{y\mid y\le-5 \text{ or } \frac{9}{2}\le y\le6\right\}\quad\text{or}\quad(-\infty,-5],\ \left[-\frac{9}{2},6\right]$$

EXERCISE SET 1.6

Determine the solution set of each inequality.

1. $x^2+5x+6>0$ 2. $x^2+3x-4\le0$ 3. $2x^2-x-1<0$ 4. $3x^2-4x-4\ge0$

5. $4x-2x^2<0$ 6. $r^2+4r\ge0$ 7. $\dfrac{x+5}{x+3}\le0$ 8. $\dfrac{x-6}{x+4}\ge0$

9. $\dfrac{2r+1}{r-3}\le0$ 10. $\dfrac{x-1}{2x-3}\ge0$ 11. $\dfrac{3s+2}{2s-1}\ge0$ 12. $\dfrac{4x+5}{x^2}\le0$

13. $(x+2)(3x-2)(x-1)>0$ 14. $(x-4)(2x+5)(2-x)\le0$

Indicate the solution set of each inequality on a real number line.

15. $x^2 + x - 6 > 0$

16. $x^2 - 3x - 10 \geq 0$

17. $2x^2 - 3x - 5 < 0$

18. $3x^2 - 4x - 4 \leq 0$

19. $\dfrac{2r + 3}{2r - 1} < 0$

20. $\dfrac{3x + 2}{2x - 3} \geq 0$

21. $\dfrac{x - 1}{x + 1} \geq 0$

22. $\dfrac{2x - 1}{x + 2} \leq 0$

23. $6x^2 + 8x + 2 \geq 0$

24. $2x^2 + 5x + 2 \leq 0$

25. $(y - 3)(2 - y)(2y + 4) \geq 0$

26. $(2x + 5)(3x - 2)(x + 1) < 0$

27. $(x - 3)(1 + 2x)(3x + 5) > 0$

28. $(1 - 2x)(2x + 1)(x - 3) \leq 0$

In Exercises 29–32, find the values of x for which the given expression has real values.

29. $\sqrt{(x - 2)(x + 1)}$

30. $\sqrt{(2x + 1)(x - 3)}$

31. $\sqrt{2x^2 + 7x + 6}$

32. $\sqrt{2x^2 + 3x + 1}$

33. A manufacturer of solar heaters finds that when x units are made and sold, the profit (in thousands of dollars) is given by $x^2 - 50x - 5000$. For what values of x will the firm show a loss?

34. A ball thrown directly upward from level ground at an initial velocity of 40 feet per second attains a height d given by $d = 40t - 16t^2$ after t seconds. During what time interval is the ball at a height of at least 16 feet?

TERMS AND SYMBOLS

set (p. 2)
element, member (p. 2)
{ } (p. 2)
subset (p. 2)
natural numbers (p. 2)
integers (p. 2)
rational numbers (p. 2)
irrational numbers (p. 2)
real number system (p. 3)
real number line (p. 3)
origin (p. 3)
nonnegative (p. 3)
$<, >, \leq, \geq$ (p. 3)
inequality symbols (p. 4)
inequalities (p. 4)
absolute value (p. 5)
| | (p. 5)
$\overline{AB}$ (p. 6)
factor (p. 10)
variable (p. 12)
algebraic expression (p. 13)
constant (p. 13)

algebraic operations (p. 13)
evaluate (p. 13)
base (p. 13)
exponent (p. 13)
power (p. 13)
polynomial (p. 14)
monomial (p. 14)
coefficient (p. 14)
degree of a monomial (p. 15)
degree of a polynomial (p. 15)
constant term (p. 15)
leading coefficient (p. 15)
zero polynomial (p. 15)
like terms (p. 15)
factoring (p. 17)
prime polynomial (p. 22)
irreducible polynomial (p. 22)
nth root (p. 26)
square root (p. 26)

cube root (p. 26)
principal nth root (p. 26)
radical sign (p. 28)
principal square root (p. 28)
radical form (p. 28)
rationalizing the denominator (p. 30)
imaginary unit i (p. 31)
imaginary number (p. 32)
complex number (p. 32)
real part (p. 32)
imaginary part (p. 32)
equation (p. 37)
left-hand side (p. 37)
right-hand side (p. 37)
solution (p. 37)
root (p. 37)
solution set (p. 37)
identity (p. 37)
conditional equation (p. 37)
equivalent equation (p. 37)

first-degree equation in one unknown (p. 38)
linear equation (p. 38)
solution of an inequality (p. 38)
linear inequality (p. 38)
graph of an inequality (p. 39)
interval (p. 39)
interval notation (p. 39)
open interval (p. 40)
closed interval (p. 40)
half-open interval (p. 40)
set-builder notation (p. 40)
$\infty, -\infty$ (p. 40)
infinite interval (p. 40)
second-degree inequality (p. 48)
critical value (p. 48)

KEY IDEAS FOR REVIEW

☐ A set is simply a collection of objects or numbers.

☐ The real number system is composed of the rational and irrational numbers. The rational numbers are those that can be written as the ratio of two integers, p/q, with

$q \neq 0$; the irrational numbers cannot be written as a ratio of integers.

☐ The real number system satisfies a number of important properties. These are

closure   commutativity   associativity
identities   inverses   distributivity

□ If two numbers are identical, we say that they are equal. Equality satisfies these basic properties:

reflexive property   symmetric property
transitive property   substitution property

□ There is a one-to-one correspondence between the set of all real numbers and the set of all points on the real number line. That is, for every point on the line there is a real number and for every real number there is a point on the line.

□ Algebraic statements using inequality symbols have straightforward geometric interpretations using the real number line. For example, $a < b$ says that a lies to the left of b on the real number line.

□ Absolute value specifies distance independent of the direction. Three important properties of absolute value are

$$|a| \geq 0 \quad |a| = |-a| \quad |a - b| = |b - a|$$

□ The distance between points A and B whose coordinates are a and b, respectively, is given by

$$\overline{AB} = |b - a|$$

□ Algebraic expressions of the form

$$P = a_n x^n + a_{n-1} x^{n-1} + \cdots + a_1 x + a_0$$

are called polynomials.

□ To add (subtract) polynomials, simply add (subtract) like terms. To multiply polynomials, form all possible products using the rule for exponents: $a^m a^n = a^{m+n}$.

□ A polynomial is said to be factored when it is written as a product of polynomials of lower degree.

□ The rules for positive integer exponents also apply to zero and negative integer exponents and to rational exponents.

□ Radical notation is simply another way of writing a rational exponent. That is, $\sqrt[n]{b} = b^{1/n}$.

□ If n is even and b is positive, there are two real numbers a such that $b^{1/n} = a$. Under these circumstances, we insist that the nth root be positive. That is, $\sqrt[n]{b}$ is a positive number if n is even and b is positive. Thus, $\sqrt{16} = 4$.

□ We must write $\sqrt{x^2} = |x|$ to insure that the result is a positive number.

□ Complex numbers were created because there are no real numbers that satisfy such simple polynomial equations as $x^2 + 5 = 0$.

□ Using the imaginary unit $i = \sqrt{-1}$, a complex number is of the form $a + bi$, where a and b are real numbers.

□ The real number system is a subset of the complex number system.

□ The linear equation $ax + b = 0$, $a \neq 0$, has the solution $-b/a$.

□ Inequalities can be operated upon in the same manner as statements involving an equals sign, with one important exception. When an inequality is multiplied or divided by a negative number, the sense of the inequality is reversed.

□ The solution set of a linear inequality can be indicated by graphing on a real number line, by set-builder notation, or by interval notation.

□ If a second-degree inequality can be written in the factored form

$$(ax + b)(cx + d) < 0$$

or

$$(ax + b)(cx + d) > 0$$

then the solution set is easily found. First, on the real number line, determine the intervals in which each factor is positive and the intervals in which each is negative. If the product of the factors is negative (< 0), then the solution set consists of the intervals in which the factors are opposite in sign; if the product is positive (> 0), the solution set consists of the intervals in which the factors are of the same sign.

REVIEW EXERCISES

Solutions to exercises whose numbers are in color are in the Solutions section in the back of the book.

1.1 For Exercises 1–4 determine whether the statement is true (T) or false (F).

1. $\sqrt{7}$ is a real number.

2. -35 is a natural number.

3. -14 is not an integer.

4. 0 is an irrational number.

In Exercises 5–8, provide a counterexample to the given statement.

5. The sum of two irrational numbers is an irrational number.

6. The product of two irrational numbers is an irrational number.

7. If a is a nonnegative real number, then $\sqrt{a}$ is irrational.

8. If a and b are real numbers such that $|a| = |b|$, then $a = b$.

In Exercises 9–11 sketch the given set of numbers on a real number line.

9. The negative real numbers.

10. The real numbers x such that $x > 4$.

11. The real numbers x such that $-1 \leq x < 1$.

12. Find the value of $|-3| - |1 - 5|$.

13. Find $\overline{PQ}$ if the coordinates of P and Q are $\frac{3}{2}$ and 6, respectively.

1.2 14. A salesperson receives $3.25x + 0.15y$ dollars, where x is the number of hours worked and y is the number of miles of automobile usage. Find the amount due the salesperson if $x = 12$ hours and $y = 80$ miles.

15. Which of the following expressions are not polynomials?
    (a) $-2xy^2 + x^2y$ (b) $3b^2 + 2b - 6$
    (c) $x^{-1/2} + 5x^2 - x$ (d) $7.5x^2 + 3x - \frac{1}{2}x^0$

In Exercises 16 and 17 indicate the leading coefficient and the degree of each polynomial.

16. $-0.5x^7 + 6x^3 - 5$ 17. $2x^2 + 3x^4 - 7x^5$

In Exercises 18–20 perform the indicated operations.

18. $(3a^2b^2 - a^2b + 2b - a) - (2a^2b^2 + 2a^2b - 2b - a)$

19. $x(2x - 1)(x + 2)$ 20. $3x(2x + 1)^2$

In Exercises 21–26 factor each expression.

21. $2x^2 - 2$

22. $x^2 - 25y^2$

23. $2a^2 + 3ab + 6a + 9b$

24. $4x^2 + 19x - 5$

25. $x^8 - 1$

26. $27r^6 + 8s^6$

1.3 In Exercises 27–33 simplify, using only positive exponents to express the answers. All variables are positive numbers.

27. $(2a^2b^{-3})^{-3}$

28. $2(a^2 - 1)^0$

29. $\left(\dfrac{x^3}{y^{-6}}\right)^{-4/3}$

30. $\dfrac{x^{3+n}}{x^n}$

31. $\sqrt{80}$

32. $\dfrac{2}{\sqrt{12}}$

33. $\dfrac{\sqrt{x}}{\sqrt{x} + \sqrt{y}}$

34. Solve for x and for y:
    $$(x - 2) + (2y - 1)i = -4 + 7i$$

35. Simplify i^{47}.

In Exercises 36–38 perform the indicated operations and write all answers in the form $a + bi$.

36. $2 + (6 - i)$ 37. $(2 + i)^2$

38. $(4 - 3i)(2 + 3i)$

1.4 In Exercises 39–42 solve for x.

39. $3x - 5 = 3$

40. $2(2x - 3) - 3(x + 1) = -9$

41. $\dfrac{2 - x}{3 - x} = 4$

42. $k - 2x = 4kx$

43. Indicate whether the statement is true (T) or false (F): The equation $3x^2 = 9$ is an identity.

44. Indicate whether the statement is true (T) or false (F): $x = 3$ is a solution of the equation $3x - 1 = 10$.

45. Solve and graph $3 \leq 2x + 1$.

46. Solve and graph $-4 < -2x + 1 \leq 10$.

In Exercises 47–49 solve, and express the solution set in interval notation.

47. $2(a + 5) > 3a + 2$

48. $\dfrac{-1}{2x - 5} < 0$

49. $\dfrac{2x}{3} + \dfrac{1}{2} \geq \dfrac{x}{2} - 1$

1.5 50. Solve $|3x + 2| = 7$ for x.

51. Solve and graph $|4x - 1| = 5$.

52. Solve and graph $|2x + 1| > 7$.

53. Solve $|2 - 5x| < 1$ and write the solution in interval notation.

54. Solve $|3x - 2| \geq 6$ and write the solution in interval notation.

1.6 55. Write the solution set of the inequality $x^2 + 4x - 5 \geq 0$ in interval notation.

56. Write the solution set for $\dfrac{2x + 1}{x + 5} \geq 0$ in interval notation.

57. Write the solution set for
$$(3 - x)(2x + 3)(x + 2) < 0$$
in interval notation.

App.

In Exercises 58–61 identify the property of the real number system that justifies the statement. All variables represent real numbers.

58. $(3a) + (-3a) = 0$ inverse

59. $(3 + 4)x = 3x + 4x$

60. $2x + 2y + z = 2x + z + 2y$

61. $9x \cdot 1 = 9x$ identites

PROGRESS TEST 1A

In Problems 1–4 determine whether the statement is true (T) or false (F).

1. -1.36 is an irrational number.

2. π is equal to $\frac{22}{7}$.

3. $\sqrt{4}$ is a real number.

4. $\sqrt{x^2} = x$ for all real numbers x.

In Problems 5 and 6 sketch the given set of numbers on a real number line.

5. The integers that are greater than -3 and less than or equal to 3.

6. The real numbers x such that $-2 \leq x < \frac{1}{2}$.

7. Find the value of $|2 - 3| - |4 - 2|$.

8. Find $\overline{AB}$ if the coordinates of A and B are -6 and -4, respectively.

9. The area of a region is given by the expression $3x^2 - xy$. Find the area when $x = 5$ meters and $y = 10$ meters.

10. Evaluate the expression $\dfrac{-|y - 2x|}{|xy|}$ when $x = 3$ and $y = -1$.

11. Which of the following expressions are not polynomials?
    (a) x^5 (b) $5x^{-4}y + 3x^2 - y$
    (c) $4x^3 + x$ (d) $2x^2 + 3x^0$

In Problems 12 and 13 indicate the leading coefficients and the degree of each polynomial.

12. $-2.2x^5 + 3x^3 - 2x$ 13. $14x^6 - 2x + 1$

In Problems 14 and 15 perform the indicated operations.

14. $3xy + 2x + 3y + 2 - (1 - y - x + xy)$

15. $(a + 2)(3a^2 - a + 5)$

In Problems 16 and 17 factor each expression.

16. $8a^3b^5 - 12a^5b^2 + 16a^2b$

17. $4 - 9x^2$

In Problems 18–21 simplify, and use only positive exponents to express the answers.

18. $\left(\dfrac{x^{7/2}}{x^{2/3}}\right)^{-6}$

19. $\dfrac{y^{2n}}{y^{n-1}}$

20. $\dfrac{-1}{(x - 1)^0}$

21. $(2a^2b^{-1})^2$

22. For what values of x is $\dfrac{1}{\sqrt{x - 2}}$ a real number?

In Problems 23 and 24 perform the indicated operations and write all answers in the form $a + bi$.

23. $(2 - i) + (-3 + i)$ 24. $(5 + 2i)(2 - 3i)$

In Problems 25 and 26 solve for y.

25. $5 - 4y = 2$ 26. $\dfrac{2 + 5y}{3y - 1} = 6$

27. Indicate whether the statement is true (T) or false (F): The equation $(2x - 1)^2 = 4x^2 - 4x + 1$ is an identity.

28. Solve $-1 \leq 2x + 3 < 5$ and graph the solution set.

In Problems 29 and 30 solve, and express the solution set in interval notation.

29. $3(2a - 1) - 4(a + 2) \le 4$

30. $-2 \le 2 - x \le 6$

31. Solve $|4x - 1| = 9$.

32. Solve $|2x - 1| \le 5$ and graph the solution set.

33. Solve $|1 - 3x| < 5$ and write the solution in interval notation.

In Problems 34 and 35 write the solution set in interval notation.

34. $-2x^2 + 3x - 1 \le 0$

35. $(x - 1)(2 - 3x)(x + 2) \le 0$

PROGRESS TEST 1B

In Problems 1–4 determine whether the statement is true (T) or false (F).

1. 19.6 is a real number.

2. π is equal to 3.14.

3. $\sqrt{5}$ is a rational number.

4. If a and b are real numbers, then $|a - b| = |b - a|$.

In Problems 5 and 6 sketch the given set of numbers on a real number line.

5. The natural numbers that are less than 5.

6. The real numbers x such that $\frac{3}{2} < x < 3$.

7. Find the value of $\dfrac{|2 - 5| + |1 - 5|}{|-7|}$.

8. Find $\overline{AB}$ if the coordinates of A and B are -2 and 5, respectively.

9. The area of a trapezoid is given by the formula $A = \frac{1}{2}h(b + b')$. Find the area if $h = 4$ meters, $b = 3$ meters, and $b' = 4$ meters.

10. Evaluate the expression $|x|/|x - y|$ when $x = -2$ and $y = -3$.

11. Which of the following expressions are not polynomials?
    (a) $3x^2 + x^{-1} - 2$ (b) $2x^3 - xy^2 + x$
    (c) $2x^2y^2 + xy - 4$ (d) $x^2y + x^{1/2}y + 2$

In Problems 12 and 13 indicate the leading coefficient and the degree of each polynomial.

12. $-3x^3 + 4x^5$ 13. $1.5x^{10} - x^9 + 17x^8$

In Problems 14 and 15 perform the indicated operations.

14. $(2s^2t^3 - st^2 + st - s + t)$
    $\qquad - (3s^2t^2 - 2s^2t - 4st^2 - t + 3)$

15. $(b + 3)(-3b^2 + 2b + 4)$

In Problems 16 and 17 factor each expression.

16. $5r^3s^4 - 40r^4s^3t$ 17. $2x^2 + 7x - 4$

In Problems 18–21 simplify, using only positive exponents to express the answers.

18. $\dfrac{4x^{-3}}{x^{-2}}$ 19. $(b^2)^5(b^3)^6$

20. $\left(\dfrac{x^8}{y^{12}}\right)^{3/4}$ 21. $\dfrac{2(x + 2)^0}{-2}$

22. For what values of x is $\sqrt{2 - x}$ a real number?

In Problems 23 and 24 perform the indicated operations and write all answers in the form $a + bi$.

23. $(4 - 2i) - \left(2 - \dfrac{1}{2}i\right)$ 24. $(3 - 2i)(2 - i)$

In Problems 25 and 26 solve for x.

25. $3(2x + 5) = 5 - (3x - 1)$

26. $3x - k^2 = -kx$

27. Indicate whether the statement is true (T) or false (F): $x = -1$ is a solution of the equation $\dfrac{x - 1}{x + 1} = 0$.

28. Solve $-9 \le 1 - 5x \le -4$ and graph the solution set.

In Problems 29 and 30 solve, and express the solution set in interval notation.

29. $\dfrac{x}{4} - \dfrac{1}{2} \le \dfrac{1}{2} - x$ 30. $\dfrac{-2}{3 - x} \ge 0$

31. Solve $|1 - 3x| = 7$.

32. Solve $\dfrac{|x - 4|}{2} \ge 1$ and graph the solution set.

33. Solve $|5x + 2| > 3$ and write the solution set in interval notation.

In Problems 34 and 35 write the solution set in interval notation.

34. $\dfrac{x^2}{x + 5} < 0$

35. $(3x - 2)(x + 4)(1 - x) > 0$

CHAPTER 2

FUNCTIONS AND GRAPHS

What is the effect of increased fertilization on the growth of an azalea? If the minimum wage is increased, what will be the impact on the number of unemployed workers? When a submarine dives, can we calculate the water pressure against the hull at a given depth?

Each of the questions posed above seeks a relationship between phenomena. The search for relationships, or correspondence, is a central activity in our attempts to understand the universe; it is used in mathematics, engineering, the physical and biological sciences, the social sciences, and business and economics.

The concept of a function has been developed as a means of organizing and assisting in the study of relationships. Since graphs are powerful means of exhibiting relationships, we begin with a study of the Cartesian, or rectangular, coordinate system. We will then formally define a function and will provide a number of ways of viewing the function concept. Function notation will be introduced to provide a convenient means of writing functions.

Much of the material in this chapter focuses on the graphs of functions. The information available at a glance from a graph is so impressive that it is vital for a student planning to study advanced mathematics to be familiar with techniques for quickly sketching the graphs of those functions that occur most frequently.

2.1

THE RECTANGULAR COORDINATE SYSTEM

In Chapter 1 we associated the system of real numbers with points on the real number line. That is, we saw that there is a one-to-one correspondence between the system of real numbers and points on the real number line.

We will now develop an analogous way to handle points in a plane. We begin by drawing a pair of perpendicular lines intersecting at a point O called the **origin.** One of the lines, called the **x-axis,** is usually drawn in a horizontal position. The other line, called the **y-axis,** is usually drawn vertically.

If we think of the x-axis as a real number line, we may mark off some convenient unit of length, with positive numbers to the right of the origin and negative numbers to the left of the origin. Similarly, we may think of the y-axis as a real number line. Again, we may mark off a convenient unit of length (usually the same as the unit of length on the x-axis) with the upward direction representing positive numbers and the downward direction negative numbers. The x and y axes are called **coordinate axes,** and together they constitute a **rectangular** or **Cartesian coordinate system.** The coordinate axes divide the plane into four **quadrants,** which we label I, II, III, and IV as in Figure 1.

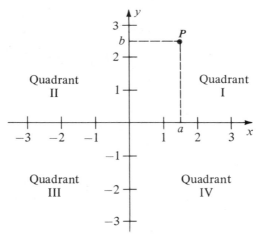

FIGURE 1

By using the coordinate axes, we can outline a procedure for labeling a point P in the plane. From P, draw a perpendicular to the x-axis and note that it meets the x-axis at $x = a$. Now draw a perpendicular from P to the y-axis and note that it meets the y-axis at $y = b$. We say that the **coordinates** of P are given by the **ordered pair** (a, b). The term "ordered pair" means that the order is significant; that is, the ordered pair (a, b) is different from the ordered pair (b, a).

The first number of the ordered pair (a, b) is sometimes called the **abscissa** or **x-coordinate** of P. The second number is called the **ordinate** or **y-coordinate** of P.

We have now developed a procedure for associating with each point P in the plane a unique ordered pair of real numbers (a, b). We usually write the point P as $P(a, b)$. Conversely, every ordered pair of real numbers (a, b) determines a unique point P in the plane. The point P is located at the intersection of the lines perpendicular to the x-axis and to the y-axis at the points on the axes having coordinates a and b, respectively. We have thus established a one-to-one correspondence between the set of all points in the plane and the set of all ordered pairs of real numbers.

We have indicated a number of points in Figure 2. Note that all points on the x-axis have a y-coordinate of 0 and all points on the y-axis have an x-coordinate

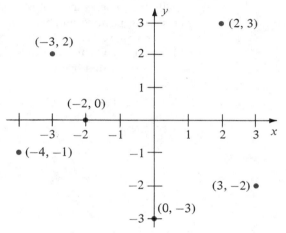

FIGURE 2

of 0. It is important to observe that the x-coordinate of a point P is the distance of P from the y-axis; the y-coordinate is its distance from the x-axis. The point $(2, 3)$ in Figure 2 is 2 units from the y-axis and 3 units from the x-axis.

THE DISTANCE FORMULA

There is a useful formula that gives the distance PQ between two points $P(x_1, y_1)$ and $Q(x_2, y_2)$. In Figure 3a we have shown the x-coordinate of a point as the distance of the point from the y-axis, and the y-coordinate as its distance from the x-axis. Thus we labeled the horizontal segments x_1 and x_2 and the vertical segments y_1 and y_2. In Figure 3b we use the lengths from Figure 3a to indicate that $\overline{PR} = x_2 - x_1$ and $\overline{QR} = y_2 - y_1$. Since triangle PRQ is a right triangle, we can apply the Pythagorean theorem.

$$d^2 = (x_2 - x_1)^2 + (y_2 - y_1)^2$$

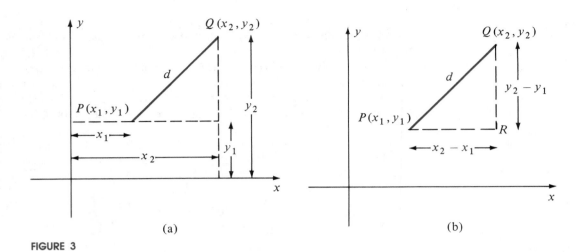

FIGURE 3

Although the points in Figure 3 are both in quadrant I, the same result will be obtained for any two points. Since distance cannot be negative, we have

---

The Distance Formula The distance $\overline{PQ}$ between the points $P(x_1, y_1)$ and $Q(x_2, y_2)$ in the plane is

$$\overline{PQ} = \sqrt{(x_2 - x_1)^2 + (y_2 - y_1)^2}$$

It is also clear from the distance formula that $\overline{PQ} = \overline{QP}$.

EXAMPLE 1
Find the distance between the points $P(-2, -3)$ and $Q(1, 2)$.

SOLUTION
Using the distance formula, we have
$$\overline{PQ} = \sqrt{[1 - (-2)]^2 + [2 - (-3)]^2} = \sqrt{3^2 + 5^2} = \sqrt{34}$$

PROGRESS CHECK
Find the distance between the points $P(-3, 2)$ and $Q(4, -2)$.

ANSWER
$\sqrt{65}$

EXAMPLE 2
Show that the triangle with vertices $A(-2, 3)$, $B(3, -2)$, and $C(6, 1)$ is a right triangle.

SOLUTION
It is a good idea to draw a diagram as in Figure 4. We compute the lengths of the three sides.

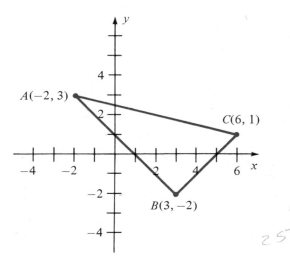

FIGURE 4

$$\overline{AB} = \sqrt{(3 + 2)^2 + (-2 - 3)^2} = \sqrt{50}$$
$$\overline{BC} = \sqrt{(6 - 3)^2 + (1 + 2)^2} = \sqrt{18}$$
$$\overline{AC} = \sqrt{(6 + 2)^2 + (1 - 3)^2} = \sqrt{68}$$

If the Pythagorean theorem holds, then triangle ABC is a right triangle. We see that

$$(\overline{AC})^2 = (\overline{AB})^2 + (\overline{BC})^2 \quad \text{since } 68 = 50 + 18$$

and we conclude that triangle ABC is a right triangle whose hypotenuse is AC.

GRAPHS OF EQUATIONS

By the **graph of an equation in two variables** x and y we mean the set of all points $P(x, y)$ whose coordinates satisfy the given equation. We say that the ordered pair (a, b) is a **solution** of the equation if substitution of a for x and b for y yields a true statement.

To graph $y = x^2 - 4$, an equation in the variables x and y, we note that we can obtain solutions by assigning arbitrary values to x and computing corresponding values of y. Thus, if $x = 3$, then $y = 3^2 - 4 = 5$, and the ordered pair $(3, 5)$ is a solution of the equation. Table 1 shows a number of solutions. We next plot the points corresponding to these ordered pairs. Since the equation has an infinite number of solutions, the plotted points represent only a portion of the graph. We assume that the curve behaves nicely between the plotted points and connect these points by a smooth curve (Figure 5). We must plot enough points to feel reasonably certain of the outline of the curve.

TABLE 1

x	-3	-2	-1	0	1	2	$\dfrac{5}{2}$
y	5	0	-3	-4	-3	0	$\dfrac{9}{4}$

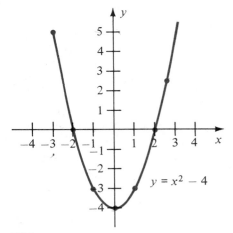

FIGURE 5

The abscissa of a point at which a graph meets the x-axis is called an **x-intercept.** Since the graph in Figure 5 meets the x-axis at the points $(2, 0)$ and $(-2, 0)$, we see that 2 and -2 are the x-intercepts. Similarly, we define the **y-intercept** as the ordinate of a point at which the graph meets the y-axis. In Figure 5 the y-intercept is -4. Intercepts are often easy to calculate and are useful in sketching a graph.

EXAMPLE 3

Sketch the graph of the equation $y = 2x + 1$. Determine the x- and y-intercepts, if any.

SOLUTION

We form a short table of values and sketch the graph in Figure 6. The graph appears to be a straight line that intersects the x-axis at $(-\frac{1}{2}, 0)$ and the y-axis at $(0, 1)$. The x-intercept is $-\frac{1}{2}$ and the y-intercept is 1. Alternatively, we can find

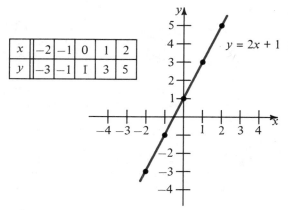

x	−2	−1	0	1	2
y	−3	−1	1	3	5

FIGURE 6

the y-intercept algebraically by letting $x = 0$ so that

$$y = 2x + 1 = 2(0) + 1 = 1$$

and the x-intercept by letting $y = 0$ so that

$$y = 2x + 1$$
$$0 = 2x + 1$$
$$x = -\tfrac{1}{2}$$

SYMMETRY

If we folded the graph of Figure 7a along the x-axis, the top and bottom portions would exactly match, which is what we intuitively mean when we speak of symmetry about the x-axis. We would like to develop a means of testing for symmetry that doesn't rely upon examining a graph. We can then use information about symmetry to help in sketching the graph.

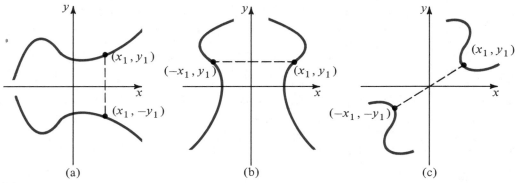

(a)   (b)   (c)

FIGURE 7

Returning to Figure 7a, we see that every point (x_1, y_1) on the portion of the curve above the x-axis is reflected in a point $(x_1, -y_1)$ that lies on the portion of the curve below the x-axis. Similarly, using the graph of Figure 7b, we can argue that symmetry about the y-axis occurs if, for every point (x_1, y_1) on the curve, $(-x_1, y_1)$ also lies on the curve. Finally, using the graph sketched in Figure 7c, we see that symmetry about the origin occurs if, for every point (x_1, y_1) on the curve, $(-x_1, -y_1)$ also lies on the curve. We now summarize these results.

Tests for Symmetry	The graph of an equation is **symmetric with respect to the** (i) **x-axis** if replacing y with $-y$ results in an equivalent equation; (ii) **y-axis** if replacing x with $-x$ results in an equivalent equation; (iii) **origin** if replacing x with $-x$ and y with $-y$ results in an equivalent equation.

EXAMPLE 4

Use intercepts and symmetry to assist in graphing the equations.
(a) $y = 1 - x^2$ (b) $x = y^2 + 1$

SOLUTION

(a) To determine the intercepts, set $x = 0$ to yield $y = 1$ as the y-intercept. Setting $y = 0$, we have $x^2 = 1$ or $x = \pm 1$ as the x-intercepts.

To test for symmetry, replace x with $-x$ in the equation $y = 1 - x^2$ to obtain

$$y = 1 - (-x)^2 = 1 - x^2$$

Since the equation is unaltered, the curve is symmetric with respect to the y-axis. Now, replacing y with $-y$, we have

$$-y = 1 - x^2$$

which is *not* equivalent to the original equation. The curve is therefore not symmetric with respect to the x-axis. Finally, replacing x with $-x$ and y with $-y$ repeats the last result and shows that the curve is not symmetric with respect to the origin.

We can now form a table of values for $x \geq 0$ and use symmetry with respect to the y-axis to help sketch the graph of the equation (see Figure 8a).

(b) The y-intercepts occur where $x = 0$. Since this leads to the equation $y^2 = -1$, which has no real roots, there are no y-intercepts. Setting $y = 0$, we have $x = 1$ as the x-intercept.

Replacing x with $-x$ in the equation $x = y^2 + 1$ gives us

$$-x = y^2 + 1$$

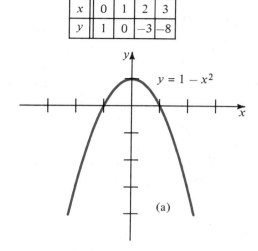

x	0	1	2	3
y	1	0	−3	−8

y	0	1	2	3
x	1	2	5	10

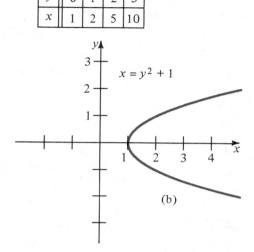

(a)

(b)

FIGURE 8

which is *not* an equivalent equation. The curve is therefore not symmetric with respect to the y-axis. Replacing y with −y, we find that

$$x = (-y)^2 + 1 = y^2 + 1$$

which is the same as the original equation. Thus, the curve is symmetric with respect to the x-axis. Replacing x with −x and y with −y also results in the equation

$$-x = y^2 + 1$$

and demonstrates that the curve is not symmetric with respect to the origin. We next form the table of values shown in Figure 8b by assigning nonnegative values to y and calculating the corresponding values of x from the equation; symmetry enables us to sketch the lower half of the graph without plotting points.

Solving the given equation for y yields $y = \pm\sqrt{x - 1}$, which confirms the symmetry about the x-axis. We can think of the upper half of Figure 8b as the graph of the equation $y = \sqrt{x - 1}$ and the lower half as the graph of the equation $y = -\sqrt{x - 1}$.

EXAMPLE 5

Without sketching the graph, determine symmetry with respect to the x-axis, the y-axis, and the origin.

(a) $x^2 + 4y^2 - y = 1$ (b) $xy = 5$ (c) $y^2 = \dfrac{x^2 + 1}{x^2 - 1}$

SOLUTION

(a) Replacing x with $-x$ in the equation, we have

$$(-x)^2 + 4y^2 - y = 1$$
$$x^2 + 4y^2 - y = 1$$

Since the equation is unaltered, the curve is symmetric with respect to the y-axis. Next, replacing y with $-y$, we have

$$x^2 + 4(-y)^2 - (-y) = 1$$
$$x^2 + 4y^2 + y = 1$$

which is *not* an equivalent equation. Replacing x with $-x$ and y with $-y$ repeats the last result. The curve is therefore not symmetric with respect to either the x-axis or the origin.

(b) Replacing x with $-x$, we have $-xy = 5$, which is *not* an equivalent equation. Replacing y with $-y$, we again have $-xy = 5$. Thus the curve is not symmetric with respect to either axis. However, replacing x with $-x$ and y with $-y$ gives us

$$(-x)(-y) = 5$$

which is equivalent to $xy = 5$. We conclude that the curve is symmetric with respect to the origin.

(c) Since x and y both appear to the second power only, all tests will lead to an equivalent equation. The curve is therefore symmetric with respect to both axes and the origin.

PROGRESS CHECK

Without graphing, determine symmetry with respect to the coordinate axes and the origin.

(a) $x^2 - y^2 = 1$ (b) $x + y = 10$ (c) $y = x + \dfrac{1}{x}$

ANSWERS

(a) Symmetric with respect to the x-axis, the y-axis, and the origin.
(b) Not symmetric with respect to either axis or the origin.
(c) Symmetric with respect to the origin only.

Note that in Example 5c and in (a) of the last Progress Check, the curves are symmetric with respect to both the x- and y-axes, as well as the origin. In fact, we have the following general rule.

A curve that is symmetric with respect to both coordinate axes is also symmetric with respect to the origin. However, a curve that is symmetric with respect to the origin need not be symmetric with respect to the coordinate axes.

The curve in Figure 7c illustrates the last point. The curve is symmetric with respect to the origin but not with respect to the coordinate axes.

EXERCISE SET 2.1

In each of Exercises 1 and 2 plot the given points on the same coordinate axes.

1. $(2, 3)$, $(-3, -2)$, $\left(-\frac{1}{2}, \frac{1}{2}\right)$, $\left(0, \frac{1}{4}\right)$, $\left(-\frac{1}{2}, 0\right)$, 2. $(-3, 4)$, $(5, -2)$, $(-1, -3)$, $\left(-1, \frac{3}{2}\right)$, $(0, 1.5)$

$(3, -2)$

In Exercises 3–8 find the distance between each pair of points.

3. $(5, 4)$, $(2, 1)$

4. $(-4, 5)$, $(-2, 3)$

5. $(-1, -5)$, $(-5, -1)$

6. $(-3, 0)$, $(2, -4)$

7. $\left(\frac{2}{3}, \frac{3}{2}\right)$, $(-2, -4)$

8. $\left(-\frac{1}{2}, 3\right)$, $\left(-1, -\frac{3}{4}\right)$

In Exercises 9–12 find the length of the shortest side of the triangle determined by the three given points.

9. $A(6, 2)$, $B(-1, 4)$, $C(0, -2)$

10. $P(2, -3)$, $Q(4, 4)$, $R(-1, -1)$

11. $R\left(-1, \frac{1}{2}\right)$, $S\left(-\frac{3}{2}, 1\right)$, $T(2, -1)$

12. $F(-5, -1)$, $G(0, 2)$, $H(1, -2)$

In Exercises 13–16 determine if the given points form a right triangle. (*Hint:* A triangle is a right triangle if and only if the lengths of the sides satisfy the Pythagorean theorem.)

13. $(1, -2)$, $(5, 2)$, $(2, 1)$

14. $(2, -3)$, $(-1, -1)$, $(3, 4)$

15. $(-4, 1)$, $(1, 4)$, $(4, -1)$

16. $(1, -1)$, $(-6, 1)$, $(1, 2)$

In Exercises 17–20 show that the points lie on the same line. (*Hint:* Three points are collinear if and only if the sum of the lengths of two sides equals the length of the third side.)

17. $(-1, 2)$, $(1, 1)$, $(5, -1)$

18. $(-1, -4)$, $(1, 10)$, $(0, 3)$

19. $(-1, 2)$, $(1, 5)$, $\left(-2, \frac{1}{2}\right)$

20. $(-1, -5)$, $(1, 1)$, $(-2, -8)$

21. Find the perimeter of the quadrilateral whose vertices are $(-2, -1)$, $(-4, 5)$, $(3, 5)$, $(4, -2)$.

22. Show that the points $(-2, -1)$, $(2, 2)$, $(5, -2)$ are the vertices of an isosceles triangle.

23. Show that the points $(9, 2)$, $(11, 6)$, $(3, 5)$, and $(1, 1)$ are the vertices of a parallelogram.

24. Show that the point $(-1, 1)$ is the midpoint of the line segment whose endpoints are $(-5, -1)$ and $(3, 3)$.

25. The points $A(2, 7)$, $B(4, 3)$, and $C(x, y)$ determine a right triangle whose hypotenuse is AB. Find x and y. (*Hint:* There is more than one answer.)

26. The points $A(2, 6)$, $B(4, 6)$, $C(4, 8)$, and $D(x, y)$ form a rectangle. Find x and y.

In Exercises 27–32 determine the intercepts and sketch the graph of the given equation.

27. $y = 2x + 4$

28. $y = -2x + 5$

29. $y = \sqrt{x}$

30. $y = \sqrt{x - 1}$

31. $y = |x + 3|$

32. $y = 2 - |x|$

In Exercises 33–38 determine the intercepts and use symmetry to assist in sketching the graph of the given equation.

33. $y = 3 - x^2$

34. $y = 3x - x^2$

35. $y = x^3 + 1$

36. $x = y^3 - 1$

37. $x = y^2 - 1$

38. $y = 3x$

Without graphing, determine whether each curve is symmetric with respect to the x-axis, the y-axis, the origin, or none of these.

39. $3x + 2y = 5$
40. $y = 4x^2$
41. $y^2 = x - 4$
42. $x^2 - y = 2$
43. $y^2 = 1 + x^3$
44. $y = (x - 2)^2$
45. $y^2 = (x - 2)^2$
46. $y^2x + 2x = 4$
47. $y^2x + 2x^2 = 4x^2y$
48. $y^3 = x^2 - 9$
49. $y = \dfrac{x^2 + 4}{x^2 - 4}$
50. $y = \dfrac{1}{x^2 + 1}$
51. $y^2 = \dfrac{x^2 + 1}{x^2 - 1}$
52. $4x^2 + 9y^2 = 36$
53. $xy = 4$

2.2
FUNCTIONS AND
FUNCTION NOTATION

The equation

$$y = 2x + 3$$

assigns a value to y for every value of x. If we let X denote the set of values that we can assign to x, and let Y denote the set of values that the equation assigns to y, we can show the correspondence schematically as in Figure 9. The equation can be thought of as a rule defining the correspondence between the sets X and Y.

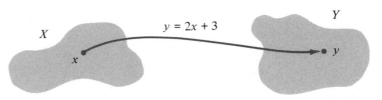

FIGURE 9

We are particularly interested in the situation where, for each element x in X, there corresponds one and only one element y in Y; that is, the rule assigns exactly one y for a given x. This type of correspondence plays a fundamental role in mathematics and is given a special name.

Function, Domain, Image, and Range

A **function** is a rule that, for each x in a set X, assigns exactly one y in a set Y. The element y is called the **image** of x. The set X is called the **domain** of the function and the set of all images is called the **range** of the function.

We can think of the rule defined by the equation $y = 2x + 3$ as a function machine (see Figure 10). Each time we drop a value of x from the domain into the input hopper, exactly one value of y falls out of the output chute. If we drop in $x = 5$, the function machine follows the rule and produces $y = 13$. Since we are free to choose the values of x that we drop into the machine, we call x the **independent variable;** the value of y that drops out depends upon the choice of x, so y is called the **dependent variable.** We say that the dependent variable is

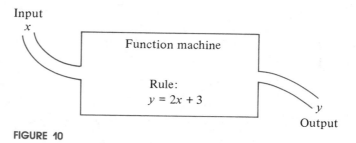

FIGURE 10

a function of the independent variable; that is, *the output is a function of the input*.

Let's look at a few schematic presentations. The correspondence in Figure 11a is a function; for each x in X there is exactly one corresponding value of y in Y. The fact that y_1 is the image of both x_1 and x_2 does not violate the definition of a function. However, the correspondence in Figure 11b is not a function, since x_1 has two images, y_1 and y_2, assigned to it, thus violating the definition of a function.

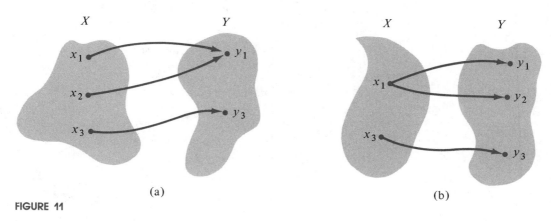

(a)   (b)

FIGURE 11

VERTICAL LINE TEST

There is a simple graphic way to test whether a correspondence determines a function. When we draw vertical lines on the graph of Figure 12a, we see that no vertical line intersects the graph at more than one point. This means that the correspondence used in sketching the graph assigns exactly one y-value for each x-value and therefore determines y as a function of x. When we draw vertical lines on the graph of Figure 12b, however, some vertical lines intersect the graph at two points. Since the correspondence graphed in Figure 12b assigns the values y_1 and y_2 to x_1, it does not determine y as a function of x. Thus, *not every equation or correspondence in the variables x and y determines y as a function of x*.

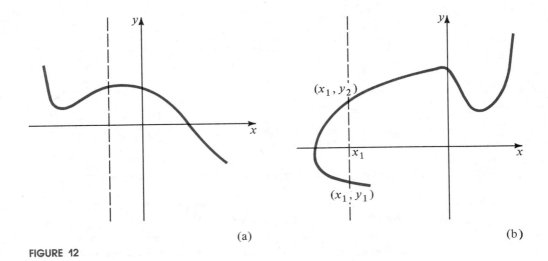

(a)   (b)

FIGURE 12

Vertical Line Test	A graph determines y as a function of x if and only if no vertical line meets the graph at more than one point.

EXAMPLE 1

Which of the following graphs determine y as a function of x?

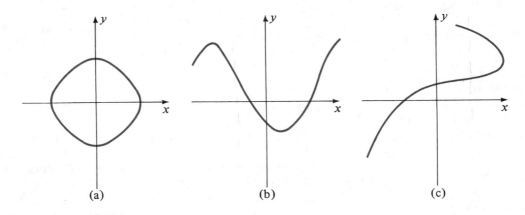

(a)   (b)   (c)

SOLUTION

(a) Not a function. Some vertical lines meet the graph in more than one point.

(b) A function. Passes the vertical line test.

(c) Not a function. Fails the vertical line test.

DOMAIN AND RANGE

We have defined the domain of a function as the set of values assumed by the independent variable. In more advanced courses in mathematics, the domain may include complex numbers. In this book we will restrict the domain of a function to those real numbers for which the image is also a real number, and we say that the function **is defined at** such values. When a function is defined by an equation, we must always be alert to two potential problems.

(a) *Division by zero.* For example, the domain of the function

$$y = \frac{2}{x - 1}$$

is the set of all real numbers other than $x = 1$. When $x = 1$, the denominator is 0, and division by 0 is not defined.

(b) *Even roots of negative numbers.* For example, the function

$$y = \sqrt{x - 1}$$

is defined only for $x \geq 1$, since we exclude the square root of negative numbers. Hence the domain of the function consists of all real numbers $x \geq 1$.

The range of a function is, in general, not as easily determined as is the domain. The range is the set of all y-values that occur in the correspondence; that is, it is the set of all outputs of the function. For our purposes, it will suffice to determine the range by examining the graph.

EXAMPLE 2

Graph the equation $y = \sqrt{x}$. If the correspondence determines a function, find the domain and range.

SOLUTION

We obtain the graph of the equation by plotting points and connecting them to form a smooth curve. Applying the vertical line test to the graph as shown in Figure 13, we see that the equation determines a function. The domain of the function is the set $\{x \mid x \geq 0\}$ and the range is the set $\{y \mid y \geq 0\}$.

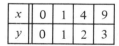

x	0	1	4	9
y	0	1	2	3

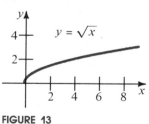

$y = \sqrt{x}$

FIGURE 13

PROGRESS CHECK

Graph the equation $y = x^2 - 4$, $-3 \leq x \leq 3$. If the correspondence determines a function, find the domain and range.

ANSWER

The graph is that portion of the curve shown in Figure 5 that lies between $x = -3$ and $x = 3$. The domain is $\{x \mid -3 \leq x \leq 3\}$; the range is $\{y \mid -4 \leq y \leq 5\}$.

FUNCTION NOTATION

It is customary to designate a function by a letter of the alphabet, such as f, g, F, or C. We then denote the output corresponding to x by $f(x)$, which is read "f of x." Thus,

$$f(x) = 2x + 3$$

specifies a rule f for determining an output $f(x)$ for a given value of x. To find $f(x)$ when $x = 5$, we simply substitute 5 for x and obtain

$$f(5) = 2(5) + 3 = 13 \qquad 2k + 3$$

The notation $f(5)$ is a convenient way of specifying "the value of the function f that corresponds to $x = 5$." The symbol f represents the function or rule; the notation $f(x)$ represents the output produced by the rule. For convenience, however, we will at times join in the common practice of designating the function f by $f(x)$.

EXAMPLE 3
(a)   If $f(x) = 2x^2 - 2x + 1$, find $f(-1)$.
(b)   If $f(t) = 3t^2 - 1$, find $f(2a)$.

SOLUTION
(a)   We substitute -1 for x.
$$f(-1) = 2(-1)^2 - 2(-1) + 1 = 5$$
(b)   We substitute $2a$ for t.
$$f(2a) = 3(2a)^2 - 1 = 3(4a^2) - 1 = 12a^2 - 1$$

PROGRESS CHECK
(a)   If $f(u) = u^3 + 3u - 4$, find $f(-2)$.
(b)   If $f(t) = t^2 + 1$, find $f(t - 1)$.

ANSWERS
(a) -18 (b) $t^2 - 2t + 2$

EXAMPLE 4
Let the function f be defined by $f(x) = x^2 - 1$. Find
(a) $f(-2)$ (b) $f(a)$ (c) $f(a + h)$ (d) $f(a + h) - f(a)$

SOLUTION
(a) $f(-2) = (-2)^2 - 1 = 4 - 1 = 3$
(b) $f(a) = a^2 - 1$
(c) $f(a + h) = (a + h)^2 - 1 = a^2 + 2ah + h^2 - 1$
(d) $f(a + h) - f(a) = (a + h)^2 - 1 - (a^2 - 1)$
$$= a^2 + 2ah + h^2 - 1 - a^2 + 1$$
$$= 2ah + h^2$$

WARNING
(a)   Note that $f(a + 3) \neq f(a) + f(3)$. Function notation is not to be confused with the distributive law.
(b)   Note that $f(x^2) \neq f \cdot x^2$. The use of parentheses in function notation does not imply multiplication.

EXAMPLE 5

A newspaper makes this offer to its advertisers: The first column inch will cost $40, and each subsequent column inch will cost $30. If T is the total cost of running an ad whose length is n column inches, and the minimum space is 1 column inch,

(a) express T as a function of n;

(b) find T when $n = 4$.

SOLUTION

(a) The equation

$$T = 40 + 30(n - 1)$$
$$= 10 + 30n$$

gives the correspondence between n and T. In function notation,

$$T(n) = 10 + 30n, \quad n \geq 1$$

(b) When $n = 4$,

$$T(4) = 10 + 30(4) = 130$$

EXERCISE SET 2.2

In Exercises 1–6 graph the equation. If the graph determines y as a function of x, find the domain and use the graph to determine the range of the function.

1. $y = 2x - 3$

2. $y = x^2 + x, \quad -2 \leq x \leq 1$

3. $x = y + 1$

4. $x = y^2 - 1$

5. $y = \sqrt{x - 1}$

6. $y = |x|$

In Exercises 7–12 determine the domain of the function defined by the given rule.

7. $f(x) = \sqrt{2x - 3}$

8. $f(x) = \sqrt{5 - x}$

9. $f(x) = \dfrac{1}{\sqrt{x - 2}}$

10. $f(x) = \dfrac{-2}{x^2 + 2x - 3}$

11. $f(x) = \dfrac{\sqrt{x - 1}}{x - 2}$

12. $f(x) = \dfrac{x}{x^2 - 4}$

In Exercises 13–16 find the number (or numbers) whose image is 2.

13. $f(x) = 2x - 5$

14. $f(x) = x^2$

15. $f(x) = \dfrac{1}{x - 1}$

16. $f(x) = \sqrt{x - 1}$

Given the function f defined by $f(x) = 2x^2 + 5$, determine the following.

17. $f(0)$

18. $f(-2)$

19. $f(a)$

20. $f(3x)$

21. $3f(x)$

22. $-f(x)$

Given the function g defined by $g(x) = x^2 + 2x$, determine the following.

23. $g(-3)$

24. $g\left(\dfrac{1}{x}\right)$

25. $\dfrac{1}{g(x)}$

26. $g(-x)$

27. $g(a + h)$

28. $\dfrac{g(a + h) - g(a)}{h}$

Given the function F defined by $F(x) = \dfrac{x^2 + 1}{3x - 1}$, determine the following.

 29. $F(-2.73)$

 30. $F(16.11)$

31. $\dfrac{1}{F(x)}$

32. $F(-x)$

33. $2F(2x)$

34. $F(x^2)$

Given the function r defined by $r(t) = \dfrac{t - 2}{t^2 + 2t - 3}$, determine the following.

35. $r(-8.27)$

36. $r(2.04)$

37. $r(2a)$

38. $2r(a)$

39. $r(a + 1)$

40. $r(1 + h)$

41. If x dollars are borrowed at 7% simple annual interest, express the interest I at the end of 4 years as a function of x.

42. Express the area A of an equilateral triangle as a function of the length s of its side.

43. Express the diameter d of a circle as a function of its circumference C.

44. Express the perimeter P of a square as a function of its area A.

2.3
GRAPHS OF
FUNCTIONS

We have used the graph of an equation to help us find out whether or not the equation determines a function. It is not surprising, therefore, that the **graph of a function** f is defined as the graph of the equation $y = f(x)$. For example, the graph of the function f defined by the rule $f(x) = \sqrt{x}$ is the graph of the equation $y = \sqrt{x}$, which was sketched in Figure 13.

"SPECIAL" FUNCTIONS
AND THEIR GRAPHS

There are a number of "special" functions that a calculus instructor is likely to use to demonstrate a point. The instructor will sketch the graph of the function, since the graph shows at a glance many characteristics of the function. For example, information about symmetry, domain, and range is available from a graph. In fact, we have already used the graphs of some of these functions to illustrate these characteristics.

You should become thoroughly acquainted with the following functions and their graphs. For each function we will form a table of values, sketch the graph of the function, and discuss symmetry, domain, and range.

$f(x) = x$ **Identity function**

The domain of f is the set of all real numbers. We form a table of values and use it to sketch the graph of $y = x$ in Figure 14. The graph is symmetric with respect to the origin (note that $-y = -x$ is equivalent to $y = x$). The range of f is seen to be the set of all real numbers.

$f(x) = -x$ **Negation function**

The domain of f is the set of all real numbers. A table of values is used to sketch the graph of $y = -x$ in Figure 15. The graph is symmetric with respect to the origin (note that $-y = x$ is equivalent to $y = -x$). The range of f is seen to be the set of all real numbers.

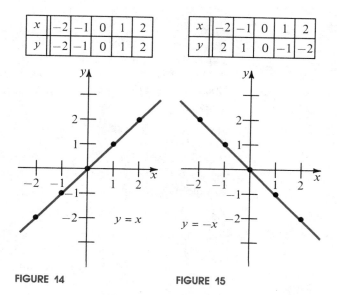

x	-2	-1	0	1	2
y	-2	-1	0	1	2

x	-2	-1	0	1	2
y	2	1	0	-1	-2

FIGURE 14   FIGURE 15

$f(x) = |x|$ **Absolute value function**

The domain of f is the set of all real numbers. A table of values allows us to sketch the graph in Figure 16. The graph is symmetric with respect to the y-axis. Since the graph always lies above the x-axis, the range of f is the set of all nonnegative real numbers.

$f(x) = c$ **Constant function**

The domain of f is the set of all real numbers. In fact, the value of f is the same for all values of x (see Figure 17). The range of f is the set $\{c\}$. The graph is symmetric with respect to the y-axis (note that $y = c$ is unaltered when x is replaced with $-x$).

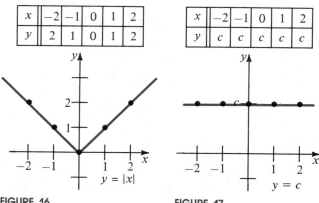

x	-2	-1	0	1	2
y	2	1	0	1	2

x	-2	-1	0	1	2
y	c	c	c	c	c

FIGURE 16   FIGURE 17

f(x) = x²   **Squaring function**

The domain of f is the set of all real numbers. The graph in Figure 18 is called a **parabola** and illustrates the general shape of all second-degree polynomials. The graph of f is symmetric with respect to the y-axis (note that $y = (-x)^2 = x^2$). Since $y \geq 0$ for all values of x, the range is the set of all nonnegative real numbers.

f(x) = √x̄   **Square root function**

Since $\sqrt{x}$ is not defined for $x < 0$, the domain is the set of nonnegative real numbers. The graph in Figure 19 always lies above the x-axis, so the range of f is $\{y \mid y \geq 0\}$, that is, the set of all nonnegative real numbers. The graph is not symmetric with respect to either axis or the origin.

f(x) = x³   **Cubing function**

The domain is the set of all real numbers. Since the graph in Figure 20 extends indefinitely both upward and downward with no gaps, the range is also the set of all real numbers. The graph is symmetric with respect to the origin (note that $-y = (-x)^3 = -x^3$ is equivalent to $y = x^3$).

x	−2	−1	0	1	2
y	−8	−1	0	1	8

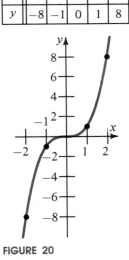

x	−2	−1	0	1	2
y	4	1	0	1	4

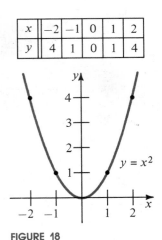

FIGURE 18

x	0	1	4	9
y	0	1	2	3

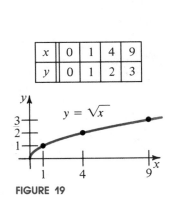

FIGURE 19

FIGURE 20

PIECEWISE-DEFINED FUNCTIONS

Thus far we have defined each function by means of an equation. A function can also be defined by a table, by a graph, or by several equations. When a function is defined in different ways over different parts of its domain, it is said to be a **piecewise-defined function.** We illustrate this idea by several examples.

EXAMPLE 1

Sketch the graph of the function f defined by

$$f(x) = \begin{cases} x^2 & \text{if} \quad -2 \le x \le 2 \\ 2x + 1 & \text{if} \quad x > 2 \end{cases}$$

SOLUTION

We form a table of points to be plotted, being careful to use the first equation when $-2 \le x \le 2$ and the second equation when $x > 2$.

x	-2	-1	0	1	2	3	4	5
y	4	1	0	1	4	7	9	11

Note that the graph in Figure 21 has a gap. Also note that the point (2, 5) has been marked with an open circle to indicate that it is not on the graph of the function.

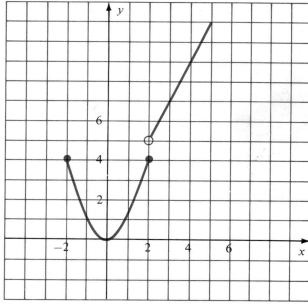

FIGURE 21

EXAMPLE 2

Sketch the graph of the function $f(x) = |x + 1|$.

SOLUTION
We apply the definition of absolute value to obtain

$$y = |x + 1| = \begin{cases} x + 1 & \text{if } x + 1 \geq 0 \\ -(x + 1) & \text{if } x + 1 < 0 \end{cases}$$

or

$$y = \begin{cases} x + 1 & \text{if } x \geq -1 \\ -x - 1 & \text{if } x < -1 \end{cases}$$

From this example it is easy to see that a function involving absolute value will usually be a piecewise-defined function. As usual, we form a table of values, being careful to use $y = x + 1$ when $x \geq -1$ and $y = -x - 1$ when $x < -1$. It is a good idea to include the value $x = -1$ in the table.

x	-3	-2	-1	0	1	2	3
y	2	1	0	1	2	3	4

The points are joined by a smooth curve (Figure 22), which consists of two rays or half-lines intersecting at $(-1, 0)$.

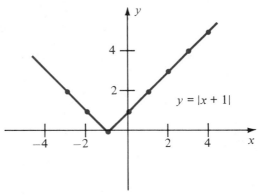

FIGURE 22

EXAMPLE 3
The commission earned by a door-to-door cosmetics salesperson is determined as shown in the accompanying table.
(a) Express the commission C as a function of sales s.
(b) Find the commission if the weekly sales are $425.
(c) Sketch the graph of the function.

Weekly sales	Commission
less than $300	20% of sales
$300 or more but less than $400	$60 + 35% of sales over $300
$400 or more	$95 + 60% of sales over $400

SOLUTION

(a)   The function C can be described by three equations.

$$C(s) = \begin{cases} 0.20s & \text{if} \quad s < 300 \\ 60 + 0.35(s - 300) & \text{if} \quad 300 \le s < 400 \\ 95 + 0.60(s - 400) & \text{if} \quad s \ge 400 \end{cases}$$

(b)   When $s = 425$, we must use the third equation and substitute to determine $C(425)$.

$$\begin{aligned} C(425) &= 95 + 0.60(425 - 400) \\ &= 95 + 0.60(25) \\ &= 110 \end{aligned}$$

The commission on sales of $425 is $110.

(c)   The graph of the function C consists of three line segments (Figure 23).

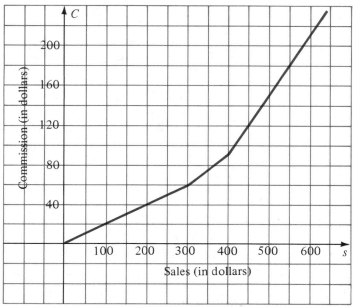

FIGURE 23

INCREASING AND DECREASING FUNCTIONS

When we apply the terms *increasing* and *decreasing* to the graph of a function, we assume that we are viewing the graph from left to right. The straight line of Figure 24a is increasing, since the values of y increase as we move from left to right; similarly, the graph in Figure 24b is decreasing, since the values of y decrease as we move from left to right.

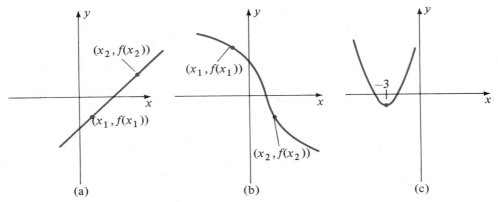

(a)   (b)   (c)

FIGURE 24

One portion of the graph pictured in Figure 24c is decreasing and another is increasing. Since this is the most common situation, we define increasing and decreasing on an interval.

If x_1 and x_2 are in the interval $[a, b]$ in the domain of a function f, then
f is **increasing** on $[a, b]$ if $f(x_1) < f(x_2)$ whenever $x_1 < x_2$
f is **decreasing** on $[a, b]$ if $f(x_1) > f(x_2)$ whenever $x_1 < x_2$
f is **constant** on $[a, b]$ if $f(x_1) = f(x_2)$ for all x_1, x_2

Returning to Figure 24c, note that the function is decreasing when $x \leq -3$ and increasing when $x \geq -3$; that is, the function is decreasing on the interval $(-\infty, -3]$ and increasing on the interval $[-3, \infty)$. The graph shows that the function has a minimum value at the point $x = -3$. Finding such points is very useful in sketching graphs and is an important technique taught in calculus courses.

It is important to become accustomed to the notation used in Figure 24, where the y-coordinate at the point $x = x_1$ is denoted by $f(x_1)$.

EXAMPLE 4
Use the graph of the function $f(x) = x^3 - 3x + 2$, shown in Figure 25, to determine where the function is increasing and where it is decreasing.

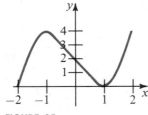

FIGURE 25

SOLUTION
From the graph we see that there are turning points at $(-1, 4)$ and at $(1, 0)$. We

conclude that

$$f \text{ is increasing on the intervals } (-\infty, -1] \text{ and } [1, \infty)$$
$$f \text{ is decreasing on the interval } [-1, 1]$$

EXAMPLE 5

The function f is defined by

$$f(x) = \begin{cases} |x| & \text{if } x \le 2 \\ -3 & \text{if } x > 2 \end{cases}$$

Use a graph to find the values of x for which the function is increasing, decreasing, and constant.

SOLUTION

Note that the piecewise-defined function f is composed of the absolute value function when $x \le 2$ and a constant function when $x > 2$. We can therefore sketch the graph of f immediately as shown in Figure 26. From the graph in Figure 26 we determine that

$$f \text{ is increasing on the interval } [0, 2]$$
$$f \text{ is decreasing on the interval } (-\infty, 0]$$
$$f \text{ is constant and has value } -3 \text{ on the interval } (2, \infty)$$

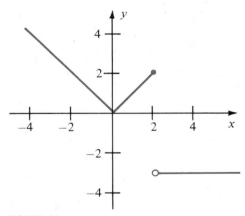

FIGURE 26

PROGRESS CHECK

The function f is defined by

$$f(x) = \begin{cases} 2x + 1 & \text{if } x < -1 \\ 0 & \text{if } -1 \le x \le 3 \\ -2x + 1 & \text{if } x > 3 \end{cases}$$

Use a graph to find the values of x for which the function is increasing, decreasing, and constant.

ANSWER
Increasing on the interval $(-\infty, -1)$. Constant on $[-1, 3]$.
Decreasing on $(3, \infty)$.

EXERCISE SET 2.3
In Exercises 1–20 sketch the graph of the function and state where it is increasing, decreasing, and constant.

1. $f(x) = 3x + 1$
2. $f(x) = 3 - 2x$
3. $f(x) = x^2 + 1$
4. $f(x) = 9 - x^2$
5. $f(x) = 4x - x^2$
6. $f(x) = |x| - 3$
7. $f(x) = |2x + 1|$
8. $f(x) = |1 - x|$
9. $f(x) = \sqrt{2x}$
10. $f(x) = 2x^3$
11. $f(x) = -3$
12. $f(x) = x^3 + 1$

13. $f(x) = \begin{cases} 2x, & x > -1 \\ -x - 1, & x \le -1 \end{cases}$

14. $f(x) = \begin{cases} x + 1, & x > 2 \\ 1, & -1 \le x \le 2 \\ -x + 1, & x < -1 \end{cases}$

15. $f(x) = \begin{cases} x, & x < 2 \\ 2, & x \ge 2 \end{cases}$

16. $f(x) = \begin{cases} -x, & x \le -2 \\ x^2, & -2 < x \le 2 \\ -x, & 3 \le x \le 4 \end{cases}$

17. $f(x) = \begin{cases} -x^2, & -3 < x < 1 \\ 0, & 1 \le x \le 2 \\ -3x, & x > 2 \end{cases}$

18. $f(x) = \begin{cases} 2 & \text{if } x \text{ is an integer} \\ -1 & \text{if } x \text{ is not an integer} \end{cases}$

19. $f(x) = \begin{cases} -2, & x < -2 \\ -1, & -2 \le x \le -1 \\ 1, & x > -1 \end{cases}$

20. $f(x) = \begin{cases} \dfrac{x^2 - 1}{x - 1}, & x \ne 1 \\ 3, & x = 1 \end{cases}$

21. The telephone company charges a fee of $6.50 per month for the first 100 message units and an additional fee of 6 cents for each of the next 100 message units. A reduced rate of 5 cents is charged for each message unit after the first 200 units. Express the monthly charge C as a function of the number of message units u.

22. The annual dues of a union are as shown in the table.

Employee's annual salary	Annual dues
less than $8000	$60
$8000 or more but less than $15,000	$60 +1% of the salary in excess of $8000
$15,000 or more	$130 + 2% of the salary in excess of $15,000

Express the annual dues d as a function of the salary S.

23. A tour operator who runs charter flights to Rome has established the following pricing schedule. For a group of no more than 100 people, the round trip fare per person is $300, with a minimum rental of $30,000 for the plane. For a group that has more than 100 but fewer than 150 people, the fare per person for all passengers will be reduced by $1 for each passenger in excess of 100. Write the tour operator's total revenue R as a function of the number of people x in the group.

24. A firm packages and ships 1-pound jars of instant coffee. The cost C of shipping is 40 cents for the first pound and 25 cents for each additional pound.
    (a) Write C as a function of the weight w (in pounds) for $0 < w \le 30$.
    (b) What is the cost of shipping a package containing 24 jars of instant coffee?

25. The daily rates of a car rental firm are $14 plus 8 cents per mile.
    (a) Express the cost C of renting a car as a function of the number of miles m traveled.
    (b) What is the domain of the function?
    (c) How much would it cost to rent a car for a 100-mile trip?

26. In a wildlife preserve, the population P of eagles depends on the population x of its basic food supply, rodents. Suppose that P is given by

$$P(x) = 0.002x + 0.004x^2$$

Find the eagle population when the rodent population is
    (a) 500   (b) 2000

2.4
AIDS IN GRAPHING

You now know the shape of the graph of the function $f(x) = |x|$. Can this knowledge be used to sketch the graph of the function $f(x) = |x| + 1$? Or, knowing the graph of $f(x) = x^2$, can you quickly sketch the graph of $f(x) = (x - 1)^2$? Of $f(x) = 3x^2$?

This section discusses techniques that will enable you to answer yes to each of the above questions. You will see that we can expand upon our knowledge of the graphs of the basic functions, the "special" functions listed in Section 2.3, to quickly sketch the graphs of a wide variety of functions. Since these functions are frequently encountered in calculus courses, the ability to sketch their graphs without using a table of values will give you a big boost toward succeeding in the study of calculus.

REFLECTIONS

Let's compare the graphs of the functions $f(x)$ and $-f(x)$. For any value of x, say $x = x_0$, the point $(x_0, f(x_0))$ lies on the graph of $f(x)$ and the point $(x_0, -f(x_0))$ lies on the graph of $-f(x)$. We see that, for every value of x, the y-coordinate on the graph of $-f(x)$ is the negative of the y-coordinate on the graph of $f(x)$. The graph of $-f(x)$ is called a **reflection about the x-axis** of the graph of $f(x)$. We illustrate with several examples.

EXAMPLE 1
Sketch the graphs of the functions.
    (a) $f(x) = -|x|$ (b) $f(x) = -x^2$ (c) $f(x) = -\sqrt{x}$ (d) $f(x) = -x^3$

SOLUTION
See Figure 27.

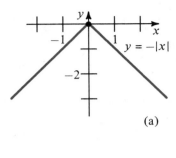

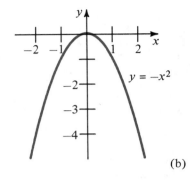

(a)

(b)

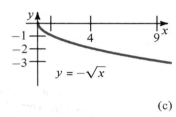

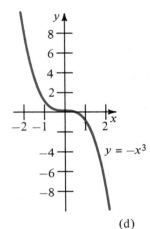

(c)

(d)

FIGURE 27

VERTICAL SHIFTS

The value of the function $f(x) + c$, where c is a constant, is obtained by adding c to the value of $f(x)$ for all x. This leads to the following result.

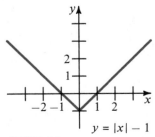

FIGURE 28

The graph of $f(x) + c$ is the graph of $f(x)$ shifted vertically c units. If $c > 0$, the shift is upward; if $c < 0$, the shift is downward.

EXAMPLE 2
Sketch the graph of the function $f(x) = |x| - 1$.

SOLUTION
The graph is that of $f(x) = |x|$ shifted downward 1 unit as in Figure 28.

EXAMPLE 3
Sketch the graph of the function $f(x) = 1 - x^2$.

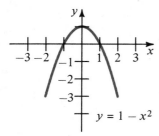

FIGURE 29

SOLUTION

The graph is that of $f(x) = -x^2$ shifted upward 1 unit as in Figure 29.

HORIZONTAL SHIFTS

Assuming that we can quickly sketch the graph of

$$y = f(x) \tag{1}$$

we would like to use this knowledge to sketch the graph of

$$y = f(x + c) \tag{2}$$

where c is a constant. Focusing on the value $x = x_0$, we see that the point $(x_0, f(x_0))$ lies on the graph of Equation (1). Similarly, the point $(x_0 - c, \ f(x_0))$ lies on the graph of Equation (2). This shows that Equation (2) attains the y-coordinate $f(x_0)$ at $x_0 - c$, whereas Equation (1) attains this y-coordinate at x_0. Thus, the graph of Equation (2) is that of Equation (1) shifted horizontally c units. In summary,

| **Horizontal Shifts** | The graph of $f(x + c)$, where c is a constant, is the graph of $f(x)$ shifted horizontally c units. If $c > 0$, the shift is to the left; if $c < 0$, the shift is to the right. |

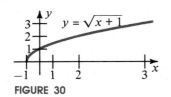

FIGURE 30

EXAMPLE 4

Sketch the graph of the function $f(x) = \sqrt{x + 1}$.

SOLUTION

If we let $g(x) = \sqrt{x}$, then $f(x) = \sqrt{x + 1} = g(x + 1)$. The graph of $f(x)$ is that of the square root function shifted left 1 unit as in Figure 30.

EXAMPLE 5

Sketch the graph of the function $f(x) = (x - 2)^3$.

SOLUTION

The graph is that of $y = x^3$ shifted right 2 units as in Figure 31.

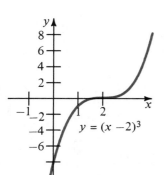

FIGURE 31

VERTICAL STRETCHING AND SHRINKING

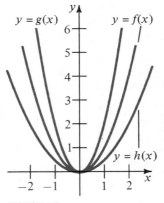

FIGURE 32

We now want to tackle the problem of graphing $c \cdot f(x)$, where c is a constant, assuming we already know the appearance of the graph of $f(x)$. If $c < 0$, then we may treat this as the graph of $-f(x)$ multiplied by $|c|$. For this reason we will restrict our results to the case where $c > 0$.

EXAMPLE 6
Sketch the graphs of the functions $f(x) = x^2$, $g(x) = 2x^2$, and $h(x) = \frac{1}{2}x^2$ on the same coordinate axes.

SOLUTION
See Figure 32. The graph of $g(x) = 2x^2$ rises more rapidly than that of $f(x) = x^2$ while the graph of $h(x) = \frac{1}{2}x^2$ rises less rapidly.

For every value of x, the y-coordinate on the graph of $c \cdot f(x)$ is precisely c times the y-coordinate on the graph of $f(x)$. This permits us to summarize in this way.

Stretching and Shrinking	When $c > 1$, the graph of $c \cdot f(x)$ is "stretched" vertically away from the x-axis, whereas for $0 < c < 1$ the graph is flattened, or "shrunk" vertically, toward the x-axis.

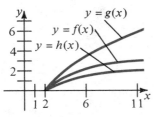

FIGURE 33

EXAMPLE 7
Sketch the graphs of $f(x) = \sqrt{x - 2}$, $g(x) = 2\sqrt{x - 2}$, and $h(x) = \frac{1}{2}\sqrt{x - 2}$ on the same coordinate axes.

SOLUTION
The graph of $f(x) = \sqrt{x - 2}$ is the same as that of $y = \sqrt{x}$ shifted right 2 units. The graph of $g(x)$ rises more rapidly and the graph of $h(x)$ rises less rapidly. See Figure 33.

EXERCISE SET 2.4
In Exercises 1–12 sketch the graph of the function by shifting or reflecting a known graph. Where needed, perform both operations.

1. $f(x) = x + 2$
2. $f(x) = \sqrt{x} - 2$
3. $f(x) = 1 - x^3$
4. $f(x) = 2 - |x|$
5. $f(x) = (x + 2)^2$
6. $f(x) = |x - 1|$
7. $f(x) = \sqrt{x - 2}$
8. $f(x) = (x + 1)^3$
9. $f(x) = (x - 1)^2 + 2$
10. $f(x) = \sqrt{x + 2} - 1$
11. $f(x) = (x + 2)^3 - 1$
12. $f(x) = \sqrt{x - 2} + 2$

In Exercises 13–20 sketch the graphs of the given functions on the same coordinate axes.

13. $f(x) = \sqrt{x}$, $g(x) = \frac{1}{2}\sqrt{x}$, $h(x) = \frac{1}{4}\sqrt{x}$

14. $f(x) = \frac{1}{2}x^2$, $g(x) = \frac{1}{3}x^2$, $h(x) = \frac{1}{4}x^2$

15. $f(x) = 2x^2$, $g(x) = -2x^2$

16. $f(x) = 2x^2 + 1$, $g(x) = 1 - 2x^2$

17. $f(x) = x^3$, $g(x) = 2x^3$

18. $f(x) = -x^3$, $g(x) = -2x^3$

19. $f(x) = \sqrt{x - 2}$, $g(x) = 2\sqrt{x - 2}$

20. $f(x) = \sqrt{x}$, $g(x) = \sqrt{x - 2}$, $h(x) = \sqrt{x + 2}$

2.5
LINEAR FUNCTIONS

The polynomial function of the first degree

$$f(x) = ax + b$$

is called a **linear function.** In this section we will show that the graph of a linear function is a straight line. We will also show that there is a property unique to the straight line that differentiates it from all other curves.

SLOPE OF THE STRAIGHT LINE

In Figure 34 we have drawn a straight line L that is not vertical. We have indicated the distinct points $P_1(x_1, y_1)$ and $P_2(x_2, y_2)$ on L. The increments or changes $x_2 - x_1$ and $y_2 - y_1$ in the x and y coordinates, respectively, from P_1 to P_2 are also indicated. Note that the increment $x_2 - x_1$ cannot be zero, since L is not vertical.

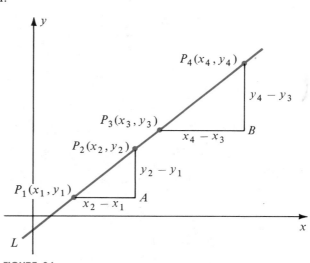

FIGURE 34

If $P_3(x_3, y_3)$ and $P_4(x_4, y_4)$ are another pair of points on L, the increments $x_4 - x_3$ and $y_4 - y_3$ will, in general, be different from the increments obtained by using P_1 and P_2. However, since triangles P_1AP_2 and P_3BP_4 are similar, the corresponding sides are in proportion; that is, the ratios

$$\frac{y_4 - y_3}{x_4 - x_3} \quad \text{and} \quad \frac{y_2 - y_1}{x_2 - x_1}$$

are the same. This ratio is called the **slope of the line** L and is denoted by m.

Slope of a Line	The slope of a line that is not vertical is given by

$$m = \frac{y_2 - y_1}{x_2 - x_1}$$

where $P_1(x_1, y_1)$ and $P_2(x_2, y_2)$ are any two distinct points on the line.

For a vertical line, $x_1 = x_2$ so that $x_2 - x_1 = 0$. Since we cannot divide by 0, we say that a vertical line has no slope.

The property of constant slope characterizes the straight line; that is, no other curve has this property. In fact, to define slope for a curve other than a straight line is not a trivial task; it requires use of the concept of limit, which is fundamental to calculus. (See Appendix B.)

EXAMPLE 1

Find the slope of the line that passes through the points $(4, 2)$, $(1, -2)$.

SOLUTION

We may choose either point as (x_1, y_1) and the other as (x_2, y_2). Our choice is

$$(x_1, y_1) = (4, 2) \quad \text{and} \quad (x_2, y_2) = (1, -2)$$

Then

$$m = \frac{y_2 - y_1}{x_2 - x_1} = \frac{-2 - 2}{1 - 4} = \frac{-4}{-3} = \frac{4}{3}$$

The student should verify that reversing the choice of P_1 and P_2 produces the same result for the slope m. We may choose either point as P_1 and the other as P_2, but we must use this choice consistently once it has been made.

Slope is a means of measuring the steepness of a line. That is, slope specifies the number of units we must move up or down to reach the line after moving 1 unit to the left or right of the line. In Figure 35 we have displayed several lines with positive and negative slopes. We can summarize this way.

Let m be the slope of a line L.
(a) When $m > 0$, the line is the graph of an increasing function.
(b) When $m < 0$, the line is the graph of a decreasing function.
(c) When $m = 0$, the line is the graph of a constant function.
(d) Slope does not exist for a vertical line, and a vertical line is not the graph of a function.

THE PIRATE TREASURE (PART I)

Five pirates traveling with a slave found a chest of gold coins. The pirates agreed to divide the coins among themselves the following morning.

During the night Pirate 1 awoke and, not trusting his fellow pirates, decided to remove his share of the coins. After dividing the coins into five equal lots, he found that one coin remained. The pirate took his lot and gave the remaining coin to the slave to ensure his silence.

Later that night Pirate 2 awoke and decided to remove his share of the coins. After dividing the remaining coins into five equal lots, he found one coin left over. The pirate took his lot and gave the extra coin to the slave.

That same night the process was repeated by Pirates 3, 4, and 5. Each time there remained one coin, which was given to the slave.

In the morning these five compatible pirates divided the remaining coins into five equal lots. Once again a single coin remained.

Question: What is the minimum number of coins there could have been in the chest? (For help, see Part II on page 95.)

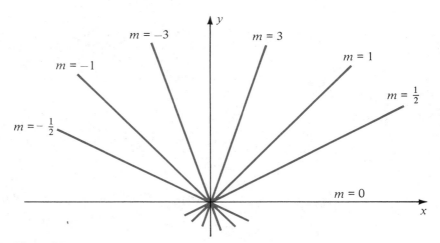

FIGURE 35

EQUATIONS OF THE STRAIGHT LINE

We can apply the concept of slope to develop two important forms of the equation of a straight line. In Figure 36 the point $P_1(x_1, y_1)$ lies on a line L whose slope is m. If $P(x, y)$ is any other point on L, then we may use P and P_1 to compute m; that is,

$$m = \frac{y - y_1}{x - x_1}$$

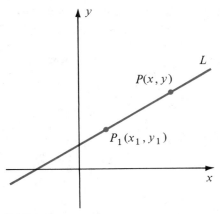

FIGURE 36

This can be written in the form

$$y - y_1 = m(x - x_1)$$

Since (x_1, y_1) satisfies this equation, every point on L satisfies this equation. Conversely, any point satisfying this equation must lie on the line L, since there is only one line through $P_1(x_1, y_1)$ with slope m. This equation is called the **point-slope form** of a line.

Point-Slope Form

$$y - y_1 = m(x - x_1)$$

is an equation of the line with slope m that passes through the point (x_1, y_1).

EXAMPLE 2

Find an equation of the line that passes through the points $(6, -2)$ and $(-4, 3)$.

SOLUTION

We first find the slope. Letting $(x_1, y_1) = (6, -2)$ and $(x_2, y_2) = (-4, 3)$, then

$$m = \frac{y_2 - y_1}{x_2 - x_1} = \frac{3 - (-2)}{-4 - 6} = \frac{5}{-10} = -\frac{1}{2}$$

Next, the point-slope form is used with $m = -\frac{1}{2}$ and $(x_1, y_1) = (6, -2)$.

$$y - y_1 = m(x - x_1)$$

$$y - (-2) = -\frac{1}{2}(x - 6)$$

$$y = -\frac{1}{2}x + 1$$

The student should verify that using the point $(-4, 3)$ and $m = -\frac{1}{2}$ in the point-slope form will yield the same equation.

PROGRESS CHECK

Find an equation of the line that passes through the points $(-5, 0)$ and $(2, -5)$.

ANSWER

$7y = -5x - 25$

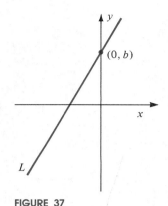

FIGURE 37

There is another form of the equation of the straight line that is very useful. In Figure 37 the line L meets the y-axis at the point $(0, b)$, and is assumed to have slope m. Then we can let $(x_1, y_1) = (0, b)$ and use the point-slope form.

$$y - y_1 = m(x - x_1)$$
$$y - b = m(x - 0)$$
$$y = mx + b$$

Recalling that b is the y-intercept, we call this equation the **slope-intercept form** of the line.

Slope-Intercept Form	The graph of the equation $$y = mx + b$$ is a straight line with slope m and y-intercept b.

The last result leads to the important conclusion mentioned in the introduction to this section. Since the graph of $y = mx + b$ is the graph of the function $f(x) = mx + b$, we have shown that the *graph of a linear function is a straight line*.

EXAMPLE 3

Find the slope and y-intercept of the line $y - 3x + 1 = 0$.

SOLUTION

The equation must be placed in the form $y = mx + b$. Solving for y gives

$$y = 3x - 1$$

and we find that $m = 3$ is the slope and $b = -1$ is the y-intercept.

PROGRESS CHECK

Find the slope and y-intercept of the line $2y + x - 3 = 0$.

ANSWER

slope $= m = -\dfrac{1}{2}$; y-intercept $= b = \dfrac{3}{2}$

HORIZONTAL AND VERTICAL LINES

In Figure 38a we have drawn a horizontal line through the point (a, b). Every point on this line has the form (x, b) since the y-coordinate remains constant. If $P(x_1, b)$ and $Q(x_2, b)$ are any two distinct points on the line, then the slope is

$$m = \frac{b - b}{x_2 - x_1} = 0$$

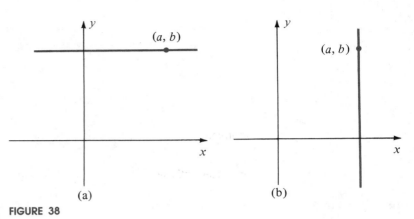

(a)   (b)

FIGURE 38

We have established the following.

Horizontal Lines

The equation of the horizontal line through the point (a, b) is

$$y = b$$

The slope of a horizontal line is 0.

In Figure 38b, every point on the vertical line through the point (a, b) has the form (a, y) since the x-coordinate remains constant. The slope computation using any two points $P(a, y_1)$ and $Q(a, y_2)$ on the line produces

$$m = \frac{y_2 - y_1}{a - a} = \frac{y_2 - y_1}{0}$$

Since we cannot divide by 0, slope is not defined for a vertical line.

Vertical Lines

The equation of the vertical line through the point (a, b) is

$$x = a$$

A vertical line has no slope.

EXAMPLE 4

Find the equations of the horizontal and vertical lines through $(-4, 7)$.

SOLUTION

The horizontal line has the equation $y = 7$. The vertical line has the equation $x = -4$.

WARNING Don't confuse "no slope" and "zero slope." A horizontal line has zero slope. A vertical line has no slope, by which we mean that the slope is undefined.

PARALLEL AND PERPENDICULAR LINES

The concept of slope of a line can be used to determine when two lines are parallel or perpendicular. Since parallel lines have the same "steepness," we intuitively recognize that they must have the same slope.

Two lines with slopes m_1 and m_2 are parallel if and only if

$$m_1 = m_2$$

The criterion for perpendicular lines can be stated in this way.

Two lines with slopes m_1 and m_2 are perpendicular if and only if

$$m_2 = -\frac{1}{m_1}$$

These two theorems do not apply to vertical lines, since the slope of a vertical line is undefined. The proofs of these theorems are geometric in nature and are outlined in Exercises 54 and 56.

EXAMPLE 5

Given the line $y = 3x - 2$, find an equation of the line passing through the point $(-5, 4)$ that is (a) parallel to the given line and (b) perpendicular to the given line.

SOLUTION

We first note that the line $y = 3x - 2$ has slope $m_1 = 3$.

(a) Every line parallel to the line $y = 3x - 2$ must have slope $m_2 = m_1 = 3$. We therefore seek a line with slope 3 that passes through the point $(-5, 4)$. Using the point-slope formula,

$$y - y_1 = m(x - x_1)$$
$$y - 4 = 3(x + 5)$$
$$y = 3x + 19$$

(b)   Every line perpendicular to the line $y = 3x - 2$ has slope $m_2 = -1/m_1 = -1/3$. The line we seek has slope $-\frac{1}{3}$ and passes through the point $(-5, 4)$. We can again apply the point-slope formula to obtain

$$y - y_1 = m(x - x_1)$$
$$y - 4 = -\frac{1}{3}(x + 5)$$
$$y = -\frac{1}{3}x + \frac{7}{3}$$

The three lines are shown in Figure 39.

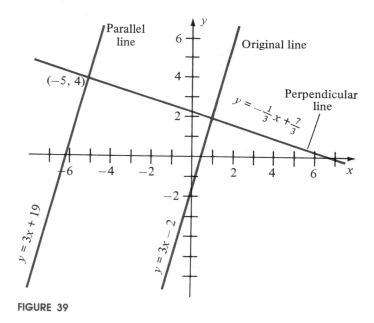

FIGURE 39

<table>
<tr><td>**GENERAL FIRST-DEGREE**
EQUATION</td><td>The **general first-degree equation** in x and y can always be written in the form</td></tr>
</table>

$$Ax + By + C = 0$$

where A, B, and C are constants and A and B are not both zero. We can rewrite

THE PIRATE TREASURE (PART II)

First, note that any number that is a multiple of 5 can be written in the form $5n$, where n is an integer. Since the number of coins found in the chest by Pirate 1 was one more than a multiple of 5, we can write the original number of coins C in the form $C = 5n + 1$, where n is a positive integer. Now, Pirate 1 removed his lot of n coins and gave one to the slave. The remaining coins can be calculated as

$$5n + 1 - (n + 1) = 4n$$

and since this is also one more than a multiple of 5, we can write $4n = 5p + 1$, where p is a positive integer. Repeating the process, we have the following sequence of equations.

$C = 5n + 1$	found by Pirate 1
$4n = 5p + 1$	found by Pirate 2
$4p = 5q + 1$	found by Pirate 3
$4q = 5r + 1$	found by Pirate 4
$4r = 5s + 1$	found by Pirate 5
$4s = 5t + 1$	found next morning

Solving for s in the last equation and substituting successively in the preceding equations leads to the requirement that

$$1024n - 3125t = 2101 \tag{1}$$

where n and t are positive integers. Equations, such as this, that require integer solutions are called Diophantine equations, and there is an established procedure for solving them that is studied in courses in number theory.

You might want to try to solve Equation (1) using a computer program. Since

$$n = \frac{3125t + 2101}{1024}$$

you can substitute successive integer values for t until you produce an integer result for n. The accompanying BASIC program does just that.

```
10  FOR K = 1 TO 3200
20  X = (3125*K
    + 2101)/1024
30  I = INT(X)
40  IF X = I THEN GO
    TO 60
50  NEXT K
60  PRINT ''MINIMUM
    NUMBER OF
    COINS = '';
    5*I + 1
70  END
```

this equation as

$$By = -Ax - C$$

If $B \neq 0$, the equation becomes

$$y = -\frac{A}{B}x - \frac{C}{B}$$

which we recognize as having a straight line graph with slope $-A/B$ and y-intercept $-C/B$. If $B = 0$, the original equation becomes $Ax + C = 0$, whose graph is a vertical line.

The General First-Degree Equation	• The graph of the general first-degree equation $$Ax + By + C = 0$$ is a straight line. • If $B = 0$, the graph is a vertical line. • If $A = 0$, the graph is a horizontal line.

EXERCISE SET 2.5

In Exercises 1–6 determine the slope of the line through the given points. State whether the line is the graph of an increasing function, a decreasing function, or a constant function.

1. $(2, 3), (-1, -3)$ 　　　　　　　　　　　　　2. $(1, 2), (-2, 5)$

3. $(-2, 3), (0, 0)$ 　　　　　　　　　　　　　4. $(2, 4), (-3, 4)$

5. $\left(\frac{1}{2}, 2\right), \left(\frac{3}{2}, 1\right)$ 　　　　　　　　　　6. $(-4, 1), (-1, -2)$

7. Use slopes to show that the points $A(-1, -5)$, $B(1, -1)$, and $C(3, 3)$ are collinear (lie on the same line).

8. Use slopes to show that the points $A(-3, 2), B(3, 4)$, $C(5, -2)$, and $D(-1, -4)$ are the vertices of a parallelogram.

In Exercises 9–12 determine an equation of the line with the given slope m that passes through the given point.

9. $m = 2, (-1, 3)$ 　　10. $m = -\frac{1}{2}, (1, -2)$ 　　11. $m = 3, (0, 0)$ 　　12. $m = 0, (-1, 3)$

In Exercises 13–18 determine an equation of the line through the given points.

13. $(2, 4), (-3, -6)$ 　　　　　　　　　　　14. $(-3, 5), (1, 7)$

15. $(0, 0), (3, 2)$ 　　　　　　　　　　　　16. $(-2, 4), (3, 4)$

17. $\left(-\frac{1}{2}, -1\right), \left(\frac{1}{2}, 1\right)$ 　　　　　　　18. $(-8, -4), (3, -1)$

In Exercises 19–24 determine an equation of the line with the given slope m and the given y-intercept b.

19. $m = 3, b = 2$ 　　20. $m = -3, b = -3$ 　　21. $m = 0, b = 2$ 　　22. $m = -\frac{1}{2}, b = \frac{1}{2}$

23. $m = \frac{1}{3}, b = -5$ 　　24. $m = -2, b = -\frac{1}{2}$

In Exercises 25–30 determine the slope m and y-intercept b of the given line.

25. $3x + 4y = 5$ 　　26. $2x - 5y + 3 = 0$ 　　27. $y - 4 = 0$ 　　28. $x = -5$

29. $3x + 4y + 2 = 0$ 　　30. $x = -\frac{1}{2}y + 3$

In Exercises 31–36 write an equation of (a) the horizontal line passing through the given point and (b) the vertical line passing through the given point.

31. $(-6, 3)$ 　　32. $(-5, -2)$ 　　33. $(-7, 0)$ 　　34. $(0, 5)$

35. $(9, -9)$ 　　36. $\left(-\frac{3}{2}, 1\right)$

In Exercises 37–40 determine the slope of (a) every line that is parallel to the given line and (b) every line that is perpendicular to the given line.

37. $y = -3x + 2$ 38. $2y - 5x + 4 = 0$ 39. $3y = 4x - 1$ 40. $5y + 4x = -1$

In Exercises 41–44 determine an equation of the line through the given point that (a) is parallel to the given line; (b) is perpendicular to the given line.

41. $(1, 3); y = -3x + 2$

42. $(-1, 2); 3y + 2x = 6$

43. $(-3, 2); 3x + 5y = 2$

44. $(-1, -3); 3y + 4x - 5 = 0$

45. The Celsius (C) and Fahrenheit (F) temperature scales are related by a linear equation. Water boils at 212°F or 100°C, and freezes at 32°F or 0°C.
   (a) Write a linear equation expressing F in terms of C.
   (b) What is the Fahrenheit temperature when the Celsius temperature is 20°?

46. The college bookstore sells a textbook that costs $10 for $13.50, and a textbook that costs $12 for $15.90. If the markup policy of the bookstore is linear, write a linear function that relates sales price S and cost C. What is the cost of a book that sells for $22?

47. An appliance manufacturer finds that it had sales of $200,000 five years ago and sales of $600,000 this year. If the growth in sales is assumed to be linear, what will the sales amount be five years from now?

48. A product that cost $2.50 three years ago sells for $3 this year. If price increases are assumed to be linear, how much will the product cost six years from now?

49. Find a real number c such that $P(-2, 2)$ is on the line $3x + cy = 4$.

50. Find a real number c such that the line $cx - 5y + 8 = 0$ has x-intercept 4.

51. If the points $(-2, -3)$ and $(-1, 5)$ are on the graph of a linear function f, find $f(x)$.

52. If $f(1) = 4$ and $f(-1) = 3$ and the function f is linear, find $f(x)$.

53. Prove that the linear function $f(x) = ax + b$ is an increasing function if $a > 0$ and is a decreasing function if $a < 0$.

54. In the accompanying figure, lines L_1 and L_2 are parallel. Points A and D are selected on lines L_1 and L_2, respectively. Lines parallel to the x-axis are constructed through A and D that intersect the y-axis at points B and E. Supply a reason for each of the steps in the following proof.

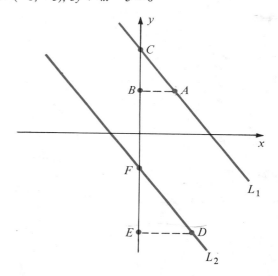

(a) Angles ABC and DEF are equal.
(b) Angles ACB and DFE are equal.
(c) Triangles ABC and DEF are similar.
(d) $\dfrac{\overline{CB}}{\overline{BA}} = \dfrac{\overline{FE}}{\overline{ED}}$
(e) $m_1 = \dfrac{\overline{CB}}{\overline{BA}}, \quad m_2 = \dfrac{\overline{FE}}{\overline{ED}}$
(f) $m_1 = m_2$
(g) Parallel lines have the same slope.

55. Prove that if two lines have the same slope, they are parallel.

56. In the accompanying figure, lines perpendicular to each other, with slopes m_1 and m_2, intersect at a point Q. A perpendicular from Q to the x-axis intersects the x-axis at the point C. Supply a reason for each of the steps in the following proof.
   (a) Angles CAQ and BQC are equal.
   (b) Triangles ACQ and BCQ are similar.

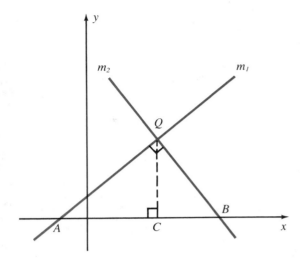

(c) $\dfrac{\overline{CQ}}{\overline{AC}} = \dfrac{\overline{CB}}{\overline{CQ}}$

(d) $m_1 = \dfrac{\overline{CQ}}{\overline{AC}}, \quad m_2 = -\dfrac{\overline{CQ}}{\overline{CB}}$

(e) $m_2 = -\dfrac{1}{m_1}$

57. Prove that if two lines have slopes m_1 and m_2 such that $m_2 = -1/m_1$, the lines are perpendicular.

58. If x_1 and x_2 are the abscissas of two points on the graph of the function $y = f(x)$, show that the slope m of the line connecting the two points can be written as

$$m = \dfrac{f(x_2) - f(x_1)}{x_2 - x_1}$$

2.6 QUADRATIC EQUATIONS

The second-degree polynomial equation of the form

$$ax^2 + bx + c = 0$$

where a, b, and c are real numbers and $a \neq 0$, is known as a **quadratic equation.** This type of equation occurs frequently, and we will devote this section to studying methods for finding solutions, or **roots,** of quadratic equations, methods for analyzing the nature of the roots, and applications of quadratics to practical problems. In the next section we will relate our findings to quadratic functions and their graphs.

SOLVING BY FACTORING

If we can factor the left-hand side of the quadratic equation

$$ax^2 + bx + c = 0, \quad a \neq 0$$

into two linear factors, then we can solve the equation quickly. For example, the quadratic equation

$$3x^2 + 5x - 2 = 0$$

can be written as

$$(3x - 1)(x + 2) = 0$$

Since the product of two real numbers can be zero only if one or more of the factors are zero, we can set each factor equal to zero.

$$3x - 1 = 0 \quad \text{or} \quad x + 2 = 0$$

$$x = \dfrac{1}{3} \qquad\qquad x = -2$$

The solutions of the given quadratic equation are $\tfrac{1}{3}$ and -2.

PROGRESS CHECK

Solve each of the given equations by factoring.

(a) $3x^2 - 4x = 0$ (b) $2x^2 - 3x - 2 = 0$

ANSWERS

(a) $0, \dfrac{4}{3}$ (b) $-\dfrac{1}{2}, 2$

COMPLETING THE SQUARE

A quadratic equation of the form

$$a(x - h)^2 + k = 0 \tag{1}$$

where a, h, and k are constants, can be easily solved. For example, from

$$2(x - 5)^2 + 8 = 0$$

we conclude that

$$(x - 5)^2 = -4$$
$$x - 5 = \pm\sqrt{-4} = \pm 2i$$
$$x = 5 \pm 2i$$

The solutions of the given equations are the complex numbers $5 + 2i$ and $5 - 2i$.

A technique known as **completing the square** permits us to rewrite *any* quadratic equation in the form of Equation (1). Beginning with the expression $x^2 + dx$, we seek a constant h^2 to complete the square so that

$$x^2 + dx + h^2 = (x - h)^2$$

Expanding and solving, we have

$$x^2 + dx + h^2 = x^2 - 2hx + h^2$$
$$dx = -2hx$$
$$h = -\frac{d}{2}$$

so h^2 is the square of half the coefficient of x.

EXAMPLE 1

Complete the square for each of the following.

(a) $x^2 - 6x$ (b) $x^2 + 3x$

SOLUTION

(a) The coefficient of x is -6 so that $h = 3$ and $h^2 = 9$. Then

$$(x^2 - 6x + 9) = (x - 3)^2$$

(b) The coefficient of x is 3 and $h^2 = \left(-\dfrac{3}{2}\right)^2 = \dfrac{9}{4}$.

Then

$$x^2 + 3x + \frac{9}{4} = \left(x + \frac{3}{2}\right)^2$$

We are now in a position to use this method to solve a quadratic equation.

EXAMPLE 2
Solve the quadratic equation $2x^2 - 10x + 1 = 0$ by completing the square.

SOLUTION
We now outline and explain each step of the process.

Completing the Square	
Step 1. Rewrite the equation with the constant term on the right-hand side.	*Step 1.* $2x^2 - 10x \quad = -1$
Step 2. Factor out the coefficient a of x^2.	*Step 2.* $2(x^2 - 5x \quad) = -1$
Step 3. Complete the square, $$x^2 + dx + h^2 = (x - h)^2$$ where $h^2 = \left(-\dfrac{d}{2}\right)^2$. Balance the equation by adding ah^2 to the right-hand side. Simplify.	*Step 3.* $$2\left(x^2 - 5x + \frac{25}{4}\right) = -1 + \frac{25}{2}$$ $$2\left(x - \frac{5}{2}\right)^2 = \frac{23}{2}$$
Step 4. Solve for x.	*Step 4.* $$\left(x - \frac{5}{2}\right)^2 = \frac{23}{4}$$ $$x - \frac{5}{2} = \pm\frac{\sqrt{23}}{2}$$ $$x = \frac{5 \pm \sqrt{23}}{2}$$

PROGRESS CHECK
Solve by completing the square.
(a) $x^2 - 3x + 2 = 0$ (b) $3x^2 - 4x + 2 = 0$

ANSWERS
(a) 1, 2   (b) $\dfrac{2 \pm i\sqrt{2}}{3}$

THE QUADRATIC FORMULA

We can apply the method of completing the square to the general quadratic equation

$$ax^2 + bx + c = 0, \quad a > 0$$

Following the steps of the method, we have

$$ax^2 + bx = -c$$

$$a\left(x^2 + \frac{b}{a}x \quad\right) = -c$$

$$a\left[x^2 + \frac{b}{a}x + \left(\frac{b}{2a}\right)^2\right] = a\left(\frac{b}{2a}\right)^2 - c$$

$$a\left(x + \frac{b}{2a}\right)^2 = \frac{b^2}{4a} - c$$

$$\left(x + \frac{b}{2a}\right)^2 = \frac{b^2}{4a^2} - \frac{c}{a} = \frac{b^2 - 4ac}{4a^2}$$

$$x + \frac{b}{2a} = \pm\sqrt{\frac{b^2 - 4ac}{4a^2}} = \frac{\pm\sqrt{b^2 - 4ac}}{2a}$$

$$x = -\frac{b}{2a} \pm \frac{\sqrt{b^2 - 4ac}}{2a}$$

$$x = \frac{-b \pm \sqrt{b^2 - 4ac}}{2a}$$

This result is a *formula* that gives us the solutions for *any* quadratic equation.

Quadratic Formula

$$x = \frac{-b \pm \sqrt{b^2 - 4ac}}{2a}, \quad a > 0$$

EXAMPLE 3

Solve $-5x^2 + 3x = 2$ by the quadratic formula.

SOLUTION

We first rewrite the given equation as $5x^2 - 3x + 2 = 0$ so that $a > 0$ and the right-hand side equals 0. Then $a = 5$, $b = -3$, and $c = 2$. Substituting in the quadratic formula, we have

$$x = \frac{-b \pm \sqrt{b^2 - 4ac}}{2a}$$

$$= \frac{-(-3) \pm \sqrt{(-3)^2 - 4(5)(2)}}{2(5)}$$

$$= \frac{3 \pm \sqrt{-31}}{10} = \frac{3 \pm i\sqrt{31}}{10}$$

PROGRESS CHECK

Solve by use of the quadratic formula.

(a) $x^2 - 8x = -10$ (b) $4x^2 - 2x + 1 = 0$

ANSWERS

(a) $4 \pm \sqrt{6}$ (b) $\dfrac{1 \pm i\sqrt{3}}{4}$

WARNING   There are a number of errors that students make in using the quadratic formula.

(a)   To solve $x^2 - 3x = -4$, you must write the equation in the form $x^2 - 3x + 4 = 0$ to properly identify a, b, and c.

(b)   The quadratic formula is

$$x = \frac{-b \pm \sqrt{b^2 - 4ac}}{2a}$$

Note that

$$x \neq -b \pm \frac{\sqrt{b^2 - 4ac}}{2a}$$

since the term $-b$ must also be divided by $2a$.

Now that you have a formula that works for *any* quadratic equation, you may be tempted to use it all the time. However, if you see an equation of the form

$$x^2 = 15$$

it is certainly easier to immediately supply the answer: $x = \pm\sqrt{15}$. Similarly, if you are faced with

$$x^2 + 3x + 2 = 0$$

it is faster to solve if you see that

$$x^2 + 3x + 2 = (x + 1)(x + 2)$$

The method of completing the square is generally not used for solving quadratic equations once you have learned the quadratic formula. The *technique* of completing the square is helpful in a variety of applications, and we will use it in a later chapter when we graph second-degree equations.

THE DISCRIMINANT

By analyzing the quadratic formula

$$x = \frac{-b \pm \sqrt{b^2 - 4ac}}{2a}$$

we can learn a great deal about the roots of the quadratic equation

$$ax^2 + bx + c = 0, \quad a > 0$$

The key to the analysis is the **discriminant** $b^2 - 4ac$ found under the radical.
(a) If $b^2 - 4ac$ is negative, we have the square root of a negative number, and the roots of the quadratic equation are complex numbers.
(b) If $b^2 - 4ac$ is positive, we have the square root of a positive number, and the roots of the quadratic equation will be real numbers.
(c) If $b^2 - 4ac = 0$, then $x = -b/2a$, which we call a **double root** or **repeated root** of the quadratic equation. For example, if $x^2 - 10x + 25 = 0$, then the discriminant is 0 and $x = 5$. But

$$x^2 - 10x + 25 = (x - 5)(x - 5) = 0$$

We call $x = 5$ a double root because the factor $(x - 5)$ is a double factor of $x^2 - 10x + 25 = 0$. This hints at the importance of the relationship between roots and factors, a relationship that we will explore in a later chapter on roots of polynomial equations.

   If the roots of the quadratic equation are real and a, b, and c are rational numbers, the discriminant enables us to determine whether the roots are rational or irrational. Since $\sqrt{k}$ is a rational number only if k is a perfect square, we see that the quadratic formula produces a rational result only if $b^2 - 4ac$ is a perfect square. We summarize as follows.

The quadratic equation $ax^2 + bx + c = 0$, $a > 0$, has exactly two roots, the nature of which is determined by the discriminant $b^2 - 4ac$.

Discriminant		Roots
Negative		Two complex roots
0		A double root
Positive		Two real roots
a, b, c $\Big\{$	A perfect square	Rational roots
rational	Not a perfect square	Irrational roots

EXAMPLE 4
Without solving, determine the nature of the roots of the quadratic equation $3x^2 - 4x + 6 = 0$.

SOLUTION
We evaluate $b^2 - 4ac$ using $a = 3$, $b = -4$, and $c = 6$.

$$b^2 - 4ac = (-4)^2 - 4(3)(6) = 16 - 72 = -56$$

The discriminant is negative and the equation has two complex roots.

EXAMPLE 5
Without solving, determine the nature of the roots of the equation

$$2x^2 - 7x = -1$$

SOLUTION
We rewrite the equation in the standard form

$$2x^2 - 7x + 1 = 0$$

and then substitute $a = 2$, $b = -7$, and $c = 1$ in the discriminant. Thus,

$$b^2 - 4ac = (-7)^2 - 4(2)(1) = 49 - 8 = 41$$

The discriminant is positive and is not a perfect square; thus, the roots are real, unequal, and irrational.

PROGRESS CHECK
Without solving, determine the nature of the roots of the quadratic equation by using the discriminant.

(a) $4x^2 - 20x + 25 = 0$ (b) $5x^2 - 6x = -2$
(c) $10x^2 = x + 2$ (d) $x^2 + x - 1 = 0$

ANSWERS
(a) a real, double root   (b) 2 complex roots
(c) 2 real, rational roots   (d) 2 real, irrational roots

FORMS LEADING TO QUADRATICS

Certain types of equations can be transformed into quadratic equations, which can be solved by the methods discussed in this section. One form that leads to a quadratic equation is the **radical equation,** such as

$$x - \sqrt{x - 2} = 4$$

which is solved in Example 6. To solve the equation we isolate the radical and raise both sides to suitable powers. The following is the key to the solution of such equations.

If P and Q are algebraic expressions, then the solution set of the equation

$$P = Q$$

is a subset of the solution set of the equation

$$P^n = Q^n$$

where n is a natural number.

This suggests that we can solve radical equations if we observe a precaution.

> If both sides of an equation are raised to the same power, the solutions of the resulting equation must be checked to see that they satisfy the original equation.

EXAMPLE 6

Solve $x - \sqrt{x - 2} = 4$.

SOLUTION

Solving Radical Equations	
Step 1. When possible, isolate the radical on one side of the equation.	*Step 1.* $\quad x - 4 = \sqrt{x - 2}$
Step 2. Raise both sides of the equation to a suitable power to eliminate the radical.	*Step 2.* Squaring both sides, $$x^2 - 8x + 16 = x - 2$$
Step 3. Solve for the unknown.	*Step 3.* $$x^2 - 9x + 18 = 0$$ $$(x - 3)(x - 6) = 0$$ $$x = 3 \qquad x = 6$$
Step 4. Check each solution by substituting in the *original* equation.	*Step 4.* $$\text{checking } x = 3 \qquad \text{checking } x = 6$$ $$3 - \sqrt{3 - 2} \overset{?}{=} 4 \qquad 6 - \sqrt{6 - 2} \overset{?}{=} 4$$ $$3 - 1 \overset{?}{=} 4 \qquad 6 - \sqrt{4} \overset{?}{=} 4$$ $$2 \neq 4 \qquad 4 \overset{\le}{=} 4$$

We conclude that 6 is a solution of the original equation and 3 is not a solution of the original equation. We say that 3 is an **extraneous solution** that was introduced when we raised each side of the original equation to the second power.

PROGRESS CHECK

Solve $x - \sqrt{1 - x} = -5$.

ANSWER

-3

Although the equation

$$x^4 - x^2 - 2 = 0$$

is not a quadratic in the unknown x, it is a quadratic in the unknown x^2; that is,

$$(x^2)^2 - (x^2) - 2 = 0$$

This may be seen more clearly by replacing x^2 by a new unknown u such that $u = x^2$. Substituting, we have

$$u^2 - u - 2 = 0$$

which is a quadratic equation in the unknown u. Solving,

$$(u + 1)(u - 2) = 0$$
$$u = -1 \quad \text{or} \quad u = 2$$

Since $x^2 = u$, we must next solve the equations

$$x^2 = -1 \quad \text{and} \quad x^2 = 2$$
$$x = \pm i \qquad\qquad x = \pm\sqrt{2}$$

The original equation has four solutions: i, $-i$, $\sqrt{2}$, and $-\sqrt{2}$.

The technique we have used is called a **substitution of variable.** Although simple in concept, this is a powerful method that is commonly used in calculus.

PROGRESS CHECK

Indicate an appropriate substitution of variable and solve each of the following equations.

(a) $\quad 3x^4 - 10x^2 - 8 = 0$ (b) $\quad 4x^{2/3} + 7x^{1/3} - 2 = 0$

(c) $\quad \dfrac{2}{x^2} + \dfrac{1}{x} - 10 = 0$ (d) $\quad \left(1 + \dfrac{2}{x}\right)^2 - 8\left(1 + \dfrac{2}{x}\right) + 15 = 0$

ANSWERS

(a) $\quad u = x^2;\ \pm 2,\ \pm\dfrac{i\sqrt{6}}{3}$ (b) $\quad u = x^{1/3};\ \dfrac{1}{64},\ -8$

(c) $\quad u = \dfrac{1}{x};\ -\dfrac{2}{5},\ \dfrac{1}{2}$ (d) $\quad u = 1 + \dfrac{2}{x};\ 1,\ \dfrac{1}{2}$

APPLICATIONS

As your knowledge of mathematical techniques and ideas grows, you will be capable of tackling an ever-wider variety of applications. For example, the two practical problems that follow can be expressed as quadratic equations, which you now know how to solve.

One word of caution: It is possible to arrive at a solution that is meaningless. For example, a negative solution that represents hours worked or the age of an individual is meaningless and must be rejected.

EXAMPLE 7

The length of a pool is 3 times its width, and the pool is surrounded by a grass walk 4 feet wide. If the total area covered and enclosed by the walk is 684 square feet, find the dimensions of the pool.

SOLUTION

A diagram such as Figure 40 is useful in solving geometric problems. If we let $x =$ width of the pool, then $3x =$ length of the pool, and the region enclosed by the walk has length $3x + 8$ and width $x + 8$. The total area is the product of the length and width, so

$$\text{length} \times \text{width} = 684$$
$$(3x + 8)(x + 8) = 684$$
$$3x^2 + 32x + 64 = 684$$
$$3x^2 + 32x - 620 = 0$$
$$(3x + 62)(x - 10) = 0$$

$$x = 10 \quad \text{Reject } x = -\frac{62}{3}.$$

The dimensions of the pool are 10 feet by 30 feet.

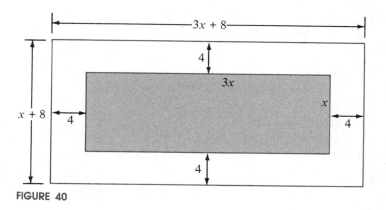

FIGURE 40

EXAMPLE 8

A number of students rented a car for $160 for a one-week camping trip. If another student had joined the original group, each person's share of expenses would have been reduced by $8. How many students were there in the original group?

SOLUTION

Let $n =$ the number of students in the original group. Then the cost per student was $160/n$ and the cost per student for the augmented group would have been

160/(n + 1). So

$$\text{original cost per student} = \text{reduced cost per student} + 8$$

$$\frac{160}{n} = \frac{160}{n + 1} + 8$$

$$160(n + 1) = 160n + 8n(n + 1)$$

$$8n^2 + 8n - 160 = 0$$

$$n^2 + n - 20 = 0$$

$$(n - 4)(n + 5) = 0$$

$$n = 4 \qquad n = -5$$

We reject the solution $n = -5$ and conclude that there were 4 students in the original group.

EXERCISE SET 2.6

Solve the given equation.

1. $3x^2 - 27 = 0$
2. $4x^2 - 64 = 0$
3. $9x^2 + 64 = 0$
4. $81x^2 + 25 = 0$
5. $(2r + 5)^2 = 8$
6. $(3x - 4)^2 = -6$
7. $(3x - 5)^2 - 8 = 0$
8. $(4t + 1)^2 - 3 = 0$

Solve by factoring.

9. $x^2 - 3x + 2 = 0$
10. $x^2 - 6x + 8 = 0$
11. $x^2 + x - 2 = 0$
12. $3r^2 - 4r + 1 = 0$
13. $x^2 + 6x = -8$
14. $x^2 + 6x + 5 = 0$
15. $y^2 - 4y = 0$
16. $2x^2 - x = 0$
17. $2x^2 - 5x = -2$
18. $2s^2 - 5s - 3 = 0$

Solve by completing the square.

19. $x^2 - 2x = 8$
20. $t^2 - 2t = 15$
21. $2r^2 - 7r = 4$
22. $9x^2 + 3x = 2$
23. $3x^2 + 8x = 3$
24. $2y^2 + 4y = 5$
25. $2y^2 + 2y = -1$
26. $3x^2 - 4x = -3$
27. $4x^2 - x = 3$
28. $2x^2 + x = 2$
29. $3x^2 + 2x = -1$
30. $3u^2 - 3u = -1$

31–42. Solve Exercises 19–30 by using the quadratic formula.

Solve by any method.

43. $2x^2 + 2x - 5 = 0$
44. $2t^2 + 2t + 3 = 0$
45. $3x^2 + 4x - 4 = 0$
46. $4u^2 - 1 = 0$
47. $x^2 + 2 = 0$
48. $4x^3 + 2x^2 + 3x = 0$

Solve for the indicated variable in terms of the remaining variables.

49. $a^2 + b^2 = c^2$, for b
50. $s = \frac{1}{2}gt^2$, for t
51. $V = \frac{1}{3}\pi r^2 h$, for r
52. $A = \pi r^2$, for r
53. $s = \frac{1}{2}gt^2 + vt$, for t
54. $F = g\frac{m_1 m_2}{d^2}$, for d

Without solving, determine the nature of the roots of each quadratic equation.

55. $x^2 - 2x + 3 = 0$
56. $3x^2 + 2x - 5 = 0$
57. $4x^2 - 12x + 9 = 0$
58. $2x^2 + x + 5 = 0$
59. $-3x^2 + 2x + 5 = 0$
60. $-3y^2 + 2y - 5 = 0$
61. $3x^2 + 2x = 0$
62. $4x^2 + 20x + 25 = 0$

Determine a value or values for k that will result in the quadratic having a double root.

63. $kx^2 - 4x + 1 = 0$ 64. $2x^2 + 3x + k = 0$ 65. $x^2 - kx - 2k = 0$ 66. $kx^2 - 4x + k = 0$

Find the solution set.

67. $x + \sqrt{x + 5} = 7$

68. $x - \sqrt{13 - x} = 1$

69. $2x + \sqrt{x + 1} = 8$

70. $3x - \sqrt{1 + 3x} = 1$

71. $\sqrt{3x + 4} - \sqrt{2x + 1} = 1$

72. $\sqrt{4 - 4x} - \sqrt{x + 4} = 3$

73. $\sqrt{2x - 1} + \sqrt{x - 4} = 4$

74. $\sqrt{5x + 1} + \sqrt{4x - 3} = 7$

Indicate an appropriate substitution of variable and solve each of the equations.

75. $3x^4 + 5x^2 - 2 = 0$

76. $2x^6 + 15x^3 - 8 = 0$

77. $\dfrac{6}{x^2} + \dfrac{1}{x} - 2 = 0$

78. $\dfrac{2}{x^4} - \dfrac{3}{x^2} - 9 = 0$

79. $2x^{2/5} + 5x^{1/5} + 2 = 0$

80. $3x^{4/3} - 4x^{2/3} - 4 = 0$

81. $2\left(\dfrac{1}{x} + 1\right)^2 - 3\left(\dfrac{1}{x} + 1\right) - 20 = 0$

82. $3\left(\dfrac{1}{x} - 2\right)^2 + 2\left(\dfrac{1}{x} - 2\right) - 1 = 0$

83. Find the width of a strip that has been mowed around a rectangular field 60 feet by 80 feet, if one-half of the lawn has not yet been mowed.

84. A 16-by-20-inch mounting board is used to mount a photograph. How wide a uniform border is there if the photograph occupies $\frac{2}{3}$ of the area of the mounting board?

85. The length of a rectangle exceeds twice its width by 4 feet. If the area of the rectangle is 48 square feet, find the dimensions.

86. The length of a rectangle is 4 centimeters less than twice its width. Find the dimensions if the area of the rectangle is 96 square centimeters.

87. The area of a rectangle is 48 square centimeters. If the length and width are each increased by 4 centimeters, the area of the newly formed rectangle is 120 square centimeters. Find the dimensions of the original rectangle.

88. The base of a triangle is 2 feet more than twice its altitude. If the area is 12 square feet, find the dimensions.

89. The smaller of two numbers is 4 less than the larger. If the sum of their squares is 58, find the numbers.

90. The sum of the reciprocals of two consecutive odd numbers is $\frac{8}{15}$. Find the numbers.

91. An investor placed an order totaling $1200 for a certain number of shares of a stock. If the price of each share of stock were $2 more, the investor would get 30 shares less for the same amount of money. How many shares did the investor buy?

92. A fraternity charters a bus for a ski trip at a cost of $360. When 6 more students join the trip, each person's cost decreases by $2. How many students were in the original group of travelers?

93. A salesman worked a certain number of days to earn $192. If he had been paid $8 more per day, he would have earned the same amount of money in two fewer days. How many days did he work?

94. A freelance photographer worked a certain number of days for a newspaper to earn $480. If she had been paid $8 less per day, she would have earned the same amount in two more days. What was her daily rate of pay?

Provide a proof of the stated theorem.

95. If r_1 and r_2 are the roots of the equation $ax^2 + bx + c = 0$, then (a) $r_1 r_2 = c/a$ and (b) $r_1 + r_2 = -b/a$.

96. If a, b, and c are rational numbers, and the discrimi-

nant of the equation $ax^2 + bx + c = 0$ is positive, then the quadratic has either two rational roots or two irrational roots.

Use the theorems of Exercise 95 to find a value or values of k so that the indicated condition is satisfied.

97. $kx^2 + 3x + 5 = 0$; sum of the roots is 6.

98. $2x^2 - 3kx - 2 = 0$; sum of the roots is -3.

99. $3x^2 - 10x + 2k = 0$; product of the roots is -4.

100. $2kx^2 + 5x - 1 = 0$; product of the roots is $\frac{1}{2}$.

101. $2x^2 - kx + 9 = 0$; one root is double the other.

102. $3x^2 - 4x + k = 0$; one root is triple the other.

103. $6x^2 - 13x + k = 0$; one root is the reciprocal of the other.

2.7 QUADRATIC FUNCTIONS

A function of the form

$$f(x) = ax^2 + bx + c \tag{1}$$

where a, b, and c are real numbers and $a \neq 0$, is called a **quadratic function.** The results we have obtained earlier in this chapter will be very helpful in studying the graph of quadratic functions.

By completing the square, it is always possible to rewrite Equation (1) in the form

$$f(x) = a(x - h)^2 + k \tag{2}$$

where h and k are constants. Here is an example.

EXAMPLE 1
Write the quadratic function

$$f(x) = 2x^2 - 4x - 1$$

in the form of Equation (2).

SOLUTION
We complete the square as follows.

$$f(x) = 2(x^2 - 2x \quad) - 1 \quad \text{Factor out } a.$$
$$= 2(x^2 - 2x + 1) - 1 - 2 \quad \text{Complete the square and balance.}$$
$$= 2(x - 1)^2 - 3$$

which is in the form of Equation (2) with $a = 2$, $h = 1$, and $k = -3$.

PROGRESS CHECK
Write each quadratic function f in the form $f(x) = a(x - h)^2 + k$.
(a) $f(x) = -3x^2 - 12x - 13$ (b) $f(x) = 2x^2 - 2x + 3$

ANSWERS
(a) $f(x) = -3(x + 2)^2 - 1$ (b) $f(x) = 2\left(x - \dfrac{1}{2}\right)^2 + \dfrac{5}{2}$

It is easier to analyze the graph of the quadratic function when written in the form of Equation (2). We have previously seen that the graph of

$$f(x) = ax^2$$

is a parabola opening from the origin. The graph of Equation (2) is that of $f(x) = ax^2$ shifted vertically k units and shifted horizontally h units. Thus the graph of Equation (2) is a parabola opening from the point (h, k), which is called the **vertex** of the parabola. If $a > 0$, the parabola opens upward from the vertex; if $a < 0$, the parabola opens downward. We can summarize in this way.

Graphing Quadratic Functions

The quadratic function

$$f(x) = ax^2 + bx + c, \quad a \neq 0$$

can be written in the form

$$f(x) = a(x - h)^2 + k$$

where h and k are constants. The graph is a parabola with vertex at (h, k), opening upward if $a > 0$ and downward if $a < 0$.

EXAMPLE 2

Sketch the graphs of the following functions.
(a) $f(x) = 2x^2 + 4x - 1$ (b) $f(x) = -2x^2 + 4x$

SOLUTION

(a) Completing the square in x, we have

$$\begin{aligned}
f(x) &= 2x^2 + 4x - 1 \\
&= 2(x^2 + 2x \quad) - 1 \\
&= 2(x^2 + 2x + 1) - 1 - 2 \\
&= 2(x + 1)^2 - 3
\end{aligned}$$

which is in the form $f(x) = a(x - h)^2 + k$ with $a = 2$, $h = -1$, and $k = -3$. The vertex of the parabola is at $(-1, -3)$ and the graph opens upward as shown in Figure 41a.

(b) Completing the square,

$$\begin{aligned}
f(x) &= -2x^2 + 4x \\
&= -2(x^2 - 2x \quad) \\
&= -2(x^2 - 2x + 1) + 2 \\
&= -2(x - 1)^2 + 2
\end{aligned}$$

Here, $a = -2$, $h = 1$, and $k = 2$. The vertex of the parabola is at $(1, 2)$ and the parabola opens downward as in Figure 41b.

INTERCEPTS AND ROOTS

Since the graph of the quadratic function of Equation (1) is the graph of the equation

$$y = ax^2 + bx + c, \quad a \neq 0$$

$$y = 2(x + 1)^2 - 3$$

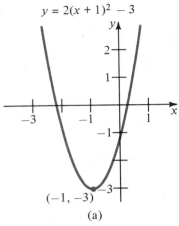

$(-1, -3)$

(a)

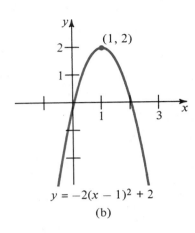

$(1, 2)$

$$y = -2(x - 1)^2 + 2$$

(b)

FIGURE 41

the graph intersects the x-axis at those points where $y = 0$. The x-intercepts are, then, those points where

$$y = 0 = ax^2 + bx + c \qquad (3)$$

which, of course, is a quadratic equation.

---

x-Intercepts of the Parabola

The x-intercepts of the parabola

$$f(x) = ax^2 + bx + c$$

are the real roots of the quadratic equation

$$ax^2 + bx + c = 0$$

and are given by the quadratic formula.

The discriminant of the quadratic equation, Equation (3), tells us the number of real roots and, therefore, the number of x-intercepts of the parabola. The possibilities are two real roots (Figure 42a), a double root (Figure 42b), and two complex roots (Figure 42c).

EXAMPLE 3
Find the vertex and all intercepts of each of the following parabolas. Sketch the graph.
(a) $f(x) = -x^2 + 3x - 2$ (b) $f(x) = x^2 + 1$ (c) $f(x) = x^2 + 2x + 1$

SOLUTION
(a) To find the vertex we must complete the square

$$f(x) = -x^2 + 3x - 2 = -\left(x - \frac{3}{2}\right)^2 + \frac{1}{4}$$

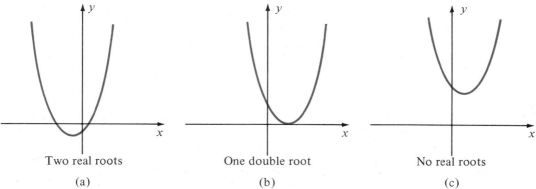

Two real roots One double root No real roots

(a) (b) (c)

FIGURE 42

The vertex is at $(\frac{3}{2}, \frac{1}{4})$ and the parabola opens downward. Setting $y = f(x) = 0$, we see that

$$x^2 - 3x + 2 = 0$$
$$(x - 1)(x - 2) = 0$$

and the x-intercepts occur at $x = 1$ and $x = 2$. Finally, the y-intercept occurs where $x = 0$ and is found by evaluating $y = f(0)$. We find that $f(0) = -2$. The graph is shown in Figure 43a.

(b) Since

$$f(x) = x^2 + 1 = (x - 0)^2 + 1$$

the vertex is at $(0, 1)$ and the parabola opens upward. To find the x-intercepts, we set $y = f(x) = 0$ so that

$$0 = x^2 + 1$$

Since this quadratic equation has no real roots, there are no x-intercepts. To find the y-intercept we set $x = 0$ and find that $y = f(0) = 1$. See Figure 43b for the graph.

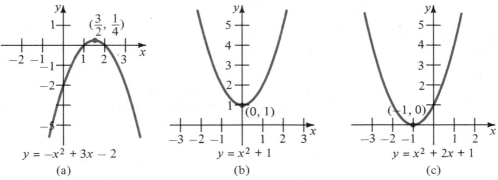

$y = -x^2 + 3x - 2$ $y = x^2 + 1$ $y = x^2 + 2x + 1$

(a) (b) (c)

FIGURE 43

(c) Completing the square,

$$f(x) = x^2 + 2x + 1 = (x + 1)^2 + 0$$

The vertex is at the point $(-1, 0)$ and the parabola opens upward. Setting $f(x) = 0$, we see that

$$x^2 + 2x + 1 = (x + 1)^2 = 0$$

so that $x = -1$ is the only x-intercept. Finally, we set $x = 0$ and find that $y = f(0) = 1$ is the y-intercept. See Figure 43c.

EXERCISE SET 2.7

In Exercises 1–10 write the quadratic function f in the form $f(x) = a(x - h)^2 + k$.

1. $f(x) = x^2 - 6x + 10$ 2. $f(x) = -x^2 - 2x - 3$ 3. $f(x) = -2x^2 + 4x - 5$ 4. $f(x) = 3x^2 + 12x + 14$

5. $f(x) = 2x^2 + 6x + 5$ 6. $f(x) = -4x^2 + 4x$ 7. $f(x) = -x^2 - x$ 8. $f(x) = 3x^2 + 18x + 9$

9. $f(x) = -2x^2 + 5$ 10. $f(x) = -\dfrac{1}{2}x^2 + 2x + 2$

In Exercises 11–18 find the vertex and all intercepts of the parabola. Sketch the graph.

11. $f(x) = 2x^2 - 4x$ 12. $f(x) = -x^2 - 2x + 3$ 13. $f(x) = -4x^2 + 4x - 1$ 14. $f(x) = x^2 + 4x + 4$

15. $f(x) = \dfrac{1}{2}x^2 + 2x + 4$ 16. $f(x) = -2x^2 + 2x - \dfrac{5}{2}$ 17. $f(x) = -\dfrac{1}{2}x^2 + 3x - 4$ 18. $f(x) = x^2 - x + 1$

2.8
COMBINING FUNCTIONS; INVERSE FUNCTIONS

Functions such as

$$f(x) = x^2 \qquad g(x) = x - 1$$

can be combined by the usual operations of addition, subtraction, multiplication, and division. Using these functions f and g, we can form

$$(f + g)(x) = f(x) + g(x) = x^2 + x - 1$$
$$(f - g)(x) = f(x) - g(x) = x^2 - (x - 1) = x^2 - x + 1$$
$$(f \cdot g)(x) = f(x) \cdot g(x) = x^2(x - 1) = x^3 - x^2$$
$$\frac{f}{g}(x) = \frac{f(x)}{g(x)} = \frac{x^2}{x - 1}$$

In each case we have combined two functions f and g to form a new function. Note, however, that the domain of the new functions need not be the same as the domain of either of the original functions. The function formed by division in the above example has as its domain the set of all real numbers x except $x = 1$, since we cannot divide by 0. On the other hand, the original functions $f(x) = x^2$ and $g(x) = x - 1$ are both defined at $x = 1$.

EXAMPLE 1
Given $f(x) = x - 4$, $g(x) = x^2 - 4$, find the following.
(a) $(f + g)(x)$ (b) $(f - g)(x)$ (c) $(f \cdot g)(x)$

(d) $\left(\dfrac{f}{g}\right)(x)$ (e)   the domain of $\left(\dfrac{f}{g}\right)(x)$

SOLUTION
(a) $(f + g)(x) = f(x) + g(x) = x - 4 + x^2 - 4 = x^2 + x - 8$
(b) $(f - g)(x) = f(x) - g(x) = x - 4 - (x^2 - 4) = -x^2 + x$
(c) $(f \cdot g)(x) = f(x) \cdot g(x) = (x - 4)(x^2 - 4) = x^3 - 4x^2 - 4x + 16$

(d) $\left(\dfrac{f}{g}\right)(x) = \dfrac{f(x)}{g(x)} = \dfrac{x - 4}{x^2 - 4}$

(e)   The domain of $\left(\dfrac{f}{g}\right)(x)$ must exclude values of x for which $x^2 - 4 = 0$.
Thus, the domain consists of the set of all real numbers except 2 and -2.

PROGRESS CHECK
Given $f(x) = 2x^2$, $g(x) = x^2 - 5x + 6$, find the following.
(a) $(f + g)(x)$ (b) $(f - g)(x)$ (c) $(f \cdot g)(x)$

(d) $\left(\dfrac{f}{g}\right)(x)$ (e)   the domain of $\left(\dfrac{f}{g}\right)(x)$

ANSWERS
(a) $3x^2 - 5x + 6$ (b) $x^2 + 5x - 6$

(c) $2x^4 - 10x^3 + 12x^2$ (d) $\dfrac{2x^2}{x^2 - 5x + 6}$

(e)   The set of all real numbers except 2 and 3.

COMPOSITE FUNCTION   There is another, important way in which two functions f and g can be combined to form a new function. In Figure 44 the function f assigns the value y in set Y to x in set X; then, function g assigns the value z in set Z to y in Y. The net effect

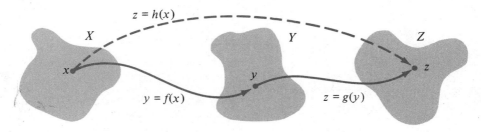

FIGURE 44

of this combination of f and g is a new function h, called the **composite function of g and f, $g \circ f$,** which assigns z in Z to x in X. We write this new function as

$$h(x) = (g \circ f)(x) = g[f(x)]$$

which is read "g of f of x."

EXAMPLE 2
Given $f(x) = x^2$, $g(x) = x - 1$, find the following.
(a) $f[g(3)]$ (b) $g[f(3)]$ (c) $f[g(x)]$ (d) $g[f(x)]$

SOLUTION
(a) We begin by evaluating $g(3)$.

$$g(x) = x - 1$$
$$g(3) = 3 - 1 = 2$$

Therefore,

$$f[g(3)] = f(2)$$

Since

$$f(x) = x^2$$

then

$$f(2) = 2^2 = 4$$

Thus,

$$f[g(3)] = 4$$

(b) Beginning with $f(3)$, we have

$$f(3) = 3^2 = 9$$

Then we find by substituting $f(3) = 9$ that

$$g[f(3)] = g(9) = 9 - 1 = 8$$

(c) Since $g(x) = x - 1$, we make the substitution

$$f[g(x)] = f(x - 1) = (x - 1)^2 = x^2 - 2x + 1$$

(d) Since $f(x) = x^2$, we make the substitution

$$g[f(x)] = g(x^2) = x^2 - 1$$

Note that $f[g(x)] \neq g[f(x)]$.

PROGRESS CHECK
Given $f(x) = x^2 - 2x$, $g(x) = 3x$, find the following.
(a) $f[g(-1)]$ (b) $g[f(-1)]$ (c) $f[g(x)]$
(d) $g[f(x)]$ (e) $(f \circ g)(2)$ (f) $(g \circ f)(2)$

ANSWERS

(a) 15   (b) 9   (c) $9x^2 - 6x$

(d) $3x^2 - 6x$ (e) 24   (f) 0

ONE-TO-ONE FUNCTIONS

An element in the range of a function may correspond to more than one element in the domain of the function. In Figure 45 we see that y in Y corresponds to both x_1 and x_2 in X. If we demand that every element in the domain be assigned to a

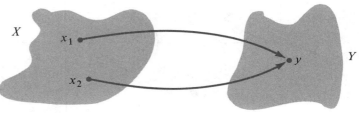

FIGURE 45

different element of the range, then the function is called **one-to-one.** More formally,

> A function f is one-to-one if $f(a) = f(b)$ only when $a = b$.

There is a simple means of determining if a function f is one-to-one by examining the graph of the function. In Figure 46a we see that a horizontal line meets the graph in more than one point. Thus, $f(a) = f(b)$ although $a \neq b$; hence

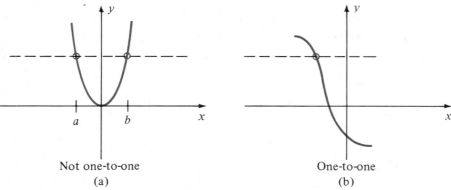

Not one-to-one
(a)

One-to-one
(b)

FIGURE 46

the function is not one-to-one. On the other hand, no horizontal line meets the graph in Figure 46b in more than one point; the graph thus determines a one-to-one function. In summary, we have the following test.

| **Horizontal Line Test** | If no horizontal line meets the graph of a function in more than one point, then the function is one-to-one. |

EXAMPLE 3

Which of the graphs in Figure 47 are graphs of one-to-one functions?

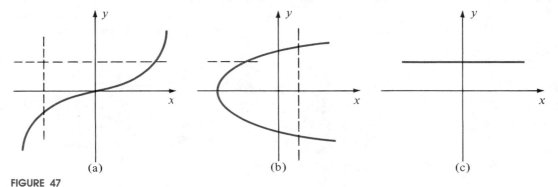

(a)   (b)   (c)

FIGURE 47

SOLUTION

(a)   No *vertical* line meets the graph in more than one point; hence, it is the graph of a function. No *horizontal* line meets the graph in more than one point; hence, it is the graph of a one-to-one function.

(b)   No *horizontal* line meets the graph in more than one point. But *vertical* lines do meet the graph in more than one point. It is therefore not the graph of a function and consequently cannot be the graph of a one-to-one function.

(c)   No *vertical* line meets the graph in more than one point; hence, it is the graph of a function. But a *horizontal* line does meet the graph in more than one point. This is the graph of a function but not of a one-to-one function.

PROGRESS CHECK

Which of the graphs in Figure 48 are graphs of one-to-one functions?

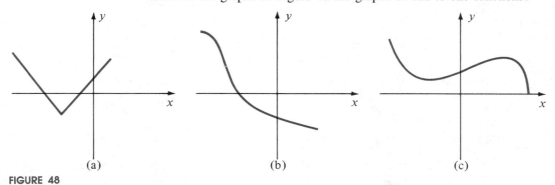

(a)   (b)   (c)

FIGURE 48

ANSWER
(b)

INVERSE FUNCTIONS

Suppose the function f in Figure 49a is a one-to-one function and that $y = f(x)$. Since f is one-to-one, we know that the correspondence is unique; that is, x in X is the *only* element of the domain for which $y = f(x)$. It is then possible to define a function g (Figure 49b) with domain Y and range X that reverses the correspondence, that is,

$$g(y) = x \quad \text{for every } x \text{ in } X$$

If we substitute $y = f(x)$, we have

$$g[f(x)] = x \quad \text{for every } x \text{ in } X \tag{1}$$

Substituting $g(y) = x$ in the equation $f(x) = y$ yields

$$f[g(y)] = y \quad \text{for every } y \text{ in } Y \tag{2}$$

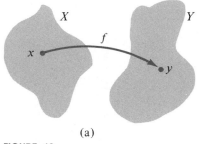

(a)

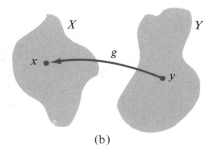

(b)

FIGURE 49

The functions f and g of Figure 49 are therefore seen to satisfy the properties of Equations (1) and (2). Such functions are called inverse functions.

Inverse Functions

If f is a one-to-one function with domain X and range Y, then the function g with domain Y and range X satisfying

$$g[f(x)] = x \quad \text{for every } x \text{ in } X$$
$$f[g(y)] = y \quad \text{for every } y \text{ in } Y$$

is called the **inverse function** of f.

It is not difficult to show that the inverse of a one-to-one function is unique (see Exercise 61).

Since the inverse (reciprocal) $1/x$ of a real number $x \neq 0$ can be written as x^{-1}, it is natural to write the inverse of a function f as f^{-1}. Thus we have

$$f^{-1}[f(x)] = x \quad \text{for every } x \text{ in } X$$
$$f[f^{-1}(y)] = y \quad \text{for every } y \text{ in } Y$$

See Figure 50 for a graphical representation.

In the following chapter we will study a very important class of inverse functions, the exponential and logarithmic functions. Always remember that *we can define the inverse function of f only if f is one-to-one.*

EXAMPLE 4

Let f be the function defined by

$$f(x) = x^2 - 4, \quad x \geq 0$$

Verify that the inverse of f is given by

$$f^{-1}(x) = \sqrt{x + 4}$$

SOLUTION

We must verify that $f[f^{-1}(x)] = x$ and $f^{-1}[f(x)] = x$. Thus,

$$f[f^{-1}(x)] = f(\sqrt{x + 4})$$
$$= (\sqrt{x + 4})^2 - 4$$
$$= x + 4 - 4 = x$$

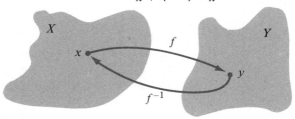

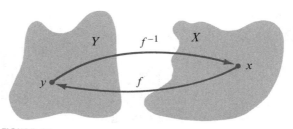

FIGURE 50

and

$$f^{-1}[f(x)] = f^{-1}(x^2 - 4)$$
$$= \sqrt{(x^2 - 4) + 4}$$
$$= \sqrt{x^2} = |x|$$

Since $x \geq 0$,

$$f^{-1}[f(x)] = |x| = x$$

We have verified that the equations defining inverse functions hold, and conclude that the inverse of f is as given. The student should verify that (a) the domain of f is the set of all nonnegative real numbers and the range of f is the set of all real numbers in the interval $[-4, \infty)$; (b) the domain of f^{-1} is the range of f and the range of f^{-1} is the domain of f.

We may also think of the function f defined by $y = f(x)$ as the set of all ordered pairs $(x, f(x))$, where x assumes all values in the domain of f. Since the inverse function reverses the correspondence, the function f^{-1} is the set of all ordered pairs $(f(x), x)$, where $f(x)$ assumes all values in the range of f. With this approach, we see that the graphs of inverse functions are related in a distinct manner. First, note that the points (a, b) and (b, a) in Figure 51a are located symmetrically with respect to the graph of the line $y = x$. That is, if we fold the

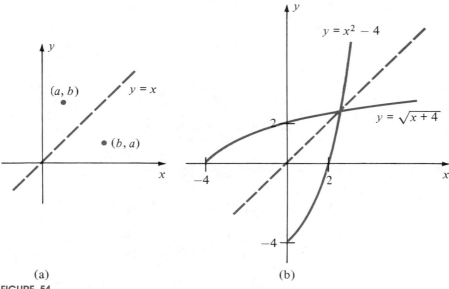

(a)

(b)

FIGURE 51

paper along the line $y = x$, the two points will coincide. And if (a, b) lies on the graph of a function f, then (b, a) must lie on the graph of f^{-1}. Thus, the graphs of a pair of inverse functions are reflections of each other about the line $y = x$. In Figure 51b we have sketched the graphs of the functions from Example 5 on the same coordinate axes to demonstrate this interesting relationship.

It is sometimes possible to find an inverse by algebraic methods, as is shown by the following example.

EXAMPLE 5
Find the inverse function of $f(x) = 2x - 3$.

SOLUTION
By definition, $f[f^{-1}(x)] = x$. Then we must have

$$f[f^{-1}(x)] = 2[f^{-1}(x)] - 3 = x$$

$$f^{-1}(x) = \frac{x + 3}{2}$$

We then verify that $f^{-1}[f(x)] = x$:

$$f^{-1}[f(x)] = \frac{2x - 3 + 3}{2} = x$$

PROGRESS CHECK
Given $f(x) = 3x + 5$, find f^{-1}.

ANSWER
$$f^{-1}(x) = \frac{x - 5}{3}$$

WARNING

(a)   In general, $f^{-1}(x) \neq \dfrac{1}{f(x)}$.

If $g(x) = x - 1$, then

$$g^{-1}(x) \neq \frac{1}{x - 1}$$

Use the methods of this section to show that

$$g^{-1}(x) = x + 1$$

(b)   The inverse function notation is *not* to be thought of as a power.

EXERCISE SET 2.8

In Exercises 1–10 $f(x) = x^2 + 1$ and $g(x) = x - 2$. Determine the following.

1. $(f + g)(x)$
2. $(f + g)(2)$
3. $(f - g)(x)$
4. $(f - g)(3)$

5. $(f \cdot g)(x)$
6. $(f \cdot g)(-1)$
7. $\left(\dfrac{f}{g}\right)(x)$
8. $\left(\dfrac{f}{g}\right)(-2)$

9. the domains of f and of g
10. the domains of $\dfrac{f}{g}$ and of $\dfrac{g}{f}$

In Exercises 11–18 $f(x) = 2x + 1$ and $g(x) = 2x^2 + x$. Determine the following.

11. $(f \circ g)(x)$
12. $(g \circ f)(x)$
13. $(f \circ g)(2)$
14. $(g \circ f)(3)$

15. $(f \circ g)(x + 1)$
16. $(f \circ f)(-2)$
17. $(g \circ f)(x - 1)$
18. $(g \circ g)(x)$

In Exercises 19–24 $f(x) = x^2 + 4$ and $g(x) = \sqrt{x + 2}$. Determine the following.

19. $(f \circ g)(x)$
20. $(g \circ f)(x)$

21. $(f \circ f)(-1)$
22. the domain of $(f \circ g)(x)$

23. the domain of $(g \circ f)(x)$
24. the domain of $(g \circ g)(x)$

In Exercises 25–28 determine $(f \circ g)(x)$ and $(g \circ f)(x)$.

25. $f(x) = x - 1$, $g(x) = x + 2$
26. $f(x) = \sqrt{x + 1}$, $g(x) = x + 2$

27. $f(x) = \dfrac{1}{x + 1}$, $g(x) = \dfrac{1}{x - 1}$
28. $f(x) = \dfrac{x + 1}{x - 1}$, $g(x) = x$

In Exercises 29–38 write the given function $h(x)$ as a composite of two functions f and g so that $h(x) = (f \circ g)(x)$. (There may be more than one answer.)

29. $h(x) = x^2 + 3$
30. $h(x) = \dfrac{1}{x + 2}$

31. $h(x) = (3x + 2)^8$
32. $h(x) = (x^3 + 2x^2 + 1)^{15}$

33. $h(x) = (x^3 - 2x^2)^{1/3}$
34. $h(x) = \left(\dfrac{x^2 + 2x}{x^3 - 1}\right)^{3/2}$

35. $h(x) = |x^2 - 4|$
36. $h(x) = |x^2 + x| - 4$

37. $h(x) = \sqrt{4 - x}$
38. $h(x) = \sqrt{2x^2 - x + 2}$

In Exercises 39–44 verify that $g = f^{-1}$ for the given functions f and g by showing that $f[g(x)] = x$ and $g[f(x)] = x$.

39. $f(x) = 2x + 4$ $g(x) = \dfrac{1}{2}x - 2$
40. $f(x) = 3x - 2$ $g(x) = \dfrac{1}{3}x + \dfrac{2}{3}$

41. $f(x) = 2 - 3x$ $g(x) = -\dfrac{1}{3}x + \dfrac{2}{3}$
42. $f(x) = x^3$ $g(x) = \sqrt[3]{x}$

43. $f(x) = \dfrac{1}{x}$ $g(x) = \dfrac{1}{x}$
44. $f(x) = \dfrac{1}{x - 2}$ $g(x) = \dfrac{1}{x} - 2$

In Exercises 45–52 find $f^{-1}(x)$. Sketch the graphs of $y = f(x)$ and $y = f^{-1}(x)$ on the same coordinate axes.

45. $f(x) = 2x + 3$
46. $f(x) = 3x - 4$
47. $f(x) = 3 - 2x$
48. $f(x) = \dfrac{1}{2}x + 1$

49. $f(x) = \dfrac{1}{3}x - 5$
50. $f(x) = 2 - \dfrac{1}{5}x$
51. $f(x) = x^3 + 1$
52. $f(x) = \dfrac{1}{x + 1}$

In Exercises 53–60 use the horizontal line test to determine whether the given function is a one-to-one function.

53. $f(x) = 2x - 1$

54. $f(x) = 3 - 5x$

55. $f(x) = x^2 - 2x + 1$

56. $f(x) = x^2 + 4x + 4$

57. $f(x) = -x^3 + 1$

58. $f(x) = x^3 - 2$

59. $f(x) = \begin{cases} 2x, & x \leq -1 \\ x^2, & -1 < x \leq 0 \\ 3x - 1, & x > 0 \end{cases}$

60. $f(x) = \begin{cases} x^2 - 4x + 4, & x \leq 2 \\ x, & x > 2 \end{cases}$

61. Prove that a one-to-one function can have at most one inverse function. (*Hint:* Assume that the functions g and h are both inverses of the function f. Show that $g(x) = h(x)$ for all real values x in the range of f.)

62. Prove that the linear function $f(x) = ax + b$ is a one-to-one function if $a \neq 0$, and is not a one-to-one function if $a = 0$.

63. Find the inverse of the linear function $f(x) = ax + b,\ a \neq 0$.

TERMS AND SYMBOLS

origin (p. 58)
x-axis (p. 58)
y-axis (p. 58)
coordinate axes (p. 58)
rectangular coordinate system (p. 58)
Cartesian coordinate system (p. 58)
quadrant (p. 58)
coordinates of a point (p. 58)
ordered pair (p. 58)
abscissa (p. 58)
x-coordinate (p. 58)
ordinate (p. 58)
y-coordinate (p. 58)
distance formula (p. 60)
graph of an equation in two variables (p. 61)
solution of an equation in two variables (p. 61)
x-intercept (p. 62)

y-intercept (p. 62)
symmetry (p. 63)
symmetry with respect to the x-axis (p. 64)
symmetry with respect to the y-axis (p. 64)
symmetry with respect to the origin (p. 64)
function (p. 68)
domain (p. 68)
range (p. 68)
independent variable (p. 68)
dependent variable (p. 68)
image (p. 68)
vertical line test (p. 70)
$f(x)$ (p. 71)
graph of a function (p. 74)
parabola (p. 76)
piecewise-defined function (p. 76)
increasing function (p. 80)

decreasing function (p. 80)
constant function (p. 80)
reflection about an axis (p. 83)
vertical shift (p. 84)
horizontal shift (p. 85)
linear function (p. 87)
slope (p. 87)
point-slope form (p. 90)
slope-intercept form (p. 91)
general first-degree equation (p. 94)
quadratic equation (p. 98)
completing the square (p. 99)
quadratic formula (p. 101)
discriminant (p. 103)
double root (p. 103)
repeated root (p. 103)
radical equation (p. 104)
extraneous solution (p. 105)

substitution of variable (p. 106)
quadratic function (p. 110)
vertex of a parabola (p. 111)
composite function (p. 116)
$f[g(x)]$ (p. 116)
$f \circ g$ (p. 116)
one-to-one function (p. 117)
horizontal line test (p. 118)
inverse function (p. 119)
f^{-1} (p. 120)

KEY IDEAS FOR REVIEW

☐ In a rectangular coordinate system, every ordered pair of real numbers (a, b) corresponds to a point in the plane, and every point in the plane corresponds to an ordered pair of real numbers.

☐ The distance $\overline{PQ}$ between points $P(x_1, y_1)$ and $Q(x_2, y_2)$ is given by the distance formula

$$\overline{PQ} = \sqrt{(x_2 - x_1)^2 + (y_2 - y_1)^2}$$

☐ An equation in two variables can be graphed by plotting points that satisfy the equation and joining the points to form a smooth curve.

☐ There are simple algebraic means for testing whether the graph of an equation is symmetric with respect to the *x*-axis, the *y*-axis, and the origin.

☐ A function is a rule that assigns exactly one element *y* of a set *Y* to each element *x* of a set *X*. The domain is the set of inputs and the range is the set of outputs.

☐ A graph represents a function if no vertical line meets the graph in more than one point.

☐ The domain of a function is the set of all real numbers for which the function is defined. Beware of division by zero and even roots of negative numbers.

☐ Function notation gives the definition of the function and also the value or expression at which to evaluate the function. If the function *f* is defined by $f(x) = x^2 + 2x$, then the notation $f(3)$ denotes the result of replacing the independent variable *x* with 3 wherever it appears.

$$f(x) = x^2 + 2x$$
$$f(3) = 3^2 + 2(3) = 15$$

☐ To graph $f(x)$, simply graph $y = f(x)$.

☐ An equation is not the only way to define a function. Sometimes a function is defined by a table or chart, or by several equations. Moreover, not every equation determines a function.

☐ As we move from left to right, the graph of an increasing function rises and the graph of a decreasing function falls. The graph of a constant function neither rises nor falls; it is horizontal.

☐ The graph of a function can have holes or gaps, and can be defined in "pieces."

☐ If we know the graph of $f(x)$, it is easy to sketch the graph of $-f(x), f(x) + c, f(x + c)$, and $c \cdot f(x)$, where *c* is a constant.

☐ Any two points on a line can be used to find its slope *m*.

$$m = \frac{y_2 - y_1}{x_2 - x_1}$$

☐ Positive slope indicates that a line is rising; negative slope indicates that a line is falling.

☐ The slope of a horizontal line is 0; the slope of a vertical line is undefined.

☐ The point-slope form of a line is $y - y_1 = m(x - x_1)$.

☐ The slope-intercept form of a line is $y = mx + b$.

☐ The equation of the horizontal line through the point (a, b) is $y = b$; the equation of the vertical line through the point (a, b) is $x = a$.

☐ Parallel lines have the same slope.

☐ The slopes of perpendicular lines are negative reciprocals of each other.

☐ The graphs of the linear function $f(x) = ax + b$ and of the general first-degree equation $Ax + By = C$ are always straight lines.

☐ The quadratic equation $ax^2 + bx + c = 0$, $a > 0$, always has two solutions, which may be found by using the quadratic formula. If $b = 0$ or if the quadratic can be factored, then faster solution methods are available.

☐ The solutions or roots of a quadratic equation may be complex numbers. The expression $b^2 - 4ac$, called the discriminant, which appears under the radical of the quadratic formula, permits the nature of the roots to be analyzed without solving the equation.

☐ Radical equations often can be transformed into quadratic equations. Since the process involves raising both sides of an equation to a power, the answers must be checked to see that they satisfy the original equation.

☐ The method called "substitution of a variable" can be used to transform certain equations into quadratics. This technique is a handy tool and will be used in other chapters of this book.

☐ Functions can be combined by the usual operations of addition, subtraction, multiplication, and division. However, the domain of the resulting function need not coincide with the domain of either of the original functions.

☐ A composite function is a function of a function.

☐ We say a function is one-to-one if every element of the range corresponds to precisely one element of the domain.

☐ No horizontal line meets the graph of a one-to-one function in more than one point.

☐ The inverse of a function, f^{-1}, reverses the correspondence defined by the function *f*. The domain of *f* becomes the range of f^{-1}, and the range of *f* becomes the domain of f^{-1}.

☐ The inverse of a function *f* is defined only if *f* is a one-to-one function.

REVIEW EXERCISES

Solutions to exercises whose numbers are in color are in the Solutions section in the back of the book.

2.1 1. Find the distance between the points $(-4, -6)$ and $(2, -1)$.

2. Find the length of the longest side of the triangle whose vertices are $A(3, -4)$, $B(-2, -6)$, and $C(-1, 2)$.

In Exercises 3 and 4 sketch the graph of the given equation by forming a table of values.

3. $y = 1 - |x|$ 4. $y = \sqrt{x - 2}$

In Exercises 5 and 6 analyze the given equation for symmetry with respect to the x-axis, y-axis, and origin.

5. $y^2 = 1 - x^3$ 6. $y^2 = \dfrac{x^2}{x^2 - 5}$

2.2 In Exercises 7 and 8 state if the graph determines y to be a function of x.

7.

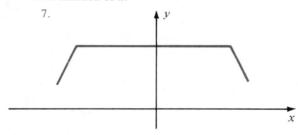

8.

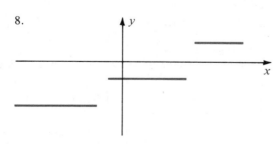

In Exercises 9 and 10 determine the domain of the given function.

9. $f(x) = \sqrt{3x - 5}$ 10. $f(x) = \dfrac{x}{x^2 + 2x + 1}$

11. If $f(x) = \sqrt{x - 1}$, find a real number whose image is 15.

12. If $f(t) = t^2 + 1$, find a real number whose image is 10.

In Exercises 13–15, $f(x) = x^2 - x$. Evaluate the following.

13. $f(-3)$ 14. $f(y - 1)$

15. $\dfrac{f(2 + h) - f(2)}{h}$, $h \neq 0$

2.3 Exercises 16–19 refer to the function f defined by

$$f(x) = \begin{cases} x - 1, & x \leq -1 \\ x^2, & -1 < x \leq 2 \\ -2, & x > 2 \end{cases}$$

16. Sketch the graph of the function f.

17. Determine where the function f is increasing, decreasing, and constant.

18. Evaluate $f(-4)$.

19. Evaluate $f(4)$.

2.4 In Exercises 20 and 21 sketch the graph of the function by shifting or reflecting a known graph, or by doing both.

20. $f(x) = |x| + 2$ 21. $f(x) = -\sqrt{x + 1}$

2.5 In Exercises 22–27 the points A and B have coordinates $(-4, -6)$ and $(-1, 3)$, respectively.

22. Find the slope of the line through A and B.

23. Find an equation of the line through the points A and B.

24. Find an equation of the line through A that is parallel to the y-axis.

25. Find an equation of the horizontal line through B.

26. Find an equation of the line through A that is parallel to the line $2x - y - 3 = 0$.

27. Find an equation of the line through B that is perpendicular to the line $2y + x - 5 = 0$.

2.6 28. Solve $x^2 - x - 20 = 0$ by factoring.

29. Solve $6x^2 - 11x + 4 = 0$ by factoring.

30. Solve $x^2 - 2x + 6 = 0$ by completing the square.

31. Solve $2x^2 - 4x + 3 = 0$ by the quadratic formula.

32. Solve $3x^2 + 2x - 1 = 0$ by the quadratic formula.

In Exercises 33–35 solve for x.

33. $49x^2 - 9 = 0$ 34. $kx^2 - 3\pi = 0$

35. $x^2 + x = 12$

In Exercises 36–38 determine the nature of the roots of the quadratic equation without solving.

36. $3r^2 = 2r + 5$ 37. $4x^2 + 20x + 25 = 0$

38. $6y^2 - 2y = -7$

In Exercises 39–41 solve the given equation.

39. $\sqrt{x + 2} = x$ 40. $3x^4 + 5x^2 - 5 = 0$

41. $\left(1 - \dfrac{2}{x}\right)^2 - 8\left(1 - \dfrac{2}{x}\right) + 15 = 0$

42. A charitable organization rented an auditorium at a cost of $420 and split the cost among the attendees. If 10 additional persons had attended the meeting, the cost per person would have decreased by $1. How many attendees were there in the original group?

2.7   In Exercises 43 and 44 find the vertex and all intercepts of the parabola.

43. $f(x) = -x^2 - 4x$ 44. $f(x) = x^2 - 5x + 7$

2.8   In Exercises 45–50 $f(x) = x + 1$ and $g(x) = x^2 - 1$. Determine the following.

45. $(f + g)(x)$ 46. $(f \cdot g)(-1)$

47. $\left(\dfrac{f}{g}\right)(x)$ 48. the domain of $\left(\dfrac{f}{g}\right)(x)$

49. $(g \circ f)(x)$ 50. $(f \circ g)(2)$

In Exercises 51–54 $f(x) = \sqrt{x} - 2$ and $g(x) = x^2$. Determine the following.

51. $(f \circ g)(x)$ 52. $(g \circ f)(x)$

53. $(f \circ g)(-2)$ 54. $(g \circ f)(-2)$

In Exercises 55 and 56 $f(x) = 2x + 4$ and $g(x) = \dfrac{x}{2} - 2$.

55. Prove that f and g are inverse functions of each other.

56. Sketch the graphs of $y = f(x)$ and $y = g(x)$ on the same coordinate axes.

PROGRESS TEST 2A

1. Find the perimeter of the triangle whose vertices are $(2, 5)$, $(-3, 1)$, and $(-3, 4)$.

2. Use symmetry to assist in sketching the graph of the equation $y = 2x^2 - 1$.

3. Analyze the equation $y = 1/x^3$ for symmetry with respect to the axes and origin.

4. Determine the domain of the function $\dfrac{1}{\sqrt{x - 1}}$.

5. If $f(x) = \sqrt{x - 1}$, find a real number whose image is 4.

6. If $f(x) = 2x^2 + 3$, find $f(2t)$.

Problems 7–9 refer to the function f defined by

$$f(x) = \begin{cases} 0, & x < -2 \\ |x|, & -2 \le x \le 3 \\ x^2 - x, & x > 3 \end{cases}$$

7. Determine where the function f is increasing, decreasing, and constant.

8. Evaluate $f(-5)$.   9. Evaluate $f(-2)$.

10. Sketch the graphs of the functions $f(x) = \sqrt{x}$ and $g(x) = \sqrt{x} - 1$ on the same coordinate axes.

11. Find an equation of the line through the points $(-3, 5)$ and $(-5, 2)$.

12. Find an equation of the vertical line through the point $(-3, 4)$.

13. Find the slope m and y-intercept b of the line whose equation is $2y - x = 4$.

14. Find an equation of the line through the point $(4, -1)$ that is parallel to the x-axis.

15. Find an equation of the line through the point $(-2, 3)$ that is perpendicular to the line $y - 3x - 2 = 0$.

16. Solve $x^2 - 5x = 14$ by factoring.

17. Solve $5x^2 - x + 4 = 0$ by completing the square.

18. Solve $12x^2 + 5x - 3 = 0$ by the quadratic formula.

In Problems 19 and 20 determine the nature of the roots of the quadratic equation without solving.

19. $6x^2 + x - 2 = 0$ 20. $3x^2 - 2x = -6$

In Problems 21 and 22 solve the given equation.

21. $x - \sqrt{4 - 3x} = -8$ 22. $3x^4 + 5x^2 - 2 = 0$

23. Sketch the graphs of $f(x) = (x + 1)^2$ and $g(x) = 2x^2 + 4x + 2$ on the same coordinate axes.

In Problems 24–26 $f(x) = \dfrac{1}{x - 1}$ and $g(x) = x^2$. Find the following.

24. $(f - g)(2)$

25. $\left(\dfrac{g}{f}\right)(x)$

26. $(g \circ f)(3)$

27. Prove that $f(x) = -3x + 1$ and $g(x) = -\frac{1}{3}(x - 1)$ are inverse functions of each other.

PROGRESS TEST 2B

1. Find the length of the shorter diagonal of the parallelogram whose vertices are $(-3, 2)$, $(-5, -4)$, $(3, -4)$, and $(5, 2)$.

2. Use symmetry to assist in sketching the graph of the equation $y^2 = -4x + 4$.

3. Analyze the equation $x^2 - xy + 2 = 0$ for symmetry with respect to the axes and the origin.

4. Determine the domain of the function

$$f(x) = \frac{x^2}{16 - x^2}$$

5. If $f(x) = x^2 - 2x$, find a real number whose image is -1.

6. If $f(x) = \sqrt{x} - 1$, find $f(4)$.

Problems 7–9 refer to the function f defined by

$$f(x) = \begin{cases} x^2 - 1, & x \le -3 \\ 10, & -3 < x \le 3 \\ \sqrt{x}, & x > 3 \end{cases}$$

7. Determine where the function f is increasing, decreasing, and constant.

8. Evaluate $f(2)$.   9. Evaluate $f(-5)$.

10. Sketch the graphs of the functions $f(x) = |x|$ and $g(x) = |x - 2|$ on the same coordinate axes.

11. Find the slope of the line through the points $(-2, -3)$ and $(-4, 6)$.

12. Find an equation of the horizontal line through the point $(-6, -5)$.

13. Find the y-intercept of the line through the points $(4, -3)$ and $(-1, 2)$.

14. Determine the slope of every line that is perpendicular to the line $6y - 2x = 5$.

15. Determine an equation of the line through the point $(3, -2)$ that is parallel to the line $3y + x - 4 = 0$.

16. Solve $6x^2 + 13x - 5 = 0$ by factoring.

17. Solve $2x^2 - 5x + 2 = 0$ by completing the square.

18. Solve $3x^2 - x = -7$ by the quadratic formula.

In Problems 19 and 20 determine the nature of the roots of the quadratic equation without solving.

19. $6z^2 - 4z = -2$ 20. $4y^2 - 20y + 25 = 0$

In Problems 21 and 22 solve the given equation.

21. $8 + \sqrt{1 - x} = 10$ 22. $\dfrac{8}{x^{4/3}} + \dfrac{9}{x^{2/3}} + 1 = 0$

23. Sketch the graphs of $f(x) = x^2$ and $g(x) = -2(x - 1)^2 + 1$ on the same coordinate axes.

In Problems 24–26 $f(x) = \dfrac{1}{\sqrt{x + 1}}$ and $g(x) = x - 1$. Find the following.

24. $(f \cdot g)(3)$ 25. $(f + g)(1)$

26. $(f \circ g)(x)$

27. Prove that $f(x) = \dfrac{1}{x}$ and $g(x) = \dfrac{1}{x}$ are inverse functions.

CHAPTER 3

ROOTS OF POLYNOMIALS

In Chapter 2 we observed that the polynomial function

$$f(x) = ax + b \tag{1}$$

is called a linear function and that the polynomial function

$$g(x) = ax^2 + bx + c, \quad a \neq 0 \tag{2}$$

is called a quadratic function. To facilitate the study of polynomial functions in general, we will use the notation

$$P(x) = a_n x^n + a_{n-1} x^{n-1} + \cdots + a_1 x + a_0, \quad a_n \neq 0 \tag{3}$$

to represent a **polynomial function of degree n.** Note that the subscript k of the coefficient a_k is the same as the exponent of x in x^k. In general, the coefficients a_k may be real or complex numbers; we will restrict our work in this chapter to real values for a_k.

If $a \neq 0$ in Equation (1), we set the polynomial function equal to zero and obtain the linear equation

$$ax + b = 0$$

which has precisely one solution, $-b/a$. If we set the polynomial function in

Equation (2) equal to zero, we have the quadratic equation

$$ax^2 + bx + c = 0$$

which has the two solutions given by the quadratic formula. If we set the polynomial function in Equation (3) equal to zero, we have the **polynomial equation of degree n**

$$a_n x^n + a_{n-1} x^{n-1} + \cdots + a_1 x + a_0 = 0 \qquad (4)$$

Our attention in this chapter will turn to finding the roots or solutions of Equation (4). These solutions are also known as the **roots,** or **zeros, of the polynomial.**

3.1
SYNTHETIC DIVISION; THE REMAINDER AND FACTOR THEOREMS

POLYNOMIAL DIVISION

To find the roots of a polynomial, it will be necessary to divide the polynomial by a second polynomial. There is a procedure for polynomial division that parallels the long division process of arithmetic. In arithmetic, if we divide an integer p by an integer $d \neq 0$, we obtain a quotient q and a remainder r, so we can write

$$\frac{p}{d} = q + \frac{r}{d} \qquad (1)$$

where

$$0 \leq r < d \qquad (2)$$

This result can also be written in the form

$$p = qd + r, \quad 0 \leq r < d \qquad (3)$$

For example,

$$\frac{7284}{13} = 560 + \frac{4}{13}$$

or

$$7284 = (560)(13) + 4$$

In the long division process for polynomials, we divide the dividend $P(x)$ by the divisor $D(x) \neq 0$ to obtain a quotient $Q(x)$ and a remainder $R(x)$. We then have

$$\frac{P(x)}{D(x)} = Q(x) + \frac{R(x)}{D(x)} \qquad (4)$$

where $R(x) = 0$ or where

$$\text{degree of } R(x) < \text{degree of } D(x) \qquad (5)$$

This result can also be written as

$$P(x) = Q(x)D(x) + R(x) \qquad (6)$$

Note that Equations (1) and (4) have the same form and that Equation (6) has the same form as Equation (3). Equation (2) requires that the remainder be less than the divisor, and the parallel requirement for polynomials in Equation (5) is that the *degree* of the remainder be less than that of the divisor.

We illustrate the long division process for polynomials by an example.

EXAMPLE 1
Divide $3x^3 - 7x^2 + 1$ by $x - 2$.

Polynomial Division	
Step 1. Arrange the terms of both polynomials by descending powers of x. If a power is missing, write the term with a zero coefficient.	$x - 2 \overline{)3x^3 - 7x^2 + 0x + 1}$
Step 2. Divide the first term of the dividend by the first term of the divisor. The answer is written above the first term of the dividend.	$\dfrac{3x^2 \qquad\qquad\qquad}{x - 2 \overline{)3x^3 - 7x^2 + 0x + 1}}$
Step 3. Multiply the divisor by the quotient obtained in Step 2 and then subtract the product.	$\begin{array}{r} 3x^2 \\ x - 2 \overline{)3x^3 - 7x^2 + 0x + 1} \\ \underline{3x^3 - 6x^2} \\ -x^2 + 0x + 1 \end{array}$
Step 4. Repeat Steps 2 and 3 until the remainder is zero or the degree of the remainder is less than the degree of the divisor.	$\begin{array}{r} 3x^2 - x \;\; -2 \;\; = Q(x) \\ x - 2 \overline{)3x^3 - 7x^2 + 0x + 1} \\ \underline{3x^3 - 6x^2} \\ -x^2 + 0x + 1 \\ \underline{-x^2 + 2x} \\ -2x + 1 \\ \underline{-2x + 4} \\ -3 = R(x) \end{array}$
Step 5. Write the answer in the form of Equation (4) or Equation (6).	$P(x) = 3x^3 - 7x^2 + 1$ $= \underbrace{(3x^2 - x - 2)}_{Q(x)} \cdot \underbrace{(x - 2)}_{D(x)} \; \underbrace{- 3}_{+ R(x)}$

PROGRESS CHECK
Divide $4x^2 - 3x + 6$ by $x + 2$.

ANSWER
$4x - 11 + \dfrac{28}{x + 2}$

SYNTHETIC DIVISION

Our work in this chapter will frequently require division of a polynomial by a first-degree polynomial $x - r$, where r is a constant. Fortunately, there is a shortcut called **synthetic division** that simplifies this task. To demonstrate synthetic division we will do Example 1 again, writing only the coefficients.

$$
\begin{array}{r}
\mathbf{3} \quad -\mathbf{1} \quad -\mathbf{2} \\
-2\overline{)3 \quad -7 \quad\; 0 \quad\; 1} \\
\underline{\mathbf{3} \quad -6} \\
-\mathbf{1} \quad\; 0 \quad\; 1 \\
\underline{-1 \quad\; 2} \\
-\mathbf{2} \quad\; 1 \\
\underline{-2 \quad\; 4} \\
-3
\end{array}
$$

Note that the boldface numerals are duplicated. We can use this to our advantage and simplify the process as follows.

$$
\begin{array}{r}
-2\underline{\big|} \quad 3 \quad -7 \quad\; 0 \quad\; 1 \\
\underline{\quad -6 \quad\; 2 \quad\; 4} \\
3 \quad -1 \quad -2 \,\big|\, -3
\end{array}
$$

$$
\underbrace{\qquad\qquad\qquad}_{\substack{\text{coefficients} \\ \text{of the} \\ \text{quotient}}} \quad \underset{\text{remainder}}{\big|}
$$

In the third row we copied the leading coefficient (3) of the dividend, multiplied it by the divisor (-2), and wrote the result (-6) in the second row under the next coefficient. The numbers in the second column were subtracted to obtain $-7 - (-6) = -1$. The procedure is repeated until the third row is of the same length as the first row.

Since subtraction is more apt to produce errors than is addition, we can modify this process slightly. If the divisor is $x - r$, we will write r instead of $-r$ in the box and use addition in each step instead of subtraction. Repeating our example, we have

$$
\begin{array}{r}
2\underline{\big|} \quad 3 \quad -7 \quad\; 0 \quad\; 1 \\
\underline{\quad\; 6 \quad -2 \quad -4} \\
3 \quad -1 \quad -2 \,\big|\, -3
\end{array}
$$

EXAMPLE 2

Divide $4x^3 - 2x + 5$ by $x + 2$ using synthetic division.

Synthetic Division	
Step 1. If the divisor is $x - r$, write r in the box. Arrange the coefficients of the dividend by descending powers of x, supplying a zero coefficient for every missing power.	*Step 1.* $\;-2\underline{\big|}\; 4 \quad\; 0 \quad -2 \quad\; 5$

Step 2. Copy the leading coefficient in the third row.

Step 2.

$$\underline{-2}\,\lfloor\ \ 4\quad\ \ 0\quad -2\qquad 5}$$

$$4$$

Step 3. Multiply the last entry in the third row by the number in the box and write the result in the second row under the next coefficient. Add the numbers in that column.

Step 3.

$$\underline{-2}\,\lfloor\ \ 4\quad\ \ 0\quad -2\qquad 5}$$
$$-8$$
$$4\quad -8$$

Step 4. Repeat Step 3 until there is an entry in the third row for each entry in the first row. The last number in the third row is the remainder; the other numbers are the coefficients of the quotient in descending order.

Step 4.

$$\underline{-2}\,\lfloor\ \ 4\quad\ \ \ 0\quad\ -2\qquad\ \ 5}$$
$$-8\quad\ 16\quad -28$$
$$4\quad -8\quad\ 14\ \lfloor\,-23$$

$$\frac{4x^3 - 2x + 5}{x + 2}$$

$$= 4x^2 - 8x + 14 - \frac{23}{x + 2}$$

PROGRESS CHECK

Use synthetic division to obtain the quotient $Q(x)$ and the constant remainder R when $2x^4 - 10x^2 - 23x + 6$ is divided by $x - 3$.

ANSWER

$Q(x) = 2x^3 + 6x^2 + 8x + 1; R = 9$

WARNING

(a)   Synthetic division can be used only when the divisor is a linear factor. Don't forget to write a zero for the coefficient of each missing term.

(b)   When dividing by $x - r$, place r in the box. For example, when the divisor is $x + 3$, place -3 in the box since $x + 3 = x - (-3)$. Similarly, when the divisor is $x - 3$, place $+3$ in the box since $x - 3 = x - (+3)$.

THE REMAINDER THEOREM

From our work with the division process we may surmise that division of a polynomial $P(x)$ by $x - r$ results in a quotient $Q(x)$ and a constant remainder R such that

$$P(x) = (x - r) \cdot Q(x) + R$$

Since this identity holds for all real values of x, it must hold when $x = r$. Consequently,

$$P(r) = (r - r) \cdot Q(r) + R$$
$$P(r) = 0 \cdot Q(r) + R$$

or

$$P(r) = R$$

We have proved the Remainder Theorem.

Remainder Theorem	If a polynomial $P(x)$ is divided by $x - r$, the remainder is $P(r)$.

EXAMPLE 1
Determine the remainder when $P(x) = 2x^3 - 3x^2 - 2x + 1$ is divided by $x - 3$.

SOLUTION
By the Remainder Theorem, the remainder is $R = P(3)$. We then have

$$R = P(3) = 2(3)^3 - 3(3)^2 - 2(3) + 1 = 22$$

We may verify this result by using synthetic division.

$$
\begin{array}{r|rrrr}
3 & 2 & -3 & -2 & 1 \\
  &   & 6 & 9 & 21 \\
\hline
  & 2 & 3 & 7 & \mathbf{22}
\end{array}
$$

The numeral in boldface is the remainder, so we have verified that $R = 22$.

PROGRESS CHECK
Determine the remainder when $3x^2 - 2x - 6$ is divided by $x + 2$.

ANSWER
10

FACTOR THEOREM

Let's assume that a polynomial $P(x)$ can be written as a product of polynomials, that is,

$$P(x) = D_1(x)D_2(x) \ . \ . \ . \ D_n(x)$$

where $D_i(x)$ is a polynomial of degree greater than zero. Then $D_i(x)$ is called a **factor** of $P(x)$. If we focus on $D_1(x)$ and let

$$Q(x) = D_2(x)D_3(x) \ . \ . \ . \ D_n(x)$$

then

$$P(x) = D_1(x)Q(x)$$

which suggests the following formal definition.

The polynomial $D(x)$ is a factor of a polynomial $P(x)$ if division of $P(x)$ by $D(x)$ results in a remainder of zero.

We can now combine this rule and the Remainder Theorem to prove the Factor Theorem.

Factor Theorem	A polynomial $P(x)$ has a factor $x - r$ if and only if $P(r) = 0$.

If $x - r$ is a factor of $P(x)$, then division of $P(x)$ by $x - r$ must result in a remainder of 0. By the Remainder Theorem, the remainder is $P(r)$, and hence $P(r) = 0$. Conversely, if $P(r) = 0$, then the remainder is 0 and $P(x) = (x - r) Q(x)$ for some polynomial $Q(x)$ of degree one less than that of $P(x)$. By definition, $x - r$ is then a factor of $P(x)$.

EXAMPLE 3
Show that $x + 2$ is a factor of

$$P(x) = x^3 - x^2 - 2x + 8$$

SOLUTION
By the Factor Theorem, $x + 2$ is a factor if $P(-2) = 0$. Using synthetic division to evaluate $P(-2)$,

$$
\begin{array}{r|rrrr}
-2 & 1 & -1 & -2 & 8 \\
   &   & -2 & 6 & -8 \\
\hline
   & 1 & -3 & 4 & 0
\end{array}
$$

we see that $P(-2) = 0$. Alternatively, we can evaluate

$$P(-2) = (-2)^3 - (-2)^2 - 2(-2) + 8 = 0$$

We conclude that $x + 2$ is a factor of $P(x)$.

PROGRESS CHECK
Show that $x - 1$ is a factor of $P(x) = 3x^6 - 3x^5 - 4x^4 + 6x^3 - 2x^2 - x + 1$.

EXERCISE SET 3.1
In Exercises 1–10 use polynomial division to find the quotient $Q(x)$ and the remainder $R(x)$ when the first polynomial is divided by the second polynomial.

1. $x^2 - 7x + 12, \quad x - 5$
2. $x^2 + 3x + 3, \quad x + 2$
3. $2x^3 - 2x, \quad x^2 + 2x - 1$
4. $3x^3 - 2x^2 + 4, \quad x^2 - 2$
5. $3x^4 - 2x^2 + 1, \quad x + 3$
6. $x^5 - 1, \quad x^2 - 1$
7. $2x^3 - 3x^2, \quad x^2 + 2$
8. $3x^3 - 2x - 1, \quad x^2 - x$
9. $x^4 - x^3 + 2x^2 - x + 1, \quad x^2 + 1$
10. $2x^4 - 3x^3 - x^2 - x - 2, \quad x - 2$

In Exercises 11–20 use synthetic division to find the quotient $Q(x)$ and the constant remainder R when the first polynomial is divided by the second polynomial.

11. $x^3 - x^2 - 6x + 5, \quad x + 2$
12. $2x^3 - 3x^2 - 4, \quad x - 2$
13. $x^4 - 81, \quad x - 3$
14. $x^4 - 81, \quad x + 3$

15. $3x^3 - x^2 + 8$, $x + 1$

16. $2x^4 - 3x^3 - 4x - 2$, $x - 1$

17. $x^5 + 32$, $x + 2$

18. $x^5 + 32$, $x - 2$

19. $6x^4 - x^2 + 4$, $x - 3$

20. $8x^3 + 4x^2 - x - 5$, $x + 3$

In Exercises 21–26 use the Remainder Theorem and synthetic division to find $P(r)$.

21. $P(x) = x^3 - 4x^2 + 1$, $r = 2$

22. $P(x) = x^4 - 3x^2 - 5x$, $r = -1$

23. $P(x) = x^5 - 2$, $r = -2$

24. $P(x) = 2x^4 - 3x^3 + 6$, $r = 2$

25. $P(x) = x^6 - 3x^4 + 2x^3 + 4$, $r = -1$

26. $P(x) = x^6 - 2$, $r = 1$

In Exercises 27–32 use the Remainder Theorem to determine the remainder when $P(x)$ is divided by $x - r$.

27. $P(x) = x^3 - 2x^2 + x - 3$, $x - 2$

28. $P(x) = 2x^3 + x^2 - 5$, $x + 2$

29. $P(x) = -4x^3 + 6x - 2$, $x - 1$

30. $P(x) = 6x^5 - 3x^4 + 2x^2 + 7$, $x + 1$

31. $P(x) = x^5 - 30$, $x + 2$

32. $P(x) = x^4 - 16$, $x - 2$

In Exercises 33–40 use the Factor Theorem to decide whether or not the first polynomial is a factor of the second polynomial.

33. $x - 2$, $x^3 - x^2 - 5x + 6$

34. $x - 1$, $x^3 + 4x^2 - 3x + 1$

35. $x + 2$, $x^4 - 3x - 5$

36. $x + 1$, $2x^3 - 3x^2 + x + 6$

37. $x + 3$, $x^3 + 27$

38. $x + 2$, $x^4 + 16$

39. $x + 2$, $x^4 - 16$

40. $x - 3$, $x^3 + 27$

In Exercises 41–44 use synthetic division to determine the value of k or r as requested.

41. Determine the values of r for which division of $x^2 - 2x - 1$ by $x - r$ has a remainder of 2.

42. Determine the values of r for which

$$\frac{x^2 - 6x - 1}{x - r}$$

has a remainder of -9.

43. Determine the values of k for which $x - 2$ is a factor of $x^3 - 3x^2 + kx - 1$.

44. Determine the values of k for which $2k^2x^3 + 3kx^2 - 2$ is divisible by $x - 1$.

45. Use the Factor Theorem to show that $x - 2$ is a factor of $P(x) = x^8 - 256$.

46. Use the Factor Theorem to show that $P(x) = 2x^4 + 3x^2 + 2$ has no factor of the form $x - r$, where r is a real number.

47. Use the Factor Theorem to show that $x - y$ is a factor of $x^n - y^n$, where n is a natural number.

3.2
GRAPHS OF POLYNOMIAL FUNCTIONS

The graph of the first-degree polynomial function

$$P(x) = ax + b \tag{1}$$

is always a straight line. The graph of the second-degree polynomial function

$$P(x) = ax^2 + bx + c, \quad a \neq 0 \tag{2}$$

is always a parabola. For $n > 2$ the graph of the polynomial function

$$P(x) = a_n x^n + a_{n-1} x^{n-1} + \cdots + a_1 x + a_0, \quad a_n \neq 0 \tag{3}$$

does not have a name, nor is it a simple matter to describe its shape. Still, there are certain general patterns of behavior that we can explore. Taken together with a neat scheme for forming a table of values, these observations will aid greatly in sketching the graphs of polynomial functions of degree greater than 2.

GENERAL BEHAVIOR FOR LARGE $|x|$

By factoring out x^n we can rewrite Equation (3) as follows.

$$P(x) = a_n x^n + a_{n-1} x^{n-1} + \cdots + a_1 x + a_0$$

$$= x^n \left(a_n + \frac{a_{n-1}}{x} + \frac{a_{n-2}}{x^2} + \cdots + \frac{a_1}{x^{n-1}} + \frac{a_0}{x^n} \right), \quad x \neq 0 \qquad (4)$$

When the equation is written this way, it is easy to see how $P(x)$ behaves when $|x|$ assumes large values, that is, when x takes on large positive and large negative values. Note that the expression

$$\frac{a_{n-k}}{x^k}$$

where k is a positive integer and a_{n-k} is a constant, will take on values closer and closer to zero as $|x|$ becomes larger and larger, since as the denominator of a fraction grows large, the value of the fraction becomes small. For sufficiently large values of $|x|$, we can ignore the contribution of all terms of the form a_{n-k}/x^k. What remains, then, is the term $a_n x^n$. In summary,

---

For large values of $|x|$, the polynomial function

$$P(x) = a_n x^n + a_{n-1} x^{n-1} + \cdots + a_1 x + a_0, \quad a_n \neq 0$$

is dominated by its lead term $a_n x^n$.

We can make practical use of this last result. In Figures 1a and 1b we have sketched the graphs of polynomial functions of degrees 3 and 4, respectively. Note that as $|x|$ increases, the "ends" of the graph of the third-degree polynomial function extend indefinitely in opposite directions, whereas the "ends" of the graph of the fourth-degree polynomial function extend indefinitely in the same direction.

The behavior exhibited in Figure 1a will be true for all third-degree polynomial functions and, in fact, will be true for *all polynomial functions of odd degree*. To see this, simply note that the lead term $a_n x^n$ of any polynomial function of *odd* degree will assume opposite signs for positive and negative values of x. By our earlier result, this term dominates for large $|x|$, so that one end of the graph will extend upward and the other downward.

We can use a similar argument to show that the behavior exhibited in Figure 1b is typical for all polynomial functions of even degree. For n that is *even*, the

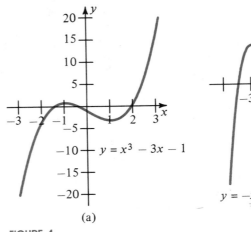

$$y = x^3 - 3x - 1$$

(a)

$$y = -x^4 - 4x^3 + 2x^2 + 12x - 4$$

(b)

FIGURE 1

lead term $a_n x^n$ will have the same sign for both positive and negative values of x. Since this term dominates for large $|x|$, the ends of the curve will both extend upward or will both extend downward.

Graphs of Polynomial Functions	The graph of the polynomial function $$P(x) = a_n x^n + a_{n-1} x^{n-1} + \cdots + a_1 x + a_0, \quad a_n \neq 0$$ has these characteristics. • If n is odd and $a_n > 0$, the graph extends upward for large positive values of x and downward for large negative values of x; if $a_n < 0$, the behavior is reversed. • If n is even and $a_n > 0$, the graph extends upward for both large positive and large negative values of x; if $a_n < 0$, the graph extends downward at both ends.

EXAMPLE 1

Without sketching the graph, determine the behavior of the graph of each of the following polynomial functions for large values of $|x|$.

(a) $P(x) = -3x^5 + 26x^2 - 5$ (b) $P(x) = -\dfrac{1}{2}x^6 + 209x^3 + 16x + 2$

SOLUTION

(a) We need only concern ourselves with the lead term, $-3x^5$. Since the degree is odd and the lead coefficient is negative, the graph will extend downward for large positive values of x and upward for large negative values of x.

(b)   The dominant term for large $|x|$ is $-\dfrac{1}{2}x^6$. Since the degree is even, the graph moves in the same direction at both ends. The negative lead coefficient tells us that this direction is downward.

TABLE OF VALUES

Now that we know the behavior of the graph of a polynomial function at the "ends" where $|x|$ is large, we need guidance as to what happens for intermediate values of x. The Remainder Theorem can be used to find the coordinates of points on the graph. Recall that if a polynomial $P(x)$ is divided by $x - r$, the remainder is $P(r)$. Then the point $(r, P(r))$ lies on the graph of the function $P(x)$.

An efficient scheme for evaluating $P(r)$ is a streamlined form of synthetic division in which the addition is performed without writing the middle row. Given

$$P(x) = 2x^3 - 3x^2 - 2x + 1$$

we can find $P(3)$ by synthetic division in this way:

$$\begin{array}{r|rrrr} & 2 & -3 & -2 & 1 \\ 3 & 2 & 3 & 7 & 22 = P(3) \end{array}$$

Then the point $(3, 22)$ lies on the graph of $P(x)$. Repeating this procedure for a number of values of r will provide a table of values for plotting.

EXAMPLE 2

Sketch the graph of $P(x) = 2x^3 - 3x^2 - 2x + 1$.

SOLUTION

For each value r of x, the point $(r, P(r))$ lies on the graph of $y = P(x)$. We will allow x to assume integer values from -3 to $+3$ and will find $P(x)$ by using synthetic division.

	2	−3	−2	1	$(x, y) = (x, P(x))$
−3	2	−9	25	−74	$(-3, -74)$
−2	2	−7	12	−23	$(-2, -23)$
−1	2	−5	3	−2	$(-1, -2)$
0	2	−3	−2	1	$(0, 1)$
1	2	−1	−3	−2	$(1, -2)$
2	2	1	0	1	$(2, 1)$
3	2	3	7	22	$(3, 22)$

The ordered pairs shown at the right of each row are the coordinates of points on the graph shown in Figure 2.

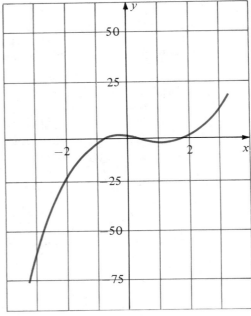

FIGURE 2

**POLYNOMIALS IN
FACTORED FORM**

When a polynomial can be written as a product of linear factors, it is a simple matter to find the x-intercepts and to determine where the graph of the polynomial lies above the x-axis and where it lies below the x-axis. The following example illustrates the procedure.

EXAMPLE 3

Sketch the graph of the polynomial

$$P(x) = x^3 + x^2 - 6x$$

SOLUTION

Factoring, we find that

$$P(x) = x(x^2 + x - 6)$$
$$= x(x + 3)(x - 2)$$

Since $P(x) = 0$ when $x = 0$, when $x = -3$, and when $x = 2$, these values are the x-intercepts. The x-intercepts divide the x-axis into the intervals

$$(-\infty, -3), \ (-3, 0), \ (0, 2), \ \text{and} \ (2, \infty)$$

To find the signs of $P(x)$ in each of these intervals, we use the method described in Section 1.6, which requires that we analyze the sign of each factor in each interval as in Figure 3. From Figure 3 we conclude that the graph of $P(x)$ lies above the x-axis in the intervals $(-3, 0)$ and $(2, \infty)$, since $P(x) > 0$ in these intervals. Similarly, the graph lies below the x-axis in the intervals $(-\infty, -3)$ and

x	$-$	$-$	$-$	$-$	$-$	$-$	$-$	$-$	0	$+$	$+$	$+$	$+$	$+$
$x + 3$	$-$	$-$	$-$	0	$+$	$+$	$+$	$+$	$+$	$+$	$+$	$+$	$+$	$+$
$x - 2$	$-$	$-$	$-$	$-$	$-$	$-$	$-$	$-$	$-$	$-$	$-$	0	$+$	$+$

$$-3 \qquad\qquad 0 \qquad\qquad 2$$

| $P(x)$ | $-$ | $-$ | $-$ | 0 | $+$ | $+$ | $+$ | $+$ | 0 | $-$ | $-$ | 0 | $+$ | $+$ | $+$ |

FIGURE 3

$(0, 2)$. Plotting a few points, we obtain the graph of Figure 4. Note that the graph extends indefinitely upward and downward as expected.

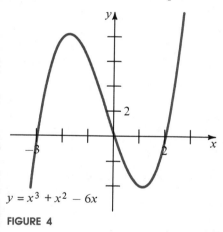

$$y = x^3 + x^2 - 6x$$

FIGURE 4

OTHER BEHAVIORAL PATTERNS

There are three additional observations that will prove helpful in sketching the graph of a polynomial function.

- No segment of the graph of a polynomial function of degree greater than 1 is a straight line.
- The graph of a polynomial function is continuous, which means that the graph can be drawn without lifting the pencil from the paper.
- The graph of a polynomial function of degree n has at most $n - 1$ turning points.

Continuity and turning points are important topics in calculus. Without the information available through methods taught in calculus courses, the graphs we sketch will be inaccurate unless we plot many points—a tedious and time-consuming task.

EXERCISE SET 3.2

In Exercises 1–8, without sketching, determine the behavior of the graph of the given polynomial function for large values of $|x|$. Use the letters U and D to indicate whether the graph extends upward or downward.

Polynomial function	Large positive values of x	Large negative values of x
1. $P(x) = x^7 - 175x^3 + 23x^2$		
2. $P(x) = -3x^8 + 22x^4 + 3$		
3. $P(x) = -8x^3 + 17x^2 - 15$		
4. $P(x) = 14x^{12} - 5x^{11} + 3x - 1$		
5. $P(x) = -5x^{10} + 16x^7 + 5$		
6. $P(x) = 2x^5 - 11x^4 - 12x^3$		
7. $P(x) = 4x^8 - 10x^6 + x^3 - 8$		
8. $P(x) = -3x^9 + 6x^6 - 2x^5 + x$		

In Exercises 9–14 use the Remainder Theorem and synthetic division to sketch the graph of the given polynomial.

9. $P(x) = x^3 + x^2 + x + 1$

10. $P(x) = 3x^4 + 5x^3 + x^2 + 5x - 2$

11. $P(x) = 2x^3 + 3x^2 - 5x - 6$

12. $P(x) = x^3 + 3x^2 - 4x - 12$

13. $P(x) = x^4 - 3x^3 + 1$

14. $P(x) = 4x^4 + 4x^3 - 9x^2 - x + 2$

In Exercises 15–20 determine the x-intercepts and the intervals wherein $P(x) > 0$ and $P(x) < 0$. Sketch the graph of $P(x)$.

15. $P(x) = (x - 3)(2x - 1)(x + 2)$

16. $P(x) = (2 - x)(x - 4)(x + 1)$

17. $P(x) = 2x^3 + 3x^2 - 5x$

18. $P(x) = x^4 - 5x^2 + 4$

19. $P(x) = x^4 - x^3 - 6x^2$

20. $P(x) = (2x + 5)(x - 1)(x + 1)(x - 3)$

3.3

THE RATIONAL ROOT THEOREM AND THE DEPRESSED EQUATION

When the coefficients of a polynomial are all integers, a systematic search for the *rational* roots is possible by using the following theorem.

Rational Root Theorem   If the coefficients of the polynomial

$$P(x) = a_n x^n + a_{n-1} x^{n-1} + \cdots + a_1 x + a_0, \quad a_n \neq 0$$

are all integers and p/q is a rational root, in lowest terms, then

(i) p is a factor of the constant term a_0, and

(ii) q is a factor of the leading coefficient a_n.

PROOF OF RATIONAL ROOT THEOREM (Optional)

Since p/q is a root of $P(x)$, then $P(p/q) = 0$. Thus,

$$a_n \left(\frac{p}{q}\right)^n + a_{n-1} \left(\frac{p}{q}\right)^{n-1} + \cdots + a_1 \left(\frac{p}{q}\right) + a_0 = 0 \qquad (1)$$

Multiplying Equation (1) by q^n, we have

$$a_n p^n + a_{n-1} p^{n-1} q + \cdots + a_1 p q^{n-1} + a_0 q^n = 0 \qquad (2)$$

or

$$a_n p^n + a_{n-1} p^{n-1} q + \cdots + a_1 p q^{n-1} = -a_0 q^n \qquad (3)$$

Taking the common factor p out of the left-hand side of Equation (3) yields

$$p(a_n p^{n-1} + a_{n-1} p^{n-2} q + \cdots + a_1 q^{n-1}) = -a_0 q^n \qquad (4)$$

Since $a_1, a_2, \ldots, a_n, p$, and q are all integers, the quantity in parentheses in the left-hand side of Equation (4) is an integer. Division of the left-hand side by p results in an integer, and we conclude that p must also be a factor of the right-hand side, $-a_0 q^n$. But p and q have no common factors since, by hypothesis, p/q is in lowest terms. Hence, p must be a factor of a_0; thus we have proved part (i) of the Rational Root Theorem.

We may also rewrite Equation (2) in the form

$$q(a_{n-1} p^{n-1} + a_{n-2} p^{n-2} q + \cdots + a_1 p q^{n-2} + a_0 q^{n-1}) = -a_n p^n \qquad (5)$$

An argument similar to the preceding one now establishes part (ii) of the theorem.

EXAMPLE 1

Find the rational roots of the equation

$$8x^4 - 2x^3 + 7x^2 - 2x - 1 = 0$$

SOLUTION

If p/q is a rational root in lowest terms, then p is a factor of 1 and q is a factor of 8. We can now list the possibilities:

> possible numerators: ± 1 (the factors of 1)
> possible denominators: $\pm 1, \pm 2, \pm 4, \pm 8$ (the factors of 8)
> possible rational roots: $\pm 1, \pm \dfrac{1}{2}, \pm \dfrac{1}{4}, \pm \dfrac{1}{8}$

Synthetic division can be used to test if these numbers are roots. Trying $x = 1$ and $x = -1$, we find that the remainder is not zero and they are therefore not roots. Trying $\frac{1}{2}$, we have

$$
\begin{array}{r|rrrrr}
\frac{1}{2} & 8 & -2 & 7 & -2 & -1 \\
  &   & 4 & 1 & 4 & 1 \\
\hline
  & 8 & 2 & 8 & 2 & 0 \\
\end{array}
$$

which demonstrates that $\frac{1}{2}$ is a root. Similarly,

$$
\begin{array}{r|rrrrr}
-\frac{1}{4} & 8 & -2 & 7 & -2 & -1 \\
  &   & -2 & 1 & -2 & 1 \\
\hline
  & 8 & -4 & 8 & -4 & 0 \\
\end{array}
$$

which shows that $-\frac{1}{4}$ is also a root. The student may verify that none of the other possible rational roots will result in a zero remainder when synthetic division is employed.

PROGRESS CHECK

Find the rational roots of the equation

$$9x^4 - 12x^3 + 13x^2 - 12x + 4 = 0$$

ANSWER

$$\frac{2}{3}, \frac{2}{3}$$

In Chapter 1 we discussed number systems and said that numbers such as $\sqrt{2}$ and $\sqrt{3}$ were irrational. The Rational Root Theorem provides a direct means of verifying that this is indeed so.

EXAMPLE 2

Prove that $\sqrt{3}$ is not a rational number.

SOLUTION

If we let $x = \sqrt{3}$, then $x^2 = 3$ or $x^2 - 3 = 0$. By the Rational Root Theorem, the only possible rational roots are ± 1, ± 3. Synthetic division can be used to show that none of these are roots. However, $\sqrt{3}$ is a root of $x^2 - 3 = 0$. Hence, $\sqrt{3}$ is not a rational number.

THE DEPRESSED EQUATION

By using the Factor Theorem we can show that there is a close relationship between the factors and the roots of the polynomial $P(x)$. By definition, r is a root of $P(x)$ if and only if $P(r) = 0$. But the Factor Theorem tells us that $P(r) = 0$ if and only if $x - r$ is a factor of $P(x)$. This leads to the following alternative statement of the Factor Theorem.

Factor Theorem	A polynomial $P(x)$ has a root r if and only if $x - r$ is a factor of $P(x)$.

SOLVING POLYNOMIAL EQUATIONS

Cardan's Formula

Cardano provided this formula for one root of the cubic equation

$$x^3 + bx + c = 0$$

$$x = \sqrt[3]{\sqrt{\frac{b^3}{27} + \frac{c^2}{4}} - \frac{c}{2}}$$
$$- \sqrt[3]{\sqrt{\frac{b^3}{27} + \frac{c^2}{4}} + \frac{c}{2}}$$

Try it for the cubics

$$x^3 - x = 0$$
$$x^3 - 1 = 0$$
$$x^3 - 3x + 2 = 0$$

The quadratic formula provides us with the solutions of a polynomial equation of second degree. How about polynomial equations of third degree? of fourth degree? of fifth degree?

The search for formulas expressing the roots of polynomial equations in terms of the coefficients of the equations intrigued mathematicians for hundreds of years. A method for finding the roots of polynomial equations of degree 3 was published around 1535 and is known as Cardan's formula despite the possibility that Girolamo Cardano stole the result from his friend Nicolo Tartaglia. Shortly afterward a method that is attributed to Ferrari was published for solving polynomial equations of degree 4.

The next 250 years were spent in seeking formulas for the roots of polynomial equations of degree 5 or higher—without success. Finally, early in the nineteenth century, the Norwegian mathematician N. H. Abel and the French mathematician Evariste Galois proved that *no such formulas exist*. Galois's work on this problem was completed a year before his death in a duel at age 20. His proof, using the new concepts of group theory, was so advanced that his teachers wrote it off as being unintelligible gibberish.

EXAMPLE 3

Find a polynomial $P(x)$ of degree 3 whose roots are -1, 1, and -2.

SOLUTION

By the Factor Theorem, $x + 1$, $x - 1$, and $x + 2$ are factors of $P(x)$. The product

$$P(x) = (x + 1)(x - 1)(x + 2) = x^3 + 2x^2 - x - 2$$

is a polynomial of degree 3 with the desired roots. Note that multiplying $P(x)$ by any nonzero real number results in another polynomial that has the same roots. For example, the polynomial

$$5 \cdot P(x) = 5x^3 + 10x^2 - 5x - 10$$

also has -1, 1, and -2 as its roots. Thus, the answer is not unique.

PROGRESS CHECK

Find a polynomial $P(x)$ of degree 3 whose roots are 2, 4, and -3.

ANSWER

$x^3 - 3x^2 - 10x + 24$

If we know that r is a root of $P(x)$, we may write

$$P(x) = (x - r)Q(x)$$

If r_1 is a root of $Q(x)$, then $Q(r_1) = 0$ and

$$P(r_1) = (r_1 - r)Q(r_1) = (r_1 - r) \cdot 0 = 0$$

which shows that r_1 is also a root of $P(x)$. We call $Q(x) = 0$ the **depressed equation,** since $Q(x)$ is of lower degree than $P(x)$. In the next example we illustrate the use of the depressed equation in finding the roots of a polynomial.

EXAMPLE 4

If 4 is a root of the polynomial $P(x) = x^3 - 8x^2 + 21x - 20$, find the other roots.

SOLUTION

Since 4 is a root of $P(x)$, $x - 4$ is a factor of $P(x)$. Therefore,

$$P(x) = (x - 4)Q(x)$$

To find the depressed equation, we compute $Q(x) = P(x)/(x - 4)$ by synthetic division.

$$
\begin{array}{r|rrrr}
4 & 1 & -8 & 21 & -20 \\
  &   & 4 & -16 & 20 \\
\hline
  & 1 & -4 & 5 & 0
\end{array}
$$

coefficients   remainder
of $Q(x)$

The depressed equation is

$$x^2 - 4x + 5 = 0$$

Using the quadratic formula, the roots of the depressed equation are found to be $2 + i$ and $2 - i$. The roots of $P(x)$ are then seen to be 4, $2 + i$, and $2 - i$.

PROGRESS CHECK

If -2 is a root of the polynomial $P(x) = x^3 - 7x - 6$, find the remaining roots.

ANSWER

$-1, 3$

We can combine the Rational Root Theorem and the depressed equation to give us even greater power in seeking the roots of a polynomial.

EXAMPLE 5

Find the rational roots of the polynomial equation

$$8x^5 + 12x^4 + 14x^3 + 13x^2 + 6x + 1 = 0$$

SOLUTION

Since the coefficients of the polynomial are all integers, we may use the Rational Root Theorem to list the possible rational roots.

> possible numerators: ± 1 (factors of 1)
> possible denominators: $\pm 1, \pm 2, \pm 4, \pm 8$ (factors of 8)
> possible rational roots: $\pm 1, \pm \frac{1}{2}, \pm \frac{1}{4}, \pm \frac{1}{8}$

TRANSCENDENTAL NUMBERS

Theorem: Every rational number p/q is algebraic.

Proof: The number p/q is a root of the equation

$$qx - p = 0$$

since

$$q\left(\frac{p}{q}\right) - p = p - p = 0$$

Further, by definition of a rational number, q and p are integers and $q \neq 0$. So p/q is a root of a polynomial equation with integer coefficients and is therefore algebraic.

A real number that is a root of some polynomial equation with integer coefficients is said to be **algebraic.** We see that $\frac{2}{3}$ is algebraic since it is the root of the equation $3x - 2 = 0$; $\sqrt{2}$ is also algebraic, since it satisfies the equation $x^2 - 2 = 0$.

Note that every real number a satisfies the equation $x - a = 0$; that is, it satisfies a polynomial equation with *real* coefficients. But to be algebraic the number a must satisfy a polynomial equation with *integer* coefficients. To show that a real number a is *not* algebraic we must demonstrate that there is *no* polynomial equation with integer coefficients that has a as one of its roots. Although this appears to be an impossible task, it was performed in 1844 when Joseph Liouville exhibited specific examples of such numbers, called **transcendental** numbers. Subsequently, Georg Cantor (1845–1918), in his brilliant work on infinite sets, provided a more general proof of the existence of transcendental numbers.

You are already familiar with a transcendental number: the number π is not a root of any polynomial equation with integer coefficients.

Trying $+1$, -1, and $+\frac{1}{2}$, we find that they are not roots. Testing $-\frac{1}{2}$ by synthetic division results in a remainder of zero.

$$
\begin{array}{r|rrrrrr}
-\frac{1}{2} & 8 & 12 & 14 & 13 & 6 & 1 \\
 & & -4 & -4 & -5 & -4 & -1 \\
\hline
 & 8 & 8 & 10 & 8 & 2 & 0
\end{array}
$$

coefficients of depressed equation

Rather than return to the original equation to continue the search, we will use the depressed equation

$$8x^4 + 8x^3 + 10x^2 + 8x + 2 = 0$$

The values $+1$, -1, and $+\frac{1}{2}$ have been eliminated, but the value $-\frac{1}{2}$ must be tried again.

$$
\begin{array}{r|rrrrr}
-\frac{1}{2} & 8 & 8 & 10 & 8 & 2 \\
 & & -4 & -2 & -4 & -2 \\
\hline
 & 8 & 4 & 8 & 4 & 0
\end{array}
$$

coefficients of depressed equation

Since the remainder is zero, $-\frac{1}{2}$ is again a root. This illustrates an important point: A rational root may be a multiple root! Applying the same technique to the resulting depressed equation

$$8x^3 + 4x^2 + 8x + 4 = 0$$

we see that $-\frac{1}{2}$ is once again a root.

$$\begin{array}{r|rrrr} -\frac{1}{2} & 8 & 4 & 8 & 4 \\ & & -4 & 0 & -4 \\ \hline & 8 & 0 & 8 & | \; 0 \end{array}$$

$$\underbrace{\qquad\qquad\qquad}$$

coefficients of depressed equation

The final depressed equation

$$8x^2 + 8 = 0 \quad \text{or} \quad x^2 + 1 = 0$$

has the roots $\pm i$. Thus, the original equation has the rational roots

$$-\frac{1}{2}, \; -\frac{1}{2}, \; -\frac{1}{2}$$

PROGRESS CHECK

Find all roots of the polynomial

$$P(x) = 9x^4 - 3x^3 + 16x^2 - 6x - 4$$

ANSWER

$\frac{2}{3}, \; -\frac{1}{3}, \; \pm\sqrt{2}i$

EXERCISE SET 3.3

In Exercises 1–10 use the Rational Root Theorem to find the rational roots of the given equation.

1. $x^3 - 2x^2 - 5x + 6 = 0$
2. $3x^3 - x^2 - 3x + 1 = 0$
3. $6x^4 - 7x^3 - 13x^2 + 4x + 4 = 0$
4. $36x^4 - 15x^3 - 26x^2 + 3x + 2 = 0$
5. $5x^6 - x^5 - 5x^4 + 6x^3 - x^2 - 5x + 1 = 0$
6. $16x^4 - 16x^3 - 29x^2 + 32x - 6 = 0$
7. $4x^4 - x^3 + 5x^2 - 2x - 6 = 0$
8. $6x^4 + 2x^3 + 7x^2 + x + 2 = 0$
9. $2x^5 - 13x^4 + 26x^3 - 22x^2 + 24x - 9 = 0$
10. $8x^5 - 4x^4 + 6x^3 - 3x^2 - 2x + 1 = 0$

In Exercises 11–16 use the given root(s) to help in finding the remaining roots of the equation.

11. $x^3 - 3x - 2 = 0; \; -1$
12. $x^3 - 7x^2 + 4x + 24 = 0; \; 3$
13. $x^3 - 8x^2 + 18x - 15 = 0; \; 5$
14. $x^3 - 2x^2 - 7x - 4 = 0; \; -1$
15. $x^4 + x^3 - 12x^2 - 28x - 16 = 0; \; -2$
16. $x^4 - 2x^2 + 1; \; 1$ is a double root

In Exercises 17–24 use the Rational Root Theorem and the depressed equation to find all roots of the given equation.

17. $4x^4 + x^3 + x^2 + x - 3 = 0$
18. $x^4 + x^3 + x^2 + 3x - 6 = 0$
19. $5x^5 - 3x^4 - 10x^3 + 6x^2 - 40x + 24 = 0$
20. $12x^4 - 52x^3 + 75x^2 - 16x - 5 = 0$
21. $6x^4 - x^3 - 5x^2 + 2x = 0$
22. $2x^4 - \frac{3}{2}x^3 + \frac{11}{2}x^2 + \frac{23}{2}x + \frac{5}{2} = 0$
23. $2x^4 - x^3 - 28x^2 + 30x - 8 = 0$
24. $12x^4 + 4x^3 - 17x^2 + 6x = 0$

In Exercises 25–28 find the integer value(s) of k for which the given equation has rational roots, and find the roots. (*Hint:* Use synthetic division.)

25. $x^3 + kx^2 + kx + 2 = 0$
26. $x^4 - 4x^3 - kx^2 + 6kx + 9 = 0$
27. $x^4 - 3x^3 + kx^2 - 4x - 1 = 0$
28. $x^3 - 4kx^2 - k^2x + 4 = 0$

29. If $P(x)$ is a polynomial with integer coefficients and the leading coefficient is $+1$ or -1, prove that the rational roots of $P(x)$ are all integers and are factors of the constant term.

30. Prove that $\sqrt{5}$ is not a rational number.

31. If p is a prime, prove that $\sqrt{p}$ is not a rational number.

3.4
RATIONAL FUNCTIONS AND THEIR GRAPHS

A function f of the form

$$f(x) = \frac{P(x)}{Q(x)}$$

where $P(x)$ and $Q(x)$ are polynomials and $Q(x) \neq 0$, is called a **rational function.** We will study the behavior of rational functions with the objective of sketching their graphs.

We first note that the polynomials $P(x)$ and $Q(x)$ are defined for all real values of x. Since we must avoid division by zero, the domain of the function f will consist of all real numbers except those for which $Q(x) = 0$.

EXAMPLE 1
Determine the domain of each function.

(a) $f(x) = \dfrac{x + 1}{x - 1}$ (b) $g(x) = \dfrac{x^2 + 9}{x^2 - 4}$ (c) $h(x) = \dfrac{x^2}{x^2 + 1}$

SOLUTION
(a) We must exclude all real values for which the denominator $x - 1 = 0$. Thus, the domain of f consists of all real numbers except $x = 1$.
(b) Since $x^2 - 4 = 0$ when $x = \pm 2$, the domain of g consists of all real numbers except $x = \pm 2$.
(c) Since $x^2 + 1 = 0$ has no real solutions, the domain of h is the set of all real numbers.

PROGRESS CHECK
Determine the domain of each function.

(a) $S(x) = \dfrac{x}{2x^2 - 3x - 2}$ (b) $T(x) = \dfrac{-1}{x^4 + x^2 + 2}$

ANSWERS
(a) $x \neq -\frac{1}{2}, 2$ (b) all real numbers

ASYMPTOTES

Let's first consider rational functions for which the numerator is a constant, for example,

$$f(x) = \frac{1}{x} \quad \text{and} \quad g(x) = \frac{1}{x^2}$$

The domain of both f and g is the set of all nonzero real numbers. Furthermore, the graph of f is symmetric with respect to the origin, since the equation $y = 1/x$ remains unchanged when x and y are replaced by $-x$ and $-y$, respectively. Similarly, the graph of g is symmetric with respect to the y-axis, since the equation $y = 1/x^2$ is unchanged when x is replaced by $-x$. We therefore need plot only those points corresponding to positive values of x and can utilize symmetry to obtain the graphs of Figure 5.

When a graph gets closer and closer to a line, we say that the line is an **asymptote** of the graph. Note the behavior of the graphs of f and g (Figure 5) as x gets closer and closer to 0. We say that the line $x = 0$ is a **vertical asymptote**

x	$\dfrac{1}{x}$	$\dfrac{1}{x^2}$
0.001	1000	1,000,000
0.01	100	10,000
0.1	10	100
1	1	1
2	0.5	0.25
4	0.25	0.06

for each of these graphs. Similarly, we note that the line $y = 0$ is a **horizontal asymptote** in both cases. We will later show that the x-axis is a horizontal asymptote for any rational function for which the numerator is a constant and the denominator is a polynomial of degree 1 or higher.

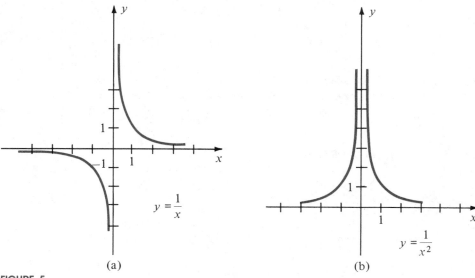

$$y = \frac{1}{x}$$

(a)

$$y = \frac{1}{x^2}$$

(b)

FIGURE 5

The determination of asymptotes is extremely helpful in the graphing of rational functions. The following theorem provides the means for finding all vertical asymptotes.

Vertical Asymptote Theorem	The graph of the rational function $$f(x) = \frac{P(x)}{Q(x)}$$ has a vertical asymptote at $x = r$ if r is a real root of $Q(x)$ but not of $P(x)$.

EXAMPLE 2

Determine the vertical asymptotes of the graph of the function

$$T(x) = \frac{2}{x^3 - 2x^2 - 3x}$$

SOLUTION

Factoring the denominator, we have

$$T(x) = \frac{2}{x(x + 1)(x - 3)}$$

and we conclude that $x = 0$, $x = -1$, and $x = 3$ are vertical asymptotes of the graph of T.

Let's examine the behavior of the function $T(x)$ of Example 2 when x is in the neighborhood of $+3$. When x is slightly more than $+3$, $x - 3$ is positive, as are x and $x + 1$; therefore, $T(x)$ is positive and growing larger and larger as x gets closer and closer to $+3$. When x is slightly less than $+3$, $x - 3$ is negative, but both x and $x + 1$ are positive; therefore, $T(x)$ is negative and growing smaller and smaller as x gets closer and closer to $+3$. This leads to the portion of the graph of $T(x)$ shown in Figure 6c. Similarly, when x is slightly more than 0, $T(x)$ is negative, and when x is slightly less than 0, $T(x)$ is positive (Figure 6b). The behavior of $T(x)$ when x is close to -1 is shown in Figure 6a. Since the numerator of $T(x)$ is constant, $T(x) \neq 0$ for any value of x and the graph of $T(x)$ does not cross the x-axis. Moreover, since $T(x)$ is of the form $k/Q(x)$, where k is a constant, the x-axis is a horizontal asymptote. Combining these observations with the portions of the graph of $T(x)$ sketched in Figure 6 leads to the graph of $T(x)$ sketched in Figure 7.

EXAMPLE 3

Sketch the graphs of the rational functions

(a) $F(x) = \dfrac{1}{x - 1}$ (b) $G(x) = \dfrac{1}{(x + 2)^2}$

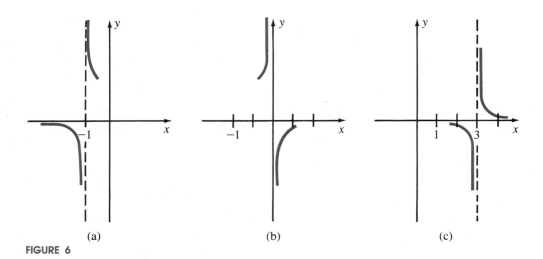

(a)   (b)   (c)

FIGURE 6

SOLUTION

The graphs are shown in Figure 8. Note that the graphs are identical to those of Figure 5 with all points moved right one unit in the case of F and moved two units left in the case of G. In both cases we say that the y-axis has been **translated.**

    The x-axis is a horizontal asymptote for each of the rational functions sketched in Figures 5, 6, and 7. In general, we can determine the existence of a horizontal asymptote by studying the behavior of a rational function as x ap-

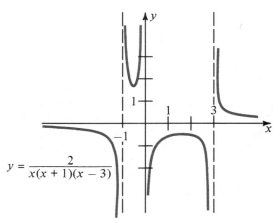

$$y = \frac{2}{x(x + 1)(x - 3)}$$

FIGURE 7

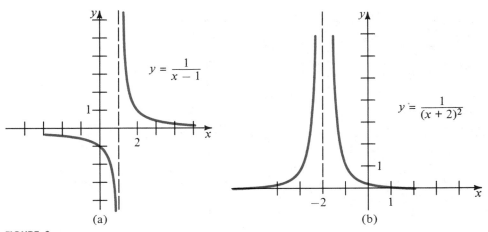

FIGURE 8

proaches $+\infty$ and $-\infty$, that is, as $|x|$ becomes very large. Recall that the expression

$$\frac{k}{x^n}$$

where n is a positive integer and k is a constant, will become very small as $|x|$ becomes very large; that is, k/x^n approaches 0 as $|x|$ approaches $+\infty$. The procedure for determining horizontal asymptotes employs the technique used earlier of factoring out the highest power of x to determine the behavior of the function as $|x|$ becomes large.

EXAMPLE 4
Determine the horizontal asymptote of the function

$$f(x) = \frac{2x^2 - 5}{3x^2 + 2x - 4}$$

SOLUTION
We illustrate the steps of the procedure.

Horizontal Asymptotes	
Step 1. Factor out the highest power of x found in the numerator; factor out the highest power of x found in the denominator.	*Step 1.* $$f(x) = \frac{x^2\left(2 - \dfrac{5}{x^2}\right)}{x^2\left(3 + \dfrac{2}{x} - \dfrac{4}{x^2}\right)}$$

Step 2. Since we are interested in large values of $|x|$, we may cancel common factors in the numerator and denominator.

Step 2.

$$f(x) = \frac{2 - \dfrac{5}{x^2}}{3 + \dfrac{2}{x} - \dfrac{4}{x^2}}, \quad x \neq 0$$

Step 3. Let $|x|$ increase. Then all terms of the form k/x^n approach 0 and may be discarded.

Step 3. The terms $-\dfrac{5}{x^2}$, $\dfrac{2}{x}$, and $-\dfrac{4}{x^2}$ approach 0 as $|x|$ approaches $+\infty$.

Step 4. If what remains is a real number c, then $y = c$ is the horizontal asymptote. Otherwise there is no horizontal asymptote.

Step 4. Discarding these terms, we have $y = \frac{2}{3}$ as the horizontal asymptote.

EXAMPLE 5

Determine the horizontal asymptote of the function

$$f(x) = \frac{2x^3 + 3x - 2}{x^2 + 5}$$

if there is one.

SOLUTION

Factoring, we have

$$f(x) = \frac{x^3\left(2 + \dfrac{3}{x^2} - \dfrac{2}{x^3}\right)}{x^2\left(1 + \dfrac{5}{x^2}\right)}$$

$$= \frac{x\left(2 + \dfrac{3}{x^2} - \dfrac{2}{x^3}\right)}{1 + \dfrac{5}{x^2}}, \quad x \neq 0$$

As $|x|$ increases, the terms $3/x^2$, $-2/x^3$, and $5/x^2$ approach zero and can be discarded. What remains is $2x$, which becomes larger and larger as $|x|$ increases. Thus, there is no horizontal asymptote, and $|y|$ becomes larger and larger as $|x|$ approaches infinity.

The following theorem can be proved by utilizing the procedure of Example 4.

Horizontal Asymptote Theorem	The graph of the rational function

$$f(x) = \frac{P(x)}{Q(x)}$$

has a horizontal asymptote if the degree of $P(x)$ is less than or equal to the degree of $Q(x)$.

Note that the graph of a rational function may have many vertical asymptotes but at most one horizontal asymptote.

PROGRESS CHECK

Determine the horizontal asymptote of the graph of each function.

(a) $f(x) = \dfrac{x - 1}{2x^2 + 1}$ (b) $g(x) = \dfrac{4x^2 - 3x + 1}{-3x^2 + 1}$

(c) $h(x) = \dfrac{3x^3 - x + 1}{2x^2 - 1}$

ANSWERS

(a) $y = 0$ (b) $y = -\frac{4}{3}$ (c) no horizontal asymptote

SKETCHING GRAPHS

We now summarize the information that can be gathered in preparation for sketching the graph of a rational function.

- symmetry with respect to the axes and the origin
- x-intercepts
- vertical asymptotes
- horizontal asymptotes
- brief table of values including points near the vertical asymptotes

EXAMPLE 6

Sketch the graph of

$$f(x) = \frac{x^2}{x^2 - 1}$$

SOLUTION

Symmetry. Replacing x with $-x$ results in the same equation, establishing symmetry with respect to the y-axis.

Intercepts. Setting the numerator equal to zero, we see that the graph of f crosses the x-axis at the point $(0, 0)$.

Vertical asymptotes. Setting the denominator equal to zero, we find that $x = 1$ and $x = -1$ are vertical asymptotes of the graph of f.

Horizontal asymptotes. We note that

$$f(x) = \frac{x^2}{x^2\left(1 - \dfrac{1}{x^2}\right)} = \frac{1}{1 - \dfrac{1}{x^2}}, \quad x \neq 0$$

As $|x|$ gets larger and larger, $1/x^2$ approaches 0 and the values of $f(x)$ approach 1. Thus, $y = 1$ is the horizontal asymptote. Plotting a few points, we sketch the graph of Figure 9.

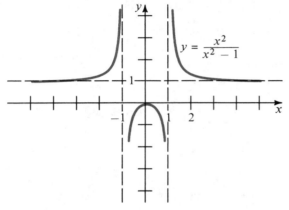

x	y
1/2	−0.33
3/4	−1.29
5/4	2.78
3/2	1.80
2	1.33

FIGURE 9

PROGRESS CHECK

Sketch the graph of

$$f(x) = \frac{x^2 - x - 6}{x^2 - 2x}$$

ANSWER

See Figure 10.

We conclude this section with an example of a rational function that is not in reduced form, that is, one in which the numerator and denominator have a common factor. Rational functions like this are often used in calculus courses to illustrate functions that have "holes" in their graphs and therefore are not continuous functions.

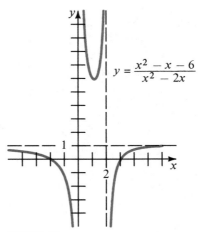

$$y = \frac{x^2 - x - 6}{x^2 - 2x}$$

FIGURE 10

EXAMPLE 7
Sketch the graph of the function

$$f(x) = \frac{x^2 - 1}{x - 1}$$

SOLUTION
We observe that

$$f(x) = \frac{x^2 - 1}{x - 1} = \frac{(x + 1)(x - 1)}{x - 1} = x + 1, \quad x \neq 1$$

Thus, the graph of the function f coincides with the straight line $y = x + 1$, with the exception that f is undefined when $x = 1$ (Figure 11).

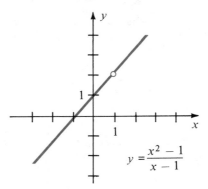

$$y = \frac{x^2 - 1}{x - 1}$$

FIGURE 11

PROGRESS CHECK
Sketch the graph of the function

$$f(x) = \frac{8 - 2x^2}{x + 2}$$

ANSWER
See Figure 12.

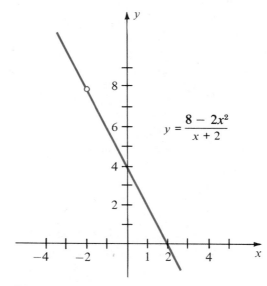

$$y = \frac{8 - 2x^2}{x + 2}$$

FIGURE 12

EXERCISE SET 3.4

In Exercises 1–6 determine the domain of the given function.

1. $f(x) = \dfrac{x^2}{x - 1}$

2. $f(x) = \dfrac{x - 1}{x^2 + x - 2}$

3. $g(x) = \dfrac{x^2 + 1}{x^2 - 2x}$

4. $g(x) = \dfrac{x^2 + 2}{x^2 - 2}$

5. $F(x) = \dfrac{x^2 - 3}{x^2 + 3}$

6. $T(x) = \dfrac{3x + 2}{2x^3 - x^2 - x}$

In Exercises 7–21 determine the vertical and horizontal asymptotes of the graph of the given function. Sketch the graph.

7. $f(x) = \dfrac{1}{x - 4}$

8. $f(x) = \dfrac{-2}{x - 3}$

9. $f(x) = \dfrac{3}{x + 2}$

10. $f(x) = \dfrac{-1}{(x - 1)^2}$

11. $f(x) = \dfrac{1}{(x + 1)^2}$

12. $f(x) = \dfrac{-1}{x^2 + 1}$

13. $f(x) = \dfrac{x + 2}{x - 2}$

14. $f(x) = \dfrac{x}{x + 2}$

15. $f(x) = \dfrac{2x^2 + 1}{x^2 - 4}$

16. $f(x) = \dfrac{x^2 + 1}{x^2 + 2x - 3}$

17. $f(x) = \dfrac{x^2 + 2}{2x^2 - x - 6}$

18. $f(x) = \dfrac{x^2 - 1}{x + 2}$

19. $f(x) = \dfrac{x^2}{4x - 4}$

20. $f(x) = \dfrac{x - 1}{2x^3 - 2x}$

21. $f(x) = \dfrac{x^3 + 4x^2 + 3x}{x^2 - 25}$

In Exercises 22–27 determine the domain and sketch the graph of the reducible function.

22. $f(x) = \dfrac{x^2 - 25}{2x - 10}$

23. $f(x) = \dfrac{2x^2 - 8}{x + 2}$

24. $f(x) = \dfrac{2x^2 + 2x - 12}{3x - 6}$

25. $f(x) = \dfrac{x^2 + 2x - 8}{2x^2 - 8x + 8}$

26. $f(x) = \dfrac{x + 2}{x^2 - x - 6}$

27. $f(x) = \dfrac{2x}{x^2 + x}$

3.5
PARTIAL FRACTIONS

Every student of algebra learns how to add fractions to arrive at

$$\frac{2}{x - 1} + \frac{3}{x + 2} = \frac{5x + 1}{(x - 1)(x + 2)}$$

Strange as it might seem, there is an important application of integral calculus that requires the *reverse* procedure.

Our objective in this section, then, is to write a rational function $P(x)/Q(x)$ as a sum of fractions, each of which is called a **partial fraction**. To begin, we note that we need only consider the situation where $P(x)/Q(x)$ is a **proper fraction**, that is, where the degree of $P(x)$ is less than the degree of $Q(x)$. If it is not, we can first divide through by $Q(x)$. For example, given the **improper fraction** (one that is not a proper fraction)

$$\frac{2x^3 - x^2 - x + 1}{x^2 - 1}$$

we can use long division to divide through by $x^2 - 1$ to obtain

$$\frac{2x^3 - x^2 - x + 1}{x^2 - 1} = 2x - 1 + \frac{x}{x^2 - 1}$$

We would then work with the proper fraction $x/(x^2 - 1)$.

The procedure for partial-fraction decomposition of the rational function $P(x)/Q(x)$ begins with the factorization of the denominator $Q(x)$. We will show in Chapter 8 that every polynomial with real coefficients can be written as a product of linear and quadratic factors with real coefficients such that the quadratic factors are irreducible (cannot be written as a product of linear factors). For example,

$$Q(x) = x^3 - x^2 + 2x - 2 = (x - 1)(x^2 + 2)$$

where the quadratic factor $x^2 + 2$ cannot be decomposed into linear factors with real coefficients since the equation $x^2 + 2 = 0$ has no real roots.

Having factored $Q(x)$ in this manner, we then collect the repeated factors so that $Q(x)$ is the product of distinct factors of the forms

$$(ax + b)^m \quad \text{and} \quad (ax^2 + bx + c)^m$$

The following rules then tell us the format of the partial-fraction decomposition of $P(x)/Q(x)$.

Rules for Partial-Fraction Decomposition

Rule 1: Linear Factors
For each distinct factor of the form $(ax + b)^m$ introduce the sum of m partial fractions

$$\frac{A_1}{ax + b} + \frac{A_2}{(ax + b)^2} + \cdots + \frac{A_m}{(ax + b)^m}$$

where $A_1, A_2, \ldots, A_m$ are constants.

Rule 2: Quadratic Factors
For each distinct factor of the form $(ax^2 + bx + c)^m$ introduce the sum of m partial fractions

$$\frac{A_1 x + B_1}{ax^2 + bx + c} + \frac{A_2 x + B_2}{(ax^2 + bx + c)^2} + \cdots + \frac{A_m x + B_m}{(ax^2 + bx + c)^m}$$

where $A_1, A_2, \ldots, A_m$ and $B_1, B_2, \ldots, B_m$ are constants.

What remains is the straightforward (if tedious) algebraic manipulation to determine the constants $A_1, A_2, \ldots, A_m$ and $B_1, B_2, \ldots, B_m$.

EXAMPLE 1

Find the partial-fraction decomposition of $\dfrac{x + 5}{x^2 + x - 2}$.

SOLUTION
Factoring the denominator gives us

$$\frac{x + 5}{x^2 + x - 2} = \frac{x + 5}{(x + 2)(x - 1)}$$

The denominator consists of linear factors. By Rule 1, each linear factor introduces one term:

Factor	$x + 2$	$x - 1$
Term	$\dfrac{A}{x + 2}$	$\dfrac{B}{x - 1}$

The partial-fraction decomposition is then

$$\frac{x + 5}{(x + 2)(x - 1)} = \frac{A}{x + 2} + \frac{B}{x - 1}$$

To solve for the constants A and B, we clear fractions by multiplying both sides by $(x + 2)(x - 1)$ to yield

$$x + 5 = A(x - 1) + B(x + 2)$$

This last equation is easily solved by substituting values of x that will make one unknown drop out. Setting $x = 1$ gives us

$$6 = 3B \quad \text{or} \quad B = 2$$

Setting $x = -2$ yields

$$3 = -3A \quad \text{or} \quad A = -1$$

Substituting these values for A and B, we arrive at the partial-fraction decomposition

$$\frac{x + 5}{(x + 2)(x - 1)} = \frac{-1}{x + 2} + \frac{2}{x - 1}$$

The student should verify that this is indeed an *identity*.

EXAMPLE 2

Find the partial-fraction decomposition of $\dfrac{x^2 - 2x - 9}{x(x + 3)^2}$.

SOLUTION
The denominator is already in factored form and consists of the linear factor x and the repeated linear factor $x + 3$. By Rule 1, the factor x introduces one term

$$\frac{A}{x}$$

and the factor $x + 3$ introduces two terms (since $m = 2$)

$$\frac{B}{x + 3} \quad \text{and} \quad \frac{C}{(x + 3)^2}$$

The partial-fraction decomposition is then

$$\frac{x^2 - 2x - 9}{x(x + 3)^2} = \frac{A}{x} + \frac{B}{x + 3} + \frac{C}{(x + 3)^2}$$

Clearing fractions by multiplying both sides by $x(x + 3)^2$ yields

$$x^2 - 2x - 9 = A(x + 3)^2 + Bx(x + 3) + Cx \tag{1}$$

Setting $x = -3$ in Equation (1) allows us to solve for C:

$$6 = -3C \quad \text{or} \quad C = -2$$

Setting $x = 0$ in Equation (1) allows us to solve for A:

$$-9 = 9A \quad \text{or} \quad A = -1$$

Expanding the right-hand side of Equation (1) and collecting terms in the powers of x, we have

$$x^2 - 2x - 9 = (A + B)x^2 + (6A + 3B + C)x + 9A \tag{2}$$

Substituting $A = -1$ and $C = -2$ in Equation (2) yields

$$x^2 - 2x - 9 = (B - 1)x^2 + (3B - 8)x - 9$$

Equating the coefficients of the terms in x^2, we have

$$1 = B - 1 \quad \text{or} \quad B = 2$$

(The same result is obtained by equating the coefficients of the terms in x.) The partial-fraction decomposition is

$$\frac{x^2 - 2x - 9}{x(x + 3)^2} = -\frac{1}{x} + \frac{2}{x + 3} - \frac{2}{(x + 3)^2}$$

WARNING For the rational function

$$\frac{2x - 1}{x^2(x^2 + 1)}$$

the factor x^2 in the denominator must be treated as the repeated *linear* factor $(x - 0)^2$. The partial-fraction decomposition has the structure

$$\frac{2x - 1}{x^2(x^2 + 1)} = \frac{A}{x} + \frac{B}{x^2} + \frac{Cx + D}{x^2 + 1}$$

EXAMPLE 3

Find the partial-fraction decomposition of $\dfrac{x^2 - 5x + 1}{2x^3 - x^2 + 2x - 1}$.

SOLUTION

The first task is to factor the denominator. By the Factor Theorem, $x - r$ is a factor of

$$Q(x) = 2x^3 - x^2 + 2x - 1$$

if and only if r is a root of $Q(x)$. The Rational Root Theorem tells us that the possible rational roots are 1, -1, $\frac{1}{2}$, and $-\frac{1}{2}$. Using the condensed form of synthetic division to test these possible roots

	2	−1	2	−1
1	2	1	3	2
−1	2	−3	5	−6
$\frac{1}{2}$	2	0	2	0

coefficients of depressed equation

we see that $Q(\frac{1}{2}) = 0$, so $x = \frac{1}{2}$ is a root of $Q(x)$. Consequently, $(x - \frac{1}{2})$, or $(2x - 1)$, is a factor of $Q(x)$, and $2x^2 + 2 = 0$ is seen to be the depressed equation.

By Rule 1, the factor $2x - 1$ will introduce a term of the form

$$\frac{A}{2x - 1}$$

and by Rule 2, the factor $x^2 + 1$ will introduce a term of the form

$$\frac{Bx + C}{x^2 + 1}$$

Then

$$\frac{x^2 - 5x + 1}{(2x - 1)(x^2 + 1)} = \frac{A}{2x - 1} + \frac{Bx + C}{x^2 + 1}$$

Multiplying by $(2x - 1)(x^2 + 1)$, we have

$$x^2 - 5x + 1 = A(x^2 + 1) + (Bx + C)(2x - 1)$$

To find A, B, and C, we first set $2x - 1 = 0$ or $x = \frac{1}{2}$:

$$-\frac{5}{4} = \frac{5}{4}A \quad \text{or} \quad A = -1$$

Equating the coefficients of x^2 yields

$$1 = A + 2B$$

Substituting $A = -1$,

$$1 = -1 + 2B$$
$$B = 1$$

Equating the coefficients of the constant term yields

$$1 = A - C$$
$$1 = -1 - C$$
$$C = -2$$

The partial-fraction decomposition is therefore

$$\frac{x^2 - 5x + 1}{2x^3 - x^2 + 2x - 1} = \frac{-1}{2x - 1} + \frac{x - 2}{x^2 + 1}$$

EXAMPLE 4

Find the partial-fraction decomposition of $\dfrac{x^2 - 2x}{(x + 2)(x^2 + 4)^2}$.

SOLUTION

The linear factor $x + 2$ introduces one term of the form

$$\frac{A}{x + 2}$$

The quadratic factor $x^2 + 4$ has no real roots. By Rule 2, this irreducible quadratic factor will introduce two terms (since $m = 2$) of the form

$$\frac{Bx + C}{x^2 + 4} \quad \text{and} \quad \frac{Dx + E}{(x^2 + 4)^2}$$

We then have to solve

$$\frac{x^2 - 2x}{(x + 2)(x^2 + 4)^2} = \frac{A}{x + 2} + \frac{Bx + C}{x^2 + 4} + \frac{Dx + E}{(x^2 + 4)^2}$$

for values of A, B, C, D, and E that will produce an identity. Multiplying by $(x + 2)(x^2 + 4)^2$ yields

$$x^2 - 2x = A(x^2 + 4)^2 + (Bx + C)(x + 2)(x^2 + 4) + (Dx + E)(x + 2) \quad (3)$$

Setting $x = -2$ enables us to solve for A:

$$8 = 64A \quad \text{or} \quad A = \frac{1}{8}$$

A methodical way to solve for B, C, D, and E is to successively equate the coefficients of the powers x^4, x^3, and x^2 in the left-hand and right-hand sides of Equation (3).

$$\text{Coefficients of } x^4: 0 = A + B \quad \text{or} \quad B = -A = -\frac{1}{8}$$

$$\text{Coefficients of } x^3: 0 = 2B + C \quad \text{or} \quad C = -2B = \frac{1}{4}$$

$$\text{Coefficients of } x^2: 1 = 8A + 4B + 2C + D$$

$$1 = 1 - \frac{1}{2} + \frac{1}{2} + D$$

$$D = 0$$

To find E, we may equate the coefficients of x or of the constant term. Choosing the latter approach,

$$0 = 16A + 8C + 2E$$
$$0 = 2 + 2 + 2E$$
$$E = -2$$

The partial-fraction decomposition is

$$\frac{x^2 - 2x}{(x + 2)(x^2 + 4)^2} = \frac{\dfrac{1}{8}}{x + 2} + \frac{-\dfrac{1}{8}x + \dfrac{1}{4}}{x^2 + 4} + \frac{-2}{(x^2 + 4)^2}$$

EXERCISE SET 3.5

Find the partial-fraction decomposition of each of the following.

1. $\dfrac{2x - 11}{(x + 2)(x - 3)}$

2. $\dfrac{1}{x^2 + 3x + 2}$

3. $\dfrac{3x - 2}{6x^2 - 5x + 1}$

4. $\dfrac{2x + 1}{x^2 - 1}$

5. $\dfrac{x^2 + x + 2}{x^3 - x}$

6. $\dfrac{3x - 14}{(x - 3)(x^2 - 4)}$

7. $\dfrac{3x - 2}{x^3 + 2x^2}$

8. $\dfrac{4x^2 - 5x + 2}{2x^3 - x^2}$

9. $\dfrac{x^2 - x + 2}{(x - 1)(x + 1)^2}$

10. $\dfrac{3x - 1}{x(x - 1)^3}$

11. $\dfrac{1 - 2x}{x^3 + 4x}$

12. $\dfrac{x^2 - 2x + 1}{x^3 + 2x^2 + 2x}$

13. $\dfrac{2x^3 - x^2 + x}{(x^2 + 3)^2}$

14. $\dfrac{x^2 + 2x + 10}{x^3 - 2x^2 + 2x - 4}$

15. $\dfrac{-x}{x^3 - 2x^2 - 4x - 1}$

16. $\dfrac{x^3 - 2x^2 + 1}{x^4 + 2x^2 + 1}$

17. $\dfrac{x^4 - x^2 - 9}{(x + 1)(x^2 + 2)^2}$

18. $\dfrac{-x^3 - x^2 + 5x + 1}{(2x - 1)(x^2 + 1)^2}$

19. $\dfrac{x^4 + x^3 + x^2 + 3x - 2}{(x + 1)(x^2 + 1)}$

20. $\dfrac{x^3 - 5x + 5}{(x - 1)^3}$

TERMS AND SYMBOLS

polynomial function of degree n (p. 129)

polynomial equation of degree n (p. 130)

roots or zeros of a polynomial (p. 130)

synthetic division (p. 132)

factor (p. 134)

depressed equation (p. 146)

rational function (p. 149)

vertical asymptote (p. 150)

horizontal asymptote (p. 150)

partial fraction (p. 159)

proper fraction (p. 159)

improper fraction (p. 159)

KEY IDEAS FOR REVIEW

☐ Polynomial division results in a quotient and a remainder, both of which are polynomials. If the remainder is not zero, then the degree of the remainder is less than the degree of the divisor.

☐ Synthetic division is a quick way to divide a polynomial by a first-degree polynomial $x - r$, where r is a real constant.

☐ **Remainder Theorem**
If a polynomial $P(x)$ is divided by $x - r$, the remainder is $P(r)$.

☐ **Factor Theorem**
A polynomial $P(x)$ has a root r if and only if $x - r$ is a factor of $P(x)$.

☐ For large values of $|x|$, a polynomial function is dominated by its lead term.

☐ For a polynomial of odd degree, the "ends" of the graph will extend indefinitely in opposite directions.

☐ For a polynomial of even degree, the "ends" of the graph will both extend upward or will both extend downward.

☐ **Rational Root Theorem**
If p/q is a rational root (in lowest terms) of the polynomial $P(x)$ with integer coefficients, then p is a factor of the constant term a_0 of $P(x)$ and q is a factor of the leading coefficient a_n of $P(x)$.

☐ If r is a real root of the polynomial $P(x)$, then the roots of the depressed equation are the other roots of $P(x)$. The depressed equation can be found by using synthetic division.

☐ If $P(x)$ has integer coefficients, then the Rational Root Theorem enables us to list all possible rational roots of $P(x)$. Synthetic division can then be used to test these potential rational roots, since r is a root if and only if the remainder is zero, that is, if and only if $P(r) = 0$.

☐ Always determine the vertical and horizontal asymptotes of a rational function before attempting to sketch its graph.

☐ A proper fraction can always be written as a sum of partial fractions whose denominators are of the form

$$(ax + b)^m \quad \text{and} \quad (ax^2 + bx + c)^m$$

REVIEW EXERCISES

Solutions to exercises whose numbers are in color are in the Solutions section in the back of the book.

3.1 In Exercises 1 and 2 use synthetic division to find the quotient $Q(x)$ and the constant remainder R when the first polynomial is divided by the second polynomial.

1. $2x^3 + 6x - 4$, $x - 1$
2. $x^4 - 3x^3 + 2x - 5$, $x + 2$

In Exercises 3 and 4 use synthetic division to find $P(2)$ and $P(-1)$.

3. $7x^3 - 3x^2 + 2$
4. $x^5 - 4x^3 + 2x$

In Exercises 5 and 6 use the Factor Theorem to show that the second polynomial is a factor of the first polynomial.

5. $2x^4 + 4x^3 + 3x^2 + 5x - 2$, $x + 2$

6. $2x^3 - 5x^2 + 6x - 2$, $x - \dfrac{1}{2}$

3.2 In Exercises 7 and 8 determine the behavior of the graph of the given polynomial function for large values of $|x|$.

7. $P(x) = -2x^5 + 27x^2 + 100$

8. $P(x) = 4x^3 - 10,000$

3.3 In Exercises 9–11 find all the rational roots of the given equation.

9. $6x^3 - 5x^2 - 33x - 18 = 0$

10. $6x^4 - 7x^3 - 19x^2 - 32x - 12 = 0$

11. $x^4 + 3x^3 + 2x^2 + x - 1 = 0$

In Exercises 12–14 use the given root to assist in finding the remaining roots of the equation.

12. $2x^3 - x^2 - 13x - 6 = 0$; -2

13. $x^3 - 2x^2 - 9x + 4 = 0$; 4

14. $2x^4 - 15x^3 + 34x^2 - 19x - 20 = 0$; $-\dfrac{1}{2}$

3.4 In Exercises 15 and 16 sketch the graph of the given function.

15. $f(x) = \dfrac{x}{x + 1}$

16. $f(x) = \dfrac{x^2}{x + 1}$

3.5 In Exercises 17–19 find the partial-fraction decomposition of the given rational function.

17. $\dfrac{8 - x}{2x^2 + 3x - 2}$

18. $\dfrac{3x^3 + 5x - 1}{(x^2 + 1)^2}$

19. $\dfrac{2x^3 - 3x^2 + 4x - 2}{(x - 1)^2}$

PROGRESS TEST 3A

1. Find the quotient and remainder when $2x^4 - x^2 + 1$ is divided by $x^2 + 2$.

2. Use synthetic division to find the quotient and remainder when $3x^4 - x^3 - 2$ is divided by $x + 2$.

3. If $P(x) = x^3 - 2x^2 + 7x + 5$, use synthetic division to find $P(-2)$.

4. Determine the remainder when $4x^5 - 2x^4 - 5$ is divided by $x + 2$.

5. Use the Factor Theorem to show that $x - 3$ is a factor of

$$2x^4 - 9x^3 + 9x^2 + x - 3$$

Problems 6–8 refer to the polynomial function

$$P(x) = -2x^9 + 3x^6 + 200$$

6. Describe the behavior of the graph of $P(x)$ for large positive values of x.

7. Describe the behavior of the graph of $P(x)$ for large negative values of x.

8. Determine the maximum number of extreme points of the graph of $P(x)$.

In Problems 9 and 10 find all rational roots of the given equation.

9. $6x^3 - 17x^2 + 14x + 3 = 0$

10. $2x^5 - x^4 - 4x^3 + 2x^2 + 2x - 1 = 0$

In Problems 11 and 12 use the given root to help in finding the remaining roots of the equation.

11. $4x^3 - 3x + 1 = 0$; -1

12. $x^4 - x^2 - 2x + 2 = 0$; 1

13. Sketch the graph of the function $f(x) = \dfrac{x^2 + 2}{x^2 - 1}$.

14. Find the partial-fraction decomposition of

$$\dfrac{x - 12}{x^2 + x - 6}$$

PROGRESS TEST 3B

1. Find the quotient and remainder when $3x^5 - x^4 - 5x^3 - x + 1$ is divided by $x^2 - x - 1$.

2. Use synthetic division to find the quotient and remainder when $-2x^3 + 3x^2 - 1$ is divided by $x - 1$.

3. If $P(x) = 2x^4 - 2x^3 + x - 4$, use synthetic division to find $P(-1)$.

4. Determine the remainder when $3x^4 - 5x^3 + 3x^2 + 4$ is divided by $x - 2$.

5. Use the Factor Theorem to show that $x + 2$ is a factor of $x^3 - 4x^2 - 9x + 6$.

Problems 6–8 refer to the polynomial function

$$P(x) = -4x^4 + 1000x - 1$$

6. Describe the behavior of the graph of $P(x)$ for large positive values of x.

7. Describe the behavior of the graph of $P(x)$ for large negative values of x.

8. Find the maximum number of turning points of the graph of $P(x)$.

In Problems 9 and 10 find all rational roots of the given equation.

9. $3x^3 + 7x^2 - 4 = 0$

10. $4x^4 - 4x^3 + x^2 - 4x - 3 = 0$

In Problems 11 and 12 use the given root(s) to help in finding the remaining roots of the equation.

11. $x^3 - x^2 - 8x - 4 = 0; \quad -2$

12. $x^4 - 3x^3 - 22x^2 + 68x - 40 = 0; \quad 2, 5$

13. Sketch the graph of the function $f(x) = \dfrac{2x}{x^2 - 1}$.

14. Find the partial-fraction decomposition of

$$\frac{5x^2 - x - 2}{x^3 + x^2}$$

CHAPTER 4

EXPONENTIAL AND LOGARITHMIC FUNCTIONS

Thus far in our study of algebra we have dealt primarily with functions that are polynomials, or sums, differences, products, quotients, or powers of polynomials. In this chapter we introduce a new type of function, the exponential function, and its inverse, the logarithmic function.

Exponential functions arise in nature and are useful in chemistry, biology, and economics, as well as in mathematics and engineering. We will study applications of exponential functions in calculating such quantities as compound interest and the growth rate of bacteria in a culture medium.

Logarithms can be viewed as another way of writing exponents. Historically, logarithms have been used to simplify calculations; in fact, the slide rule, a device long used by engineers, is based on logarithmic scales. In today's world of inexpensive hand calculators, the need for manipulating logarithms is reduced. The section on computing with logarithms will provide enough background to allow you to use this powerful tool but omits some of the detail found in older textbooks.

**4.1
EXPONENTIAL
FUNCTIONS**

The function $f(x) = 2^x$ is very different from any of the functions we have worked with thus far. Previously, we defined functions by using the basic algebraic operations (addition, subtraction, multiplication, division, powers, and roots). However, $f(x) = 2^x$ has a variable in the exponent and doesn't fall into the class of algebraic functions. Rather, it is our first example of an exponential function.

An **exponential function** has the form

$$f(x) = a^x$$

where $a > 0$, $a \neq 1$. The constant a is called the **base**, and the independent variable x may assume any real value.

**GRAPHS OF EXPONENTIAL
FUNCTIONS**

The best way to become familiar with exponential functions is to sketch their graphs.

EXAMPLE 1
Sketch the graph of $f(x) = 2^x$.

SOLUTION
We let $y = 2^x$ and we form a table of values of x and y. Then we plot these points and sketch the smooth curve as in Figure 1. Note that the x-axis is a horizontal asymptote.

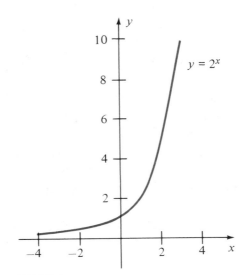

FIGURE 1

In a sense, we have cheated in our definition of $f(x) = 2^x$ and in sketching the graph in Figure 1. Since we have not explained the meaning of 2^x when x is irrational, we have no right to plot values such as $2^{\sqrt{2}}$. For our purposes, however, it will be adequate to think of $2^{\sqrt{2}}$ as the value we approach by taking successively closer approximations to $\sqrt{2}$, such as $2^{1.4}$, $2^{1.41}$, $2^{1.414}$, ... A precise definition is given in more advanced mathematics courses, where it is also shown that the laws of exponents hold for irrational exponents.

We now look at $f(x) = a^x$ when $0 < a < 1$.

EXAMPLE 2

Sketch the graph of $f(x) = \left(\dfrac{1}{2}\right)^x = 2^{-x}.$

SOLUTION

We form a table, plot points, and sketch the graph shown in Figure 2. Note that the graph of $y = 2^{-x}$ is a reflection about the y-axis of the graph of $y = 2^x$.

In Figure 3 we have sketched the graphs of

$$f(x) = 2^x \quad g(x) = 3^x \quad h(x) = \left(\frac{1}{2}\right)^x \quad k(x) = \left(\frac{1}{3}\right)^x$$

on the same coordinate axes to provide additional examples of the graphs of exponential functions.

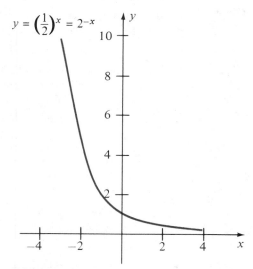

FIGURE 2

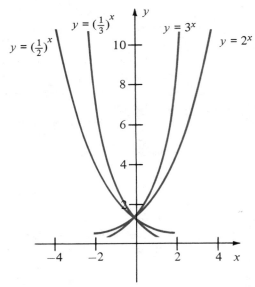

FIGURE 3

PROPERTIES OF THE EXPONENTIAL FUNCTIONS

The graphs in Figure 3 illustrate these important properties of the exponential functions. (Recall that the definition of the exponential function $f(x) = a^x$ requires that $a > 0$ and $a \neq 1$.)

Properties of the Exponential Functions

- The graph of $f(x) = a^x$ always passes through the point $(0, 1)$ since $a^0 = 1$.
- The domain of $f(x) = a^x$ consists of the set of all real numbers; the range is the set of all positive real numbers.
- If $a > 1$, a^x is an increasing function; if $a < 1$, a^x is a decreasing function.
- If $a < b$, then $a^x < b^x$ for all $x > 0$, and $a^x > b^x$ for all $x < 0$. Note in Figure 3 that $y = 3^x$ lies above $y = 2^x$ when $x > 0$ and below when $x < 0$.

Since a^x is either increasing or decreasing, it never assumes the same value twice. (Recall that $a \neq 1$.) This leads to a useful conclusion.

If $a^u = a^v$, then $u = v$.

The graphs of a^x and b^x intersect only at $x = 0$. This observation provides us with the following result.

If $a^u = b^u$ for $u \neq 0$, then $a = b$.

EXAMPLE 3
Solve for x.
(a) $3^{10} = 3^{5x}$ (b) $2^7 = (x - 1)^7$

SOLUTION
(a) Since $a^u = a^v$ implies $u = v$, we have

$$10 = 5x$$
$$2 = x$$

(b) Since $a^u = b^u$ implies $a = b$, we have

$$2 = x - 1$$
$$3 = x$$

PROGRESS CHECK
Solve for x.
(a) $2^8 = 2^{x+1}$ (b) $4^{2x+1} = 4^{11}$

ANSWERS
(a) 7   (b) 5

THE NUMBER *e*

There is an irrational number that was first designated by the letter e by the Swiss mathematician Leonhard Euler (1707–1783). The number e is the value approached by the expression

$$\left(1 + \frac{1}{m}\right)^m$$

as m gets larger. The procedure for studying the behavior of this expression as m gets larger and larger is developed in calculus courses. We will simply evaluate this expression for different values of m, as shown in Table 1.

TABLE 1

m	1	2	10	100	1000	10,000	100,000	1,000,000
$\left(1 + \dfrac{1}{m}\right)^m$	2.0	2.25	2.5937	2.7048	2.7169	2.7181	2.7182	2.71828

The function $f(x) = e^x$ is called the **natural exponential function;** we assume that this is the function referred to when someone speaks of "the exponential function." The graphs of $f(x) = e^x$ and $f(x) = e^{-x}$ are shown in Figure 4. Since $e \approx 2.71828$, the graph of $y = e^x$ falls between the graphs of $y = 2^x$ and $y = 3^x$. Table I in the Tables Appendix lists values for e^x and e^{-x}.

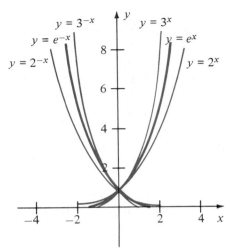

FIGURE 4

APPLICATIONS

Exponential functions occur in a wide variety of applied problems. We will look at problems dealing with population growth, such as predicting the growth of bacteria in a culture medium; radioactive decay, such as determining the half-life of strontium 90; and the interest earned when an interest rate is compounded.

Exponential Growth

The function Q defined by

$$Q(t) = q_0 e^{kt}, \quad k > 0$$

in which the variable t represents time, is called an **exponential growth model**; k is a constant and t is the independent variable. We may think of Q as the quantity of a substance available at any given time t. Note that when $t = 0$ we have

$$Q(0) = q_0 e^0 = q_0$$

which says that q_0 is the initial quantity. (It is customary to use the subscript 0 to denote an initial value.) The constant k is called the **growth constant.**

EXAMPLE 4
The number of bacteria in a culture after t hours is described by the exponential growth model

$$Q(t) = 50e^{0.7t}$$

(a) Find the initial number of bacteria, q_0, in the culture.
(b) How many bacteria are in the culture after 10 hours?

SOLUTION

(a)   To find q_0 we need to evaluate $Q(t)$ at $t = 0$.

$$Q(0) = 50e^{0.7(0)} = 50e^0 = 50 = q_0$$

Thus, there are initially 50 bacteria in the culture.

(b)   The number of bacteria in the culture after 10 hours is given by $Q(10)$.

$$Q(10) = 50e^{0.7(10)} = 50e^7 = 50(1096.6) = 54,830$$

Thus, there are 54,830 bacteria after 10 hours. (The value $e^7 = 1096.6$ can be found by using Table I in the Appendix; it can also be found by using a calculator with a "y^x" key, with $y = e = 2.71828$ and $x = 7$.)

PROGRESS CHECK

The number of bacteria in a culture after t minutes is described by the exponential growth model $Q(t) = q_0e^{0.005t}$. If there were 100 bacteria present initially, how many bacteria will be present after 1 hour has elapsed?

ANSWER

135

Exponential Decay

The model defined by the function

$$Q(t) = q_0e^{-kt}, \quad k > 0$$

is called an **exponential decay model**; k is a constant, called the **decay constant,** and t is the independent variable denoting time. Here is an application of this model.

EXAMPLE 5

A substance has a decay rate of 5% per hour. If 500 grams are present initially, how much of the substance remains after 4 hours?

SOLUTION

The general equation of an exponential decay model is

$$Q(t) = q_0e^{-kt}$$

In our model, $q_0 = 500$ grams (since the quantity available initially is 500 grams), and $k = 0.05$ (since the decay rate is 5% per hour). After 4 hours

$$Q(4) = 500e^{-0.05(4)} = 500e^{-0.2} = 500(0.8187) = 409.4$$

($e^{-0.2} = 0.8187$ is obtained from Table I in the Tables Appendix). Thus, there remain 409.4 grams of the substance.

PROGRESS CHECK

The number of grams Q of a certain radioactive substance present after t seconds is given by the exponential decay model $Q(t) = q_0 e^{-0.4t}$. If 200 grams of the substance are present initially, find how much remains after 6 seconds.

ANSWER

18.1 grams

Compound Interest

We begin by recalling the definition of simple interest. If the principal P is invested at a simple annual interest rate r, then the amount or sum S that we will have after t years is given by

$$S = P + Prt$$

In many business transactions the interest that is added to the principal at regular time intervals also earns interest. This is called the **compound interest** process.

The time period between successive additions of interest is known as the **conversion period.** If interest is compounded quarterly, the conversion period is three months; if interest is compounded semiannually, the conversion period is six months.

Suppose now that a principal P is invested at an annual interest rate r, compounded k times a year. Then each conversion period lasts $t = 1/k$ years. Thus, the amount S_1 at the end of the first conversion period is

$$S_1 = P + Prt$$
$$= P + P \cdot r \cdot \frac{1}{k} = P\left(1 + \frac{r}{k}\right)$$

The amount S_2 at the end of the second conversion period is

$$S_2 = S_1 + S_1 rt$$
$$= P\left(1 + \frac{r}{k}\right) + P\left(1 + \frac{r}{k}\right) \cdot r \cdot \frac{1}{k}$$
$$= \left[P\left(1 + \frac{r}{k}\right)\right]\left(1 + \frac{r}{k}\right)$$
$$S_2 = P\left(1 + \frac{r}{k}\right)^2$$

In this way, we see that the amount S_n after n conversion periods is given by

$$S_n = P\left(1 + \frac{r}{k}\right)^n$$

which is usually written

$$S = P(1 + i)^n$$

where $i = r/k$. Table IV in the Tables Appendix gives values of $(1 + i)^n$ for a number of values of i and n. Accurate results can also be obtained by using a calculator with a "y^x" key.

EXAMPLE 6

Suppose that $6000 is invested at an annual interest rate of 8%. What will the value of the investment be after 3 years if
(a)   interest is compounded quarterly?
(b)   interest is compounded semiannually?

SOLUTION

(a)   We are given $P = 6000$, $r = 0.08$, $k = 4$, and $n = 12$ (since there are 4 conversion periods per year for 3 years). Thus,

$$i = \frac{r}{k} = \frac{0.08}{4} = 0.02$$

and

$$S = P(1 + i)^n = 6000(1 + 0.02)^{12}$$

Table IV in the Tables Appendix, with $i = r/k = 0.02 = 2\%$ and $n = 12$, yields

$$S = 6000(1.26824179) = 7609.45$$

Thus, the sum at the end of the three-year period will be $7609.45.
(b)   We have $P = 6000$, $r = 0.08$, $k = 2$, and $n = 6$ (since there are 2 conversion periods per year for 3 years). Then

$$i = \frac{r}{k} = \frac{0.08}{2} = 0.04$$

$$S = P(1 + i)^n = 6000(1 + 0.04)^6$$

Table IV in the Tables Appendix, with $i = 0.04 = 4\%$ and $n = 6$, yields

$$S = 6000(1.26531902) = 7591.91$$

The sum at the end of the three-year period will be $7591.91, which is $17.54 less than the interest earned when compounding is quarterly rather than semiannual.

WARNING When using Table IV, be certain that n is the total number of conversion periods and i is the interest rate per conversion period. For example, an interest rate of 18% compounded monthly for 2 years leads to $n = 24$ and $i = 1\frac{1}{2}\%$.

Continuous Compounding

When P, r, and t are held fixed and the frequency of compounding is increased, the return on the investment is increased. We wish to determine the effect of making the number of conversions per year larger and larger.

Suppose a principal P is invested at an annual rate r, compounded k times per year. After t years, the number of conversions is $n = tk$. Then, the value of the investment after t years is

$$S = P\left(1 + \frac{r}{k}\right)^{tk}$$

Letting $m = k/r$, we can rewrite this equation as

$$S = P\left(1 + \frac{1}{m}\right)^{tmr}$$

or

$$S = P\left[\left(1 + \frac{1}{m}\right)^{m}\right]^{rt}$$

If the number of conversions k per year gets larger and larger, then m gets larger and larger. Since we saw in Table 1 that the expression

$$\left(1 + \frac{1}{m}\right)^{m}$$

gets closer and closer to e as m gets larger and larger, we conclude that

$$S = Pe^{rt} \tag{1}$$

As the number of conversions increases, so does the value of the investment. But there is a limit, or bound, to this value, and it is given by Equation (1). We say that Equation (1) represents the result of **continuous compounding.**

EXAMPLE 7

Suppose that $20,000 is invested at an annual interest rate of 7% compounded continuously. What is the value of the investment after 4 years?

SOLUTION

We have $P = 20,000$, $r = 0.07$, and $t = 4$, and we substitute in Equation (1).

$$S = Pe^{rt}$$
$$= 20,000e^{0.07(4)} = 20,000e^{0.28}$$
$$= 20,000(1.3231) \qquad \text{from Table I, Tables Appendix, or a calculator}$$
$$= 26,462$$

The sum available after 4 years is $26,462.

PROGRESS CHECK

Suppose that $10,000 is invested at an annual interest rate of 10% compounded continuously. What is the value of the investment after 6 years?

ANSWER
$18,221

By solving Equation (1) for P, we can determine the principal P that must be invested at continuous compounding to have a certain amount S at some future time. The values of e^{-x} from Table I in the Tables Appendix will be used in this connection.

EXAMPLE 8

Suppose that a principal P is to be invested at continuous compound interest of 8% per year to yield $10,000 in 5 years. Approximately how much should be invested?

SOLUTION

Using Equation (1) with $S = 10,000$, $r = 0.08$, and $t = 5$, we have

$$S = Pe^{rt}$$
$$10,000 = Pe^{0.08(5)} = Pe^{0.40}$$
$$P = \frac{10,000}{e^{0.40}}$$
$$= 10,000e^{-0.40}$$
$$= 10,000(0.6703) \qquad \text{from Table I, Tables Appendix, or a calculator}$$
$$= 6703$$

Thus, approximately $6703 should be invested initially.

PROGRESS CHECK

Approximately how much money should a 35-year-old woman invest now at continuous compound interest of 10% per year to obtain the sum of $20,000 upon her retirement at age 65?

ANSWER

$996

EXERCISE SET 4.1

In Exercises 1–12 sketch the graph of the given function f.

1. $f(x) = 4^x$
2. $f(x) = 4^{-x}$
3. $f(x) = 10^x$
4. $f(x) = 10^{-x}$
5. $f(x) = 2^{x+1}$
6. $f(x) = 2^{x-1}$
7. $f(x) = 2^{|x|}$
8. $f(x) = 2^{-|x|}$
9. $f(x) = 2^{2x}$
10. $f(x) = 3^{-2x}$
11. $f(x) = e^{x+1}$
12. $f(x) = e^{-2x}$

In Exercises 13–20 solve for x.

13. $2^x = 2^3$
14. $2^{x-1} = 2^4$
15. $3^x = 9^{x-2}$
16. $2^x = 8^{x+2}$
17. $2^{3x} = 4^{x+1}$
18. $3^{4x} = 9^{x-1}$
19. $e^{x-1} = e^3$
20. $e^{x-1} = 1$

In Exercises 21–24 solve for a.

21. $(a + 1)^x = (2a - 1)^x$
22. $(2a + 1)^x = (a + 4)^x$
23. $(a + 1)^x = (2a)^x$
24. $(2a + 3)^x = (3a + 1)^x$

In Exercises 25–29 use Table I in the Tables Appendix to evaluate e^x and e^{-x}.

25. The number of bacteria in a culture after t hours is described by the exponential growth model $Q(t) = 200e^{0.25t}$.
    (a) What is the initial number of bacteria in the culture?
    (b) Find the number of bacteria in the culture after 20 hours.
    (c) Use Table I in the Tables Appendix to complete the following table.

t	1	4	8	10
Q				

26. The number of bacteria in a culture after t hours is described by the exponential growth model $Q(t) = q_0e^{0.01t}$. If there were 400 bacteria present initially, how many bacteria will be present after 2 *days*?

27. At the beginning of 1975 the world population was approximately 4 billion. Suppose that the population is described by an exponential growth model, and that the rate of growth is 2% per year. Give the approximate world population in the year 2000.

28. The number of grams of potassium 42 present after t hours is given by the exponential decay model $Q(t) = q_0e^{-0.055t}$. If 400 grams of the substance were present initially, how much remains after 10 hours?

29. A radioactive substance has a decay rate of 4% per hour. If 1000 grams are present initially, how much of the substance remains after 10 hours?

 In Exercises 30–33 use Table IV in the Tables Appendix, or a calculator, to assist in the computations.

30. An investor purchases a $12,000 savings certificate paying 10% annual interest compounded semiannually. Find the amount received when the savings certificate is redeemed at the end of 8 years.

31. The parents of a newborn infant place $10,000 in an investment that pays 8% annual interest compounded quarterly. What sum is available at the end of 18 years to finance the child's college education?

32. A widow is offered a choice of two investments. Investment A pays 8% annual interest compounded quarterly, and investment B pays 9% compounded annually. Which investment will yield a greater return?

33. A firm intends to replace its present computer in 5 years. The treasurer suggests that $25,000 be set aside in an investment paying 12% compounded monthly. What sum will be available for the purchase of the new computer?

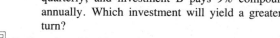 In Exercises 34–38 use Tables I and IV in the Tables Appendix, or a calculator, to assist in the computations.

34. If $5000 is invested at an annual interest rate of 9% compounded continuously, how much is available after 5 years?

35. If $100 is invested at an annual interest rate of 5.5% compounded continuously, how much is available after 10 years?

36. A principal P is to be invested at continuous compound interest of 9% to yield $50,000 in 20 years. What is the approximate value of P to be invested?

37. A 40-year-old executive plans to retire at age 65. How much should be invested at 12% annual interest compounded continuously to provide the sum of $50,000 upon retirement?

38. Investment A offers 8% annual interest compounded semiannually, and investment B offers 8% annual interest compounded continuously. If $1000 were invested in each, what would be the approximate difference in value after 10 years?

In Exercises 39 and 40 use a calculator to determine which number is greater.

39. $2^{\pi}, \pi^2$

40. $3^{\pi}, \pi^3$

4.2 LOGARITHMIC FUNCTIONS

LOGARITHMS AS EXPONENTS

We have previously noted that $f(x) = a^x$ is an increasing function if $a > 1$ and is a decreasing function if $0 < a < 1$. It is then clear that no horizontal line can meet the graph of $f(x) = a^x$ in more than one point. We conclude that the exponential function is a one-to-one function.

In Figure 5a, we see the function $f(x) = 2^x$ assigning values in the set Y for various values of x in the domain X. Since $f(x) = 2^x$ is a one-to-one function, it makes sense to seek a function f^{-1} that will return the values of the range of f back to their origin as in Figure 5b. That is,

$$f \text{ maps } 3 \text{ into } 8, \quad f^{-1} \text{ maps } 8 \text{ into } 3$$
$$f \text{ maps } 4 \text{ into } 16, \quad f^{-1} \text{ maps } 16 \text{ into } 4$$

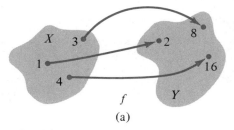

(a)

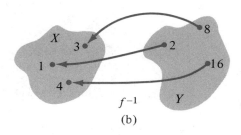

(b)

FIGURE 5

and so on. Since 2^x is always positive, we see that the domain of f^{-1} is the set of all positive real numbers. Its range is the set of all real numbers.

The function f^{-1} of Figure 5b has a special name, the **logarithmic function base 2,** which we write as $\log_2$. It is also possible to generalize and define the logarithmic function as the inverse of the exponential function with any base a such that $a > 0$ and $a \neq 1$.

Logarithmic Function Base a	$y = \log_a x$ if and only if $x = a^y$

When no base is indicated, the notation $\log x$ is interpreted to mean $\log_{10} x$.

The notation **ln x** is used to indicate logarithms to the base e. Since ln x is the inverse of the natural exponential function e^x, it is called the **natural logarithm of x**.

Natural Logarithm	$\ln x = \log_e x$

The exponential form $x = a^y$ and the logarithmic form $y = \log_a x$ are two ways of expressing the same relationship among x, y, and a. Further, it is always possible to convert from one form to the other. The natural question, then, is why bother to create a logarithmic form when we already have an equivalent exponential form? We will show later that the logarithmic function has a number of very useful properties that make its study well worthwhile.

EXAMPLE 1
Write in exponential form.

(a) $\log_3 9 = 2$ (b) $\log_2 \dfrac{1}{8} = -3$

(c) $\log_{16} 4 = \dfrac{1}{2}$ (d) $\ln 7.39 = 2$

SOLUTION
We change from the logarithmic form $\log_a x = y$ to the equivalent exponential form $a^y = x$.

(a) $3^2 = 9$ (b) $2^{-3} = \dfrac{1}{8}$ (c) $16^{1/2} = 4$ (d) $e^2 = 7.39$

PROGRESS CHECK
Write in exponential form.

(a) $\log_4 64 = 3$ (b) $\log_{10} \left(\dfrac{1}{10,000} \right) = -4$

(c) $\log_{25} 5 = \dfrac{1}{2}$ (d) $\ln 0.3679 = -1$

ANSWERS

(a) $4^3 = 64$ (b) $10^{-4} = \dfrac{1}{10,000}$

(c) $25^{1/2} = 5$ (d) $e^{-1} = 0.3679$

EXAMPLE 2
Write in logarithmic form.

(a) $36 = 6^2$ (b) $7 = \sqrt{49}$ (c) $\dfrac{1}{16} = 4^{-2}$ (d) $0.1353 = e^{-2}$

SOLUTION
Since $y = \log_a x$ if and only if $x = a^y$, the logarithmic forms are

(a) $\log_6 36 = 2$ (b) $\log_{49} 7 = \dfrac{1}{2}$

(c) $\log_4 \dfrac{1}{16} = -2$ (d) $\ln 0.1353 = -2$

PROGRESS CHECK
Write in logarithmic form.
(a) $64 = 8^2$ (b) $6 = 36^{1/2}$

(c) $\dfrac{1}{7} = 7^{-1}$ (d) $20.09 = e^3$

ANSWERS

(a) $\log_8 64 = 2$ (b) $\log_{36} 6 = \dfrac{1}{2}$

(c) $\log_7 \dfrac{1}{7} = -1$ (d) $\ln 20.09 = 3$

**LOGARITHMIC
EQUATIONS**

Logarithmic equations can often be solved by changing them to equivalent exponential forms. Here are some straightforward examples; more challenging problems will be handled in Section 4.5.

EXAMPLE 3
Solve for x.
(a) $\log_3 x = -2$ (b) $\log_x 81 = 4$

(c) $\log_5 125 = x$ (d) $\ln x = \dfrac{1}{2}$

SOLUTION
(a) Using the equivalent exponential form,

$$x = 3^{-2} = \frac{1}{9}$$

(b) Changing to the equivalent exponential form,

$$x^4 = 81 = 3^4$$

Since $a^u = b^u$ implies $a = b$, we have

$$x = 3$$

(c)   In exponential form we have

$$5^x = 125$$

Writing 125 to the base 5, we have

$$5^x = 5^3$$

and since $a^u = a^v$ implies $u = v$, we conclude that

$$x = 3$$

(d)   The equivalent exponential form is

$$x = e^{1/2} \approx 1.65$$

which we obtain from Table I in the Tables Appendix or by using a calculator with a "y^x" key.

PROGRESS CHECK
Solve for x.

(a) $\log_x 1000 = 3$ (b) $\log_2 x = 5$ (c) $x = \log_7 \dfrac{1}{49}$

ANSWERS
(a)   10   (b)   32   (c) -2

LOGARITHMIC IDENTITIES

If $f(x) = a^x$, then $f^{-1}(x) = \log_a x$. Recall that inverse functions have the property that

$$f[f^{-1}(x)] = x \quad \text{and} \quad f^{-1}[f(x)] = x$$

Substituting $f(x) = a^x$ and $f^{-1}(x) = \log_a x$, we have

$$f[f^{-1}(x)] = x \qquad f^{-1}[f(x)] = x$$
$$f(\log_a x) = x \qquad f^{-1}(a^x) = x$$
$$a^{\log_a x} = x \qquad \log_a a^x = x$$

These two identities are useful in simplifying expressions and should be remembered.

$$a^{\log_a x} = x$$
$$\log_a a^x = x$$

MEASURING AN EARTHQUAKE

Richter Scale Readings

Here's what you can anticipate from an earthquake of various Richter scale readings.

2.0 not noticed
4.5 some damage in a very limited area
6.0 hazardous; serious damage with destruction of buildings in a limited area
7.0 felt over a wide area with significant damage
8.0 great damage
8.7 maximum recorded

The great San Francisco earthquake of 1906 is estimated to have had a Richter scale reading of 8.3.

Radio and television newscasts often describe earthquakes in this way: "A minor earthquake in China registered 3.0 on the Richter scale," or, "A major earthquake in Chile registered 8.0 on the Richter scale." From statements like this we know that 3.0 is a "low" value and 8.0 is a "high" value. But just what is the Richter scale?

On the Richter scale, the magnitude R of an earthquake is defined as

$$R = \log \frac{I}{I_0}$$

where I_0 is a constant that represents a standard intensity and I is the intensity of the earthquake being measured. The Richter scale is a means of measuring a given earthquake against a "standard earthquake" of intensity I_0.

What does 3.0 on the Richter scale mean? Substituting $R = 3$ in the above equation, we have

$$3 = \log \frac{I}{I_0}$$

or, in the equivalent exponential form,

$$1000 = \frac{I}{I_0}$$

Solving for I,

$$I = 1000 \, I_0$$

which states that an earthquake with a Richter scale reading of 3.0 is 1000 times as intense as the standard! No wonder, then, that an earthquake registering 8.0 on the Richter scale is serious: it has an intensity 100,000,000 times that of the standard!

The following pair of identities can be established by converting to the equivalent exponential form.

$$\log_a a = 1$$
$$\log_a 1 = 0$$

EXAMPLE 4
Evaluate.
(a) $8^{\log_8 5}$ (b) $\log_{10} 10^{-3}$ (c) $\log_7 7$ (d) $\log_4 1$

SOLUTION
(a) 5   (b) -3 (c) 1   (d) 0

PROGRESS CHECK

Evaluate.

(a) $\log_3 3^4$ (b) $6^{\log_6 9}$ (c) $\log_5 1$ (d) $\log_8 8$

ANSWERS

(a) 4   (b) 9   (c) 0   (d) 1

GRAPHS OF THE LOGARITHMIC FUNCTIONS

To sketch the graph of a logarithmic function, we convert to the equivalent exponential form. For example, to sketch the graph of $y = \log_2 x$, we form a table of values for the equivalent exponential equation $x = 2^y$.

y	-3	-2	-1	0	1	2	3
$x = 2^y$	$\dfrac{1}{8}$	$\dfrac{1}{4}$	$\dfrac{1}{2}$	1	2	4	8

We can now plot these points and sketch a smooth curve, as in Figure 6. Note that the y-axis is a vertical asymptote. We have included the graph of $y = 2^x$ to illustrate that the graphs of a pair of inverse functions are reflections of each other about the line $y = x$.

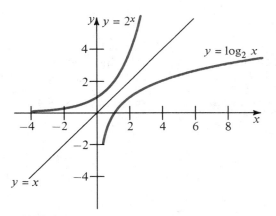

FIGURE 6

EXAMPLE 5

Sketch the graphs of $y = \log_3 x$ and $y = \log_{1/3} x$ on the same coordinate axes.

SOLUTION

The graphs are shown in Figure 7. Practical applications of logarithms generally involve a base $a > 1$.

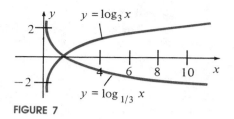

FIGURE 7

PROPERTIES OF LOGARITHMIC FUNCTIONS

The graphs in Figures 6 and 7 illustrate these important properties of logarithmic functions.

Properties of Logarithmic Functions	
	• The point $(1, 0)$ lies on the graph of the function $f(x) = \log_a x$ for all real numbers $a > 0$. This is another way of saying $\log_a 1 = 0$. • The domain of $f(x) = \log_a x$ is the set of all positive real numbers; the range is the set of all real numbers. • When $a > 1$, $f(x) = \log_a x$ is an increasing function; when $0 < a < 1$, $f(x) = \log_a x$ is a decreasing function.

These results are in accord with what we anticipate for a pair of inverse functions. As expected, the domain of the logarithmic function is the range of the corresponding exponential function, and vice versa.

Since $\log_a x$ is either increasing or decreasing, the same value cannot be assumed more than once. Thus,

> If $\log_a u = \log_a v$, then $u = v$.

Since the graphs of $\log_a u$ and $\log_b u$ intersect only at $u = 1$, we have

> If $\log_a u = \log_b u$ and $u \neq 1$, then $a = b$.

EXAMPLE 6
Solve for x.
(a) $\log_5 (x + 1) = \log_5 25$ (b) $\log_{x-1} 31 = \log_5 31$

SOLUTION

(a) Since $\log_a u = \log_a v$ implies $u = v$, then

$$x + 1 = 25$$
$$x = 24$$

(b) Since $\log_a u = \log_b u$, $u \neq 1$, implies $a = b$,

$$x - 1 = 5$$
$$x = 6$$

PROGRESS CHECK

Solve for x.

(a) $\log_2 x^2 = \log_2 9$ (b) $\log_7 14 = \log_{2x} 14$

ANSWERS

(a) $3, -3$ (b) $\dfrac{7}{2}$

EXERCISE SET 4.2

Write each of the following in exponential form.

1. $\log_2 4 = 2$

2. $\log_5 125 = 3$

3. $\log_9 \dfrac{1}{81} = -2$

4. $\log_{64} 4 = \dfrac{1}{3}$

5. $\ln 20.09 = 3$

6. $\ln \dfrac{1}{7.39} = -2$

7. $\log_{10} 1000 = 3$

8. $\log_{10} \dfrac{1}{1000} = -3$

9. $\ln 1 = 0$

10. $\log_{10} 0.01 = -2$

11. $\log_3 \dfrac{1}{27} = -3$

12. $\log_{125} \dfrac{1}{5} = -\dfrac{1}{3}$

Write each of the following in logarithmic form.

13. $25 = 5^2$

14. $27 = 3^3$

15. $10,000 = 10^4$

16. $\dfrac{1}{100} = 10^{-2}$

17. $\dfrac{1}{8} = 2^{-3}$

18. $\dfrac{1}{27} = 3^{-3}$

19. $1 = 2^0$

20. $1 = e^0$

21. $6 = \sqrt{36}$

22. $2 = \sqrt[3]{8}$

23. $64 = 16^{3/2}$

24. $81 = 27^{4/3}$

25. $\dfrac{1}{3} = 27^{-1/3}$

26. $\dfrac{1}{2} = 16^{-1/4}$

Solve for x.

27. $\log_5 x = 2$

28. $\log_{16} x = \dfrac{1}{2}$

29. $\log_{25} x = -\dfrac{1}{2}$

30. $\log_{1/2} x = 3$

31. $\ln x = 2$

32. $\ln x = -3$

33. $\ln x = -\dfrac{1}{2}$

34. $\log_4 64 = x$

35. $\log_5 \dfrac{1}{25} = x$

36. $\log_x 4 = \dfrac{1}{2}$

37. $\log_x \dfrac{1}{8} = -\dfrac{1}{3}$

38. $\log_3(x - 1) = 2$

39. $\log_5(x + 1) = 3$

40. $\log_2(x - 1) = \log_2 10$

41. $\log_{x+1} 24 = \log_3 24$

42. $\log_3 x^3 = \log_3 64$

43. $\log_{x+1} 17 = \log_4 17$

44. $\log_{3x} 18 = \log_4 18$

Evaluate.

45. $3^{\log_3 6}$

46. $2^{\log_2(2/3)}$

47. $e^{\ln 2}$

48. $e^{\ln 1/2}$

49. $\log_5 5^3$

50. $\log_4 4^{-2}$

51. $\log_8 8^{1/2}$

52. $\log_{64} 64^{-1/3}$

53. $\log_7 49$

54. $\log_7 \sqrt{7}$

55. $\log_5 5$

56. $\ln e$

57. $\ln 1$

58. $\log_4 1$

59. $\log_2 \dfrac{1}{4}$

60. $\log_{16} 4$

61. $\log 10{,}000$

62. $\log 0.001$

63. $\ln e^2$

64. $\ln e^{-2/3}$

Sketch the graph of each given function.

65. $f(x) = \log_4 x$

66. $f(x) = \log_{1/2} x$

67. $f(x) = \log 2x$

68. $f(x) = \dfrac{1}{2} \log x$

69. $f(x) = \ln \dfrac{x}{2}$

70. $f(x) = \ln 3x$

71. $f(x) = \log_3(x - 1)$

72. $f(x) = \log_3(x + 1)$

4.3
FUNDAMENTAL PROPERTIES OF LOGARITHMS

There are three fundamental properties of logarithms that have made them a powerful computational aid.

Property 1. $\quad \log_a(x \cdot y) = \log_a x + \log_a y$

Property 2. $\quad \log_a\!\left(\dfrac{x}{y}\right) = \log_a x - \log_a y$

Property 3. $\quad \log_a x^n = n \log_a x, \quad n$ a real number

These properties can be proved by using equivalent exponential forms. To prove the first property, $\log_a(x \cdot y) = \log_a x + \log_a y$, we let

$$\log_a x = u \quad \text{and} \quad \log_a y = v$$

Then the equivalent exponential forms are

$$a^u = x \quad \text{and} \quad a^v = y$$

Multiplying the left-hand and right-hand sides of these equations, we have

$$a^u \cdot a^v = x \cdot y$$

or

$$a^{u+v} = x \cdot y$$

Substituting a^{u+v} for $x \cdot y$ in $\log_a(x \cdot y)$ we have

$$\log_a(x \cdot y) = \log_a(a^{u+v})$$
$$= u + v \qquad \text{since } \log_a a^x = x.$$

Substituting for u and v,

$$\log_a(x \cdot y) = \log_a x + \log_a y$$

Properties 2 and 3 can be established in much the same way.

It is these properties of logarithms that originally made their study worthwhile. Note that the more complex operations of multiplication and division are converted to addition and subtraction, and exponentiation is converted to multiplication. We will first demonstrate these properties, and in the next section will apply them to realistic computational problems.

EXAMPLE 1
(a) $\log_{10}(225 \times 478) = \log_{10} 225 + \log_{10} 478$

(b) $\log_8\left(\dfrac{422}{735}\right) = \log_8 422 - \log_8 735$

(c) $\log_2(2)^5 = 5 \log_2 2 = 5 \cdot 1 = 5$

(d) $\log_a\left(\dfrac{x \cdot y}{z}\right) = \log_a x + \log_a y - \log_a z$

PROGRESS CHECK
Write in terms of simpler logarithmic forms.

(a) $\log_4(1.47 \times 22.3)$ (b) $\log_a \dfrac{x - 1}{\sqrt{x}}$

ANSWERS

(a) $\log_4 1.47 + \log_4 22.3$ (b) $\log_a(x - 1) - \dfrac{1}{2}\log_a x$

SIMPLIFYING LOGARITHMS

The next example illustrates rules that speed the handling of logarithmic forms.

EXAMPLE 2

Write $\log_a \dfrac{(x - 1)^{-2}(y + 2)^3}{\sqrt{x}}$ in terms of simpler logarithmic forms.

SOLUTION

Simplifying Logarithms	
Step 1. Rewrite the expression so that each factor has a positive exponent.	*Step 1.* $\log_a \dfrac{(x-1)^{-2}(y+2)^3}{\sqrt{x}}$ $= \log_a \dfrac{(y+2)^3}{(x-1)^2\sqrt{x}}$
Step 2. Apply Property 1 and Property 2 for multiplication and division of logarithms. Each factor in the numerator will yield a term with a plus sign. Each factor in the denominator will yield a term with a minus sign.	*Step 2.* $= \log_a(y+2)^3 - \log_a(x-1)^2 - \log_a \sqrt{x}$
Step 3. Apply Property 3 to simplify.	*Step 3.* $= 3\log_a(y+2) - 2\log_a(x-1) - \dfrac{1}{2}\log_a x$

PROGRESS CHECK

Simplify $\log_a \dfrac{(2x-3)^{1/2}(y+2)^{2/3}}{z^4}$.

ANSWER

$\dfrac{1}{2}\log_a(2x-3) + \dfrac{2}{3}\log_a(y+2) - 4\log_a z$

EXAMPLE 3

If $\log_a 1.5 = r$, $\log_a 2 = s$, and $\log_a 5 = t$, find the following.

(a) $\log_a 7.5$ (b) $\log_a\left[(1.5)^3 \sqrt[5]{\dfrac{2}{5}}\right]$

SOLUTION
(a) Since

$$7.5 = 1.5 \times 5$$
$$\log_a 7.5 = \log_a(1.5 \times 5)$$
$$= \log_a 1.5 + \log_a 5 \quad \text{Property 1}$$
$$= r + t \quad\quad\quad\quad\quad \text{Substitution}$$

(b)   Write this as

$$\log_a(1.5)^3 + \log_a\left(\frac{2}{5}\right)^{1/5} \qquad\qquad \text{Property 1}$$

$$= 3\log_a 1.5 + \frac{1}{5}\log_a\left(\frac{2}{5}\right) \qquad\qquad \text{Property 3}$$

$$= 3\log_a 1.5 + \frac{1}{5}[\log_a 2 - \log_a 5] \quad \text{Property 2}$$

$$= 3r + \frac{1}{5}(s - t) \qquad\qquad\qquad \text{Substitution}$$

PROGRESS CHECK

If $\log_a 2 = 0.43$ and $\log_a 3 = 0.68$, find the following.

(a) $\log_a 18$ (b) $\log_a \sqrt[3]{\dfrac{9}{2}}$

ANSWERS

(a)   1.79   (b)   0.31

WARNING

(a)   Note that

$$\log_a(x + y) \neq \log_a x + \log_a y$$

Property 1 tells us that

$$\log_a(x \cdot y) = \log_a x + \log_a y$$

Don't try to apply this property to $\log_a(x + y)$, which cannot be simplified.

(b)   Note that

$$\log_a x^n \neq (\log_a x)^n$$

By Property 3,

$$\log_a x^n = n\log_a x$$

We can also apply the properties of logarithms to combine terms involving logarithms.

EXAMPLE 4

Write as a single logarithm.
$$2\log_a x - 3\log_a(x + 1) + \log_a \sqrt{x - 1}$$

SOLUTION
$$2\log_a x - 3\log_a(x + 1) + \log_a \sqrt{x - 1}$$
$$= \log_a x^2 - \log_a(x + 1)^3 + \log_a \sqrt{x - 1} \quad \text{Property 3}$$

$$= \log_a x^2 \sqrt{x - 1} - \log_a(x + 1)^3 \qquad \text{Property 1}$$
$$= \log_a \frac{x^2 \sqrt{x - 1}}{(x + 1)^3} \qquad \text{Property 2}$$

PROGRESS CHECK

Write as a single logarithm.

$$\frac{1}{3}[\log_a(2x - 1) - \log_a(2x - 5)] + 4 \log_a x$$

ANSWER

$$\log_a x^4 \sqrt[3]{\frac{2x - 1}{2x - 5}}$$

WARNING

(a) Note that

$$\frac{\log_a x}{\log_a y} \neq \log_a(x - y)$$

Property 2 tells us that

$$\log_a\left(\frac{x}{y}\right) = \log_a x - \log_a y$$

Don't try to apply this property to $\dfrac{\log_a x}{\log_a y}$, which cannot be simplified.

(b) The expressions

$$\log_a x + \log_b x$$

and

$$\log_a x - \log_b x$$

cannot be simplified. Logarithms with different bases do not readily combine except in special cases.

CHANGE OF BASE

Sometimes it is convenient to be able to write a logarithm that is given in terms of a base a in terms of another base b, that is, to convert $\log_a x$ to $\log_b x$. (As always, we must require a and b to be positive real numbers other than 1.) For example, some calculators can compute $\log x$ but not $\ln x$, and vice versa.

To compute $\log_b x$ given $\log_a x$, let $y = \log_b x$. The equivalent exponential form is then

$$b^y = x$$

Taking logarithms to the base a of both sides of this equation, we have

$$\log_a b^y = \log_a x$$

We now apply the fundamental properties of logarithms developed earlier in this section. By Property 3,

$$y \log_a b = \log_a x$$

Solving for y,

$$y = \frac{\log_a x}{\log_a b}$$

Since $y = \log_b x$, we have

Change of Base Formula	$\log_b x = \dfrac{\log_a x}{\log_a b}$

EXAMPLE 5
A calculator has a key labeled ''log'' (for $\log_{10}$) but doesn't have a key labeled ''ln.'' The calculator is used to find that

$$\log 27 = 1.4314$$
$$\log e \approx \log 2.7183 = 0.4343$$

Find ln 27.

SOLUTION
We use the change of base formula

$$\log_b x = \frac{\log_a x}{\log_a b}$$

with $b = e$, $a = 10$, and $x = 27$. Then

$$\log_e 27 = \ln 27 = \frac{\log 27}{\log e}$$

$$= \frac{1.4314}{0.4343} = 3.2959$$

Note that for any positive number k, the conversion from $\log k$ to $\ln k$ involves division by the constant $\log e = 0.4343$. A calculator that has a ''log'' key can thereby be used efficiently to find natural logarithms.

PROGRESS CHECK

Given $\log 16 = 1.2041$ and $\log 5 = 0.6990$, find $\log_5 16$.

ANSWER

1.7226

EXERCISE SET 4.3

Express each of the following in terms of simpler logarithmic forms.

1. $\log_{10}(120 \times 36)$

2. $\log_6\left(\dfrac{187}{39}\right)$

3. $\log_3(3^4)$

4. $\log_3(4^3)$

5. $\log_a(2xy)$

6. $\ln(4xyz)$

7. $\log_a\left(\dfrac{x}{yz}\right)$

8. $\ln\left(\dfrac{2x}{y}\right)$

9. $\ln x^5$

10. $\log_3 y^{2/3}$

11. $\log_a(x^2y^3)$

12. $\log_a(xy)^3$

13. $\log_a\sqrt{xy}$

14. $\log_a\sqrt[3]{xy^4}$

15. $\ln(x^2y^3z^4)$

16. $\log_a(xy^3z^2)$

17. $\ln\left(\sqrt{x}\sqrt[3]{y}\right)$

18. $\ln\sqrt[3]{xy^2}\sqrt[4]{z}$

19. $\log_a\left(\dfrac{x^2y^3}{z^4}\right)$

20. $\ln\dfrac{x^4y^2}{z^{1/2}}$

If $\log 2 = 0.30$, $\log 3 = 0.47$, and $\log 5 = 0.70$, find the following.

21. $\log 6$

22. $\log\dfrac{2}{3}$

23. $\log 9$

24. $\log\sqrt{5}$

25. $\log 12$

26. $\log\dfrac{6}{5}$

27. $\log\dfrac{15}{2}$

28. $\log 0.3$

29. $\log\sqrt{7.5}$

30. $\log\sqrt[4]{30}$

Write each of the following as a single logarithm.

31. $2\log x + \dfrac{1}{2}\log y$

32. $3\log_a x - 2\log_a z$

33. $\dfrac{1}{3}\ln x + \dfrac{1}{3}\ln y$

34. $\dfrac{1}{3}\ln x - \dfrac{2}{3}\ln y$

35. $\dfrac{1}{3}\log_a x + 2\log_a y - \dfrac{3}{2}\log_a z$

36. $\dfrac{2}{3}\log_a x + \log_a y - 2\log_a z$

37. $\dfrac{1}{2}(\log_a x + \log_a y)$

38. $\dfrac{2}{3}(4\ln x - 5\ln y)$

39. $\dfrac{1}{3}(2\ln x + 4\ln y) - 3\ln z$

40. $\ln x - \dfrac{1}{2}(3\ln x + 5\ln y)$

41. $\dfrac{1}{2}\log_a(x - 1) - 2\log_a(x + 1)$

42. $2\log_a(x + 2) - \dfrac{1}{2}(\log_a y + \log_a z)$

43. $3\log_a x - 2\log_a(x - 1) + \dfrac{1}{2}\log_a\sqrt[3]{x + 1}$

44. $4\ln(x - 1) + \dfrac{1}{2}\ln(x + 1) - 3\ln y$

The key labeled "ln" on a calculator is used to compute ln $10 = 2.3026$, ln $6 = 1.7918$, and ln $3 = 1.0986$. In Exercises 45–50 use the first value to find the required value.

45. ln $17 = 2.8332$; find log 17

46. ln $22 = 3.0910$; find $\log_6$ 22

47. ln $141 = 4.9488$; find $\log_3$ 141

48. ln $78 = 4.3567$; find $\log_6$ 78

49. ln $245 = 5.5013$; find log 245

50. ln $7 = 1.9459$; find $\log_3$ 7

4.4
COMPUTING WITH
LOGARITHMS
(Optional)

We indicated earlier that logarithms can be used to simplify complex calculations. In this section we will demonstrate the power of logarithms in computational work.

We will use 10 as the base for our computations with logarithms since 10 is the base of our number system. We call logarithms to the base 10 **common logarithms.**

We begin with the observation that any positive real number can be written as a product of a number c, $1 \leq c < 10$, and an integer power of 10, say 10^k. This format is often referred to as **scientific notation.** Here are some examples.

$$643 = 6.43 \times 10^2 \qquad 4629 = 4.629 \times 10^3$$
$$754,000 = 7.54 \times 10^5 \qquad 1.76 = 1.76 \times 10^0$$
$$0.0423 = 4.23 \times 10^{-2} \qquad 0.0000926 = 9.26 \times 10^{-5}$$

Let's begin with the number 643 expressed in scientific notation, that is,

$$643 = 6.43 \times 10^2$$

We next take the logarithm of both sides of this equation (to the base 10) and then apply the properties of logarithms.

$$
\begin{aligned}
\log 643 &= \log(6.43 \times 10^2) \\
&= \log 6.43 + \log 10^2 &&\text{Property 1} \\
&= \log 6.43 + 2 \log 10 &&\text{Property 3} \\
&= \log 6.43 + 2 &&\log 10 = 1
\end{aligned}
$$

We can generalize this result dealing with the logarithm of a number that is written in scientific notation.

If x is a positive real number and $x = c \cdot 10^k$, then

$$\log x = \log c + k$$

The number log c is called the **mantissa** and the integer k the **characteristic** of the number x. Since

$$x = c \cdot 10^k \quad \text{where } 1 \leq c < 10$$

and since the function $f(x) = \log x$ is an increasing function, we see that

$$\log 1 \leq \log c < \log 10$$

or

$$0 \leq \log c < 1$$

We conclude that the mantissa is always a number between 0 and 1.

Table II in the Tables Appendix can be used to approximate the common logarithm of any three-digit number between 1.00 and 9.99 at intervals of 0.01. The following example shows how to proceed.

EXAMPLE 1
Find (a) $\log 73.5$ (b) $\log 0.00451$.

SOLUTION
(a) Since $73.5 = 7.35 \times 10^1$, the characteristic of 73.5 is 1. Using Table II in the Tables Appendix, we find that $\log 7.35$ is 0.8663 (approximately). Then

$$\log 73.5 = \log(7.35 \times 10^1) = \log 7.35 + \log 10^1$$
$$= 0.8663 + 1 = 1.8663$$

(b) Since $0.00451 = 4.51 \times 10^{-3}$, the characteristic of 0.00451 is -3. From Table II we have

$$\log 0.00451 = 0.6542 - 3$$

Here we have a positive mantissa and a negative characteristic. For reasons that will be clear later, we always leave the answer in this form.

PROGRESS CHECK
Find the following.
(a) $\log 69{,}700$ (b) $\log 0.000697$ (c) $\log 0.697$

ANSWERS
(a) 4.8432 (b) $0.8432 - 4$ (c) $0.8432 - 1$

WARNING Note that

$$\log 0.00547 = 0.7380 - 3$$

but

$$\log 0.00547 \neq -3.7380$$

since this is algebraically incorrect. In fact, $0.7380 - 3 = -2.2620$.

Table II in the Tables Appendix can also be used in the reverse process, that is, to find x if $\log x$ is known. Since the entries in the body of Table II are numbers between 0 and 1, we must first write the number $\log x$ in the form

$$\log x = \log c + k$$

where $\log c$ is the mantissa, $0 \le \log c < 1$, and k, the characteristic, is an integer. (This is why we insisted in Example 1b that the number be left in the form $0.6542 - 3$.)

EXAMPLE 2
Find x if
(a) $\log x = 2.8351$ (b) $\log x = -6.6478$

SOLUTION
(a)   We must write $\log x$ in the form

$$\log x = \log c + k, \quad 0 \le \log c < 1$$

so

$$\log x = 2.8351 = \underbrace{0.8351}_{\log c} + \underbrace{2}_{k}$$

We seek the mantissa 0.8351 in the body of Table II in the Tables Appendix and find that it corresponds to $\log 6.84$. Since the characteristic $k = 2$,

$$x = 6.84 \times 10^2 = 684$$

(b)   We have to proceed carefully to ensure that the mantissa is between 0 and 1. If we add and subtract 7, we have

$$\log x = (7 - 6.6478) - 7$$
$$= \underbrace{0.3522}_{\log c} - \underbrace{7}_{k}$$

We seek the mantissa 0.3522 in the body of Table II in the Tables Appendix and find

$$0.3522 = \log 2.25$$

Since the characteristic $k = -7$, we have

$$x = 2.25 \times 10^{-7} = 0.000000225$$

PROGRESS CHECK
Find x in the following.
(a) $\log x = 3.8457$ (b) $\log x = 0.6201 - 2$ (c) $\log x = -2.0487$

ANSWERS
(a)   7010   (b)   0.0417   (c)   0.00894

DATING THE LATEST ICE AGE	All organic forms of life contain radioactive carbon 14. In 1947 the chemist Willard Libby (who won the Nobel prize in 1960) found that the percentage of carbon 14 in the atmosphere equals the percentage found in the living tissues of all organic forms of life. When an organism dies, it stops replacing carbon 14 in its living tissues. Yet the carbon 14 continues decaying at the rate of 0.012% per year. By measuring the amount of carbon 14 in the remains of an organism, it is possible to estimate fairly accurately when the organism died.
$Q(t) = q_0 e^{-kt}$ $0.254 q_0 = q_0 e^{-0.00012t}$ $0.254 = e^{-0.00012t}$ $\ln 0.254 = \ln e^{-0.00012t}$ $-1.3704 = -0.00012t$ $t = 11{,}420$	In the late 1940s radiocarbon dating was used to date the last ice sheet to cover the North American and European continents. Remains of trees in the Two Creeks Forest in northern Wisconsin were found to have lost 74.6% of their carbon 14 content. The remaining carbon 14, therefore, was 25.4% of the original quantity q_0 that was present when the descending ice sheet felled the trees. The accompanying computations use the general equation of an exponential decay model to find the age t of the wood. Conclusion: The latest ice age occurred approximately 11,420 years before the measurements were taken.

The following example shows how to use logarithms to simplify computations.

EXAMPLE 3
Approximate 478×0.0345 by using logarithms.

SOLUTION
If

$$N = 478 \times 0.0345$$

then

$$\log N = \log(478 \times 0.0345)$$
$$= \log 478 + \log 0.0345 \qquad \text{Property 1}$$

Using Table II in the Tables Appendix,

$$\log 478 = 2.6794$$
$$\underline{\log 0.0345 = 0.5378 - 2}$$
$$\log N = 3.2172 - 2 \qquad \text{Adding the logarithms}$$
$$= 1.2172 = \underbrace{0.2172}_{\log c} + \underbrace{1}_{k}$$

Looking in the body of Table II, we find that the mantissa 0.2172 does not appear. However, 0.2175 does appear and corresponds to log 1.65. Thus,

$$N \approx 1.65 \times 10^1 = 16.5$$

Inexpensive calculators have reduced the importance of logarithms as a computational device. Still, many calculators cannot, for example, handle $\sqrt[5]{14.2}$ directly. If you know how to compute with logarithms and if you combine this knowledge with a calculator that can handle logarithms, you can enhance the power of your calculator. Many additional applications of logarithms occur in more advanced mathematics, especially in calculus.

EXAMPLE 4

Approximate $\dfrac{\sqrt{47.4}}{(2.3)^3}$ by using logarithms.

SOLUTION

If

$$N = \frac{\sqrt{47.4}}{(2.3)^3}$$

then

$$\log N = \frac{1}{2} \log 47.4 - 3 \log 2.3$$

$$\frac{1}{2} \log 47.4 = \frac{1}{2} (1.6758) = 0.8379$$

$$3 \log 2.3 = 3(0.3617) = 1.0851$$

$$\log N = 0.8379 - 1.0851 = -0.2472 \quad \text{Subtracting the logarithms}$$

or

$$\log N = \underbrace{0.7528}_{\log c} - \underbrace{1}_{k} \qquad\qquad \text{Adding and subtracting 1}$$

From the body of Table II in the Tables Appendix, we find $0.7528 \approx \log 5.66$, so

$$N \approx 5.66 \times 10^{-1} = 0.566$$

PROGRESS CHECK

Approximate $\dfrac{(4.64)^{3/2}}{\sqrt{7.42 \times 165}}$ by logarithms.

ANSWER

0.286

Problems in compound interest provide us with an opportunity to demonstrate the power of logarithms in computational work. In Section 4.1 we showed that the amount S available when a principal P is invested at an annual interest rate r compounded k times a year is given by

$$S = P(1 + i)^n$$

where $i = r/k$, and n is the number of conversion periods.

EXAMPLE 5
If $1000 is left on deposit at an interest rate of 8% per year compounded quarterly, how much money is in the account at the end of 6 years?

SOLUTION
We have $P = 1000$, $r = 0.08$, $k = 4$, and $n = 24$ (since there are 24 quarters in 6 years). Thus,

$$S = P(1 + i)^n = 1000\left(1 + \frac{0.08}{4}\right)^{24}$$
$$= 1000(1 + 0.02)^{24} = 1000(1.02)^{24}$$

Then

$$\log S = \log 1000 + 24 \log 1.02$$
$$= 3 + 24(0.0086) = 3.2064$$

From the body of Table II in the Tables Appendix, we find $0.2064 \approx \log 1.61$, so

$$S \approx 1.61 \times 10^3 = 1610$$

The account contains $1610 (approximately) at the end of 6 years.

PROGRESS CHECK
If $1000 is left on deposit at an interest rate of 6% per year compounded semiannually, approximately how much is in the account at the end of 4 years?

ANSWER
$1267

EXERCISE SET 4.4
Write each of the following in scientific notation.
1. 2725
2. 493
3. 0.0084
4. 0.000914
5. 716,000
6. 527,600,000
7. 296.2
8. 32.767

Compute the following logarithms by using Tables II and III in the Tables Appendix.
9. log 3.56
10. ln 3.2
11. log 37.5
12. log 85.3
13. ln 4.7
14. ln 60
15. log 74
16. log 4230

17. log 48,200   18. log 7,890,000   19. log 0.342   20. log 0.00532

Using Tables II and III in the Tables Appendix, find x.

21. $\log x = 0.4014$ 22. $\ln x = -0.5108$ 23. $\ln x = 1.0647$ 24. $\log x = 2.7332$

25. $\ln x = 2.0669$ 26. $\log x = 0.1903 - 2$ 27. $\log x = 0.4099 - 1$ 28. $\log x = 0.7024 - 2$

29. $\log x = 0.7832 - 4$ 30. $\log x = 0.9320 - 2$ 31. $\log x = -1.6599$ 32. $\log x = -3.9004$

Find an approximate answer by using logarithms.

33. $(320)(0.00321)$ 34. $(8780)(2.13)$ 35. $\dfrac{679}{321}$ 36. $\dfrac{88.3}{97.2}$

37. $(3.19)^4$ 38. $(42.3)^3(71.2)^2$ 39. $\dfrac{(87.3)^2(0.125)^3}{(17.3)^3}$ 40. $\sqrt[3]{(66.9)^4(0.781)^2}$

41. $\dfrac{\sqrt{7870}}{(46.3)^4}$ 42. $\dfrac{(7.28)^{2/3}}{\sqrt[3]{(87.3)(16.2)^4}}$ 43. $\dfrac{(32.870)(0.00125)}{(12.8)(124,000)}$

44. The period T (in seconds) of a simple pendulum of length L (in feet) is given by the formula

$$T = 2\pi\sqrt{\frac{L}{g}}$$

Using common logarithms, find the approximate value of T if $L = 4.72$ feet, $g = 32.2$, and $\pi = 3.14$.

45. Use logarithms to find the approximate amount that accumulates if $6000 is invested for 8 years in a bank paying 7% interest per year compounded quarterly.

46. Use logarithms to find the approximate sum if $8000 is invested for 6 years in a bank paying 8% interest per year compounded monthly.

47. If $10,000 is invested at 7.8% interest per year compounded semiannually, what sum is available after 5 years?

48. Which of the following offers will yield a greater return: 8% annual interest compounded annually, or 7.75% annual interest compounded quarterly?

49. Which of the following offers will yield a greater return: 9% annual interest compounded annually, or 8.75% annual interest compounded quarterly?

50. The area of a triangle whose sides are a, b, and c in length is given by the formula

$$A = \sqrt{s(s - a)(s - b)(s - c)}$$

where $s = \frac{1}{2}(a + b + c)$. Use logarithms to find the approximate area of a triangle whose sides are 12.86 feet, 13.72 feet, and 20.3 feet.

4.5
EXPONENTIAL AND LOGARITHMIC EQUATIONS

The following approach will often help in solving exponential and logarithmic equations.

- To solve an exponential equation, take logarithms of both sides of the equation.
- To solve a logarithmic equation, form a single logarithm on one side of the equation, and then convert the equation to the equivalent exponential form.

EXAMPLE 1
Solve $3^{2x-1} = 17$.

SOLUTION

Taking logarithms to the base 10 of both sides of the equation,

$$\log 3^{2x-1} = \log 17$$

$$(2x - 1)\log 3 = \log 17 \qquad \text{Property 3}$$

$$2x - 1 = \frac{\log 17}{\log 3}$$

$$2x = 1 + \frac{\log 17}{\log 3}$$

$$x = \frac{1}{2} + \frac{\log 17}{2 \log 3}$$

If a numerical value is required, Table II in the Tables Appendix, or a calculator, can be used to approximate log 17 and log 3. Also, note that we could have taken logarithms to *any* base in solving this equation.

PROGRESS CHECK

Solve $2^{x+1} = 3^{2x-3}$

ANSWER

$$\frac{\log 2 + 3 \log 3}{2 \log 3 - \log 2}$$

EXAMPLE 2

Solve $\log(2x + 8) = 1 + \log(x - 4)$.

SOLUTION

If we rewrite the equation in the form

$$\log(2x + 8) - \log(x - 4) = 1$$

then we can apply Property 2 to form a single logarithm.

$$\log \frac{2x + 8}{x - 4} = 1$$

Now we convert to the equivalent exponential form.

$$\frac{2x + 8}{x - 4} = 10^1 = 10$$

$$2x + 8 = 10x - 40$$

$$x = 6$$

PROGRESS CHECK

Solve $\log(x + 1) = 2 + \log(3x - 1)$.

ANSWER

$\dfrac{101}{299}$

EXAMPLE 3

Solve $\log_2 x = 3 - \log_2(x + 2)$.

SOLUTION

Rewriting the equation with a single logarithm, we have

$$\log_2 x + \log_2(x + 2) = 3$$
$$\log_2[x(x + 2)] = 3 \qquad \text{Why?}$$
$$x(x + 2) = 2^3 = 8 \quad \text{Equivalent exponential form}$$
$$x^2 + 2x - 8 = 0$$
$$(x - 2)(x + 4) = 0 \qquad \text{Factor}$$
$$x = 2 \quad \text{or} \quad x = -4$$

The "solution" $x = -4$ must be rejected since the original equation contains $\log_2 x$, which requires that x be positive.

PROGRESS CHECK

Solve $\log_3(x - 8) = 2 - \log_3 x$.

ANSWER

$x = 9$

EXAMPLE 4

World population is increasing at an annual rate of 2.5%. If we assume an exponential growth model, in how many years will the population double?

SOLUTION

The exponential growth model

$$Q(t) = q_0 e^{0.025t}$$

describes the population Q as a function of time t. Since the initial population is $Q(0) = q_0$, we seek the time t required for the population to double or become $2q_0$. We wish to solve the equation

$$Q(t) = 2q_0 = q_0 e^{0.025t}$$

for t. We then have

$$2q_0 = q_0 e^{0.025t}$$
$$2 = e^{0.025t} \qquad \text{Divide by } q_0$$
$$\ln 2 = \ln e^{0.025t} \qquad \text{Take natural logs of both sides}$$
$$= 0.025t \qquad \text{Since } \ln e^x = x$$
$$t = \frac{\ln 2}{0.025} = \frac{0.6931}{0.025} = 27.7$$

or approximately 28 years.

EXAMPLE 5

A trust fund invests $8000 at an annual interest rate of 8% compounded continuously. How long does it take for the initial investment to grow to $12,000?

SOLUTION

Using Equation (4) of Section 4.1,

$$S = Pe^{rt}$$

we have $S = 12,000$, $P = 8000$, $r = 0.08$, and we must solve for t. Thus,

$$12,000 = 8000e^{0.08t}$$
$$\frac{12,000}{8000} = e^{0.08t}$$
$$e^{0.08t} = 1.5$$

Taking natural logarithms of both sides, we have

$$0.08t = \ln 1.5$$
$$t = \frac{\ln 1.5}{0.08} \approx \frac{0.4055}{0.08} \quad \text{from Table III}$$
$$\approx 5.07$$

It takes approximately 5.07 years for the initial $8000 to grow to $12,000.

EXERCISE SET 4.5

In Exercises 1–31 solve for x.

1. $5^x = 18$
2. $2^x = 24$
3. $2^{x-1} = 7$
4. $3^{x-1} = 12$
5. $3^{2x} = 46$
6. $2^{2x-1} = 56$
7. $5^{2x-5} = 564$
8. $3^{3x-2} = 23.1$
9. $3^{x-1} = 2^{2x+1}$
10. $4^{2x-1} = 3^{2x+3}$
11. $2^{-x} = 15$
12. $3^{-x+2} = 103$
13. $4^{-2x+1} = 12$
14. $3^{-3x+2} = 2^{-x}$
15. $e^x = 18$
16. $e^{x-1} = 2.3$

17. $e^{2x+3} = 20$

18. $e^{-3x+2} = 40$

19. $\log x + \log 2 = 3$

20. $\log x - \log 3 = 2$

21. $\log_x(3 - 5x) = 1$

22. $\log_x(8 - 2x) = 2$

23. $\log x + \log(x - 3) = 1$

24. $\log x + \log(x + 21) = 2$

25. $\log(3x + 1) - \log(x - 2) = 1$

26. $\log(7x - 2) - \log(x - 2) = 1$

27. $\log_2 x = 4 - \log_2(x - 6)$

28. $\log_2(x - 4) = 2 - \log_2 x$

29. $\log_2(x + 4) = 3 - \log_2(x - 2)$

30. $y = \dfrac{e^x + e^{-x}}{2}$

31. $y = \dfrac{e^x - e^{-x}}{2}$

32. Suppose that world population is increasing at an annual rate of 2%. If we assume an exponential growth model, in how many years will the population double?

33. Suppose that the population of a certain city is increasing at an annual rate of 3%. If we assume an exponential growth model, in how many years will the population triple?

34. The population P of a certain city t years from now is given by

$$P = 20{,}000e^{0.05t}$$

How many years from now will the population be 50,000?

35. Potassium 42 has a decay rate of approximately 5.5% per hour. Assuming an exponential decay model, in how many hours will the original quantity of potassium 42 have been halved?

36. Consider an exponential decay model given by

$$Q = q_0 e^{-0.4t}$$

where t is in weeks. How many weeks does it take for Q to decay to $\frac{1}{4}$ of its original amount?

37. How long does it take an amount of money to double if it is invested at a rate of 8% per year compounded semiannually?

38. At what rate of annual interest, compounded semiannually, should a certain amount of money be invested so that it will double in 8 years?

39. The number N of radios that an assembly line worker can assemble daily after t days of training is given by

$$N = 60 - 60e^{-0.04t}$$

After how many days of training does the worker assemble 40 radios daily?

40. The quantity Q (in grams) of a radioactive substance that is present after t days of decay is given by

$$Q = 400e^{-kt}$$

If $Q = 300$ when $t = 3$, find k, the decay rate.

41. A person on an assembly line produces P items per day after t days of training, where

$$P = 400(1 - e^{-t})$$

How many days of training will it take this person to be able to produce 300 items per day?

42. Suppose that the number N of mopeds sold when x thousands of dollars are spent on advertising is given by

$$N = 4000 + 1000 \ln(x + 2)$$

How much advertising money must be spent to sell 6000 mopeds?

TERMS AND SYMBOLS

exponential function (p. 170)
base (p. 170)
a^x (p. 170)
e (p. 173)
exponential growth model (p. 174)

growth constant (p. 174)
exponential decay model (p. 175)
decay constant (p. 175)
compound interest (p. 176)
conversion period (p. 176)

continuous compounding (p. 178)
logarithmic function (p. 181)
$\log_a x$ (p. 182)
$\ln x$ (p. 182)

natural logarithm (p. 182)
common logarithm (p. 196)
scientific notation (p. 196)
mantissa (p. 196)
characteristic (p. 196)

KEY IDEAS FOR REVIEW

☐ An exponential function has a variable in the exponent and has a base that is a positive constant.

☐ The graph of the exponential function $f(x) = a^x$, where $a > 0$ and $a \neq 1$,
  - passes through the points $(0, 1)$ and $(1, a)$ for any value of x;
  - is increasing if $a > 1$ and decreasing if $0 < a < 1$.

☐ The domain of the exponential function is the set of all real numbers; the range is the set of all positive numbers.

☐ If $a^x = a^y$, then $x = y$ (assuming $a > 0$, $a \neq 1$).

☐ If $a^x = b^x$ for all $x \neq 0$, then $a = b$ (assuming $a > 0$, $b > 0$).

☐ Exponential functions play a key role in the following important applications:
  - Exponential growth model: $Q(t) = q_0 e^{kt}, \quad k > 0$
  - Exponential decay model: $Q(t) = q_0 e^{-kt}, \quad k > 0$
  - Compound interest: $S = P(1 + i)^n$
  - Continuous compounding: $S = Pe^{rt}$

☐ The logarithmic function $\log_a x$ is the inverse of the function a^x.

☐ The logarithmic form $y = \log_a x$ and the exponential form $x = a^y$ are two ways of expressing the same relationship. In short, logarithms are exponents. Consequently, it is always possible to convert from one form to the other.

☐ The following identities are useful in simplifying expressions and in solving equations.

$$a^{\log_a x} = x \qquad \log_a a = 1$$
$$\log_a a^x = x \qquad \log_a 1 = 0$$

☐ The graph of the logarithmic function $f(x) = \log_a x$, where $x > 0$,
  - passes through the points $(1, 0)$ and $(a, 1)$ for any $a > 0$;
  - is increasing if $a > 1$ and decreasing if $0 < a < 1$.

☐ The domain of the logarithmic function is the set of all positive real numbers; the range is the set of all real numbers.

☐ If $\log_a x = \log_a y$, then $x = y$.

☐ If $\log_a x = \log_b x$ and $x \neq 1$, then $a = b$.

☐ The fundamental properties of logarithms are as follows.

> *Property 1.* $\log_a(xy) = \log_a x + \log_a y$
>
> *Property 2.* $\log_a\left(\dfrac{x}{y}\right) = \log_a x - \log_a y$
>
> *Property 3.* $\log_a x^n = n \log_a x$

☐ The fundamental properties of logarithms, used in conjunction with tables of logarithms, are a powerful tool in performing calculations. It is these properties that make the study of logarithms worthwhile.

☐ The change of base formula is

$$\log_b x = \frac{\log_a x}{\log_a b}$$

REVIEW EXERCISES

Solutions to exercises whose numbers are in color are in the Solutions section in the back of the book.

4.1 1. Sketch the graph of $f(x) = \left(\dfrac{1}{3}\right)^x$. Label the point $(-1, f(-1))$.

  2. Solve $2^{2x} = 8^{x-1}$ for x.

  3. Solve $(2a + 1)^x = (3a - 1)^x$ for a.

  4. The sum of $8000 is invested in a certificate paying 12% annual interest compounded semiannually. What sum is available at the end of 4 years?

4.2 In Exercises 5–8 write each logarithmic form in exponential form and vice versa.

  5. $27 = 9^{3/2}$

  6. $\log_{64} 8 = \dfrac{1}{2}$

  7. $\log_2 \dfrac{1}{8} = -3$

  8. $6^0 = 1$

In Exercises 9–12 solve for x.

9. $\log_x 16 = 4$

10. $\log_5 \dfrac{1}{125} = x - 1$

11. $\ln x = -4$

12. $\log_3(x + 1) = \log_3 27$

In Exercises 13–16 evaluate the given expression.

13. $\log_3 3^5$

14. $\ln e^{-1/3}$

15. $\log_3\left(\dfrac{1}{3}\right)$

16. $e^{\ln 3}$

17. Sketch the graph of $f(x) = \log_3 x + 1$.

In Exercises 18–21 write the given expression in terms of simpler logarithmic forms.

4.3 18. $\log_a \dfrac{\sqrt{x - 1}}{2x}$

19. $\log_a \dfrac{x(2 - x)^2}{(y + 1)^{1/2}}$

20. $\ln (x + 1)^4(y - 1)^2$

21. $\log \sqrt[5]{\dfrac{y^2 z}{z + 3}}$

In Exercises 22–25 use the values $\log 2 = 0.30$, $\log 3 = 0.50$, and $\log 7 = 0.85$ to evaluate the given expression.

22. $\log 14$

23. $\log 3.5$

24. $\log \sqrt{6}$

25. $\log 0.7$

In Exercises 26–29 write the given expression as a single logarithm.

26. $\dfrac{1}{3}\log_a x - \dfrac{1}{2} \log_a y$

27. $\dfrac{4}{3}[\log x + \log(x - 1)]$

28. $\ln 3x + 2\left(\ln y - \dfrac{1}{2} \ln z\right)$

29. $2 \log_a(x + 2) - \dfrac{3}{2} \log_a(x + 1)$

In Exercises 30 and 31 use the values $\log 32 = 1.5$, $\log 8 = 0.9$, and $\log 5 = 0.7$ to find the requested value.

30. $\log_8 32$

31. $\log_5 32$

4.4 In Exercises 32–35 write the given number in scientific notation.

32. 476.5

33. 0.098

34. $26,475$

35. 77.67

In Exercises 36–38 use logarithms to calculate the value of the given expression.

36. $(0.765)(32.4)^2$

37. $\sqrt{62.3}$

38. $\dfrac{2.1}{(32.5)^{5/2}}$

39. A substance is known to have a decay rate of 6% per hour. Approximately how many hours are required for the remaining quantity to be half of the original quantity?

4.5 In Exercises 40–42 solve for x.

40. $2^{3x-1} = 14$

41. $2 \log x - \log 5 = 3$

42. $\log(2x - 1) = 2 + \log(x - 2)$

PROGRESS TEST 4A

1. Sketch the graph of $f(x) = 2^{x+1}$. Label the point $(1, f(1))$.

2. Solve $\left(\dfrac{1}{2}\right)^x = \left(\dfrac{1}{4}\right)^{2x+1}$

In Problems 3 and 4 convert from logarithmic form to exponential form or vice versa.

3. $\log_3 \dfrac{1}{9} = -2$ 4. $64 = 16^{3/2}$

In Problems 5 and 6 solve for x.

5. $\log_x 27 = 3$

6. $\log_6\left(\dfrac{1}{36}\right) = 3x + 1$

In Problems 7 and 8 evaluate the given expression.

7. $\ln e^{5/2}$

8. $\log_5 \sqrt{5}$

In Problems 9 and 10 write the given expression in terms of simpler logarithmic forms.

9. $\log_a \dfrac{x^3}{y^2 z}$

10. $\log \dfrac{x^2 \sqrt{2y - 1}}{y^3}$

In Problems 11 and 12 use the values $\log 2.5 = 0.4$ and $\log 2 = 0.3$ to evaluate the given expression.

11. $\log 5$

12. $\log 2\sqrt{2}$

In Problems 13 and 14 write the given expression as a single logarithm.

13. $2 \log x - 3 \log(y + 1)$

14. $\dfrac{2}{3}[\log_a(x + 3) - \log_a(x - 3)]$

In Problems 15 and 16 write the given number in scientific notation.

15. 0.000273

16. 5.972

In Problems 17 and 18 use logarithms to evaluate the given expression.

17. $\dfrac{72.9}{(39.4)^2}$

18. $\sqrt[3]{0.0176}$

19. The number of bacteria in a culture is described by the exponential growth model

$$Q(t) = q_0 e^{0.02t}$$

Approximately how many hours are required for the number of bacteria to double?

In Problems 20 and 21 solve for x.

20. $\log x - \log 2 = 2$

21. $\log_4(x - 3) = 1 - \log_4 x$

PROGRESS TEST 4B

1. Sketch the graph of $f(x) = \left(\dfrac{1}{2}\right)^{x-1}$. Label the point $(2, f(2))$.

2. Solve $(a + 3)^x = (2a - 5)^x$ for a.

In Problems 3 and 4 convert from logarithmic form to exponential form and vice versa.

3. $\dfrac{1}{1000} = 10^{-3}$

4. $\log_3 1 = 0$

In Problems 5 and 6 solve for x.

5. $\log_2(x - 1) = -1$

6. $\log_{2x} 27 = \log_3 27$

In Problems 7 and 8 evaluate the given expression.

7. $\log_3 3^{10}$

8. $e^{\ln 4}$

In Problems 9 and 10 write the given expression in terms of simpler logarithmic forms.

9. $\log_a(x - 1)(y + 3)^{5/4}$

10. $\ln \sqrt{xy} \sqrt[4]{2z}$

In Problems 11 and 12 use the values $\log 2.5 = 0.4$, $\log 2 = 0.3$, and $\log 6 = 0.75$ to evaluate the given expression.

11. $\log 7.5$

12. $\log 36$

In Problems 13 and 14 write the given expression as a single logarithm.

13. $\dfrac{3}{5} \ln(x - 1) + \dfrac{2}{5} \ln y - \dfrac{1}{5} \ln z$

14. $\log\dfrac{x}{y} - \log\dfrac{y}{x}$

In Problems 15 and 16 write the given number in scientific notation.

15. 22,684,321

16. 0.297

In Problems 17 and 18 use logarithms to evaluate the given expression.

17. $(0.295)(31.7)^3$

18. $\dfrac{\sqrt{42.9}}{(3.75)^2(747)}$

19. Suppose that \$500 is invested in a certificate at an annual interest rate of 12% compounded monthly. What is the value of the investment after 6 months?

In Problems 20 and 21 solve for x.

20. $\log_x(x + 6) = 2$

21. $\log(x - 9) = 1 - \log x$

CHAPTER 5

TRIGONOMETRY: THE CIRCULAR FUNCTIONS

The word "trigonometry" derives from the Greek, meaning "measurement of triangles." The conventional approach to the subject matter of trigonometry deals with relationships between the sides and angles of a triangle, reflecting the important applications of trigonometry in such fields as navigation and surveying.

The modern approach is to deal with trigonometry in terms of functional relationships. The *trigonometric functions* can be viewed as functions whose domains are angles or as functions whose domains are real numbers, in which case they are referred to as the *circular functions*. We will take the latter approach since it more clearly demonstrates the utility of the function concept. We will devote this chapter to the circular functions and will deal with both angles and triangles in the next chapter.

**5.0
REVIEW OF
GEOMETRY**

We need to recall various facts about the circle from plane geometry. A line segment joining the center of a circle to any point on the circle is called a **radius.** Since every point on the circle is the same distance from the center, the radii of a circle are all equal. Thus, in Figure 1, $\overline{OP} = \overline{OQ}$. A **chord** of a circle is a line segment joining any two points on the circle; a **diameter** of a circle is a chord that passes through the center of the circle. Note that the length of a diameter is twice that of a radius.

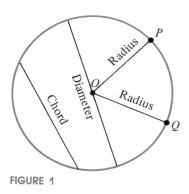

FIGURE 1

The **circumference** C of a circle is the distance around the circle and is given by

$$C = 2\pi r$$

where r is the radius of the circle. The constant π is then seen to be the ratio of the circumference of a circle to the length of its diameter. The **area** A of a circle of radius r is given by

$$A = \pi r^2$$

An **arc** of a circle is simply a part of a circle. The arc $\widehat{AB}$ of Figure 2 consists of the two endpoints A and B and the set of all points on the circle that are between A and B and are shown in color.

A **central angle** has its vertex at the center of the circle, and its sides are radii of the circle. We define the measure of a central angle to be the same as that of the arc it intercepts. Thus, in Figure 2, $\sphericalangle AOB = \widehat{AB}$. We can then show that equal arcs determine or subtend equal chords. If arc $\widehat{AB} =$ arc $\widehat{CD}$ in Figure 2, then, by definition, $\sphericalangle AOB = \sphericalangle COD$. Since $\overline{AO} = \overline{BO} = \overline{CO} = \overline{DO}$ are all radii, it follows that triangles AOB and COD are congruent. Hence, $\overline{CD} = \overline{AB}$.

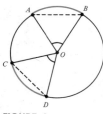

FIGURE 2

The converse can be proven in a similar manner. Thus,

> Equal arcs determine equal chords.
> Equal chords determine equal arcs.

5.1
THE WRAPPING
FUNCTION

THE UNIT CIRCLE

We begin by discussing the **unit circle**, a circle of radius 1 whose center is at the origin of a rectangular coordinate system (Figure 3a). A point $P(x, y)$ is on the unit circle if and only if the distance $\overline{OP} = 1$. Using the distance formula,

$$\overline{OP} = \sqrt{(x - 0)^2 + (y - 0)^2} = 1$$

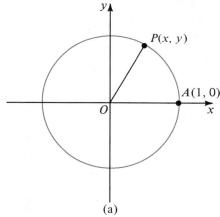

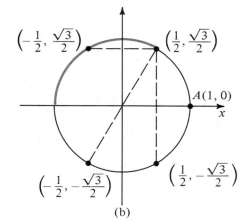

(a)   (b)

FIGURE 3

Squaring both sides, we conclude

> The equation of the unit circle is
> $$x^2 + y^2 = 1$$

Using the methods of Section 2.1, we find that the unit circle is symmetric with respect to the x-axis, the y-axis, and the origin. These symmetries will prove to be very useful. For example, you can easily verify that the point $(1/2, \sqrt{3}/2)$ lies on the unit circle. Figure 3b shows the coordinates of various other points that can be obtained from the symmetries of the circle. (See Exercises 44 and 45.)

THE WRAPPING FUNCTION

We now seek to establish a correspondence between the real numbers and points on the unit circle. The point $A(1, 0)$ lies on the unit circle, since its coordinates satisfy the equation $x^2 + y^2 = 1$. If $t \geq 0$ is any nonnegative real number, we may locate t on a real number line L whose scale is the same as that of the coordinate axes on which we have sketched the unit circle (Figure 4). We now imagine that the real number line is made of string, place its origin at $A(1, 0)$, and "wrap" the real number line about the circle in a *counterclockwise* direction. Eventually, the entire length t of the "string" is used, and it ends at a point P on the unit circle. (Note that if $t = 0$, then the point P coincides with the point A.)

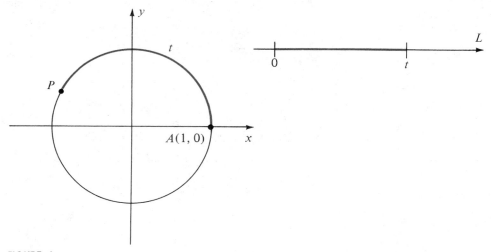

FIGURE 4

If $t < 0$ is a negative real number, we proceed in the same manner except that we wrap a string of length $|t|$ in a *clockwise* direction. In either event, the wrapping procedure is seen to be a rule that assigns a unique point P to every real number t. We may therefore refer to this rule as the **wrapping function** W and say that $W(t)$ is the point corresponding to the real number t.

Figure 5 is an illustration of the wrapping function. The real number line L has been placed tangent to the unit circle at A, with its origin coincident with A (Figure 5a). The wrapping function is seen to wrap the positive real numbers around the unit circle in a counterclockwise direction (Figure 5b), and the negative real numbers are wrapped around the unit circle in a clockwise direction (Figure 5c).

The wrapping function is not a one-to-one function. If we consider a point P on the unit circle (Figure 4), we may associate this point with the length t of the arc $\overset{\frown}{AP}$. Since the circumference of the unit circle is 2π, the length t of arc $\overset{\frown}{AP}$ is less than 2π, that is, $0 \leq t < 2\pi$. The correspondence assigning a real number to

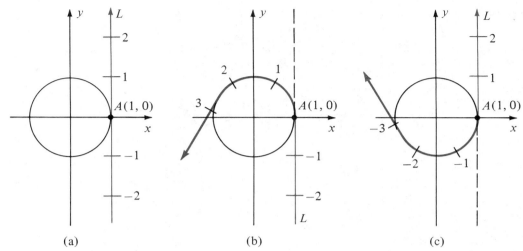

(a)   (b)   (c)

FIGURE 5

the point P is not unique, however. The point P not only corresponds to t but also corresponds to $t + 2\pi$ and to $t - 2\pi$ since an additional revolution in either direction will again terminate at P. In fact, the point P corresponds to all of the real numbers $t + 2\pi n$, where n is an integer. This gives us an important property of the wrapping function W.

The wrapping function W has the property that

$$W(t) = W(t + 2\pi n)$$

for any real number t and integer n.

EXAMPLE 1
Find two positive and two negative values of t for which

$$W(t) = W\left(\frac{\pi}{6}\right)$$

SOLUTION
Since $W\left(\dfrac{\pi}{6}\right) = W\left(\dfrac{\pi}{6} + 2\pi n\right)$, we can select any integer values of n. Thus

$$n = 1: \quad W\left(\frac{\pi}{6}\right) = W\left(\frac{\pi}{6} + 2\pi\right) = W\left(\frac{13\pi}{6}\right)$$

$$n = 2: \quad W\left(\frac{\pi}{6}\right) = W\left(\frac{\pi}{6} + 4\pi\right) = W\left(\frac{25\pi}{6}\right)$$

$$n = -1: \quad W\left(\frac{\pi}{6}\right) = W\left(\frac{\pi}{6} - 2\pi\right) = W\left(-\frac{11\pi}{6}\right)$$

$$n = -5: \quad W\left(\frac{\pi}{6}\right) = W\left(\frac{\pi}{6} - 10\pi\right) = W\left(-\frac{59\pi}{6}\right)$$

The values of t are $\dfrac{13\pi}{6}, \dfrac{25\pi}{6}, -\dfrac{11\pi}{6}, -\dfrac{59\pi}{6}$.

EXAMPLE 2
Find the real number t, $0 \le t < 2\pi$, such that $W(t)$ is equal to each of the following.

(a) $W(7\pi)$ (b) $W\left(\dfrac{10\pi}{3}\right)$ (c) $W\left(-\dfrac{\pi}{4}\right)$

SOLUTION
Given $W(s)$, we may add (or subtract) 2π to (or from) s until we obtain a number t in the interval $[0, 2\pi)$. Stated algebraically, we can find an integer multiple n of 2π such that $0 \le s + 2\pi n < 2\pi$. Then the number t we seek is $t = s + 2\pi n$.

(a) $W(7\pi) = W(7\pi - 2\pi) = W(5\pi)$

     $W(5\pi) = W(5\pi - 2\pi) = W(3\pi)$

     $W(3\pi) = W(3\pi - 2\pi) = W(\pi)$,   and $t = \pi$

or

     $W(7\pi) = W(7\pi - 6\pi) = W(\pi)$,   and $t = \pi$

(b) $W\left(\dfrac{10\pi}{3}\right) = W\left(\dfrac{10\pi}{3} - 2\pi\right) = W\left(\dfrac{4\pi}{3}\right)$,   and $t = \dfrac{4\pi}{3}$

(c) $W\left(-\dfrac{\pi}{4}\right) = W\left(-\dfrac{\pi}{4} + 2\pi\right) = W\left(\dfrac{7\pi}{4}\right)$,   and $t = \dfrac{7\pi}{4}$

PROGRESS CHECK
Find the real number t, $0 \le t < 2\pi$, such that $W(t)$ is equal to each of the following.

(a) $W\left(\dfrac{5\pi}{2}\right)$ (b) $W(29\pi)$ (c) $W\left(-\dfrac{7\pi}{3}\right)$

ANSWERS
(a) $\dfrac{\pi}{2}$ (b) π (c) $\dfrac{5\pi}{3}$

We are now prepared to summarize our results.

The Wrapping Function

• The wrapping process determines a function W whose domain is the set of all real numbers and whose range is the set of all points on the unit circle.

• Given a real number t, the wrapping function W determines a point on the circle corresponding to t as follows:

(a) If $t = 0$, $W(t) = A(1, 0)$.

(b) If $t > 0$, $W(t) = P(x, y)$ is the point reached by starting at $(1, 0)$ and wrapping t units along the unit circle in the counterclockwise direction.

(c) If $t < 0$, $W(t) = P(x, y)$ is the point reached by starting at $(1, 0)$ and wrapping $|t|$ units along the unit circle in the clockwise direction.

• The wrapping function is not one-to-one. In fact,

$$W(t) = W(t + 2\pi n), \quad n \text{ an integer}$$

so that every point on the unit circle corresponds to infinitely many real numbers.

As a result of the wrapping function, every point P on the unit circle may now be designated in two ways, by the usual rectangular coordinates (x, y) or by a real number t such that $W(t) = P(x, y)$. We illustrate by finding the rectangular coordinates of the points labeled in Figure 6. Since the unit circle has circumfer-

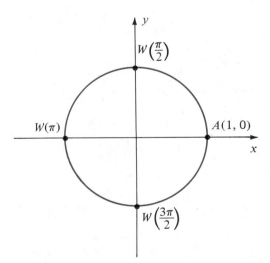

FIGURE 6

ence 2π, the point $W(\pi)$ lies halfway around the circle and has rectangular coordinates $(-1, 0)$. Similarly, we find that $W(\pi/2)$ has rectangular coordinates $(0, 1)$ and $W(3\pi/2)$ has rectangular coordinates $(0, -1)$. Since $W(t) =$

$W(t + 2\pi n)$, n an integer, we see that

$$W\left(\frac{\pi}{2}\right) = W\left(\frac{\pi}{2} + 2\pi\right) = W\left(\frac{5\pi}{2}\right)$$

also has rectangular coordinates $(0, 1)$.

EXAMPLE 3
Locate the following points on a unit circle.

(a) $W\left(\dfrac{\pi}{4}\right)$ (b) $W\left(-\dfrac{\pi}{2}\right)$ (c) $W(3\pi)$ (d) $W\left(\dfrac{9\pi}{4}\right)$

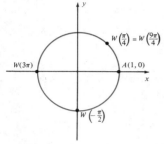

FIGURE 7

SOLUTION
The points are shown in Figure 7.

(a) $W\left(\dfrac{\pi}{4}\right)$ lies halfway between $W(0)$ and $W\left(\dfrac{\pi}{2}\right)$.

(b) $W\left(-\dfrac{\pi}{2}\right)$ can be found by moving $\dfrac{\pi}{2}$ units on the unit circle in a *clockwise* direction. Alternatively, $W\left(-\dfrac{\pi}{2}\right) = W\left(-\dfrac{\pi}{2} + 2\pi\right) = W\left(\dfrac{3\pi}{2}\right)$.

(c) $W(3\pi) = W(3\pi - 2\pi) = W(\pi)$

(d) $W\left(\dfrac{9\pi}{4}\right) = W\left(\dfrac{9\pi}{4} - 2\pi\right) = W\left(\dfrac{\pi}{4}\right)$

PROGRESS CHECK
Locate the following points on a unit circle.

(a) $W\left(\dfrac{3\pi}{4}\right)$ (b) $W\left(-\dfrac{\pi}{4}\right)$ (c) $W\left(\dfrac{5\pi}{2}\right)$ (d) $W(5\pi)$

ANSWER
The location of each point is shown in Figure 8.

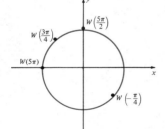

FIGURE 8

We have seen that the intercepts of the unit circle correspond to multiples of π on the real number line. In locating the point on the unit circle corresponding to a real number t, it is helpful to think of t in relation to multiples of π. Here are some examples.

EXAMPLE 4
Locate each of the following on the unit circle.

(a) $W(3)$ (b) $W(4)$ (c) $W(-2)$ (d) $W(13)$

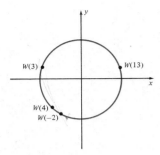

FIGURE 9

SOLUTION
The approximate location of each point is shown in Figure 9.
(a)   Since $\pi \approx 3.14$ and $W(\pi) = (-1, 0)$, $W(3)$ occurs slightly less than half-way around the unit circle in the counterclockwise direction.
(b)   Since $3\pi/2 \approx 4.71$, $W(4)$ occurs between $W(\pi)$ and $W(3\pi/2)$.
(c)   Since $-\pi/2 \approx -1.57$ and $-\pi \approx -3.14$, $W(-2)$ lies between $W(-\pi/2)$ and $W(-\pi)$.
(d)   Since $4\pi \approx 12.57$ and $W(4\pi) = W(0)$, we see that $W(13)$ occurs just beyond $W(0)$ on the unit circle in the counterclockwise direction.

VALUES OF THE WRAPPING FUNCTION

The coordinates of the point $W(t) = P(x, y)$ are, in general, not easily determined. For some "special" values of t, however, it is possible to provide a geometric argument for determining the coordinates of $W(t)$. In Figure 10, we

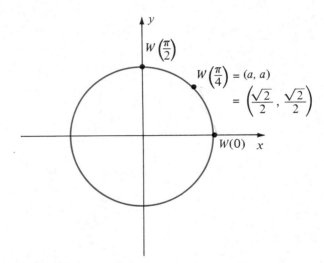

FIGURE 10

see that $W(\pi/4)$ bisects the arc between $W(0)$ and $W(\pi/2)$. Thus, $W(\pi/4)$ must lie on the line $y = x$, and we may designate the coordinates of $W(\pi/4)$ as (a, a). Since the coordinates of any point on the unit circle satisfy $x^2 + y^2 = 1$, we have

$$a^2 + a^2 = 1$$
$$2a^2 = 1$$
$$a = \sqrt{\frac{1}{2}} = \frac{\sqrt{2}}{2}$$

We conclude that

$$W\left(\frac{\pi}{4}\right) = \left(\frac{\sqrt{2}}{2}, \frac{\sqrt{2}}{2}\right)$$

EXAMPLE 5

Use the symmetries of the circle to find the coordinates of $W\left(\frac{3\pi}{4}\right)$, $W\left(\frac{5\pi}{4}\right)$, $W\left(\frac{7\pi}{4}\right)$.

SOLUTION

The points are shown in Figure 11. By using the symmetries of the circle, we determine the coordinates to be

$$W\left(\frac{3\pi}{4}\right) = \left(-\frac{\sqrt{2}}{2}, \frac{\sqrt{2}}{2}\right)$$

$$W\left(\frac{5\pi}{4}\right) = \left(-\frac{\sqrt{2}}{2}, -\frac{\sqrt{2}}{2}\right)$$

$$W\left(\frac{7\pi}{4}\right) = \left(\frac{\sqrt{2}}{2}, -\frac{\sqrt{2}}{2}\right)$$

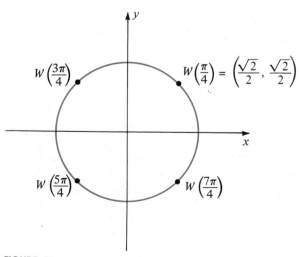

FIGURE 11

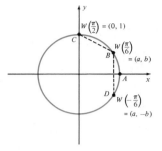

FIGURE 12

We can employ a geometric argument to find the rectangular coordinates of $W(\pi/6)$. In Figure 12 we let the point $B(a, b)$ correspond to $W(\pi/6)$. Then the length of arc $\overparen{BC}$ can be calculated by

$$\overparen{BC} = \overparen{AC} - \overparen{AB} = \frac{\pi}{2} - \frac{\pi}{6} = \frac{\pi}{3}$$

Locating the point D corresponding to $W(-\pi/6)$, we see from the symmetries of the circle that D has coordinates $(a, -b)$ and that arc $\overparen{BD}$ can be calculated in this way.

$$\overparen{BD} = \overparen{BA} + \overparen{AD} = \frac{\pi}{6} + \frac{\pi}{6} = \frac{\pi}{3}$$

Recall from plane geometry that equal arcs subtend equal chords, so that $\overline{BC} = \overline{BD}$ and, by the distance formula,

$$\sqrt{(a - 0)^2 + (b - 1)^2} = \sqrt{(a - a)^2 + (b - (-b))^2} = \sqrt{(2b)^2} = 2b$$

Squaring both sides,

$$a^2 + b^2 - 2b + 1 = 4b^2$$

or

$$a^2 + b^2 = 4b^2 + 2b - 1$$

Since (a, b) are the coordinates of a point on the unit circle, they must satisfy the equation $x^2 + y^2 = 1$. Thus, $a^2 + b^2 = 1$, and we have by substitution

$$1 = 4b^2 + 2b - 1$$
$$0 = 4b^2 + 2b - 2$$
$$0 = 2(2b - 1)(b + 1)$$
$$b = \frac{1}{2} \quad \text{or} \quad b = -1$$

Since $W(\pi/6)$ is in the first quadrant, b must be positive. Thus, we have $b = 1/2$. Since $a^2 + b^2 = 1$, we find that $a = \sqrt{3}/2$. We conclude that

$$W\left(\frac{\pi}{6}\right) = \left(\frac{\sqrt{3}}{2}, \frac{1}{2}\right)$$

A similar argument (see Exercise 46) can be used to show that

$$W\left(\frac{\pi}{3}\right) = \left(\frac{1}{2}, \frac{\sqrt{3}}{2}\right)$$

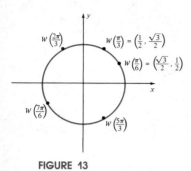

FIGURE 13

EXAMPLE 6

Use the symmetries of the circle to find $W\left(\dfrac{2\pi}{3}\right)$, $W\left(\dfrac{7\pi}{6}\right)$, $W\left(\dfrac{5\pi}{3}\right)$.

SOLUTION

The points are shown in Figure 13.

By the symmetries of the circle,

$$W\left(\frac{2\pi}{3}\right) = \left(-\frac{1}{2}, \frac{\sqrt{3}}{2}\right)$$

$$W\left(\frac{7\pi}{6}\right) = \left(-\frac{\sqrt{3}}{2}, -\frac{1}{2}\right)$$

$$W\left(\frac{5\pi}{3}\right) = \left(\frac{1}{2}, -\frac{\sqrt{3}}{2}\right)$$

PROGRESS CHECK

Use the symmetries of the circle to find the following.

(a) $W\left(\dfrac{4\pi}{3}\right)$ (b) $W\left(\dfrac{5\pi}{6}\right)$ (c) $W\left(\dfrac{11\pi}{6}\right)$

ANSWERS

(a) $\left(-\dfrac{1}{2}, -\dfrac{\sqrt{3}}{2}\right)$ (b) $\left(-\dfrac{\sqrt{3}}{2}, \dfrac{1}{2}\right)$ (c) $\left(\dfrac{\sqrt{3}}{2}, -\dfrac{1}{2}\right)$

We conclude this section with yet another observation concerning the symmetries of the circle.

If $P(a, b) = W(t)$ and $P'(a', b') = W\left(t \pm \dfrac{\pi}{2}\right)$ are points on a circle in standard position, then either

(i) $(a', b') = (-b, a)$, or
(ii) $(a', b') = (b, -a)$.

The proof of this theorem is outlined in Exercises 47–49. The student is urged to use this result in solving the accompanying Progress Check.

PROGRESS CHECK

Given $W(t) = \left(\dfrac{3}{5}, -\dfrac{4}{5}\right)$, use the symmetries of the circle to find

(a) $W\left(t + \dfrac{\pi}{2}\right)$ (b) $W\left(t - \dfrac{\pi}{2}\right)$ (c) $W\left(-t + \dfrac{\pi}{2}\right)$

ANSWERS

(a) $\left(\dfrac{4}{5}, \dfrac{3}{5}\right)$ (b) $\left(-\dfrac{4}{5}, -\dfrac{3}{5}\right)$ (c) $\left(-\dfrac{4}{5}, \dfrac{3}{5}\right)$

EXERCISE SET 5.1

Replace each given real number s by t, $0 \le t < 2\pi$, so that $W(t) = W(s)$.

1. 4π
2. $\dfrac{13\pi}{2}$
3. $\dfrac{15\pi}{7}$
4. $-\dfrac{25\pi}{4}$

5. $-\dfrac{21\pi}{2}$
6. $-\dfrac{11\pi}{2}$
7. $\dfrac{41\pi}{6}$
8. $\dfrac{11\pi}{2}$

9. -9π
10. 7π
11. $\dfrac{27\pi}{5}$
12. $-\dfrac{22\pi}{3}$

Plot the approximate positions of the given points on the unit circle.

13. $W(7\pi)$, $W\left(\dfrac{4\pi}{3}\right)$, $W\left(\dfrac{5\pi}{2}\right)$, $W\left(-\dfrac{7\pi}{4}\right)$, $W\left(-\dfrac{13\pi}{6}\right)$

14. $W\left(\dfrac{11\pi}{2}\right)$, $W\left(-\dfrac{3\pi}{4}\right)$, $W\left(\dfrac{11\pi}{6}\right)$, $W\left(-\dfrac{10\pi}{3}\right)$, $W\left(\dfrac{33\pi}{4}\right)$

15. $W(-10)$, $W(8)$, $W(3.3)$, $W(-4)$, $W(1.7)$

16. $W(14)$, $W(-0.5)$, $W(-8)$, $W(6)$, $W(3)$

Find the rectangular coordinates of the given points.

17. $W(5\pi)$
18. $W\left(\dfrac{5\pi}{2}\right)$
19. $W\left(-\dfrac{\pi}{4}\right)$
20. $W\left(-\dfrac{3\pi}{2}\right)$

21. $W\left(\dfrac{5\pi}{4}\right)$
22. $W(8\pi)$
23. $W\left(\dfrac{4\pi}{3}\right)$
24. $W\left(\dfrac{2\pi}{3}\right)$

25. $W\left(-\dfrac{2\pi}{3}\right)$
26. $W\left(-\dfrac{19\pi}{3}\right)$
27. $W\left(\dfrac{19\pi}{6}\right)$
28. $W\left(\dfrac{17\pi}{6}\right)$

29. $W\left(-\dfrac{5\pi}{6}\right)$
30. $W\left(-\dfrac{11\pi}{6}\right)$
31. $W\left(\dfrac{19\pi}{3}\right)$
32. $W\left(\dfrac{25\pi}{3}\right)$

Determine a positive and a negative value of t, $|t| < 2\pi$, for which $W(t)$ has the given rectangular coordinates.

33. $W(t) = (-1, 0)$
34. $W(t) = (0, -1)$

35. $W(t) = \left(-\dfrac{\sqrt{2}}{2}, \dfrac{\sqrt{2}}{2}\right)$
36. $W(t) = \left(\dfrac{\sqrt{2}}{2}, -\dfrac{\sqrt{2}}{2}\right)$

37. $W(t) = \left(-\dfrac{\sqrt{3}}{2}, \dfrac{1}{2}\right)$
38. $W(t) = \left(-\dfrac{1}{2}, -\dfrac{\sqrt{3}}{2}\right)$

39. $W(t) = \left(\dfrac{1}{2}, -\dfrac{\sqrt{3}}{2}\right)$
40. $W(t) = \left(-\dfrac{\sqrt{3}}{2}, -\dfrac{1}{2}\right)$

41. $W(t) = \left(-\dfrac{1}{2}, \dfrac{\sqrt{3}}{2}\right)$

42. Given $W(t) = \left(\dfrac{3}{5}, \dfrac{4}{5}\right)$, use the symmetries of the circle to find

   (a) $W(t + \pi)$ (b) $W\left(t - \dfrac{\pi}{2}\right)$

   (c) $W(-t)$ (d) $W(-t - \pi)$

43. Given $W(t) = \left(-\dfrac{4}{5}, -\dfrac{3}{5}\right)$, use the symmetries of the circle to find

   (a) $W(t - \pi)$ (b) $W\left(t + \dfrac{\pi}{2}\right)$

   (c) $W(-t)$ (d) $W(-t + \pi)$

44. If the point (a, b) is on the unit circle, show that $(a, -b)$, $(-a, b)$, and $(-a, -b)$ also lie on the unit circle.

45. If the point (a, b) is on the unit circle, show that (b, a), $(b, -a)$, $(-b, a)$, and $(-b, -a)$ also lie on the unit circle.

46. Show that $W\left(\dfrac{\pi}{3}\right) = \left(\dfrac{1}{2}, \dfrac{\sqrt{3}}{2}\right)$. $\left(Hint:\right.$ If $W\left(\dfrac{\pi}{3}\right) = (a, b)$, then $W\left(\dfrac{2\pi}{3}\right) = (-a, b)$. The arc from $A(1, 0)$ to $W\left(\dfrac{\pi}{3}\right)$ and the arc from $W\left(\dfrac{\pi}{3}\right)$ to $W\left(\dfrac{2\pi}{3}\right)$ are equal.$\left.\right)$

In Exercises 47–49, the points $P(a, b) = W(t)$ and $P'(a', b') = W\left(t \pm \dfrac{\pi}{2}\right)$ lie on a circle that is in standard position and whose center is at the origin O.

47. Show that $\dfrac{b'}{a'} = -\dfrac{a}{b}$. (*Hint:* Show that the lines OP and OP' are perpendicular and then determine their slopes.)

48. Show that $b' = \pm a$ and $a' = \pm b$. (*Hint:* The radii OP and OP' are equal in length. Use the distance formula combined with the result of Exercise 47 above to substitute alternately for b' and for a'.)

49. Show that either (i) $(a', b') = (-b, a)$ or (ii) $(a', b') = (b, -a)$. (*Hint:* Begin with the result of Exercise 48 and apply the result of Exercise 47.)

5.2
THE SINE, COSINE, AND TANGENT FUNCTIONS

There are six **trigonometric** or **circular functions** that are defined by the wrapping function. In this section we will discuss the **sine, cosine,** and **tangent** functions, which are written as **sin, cos,** and **tan,** respectively. The remaining three functions are reciprocals of these and will be described in a later section. We now define the first three circular functions.

DEFINITION AND DOMAIN

Circular Functions	If t is a real number and $W(t) = (x, y)$, then
	$$\sin t = y$$
	$$\cos t = x$$
	$$\tan t = \dfrac{y}{x}, \quad x \neq 0$$

It is often convenient to think of these definitions in this way:

$$W(t) = (\cos t, \sin t)$$

We see immediately that the domain of the sine and cosine functions is the set of all real numbers. The tangent function, however, is not defined when $x = 0$. Since $W(t) = (0, y)$ when $t = \pi/2$ and $3\pi/2$ and since $W(t) = W(t + 2\pi n)$, we see that $W(t) = (0, y)$ when $t = \pi/2 + 2\pi n$ and $t = 3\pi/2 + 2\pi n$ or, more compactly, when $t = \pi/2 + n\pi$ for all integers n.

Domain of the Circular Functions

$\sin t$:   all real values of t

$\cos t$:   all real values of t

$\tan t$:   all real values of t such that $t \neq \dfrac{\pi}{2} + n\pi$ for all integers n

---

SPECIAL VALUES OF THE CIRCULAR FUNCTIONS

Given a real number t, if we can find the rectangular coordinates of $W(t)$, we can determine the values of the circular functions. Here are some examples.

EXAMPLE 1

Evaluate $\sin t$, $\cos t$, and $\tan t$ for each of the following.

(a) $t = 0$ (b) $t = \dfrac{\pi}{6}$ (c) $t = \dfrac{\pi}{4}$

SOLUTION

(a) $W(0) = (1, 0)$. Then $\sin 0 = y = 0$, $\cos 0 = x = 1$, and $\tan 0 = \dfrac{y}{x} = 0$.

(b) $W\left(\dfrac{\pi}{6}\right) = \left(\dfrac{\sqrt{3}}{2}, \dfrac{1}{2}\right)$. Then

$$\sin \frac{\pi}{6} = y = \frac{1}{2}, \quad \cos \frac{\pi}{6} = x = \frac{\sqrt{3}}{2}$$

$$\tan \frac{\pi}{6} = \frac{y}{x} = \frac{\dfrac{1}{2}}{\dfrac{\sqrt{3}}{2}} = \frac{1}{\sqrt{3}} = \frac{\sqrt{3}}{3}$$

(c) $W\left(\dfrac{\pi}{4}\right) = \left(\dfrac{\sqrt{2}}{2}, \dfrac{\sqrt{2}}{2}\right)$. Then

$$\sin \frac{\pi}{4} = \cos \frac{\pi}{4} = \frac{\sqrt{2}}{2}$$

$$\tan \frac{\pi}{4} = 1$$

PROGRESS CHECK

Evaluate sin t, cos t, and tan t for each of the following.

(a) $t = \dfrac{\pi}{2}$ (b) $t = \dfrac{\pi}{3}$ (c) $t = \pi$

ANSWERS

(a) $\sin \dfrac{\pi}{2} = 1$, $\cos \dfrac{\pi}{2} = 0$, $\tan \dfrac{\pi}{2}$ is undefined

(b) $\sin \dfrac{\pi}{3} = \dfrac{\sqrt{3}}{2}$, $\cos \dfrac{\pi}{3} = \dfrac{1}{2}$, $\tan \dfrac{\pi}{3} = \sqrt{3}$

(c) $\sin \pi = 0$, $\cos \pi = -1$, $\tan \pi = 0$

Since we will frequently refer to the values of the sine, cosine, and tangent functions for certain "special" real numbers t, we list these in Table 1. Note that these values of t are in the interval $[0, 2\pi]$ and that we already know $W(t)$. There are no entries for tan $\pi/2$ and tan $3\pi/2$ since the tangent function is not defined for these values.

TABLE 1

t	$W(t)$	sin t	cos t	tan t
0	$(1, 0)$	0	1	0
$\dfrac{\pi}{6}$	$\left(\dfrac{\sqrt{3}}{2}, \dfrac{1}{2}\right)$	$\dfrac{1}{2}$	$\dfrac{\sqrt{3}}{2}$	$\dfrac{\sqrt{3}}{3}$
$\dfrac{\pi}{4}$	$\left(\dfrac{\sqrt{2}}{2}, \dfrac{\sqrt{2}}{2}\right)$	$\dfrac{\sqrt{2}}{2}$	$\dfrac{\sqrt{2}}{2}$	1
$\dfrac{\pi}{3}$	$\left(\dfrac{1}{2}, \dfrac{\sqrt{3}}{2}\right)$	$\dfrac{\sqrt{3}}{2}$	$\dfrac{1}{2}$	$\sqrt{3}$
$\dfrac{\pi}{2}$	$(0, 1)$	1	0	
π	$(-1, 0)$	0	-1	0
$\dfrac{3\pi}{2}$	$(0, -1)$	-1	0	

PROPERTIES OF THE CIRCULAR FUNCTIONS

In mathematics, whenever we define a new quantity or function, we then proceed to investigate its properties. We will spend the rest of this section determining some simple properties of the circular functions.

The *signs* of the circular functions in each of the four quadrants are shown in Figure 14a. These follow immediately from the definitions. For example, since both the x- and y-coordinates of any point in the third quadrant are negative, the

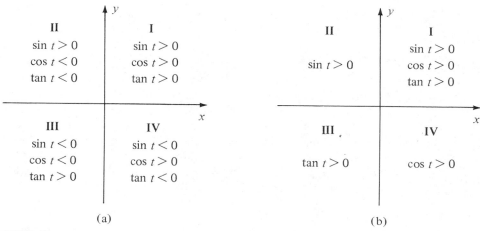

FIGURE 14

sine and cosine functions both have negative values if $W(t)$ is in the third quadrant. The tangent will be positive in the third quadrant since it is the ratio y/x of two negative values.

Figure 14b shows where each of the circular functions is positive. It is easier to remember this and to determine the negative values by inference.

EXAMPLE 2
Determine the quadrant in which $W(t)$ lies in each of the following.
(a) $\sin t > 0$ and $\tan t < 0$ (b) $\sin t < 0$ and $\cos t > 0$

SOLUTION
(a) $\sin t > 0$ in quadrants I and II; $\tan t < 0$ in quadrants II and IV. Both conditions therefore apply only in quadrant II.
(b) $\sin t < 0$ in quadrants III and IV; $\cos t > 0$ in quadrants I and IV. Both conditions therefore apply only in quadrant IV.

PROGRESS CHECK
Determine the quadrant in which $W(t)$ lies in each of the following.
(a) $\cos t < 0$ and $\tan t > 0$ (b) $\cos t < 0$ and $\sin t > 0$

ANSWERS
(a) quadrant III   (b) quadrant II

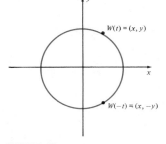

FIGURE 15

We can use the symmetries of the unit circle to find $\sin(-t)$ and $\cos(-t)$. In Figure 15 we see that $W(t)$ and $W(-t)$ correspond to points having the same x-coordinates while the y-coordinates are opposite in sign. Then $\sin t = y$ and $\sin(-t) = -y$ or $\sin(-t) = -\sin t$. Similarly, $\cos t = x = \cos(-t)$. Finally, $\tan(-t) = -y/x = -\tan t$. In summary,

$$\sin(-t) = -\sin t$$
$$\cos(-t) = \cos t$$
$$\tan(-t) = -\tan t$$

where t is any real number in the domain of the function. A function f for which $f(-x) = f(x)$ is said to be an **even function**; if $f(-x) = -f(x)$, then f is called an **odd function**. We see that sine and tangent are odd functions and cosine is an even function. Another example of an even function is $f(x) = x^2$ while $f(x) = x^3$ is an example of an odd function. From our earlier work with graphs you can see that the graph of an even function is symmetric about the y-axis while the graph of an odd function is symmetric with respect to the origin.

EXAMPLE 3

Find $\sin\left(-\dfrac{\pi}{4}\right)$ and $\cos\left(-\dfrac{\pi}{3}\right)$.

SOLUTION

$$\sin\left(-\frac{\pi}{4}\right) = -\sin\left(\frac{\pi}{4}\right) = -\frac{\sqrt{2}}{2} \qquad \cos\left(-\frac{\pi}{3}\right) = \cos\left(\frac{\pi}{3}\right) = \frac{1}{2}$$

IDENTITIES

Trigonometry often involves the use of **identities,** that is, equations that are true for *all* values in the domain of the variable. Identities are useful in simplifying equations and in providing alternative forms for computations. We can now establish two fundamental identities of trigonometry.

Since the coordinates (x, y) of every point on the unit circle satisfy the equation $x^2 + y^2 = 1$, we may substitute $x = \cos t$ and $y = \sin t$ to obtain

$$(\cos t)^2 + (\sin t)^2 = 1$$

Expressions of the form $(\sin t)^n$ occur so frequently that a special notation is used:

$$\sin^n t = (\sin t)^n \quad \text{when } n \neq -1$$

Using this notation and reordering the terms, the identity becomes

$$\sin^2 t + \cos^2 t = 1$$

Of course, we may also use this identity in the alternative forms

$$\sin^2 t = 1 - \cos^2 t$$
$$\cos^2 t = 1 - \sin^2 t$$

Since $\tan t = y/x$, $x \neq 0$, we may substitute $\sin t = y$ and $\cos t = x$ to obtain

$$\tan t = \frac{\sin t}{\cos t}$$

for all values of t in the domain of the tangent function.

EXAMPLE 4

If $\cos t = \dfrac{3}{5}$ and t is in quadrant IV, find $\sin t$ and $\tan t$.

SOLUTION

Using the identity $\sin^2 t + \cos^2 t = 1$, we have

$$\sin^2 t + \left(\frac{3}{5}\right)^2 = 1$$

$$\sin^2 t = 1 - \frac{9}{25} = \frac{16}{25}$$

$$\sin t = \pm\frac{4}{5}$$

Since t is in quadrant IV, $\sin t$ must be negative so that $\sin t = -\dfrac{4}{5}$. Then

$$\tan t = \frac{\sin t}{\cos t} = \frac{-\dfrac{4}{5}}{\dfrac{3}{5}} = -\frac{4}{3}$$

PROGRESS CHECK

If $\sin t = \dfrac{12}{13}$ and t is in quadrant II, find the following.

(a) $\cos t$ (b) $\tan t$

ANSWERS

(a) $-\dfrac{5}{13}$ (b) $-\dfrac{12}{5}$

EXAMPLE 5

Show that $1 + \tan^2 x = \dfrac{1}{\cos^2 x}$.

SOLUTION

We will use the trigonometric identities to transform the left-hand side of the equation into the right-hand side. Since $\tan x = \dfrac{\sin x}{\cos x}$, we have

$$
\begin{aligned}
1 + \tan^2 x &= 1 + \frac{\sin^2 x}{\cos^2 x} \\
&= \frac{\cos^2 x + \sin^2 x}{\cos^2 x}
\end{aligned}
$$

Since $\cos^2 x + \sin^2 x = 1$,

$$
1 + \tan^2 x = \frac{1}{\cos^2 x}
$$

PROGRESS CHECK

Use identities to transform the expression $(\tan t)(\cos t) + \sin t$ to $2 \sin t$.

EXERCISE SET 5.2

Use the rectangular coordinates of $W(t)$ to find $\sin t$, $\cos t$, $\tan t$.

1. $t = \dfrac{5\pi}{3}$
2. $t = \dfrac{3\pi}{4}$
3. $t = -5\pi$
4. $t = -\dfrac{5\pi}{4}$

5. $t = \dfrac{7\pi}{4}$
6. $t = \dfrac{7\pi}{6}$
7. $t = \dfrac{2\pi}{3}$
8. $t = \dfrac{5\pi}{6}$

9. $t = -\dfrac{\pi}{3}$
10. $t = -\dfrac{5\pi}{6}$
11. $t = \dfrac{5\pi}{4}$
12. $t = -11\pi$

Find the quadrant in which $W(t)$ lies if the following conditions hold.

13. $\sin t > 0$ and $\cos t < 0$
14. $\sin t < 0$ and $\tan t > 0$
15. $\cos t < 0$ and $\tan t > 0$
16. $\tan t < 0$ and $\sin t > 0$
17. $\sin t < 0$ and $\cos t < 0$
18. $\tan t < 0$ and $\cos t < 0$

Find the values of t in the interval $[0, 2\pi]$ that satisfy the given equation.

19. $\cos t = 0$
20. $\sin t = 1$
21. $\tan t = 1$
22. $\cos t = -1$

23. $\sin t = \dfrac{\sqrt{2}}{2}$
24. $\cos t = \dfrac{\sqrt{3}}{2}$
25. $\cos t = -\dfrac{\sqrt{3}}{2}$
26. $\sin t = -\dfrac{1}{2}$

27. $\tan t = \sqrt{3}$
28. $\sin t = \dfrac{3}{2}$
29. $\sin t = \dfrac{\sqrt{3}}{2}$
30. $\tan t = \dfrac{\sqrt{3}}{3}$

31. $\sin t = -\dfrac{1}{2}$
32. $\cos t = -\dfrac{\sqrt{2}}{2}$
33. $\sin t = 2$
34. $\cos t = \dfrac{\sqrt{2}}{2}$

35. $\sin t = \dfrac{1}{2}$
36. $\cos t = \dfrac{3}{2}$

Use trigonometric identities to find the indicated value under the given conditions.

37. $\sin t = \dfrac{3}{5}$ and $W(t)$ is in quadrant II; find $\tan t$.

38. $\tan t = -\dfrac{3}{4}$ and $W(t)$ is in quadrant II; find $\cos t$.

39. $\cos t = -\dfrac{5}{13}$ and $W(t)$ is in quadrant III; find $\sin t$.

40. $\sin t = -\dfrac{5}{13}$ and $W(t)$ is in quadrant III; find $\tan t$.

41. $\cos t = \dfrac{4}{5}$ and $\sin t < 0$; find $\sin t$.

42. $\tan t = \dfrac{12}{5}$ and $\cos t < 0$; find $\sin t$.

43. $\sin t = -\dfrac{3}{5}$ and $\tan t < 0$; find $\cos t$.

44. $\tan t = -\dfrac{5}{12}$ and $\sin t > 0$; find $\sin t$.

Use the trigonometric identities to transform the first expression into the second.

45. $(\tan t)(\cos t), \quad \sin t$

46. $\dfrac{\cos t}{\sin t}, \quad \dfrac{1}{\tan t}$

47. $\dfrac{1 - \sin^2 t}{\sin t}, \quad \dfrac{\cos t}{\tan t}$

48. $(\tan t)(\sin t) + \cos t, \quad \dfrac{1}{\cos t}$

49. $\cos t\left(\dfrac{1}{\cos t} - \cos t\right), \quad \sin^2 t$

50. $\dfrac{1 - \cos^2 t}{\sin t}, \quad \sin t$

51. $\dfrac{1 - \cos^2 t}{\cos^2 t}, \quad \tan^2 t$

52. $\dfrac{\cos^2 t}{1 - \sin t}, \quad 1 + \sin t$

53. $(\sin t - \cos t)^2, \quad 1 - 2(\sin t)(\cos t)$

54. $\dfrac{1}{1 - \sin t} + \dfrac{1}{1 + \sin t}, \quad \dfrac{2}{\cos^2 t}$

5.3
GRAPHS OF SINE, COSINE, AND TANGENT

PERIODIC FUNCTIONS

Before we tackle the problem of graphing the sine, cosine, and tangent functions, it is helpful to observe that these functions exhibit a cyclical behavior. This characteristic makes these functions especially useful in describing cyclic phenomena in a wide variety of applications. The following definition gives a name to such functions.

> A function f is **periodic** if there exists a positive number c such that
> $$f(x + c) = f(x)$$
> for all x in the domain of f. The least number c for which f is periodic is called the **period** of f.

Since $W(t + 2\pi) = W(t)$, the wrapping function is an example of a periodic function. From this observation and the definition of sine and cosine it follows immediately that

$$\sin(t + 2\pi) = \sin t \quad \text{and} \quad \cos(t + 2\pi) = \cos t$$

Thus, the sine and cosine functions are also periodic functions.

It is not difficult to show that the period of the sine and cosine functions is 2π (see Exercises 11 and 12 in Exercise Set 5.3), that is, 2π is the smallest positive real value of c such that $\sin(t + c) = \sin t$ and $\cos(t + c) = \cos t$ for all real numbers t.

**GRAPHS OF
SINE AND COSINE**

If we can graph the sine and cosine functions over the interval $[0, 2\pi]$, we can then repeat the graph for every interval of length 2π. As usual, we form a table of values, plot the corresponding points on a ty coordinate system, and sketch a smooth curve. We can make use of the results of the last section to provide us with values for plotting, as in Table 2.

TABLE 2

t	0	$\pi/6$	$\pi/4$	$\pi/3$	$\pi/2$	$3\pi/4$	π	$5\pi/4$	$3\pi/2$	$7\pi/4$	2π
$\sin t$	0	0.50	0.71	0.87	1	0.71	0	-0.71	-1	-0.71	0
$\cos t$	1	0.87	0.71	0.50	0	-0.71	-1	-0.71	0	0.71	1

We have used the approximations $\sqrt{2} \approx 1.414$ and $\sqrt{3} \approx 1.732$. With the values in the table we can sketch $y = \sin t$ over the interval $[0, 2\pi]$, as in Figure 16.

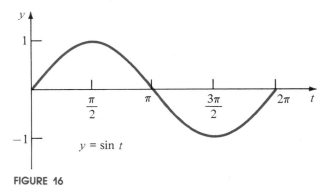

$y = \sin t$

FIGURE 16

Repeating for adjacent intervals of length 2π yields the graph in Figure 17.

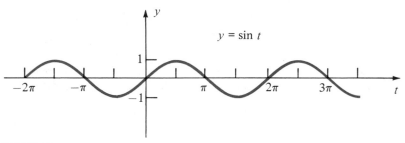

$y = \sin t$

FIGURE 17

Turning to the cosine function, we can use values given in Table 2 to sketch the graph of $y = \cos t$ as in Figure 18.

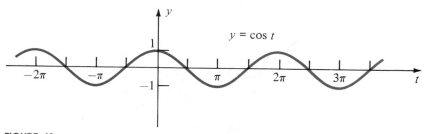

FIGURE 18

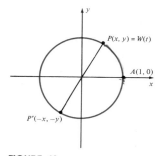

FIGURE 19

To graph the tangent function, we first establish that $\tan(t + \pi) = \tan t$ for all real values of t. If $P(x, y)$ is any point on the unit circle, then $P'(-x, -y)$ also lies on the unit circle (Figure 19), and arc $\widehat{PP'}$ is of length π. If $W(t) = P(x, y)$, then

$$\tan t = \frac{y}{x}$$

and

$$\tan(t + \pi) = \frac{-y}{-x} = \frac{y}{x}$$

so that $\tan(t + \pi) = \tan t$. It is easy to show that there is no real number c, $0 < c < \pi$, such that $\tan(t + c) = \tan t$ for all real numbers t (see Exercise 13). Hence, the tangent function has period π.

Table 3 provides us with some values of the tangent function for $-\pi/2 < t < \pi/2$. Since $\tan t$ is undefined at $\pi/2$ and at $-\pi/2$, we need to carefully consider the behavior of the graph *near* these values of t. As t increases from 0

TABLE 3

t	$-\dfrac{\pi}{2}$	$-\dfrac{\pi}{3}$	$-\dfrac{\pi}{4}$	$-\dfrac{\pi}{6}$	0	$\dfrac{\pi}{6}$	$\dfrac{\pi}{4}$	$\dfrac{\pi}{3}$	$\dfrac{\pi}{2}$
$\tan t$		-1.73	-1	-0.58	0	0.58	1	1.73	

toward $\pi/2$, the x-coordinate of $W(t)$ gets closer and closer to 0. Since $\tan t = y/x$, arbitrarily small values of x produce arbitrarily large values for the quotient y/x. We say that $\tan t$ **approaches positive infinity** as t approaches $\pi/2$. Similarly, as t decreases from 0 toward $-\pi/2$, $\tan t$ grows larger and larger in the negative sense. Accordingly, we say that $\tan t$ **approaches negative infinity** as t approaches $-\pi/2$. These considerations lead us to the graph of $\tan t$ shown in Figure 20. The vertical, dashed lines are vertical asymptotes.

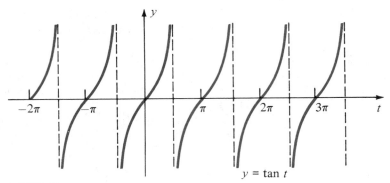

FIGURE 20

RANGE OF THE CIRCULAR FUNCTIONS

From the graphs of the sine, cosine, and tangent functions it is easy to determine the range of these functions. We list both the range and the period in Table 4.

TABLE 4

	$\sin t$	$\cos t$	$\tan t$
Range	$-1 \le y \le 1$	$-1 \le y \le 1$	all real numbers
Period	2π	2π	π

EXAMPLE 1

Sketch the graph of $f(t) = 1 + \sin t$.

SOLUTION

Rather than form a table of values and plot points, we simply note that the y-coordinate of $f(t) = 1 + \sin t$ is one unit larger than that of $\sin t$ for each value of t. In Figure 21 we have sketched $\sin t$ with dashed lines and $f(t) = 1 + \sin t$ with a solid line.

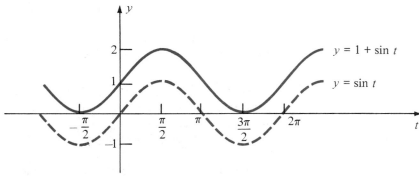

FIGURE 21

EXAMPLE 2

Sketch the graph of $f(t) = \sin t + \cos t$.

SOLUTION

Again, rather than plot points, we note that the y-coordinate of $f(t) = \sin t + \cos t$ is simply the sum of the y-coordinates of $\sin t$ and $\cos t$ for each value of t. In Figure 22 we have sketched the graphs of $\sin t$ and $\cos t$ with dashed lines, formed the sum of the y-coordinates geometrically, and then sketched a smooth curve through the resulting points.

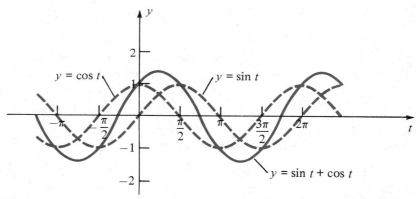

FIGURE 22

The wrapping function and the graphs of the trigonometric functions provide alternative ways of tackling problems or verifying basic results. Here is an example.

EXAMPLE 3

Verify that $\tan(-t) = -\tan t$ by using (a) the wrapping function and (b) the graph of $y = \tan t$.

SOLUTION

(a) In Figure 23a we let $W(t) = P(x, y)$ for some real number t. We see that the coordinates of $W(-t)$ are then $(x, -y)$. By definition,

$$\tan t = \frac{y}{x} \quad \text{and} \quad \tan(-t) = \frac{-y}{x} = -\frac{y}{x}$$

which establishes that $\tan(-t) = -\tan t$.

(b) In Figure 23b we see from the symmetry of the graph that the y-coordinates of the points $(t, \tan t)$ and $(-t, \tan(-t))$ are opposite in sign. From the periodic nature of the tangent function we can conclude that $\tan(-t) = -\tan t$.

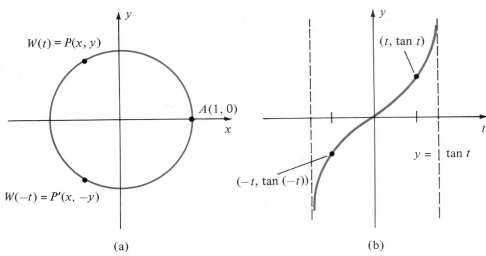

(a)   (b)

FIGURE 23

EXERCISE SET 5.3

In Exercises 1–10 sketch the graph of each given function.

1. $f(t) = 1 + \cos t$
2. $f(t) = -1 + \sin t$
3. $f(t) = 2 \sin t$
4. $f(t) = \dfrac{1}{2} \cos t$

5. $f(t) = \sin t + \dfrac{1}{2} \cos t$
6. $f(t) = 2 \sin t + \cos t$
7. $f(t) = \sin t - \cos t$
8. $f(t) = \sin(-t) + \cos t$

9. $f(t) = t + \sin t$
10. $f(t) = -t + \cos t$

11. Prove that the period of the sine function is 2π. (*Hint:* Assume $\sin(t + c) = \sin t$, $0 < c < 2\pi$, for all t. By letting $t = 0$, show that $\sin c = 0$ and, consequently, that $c = \pi$. Finally, conclude that $\sin(t + \pi) = \sin t$ does not hold for $t = \pi/2$.)

12. Prove that the period of the cosine function is 2π.

13. Prove that the period of the tangent function is π.

14. Verify that $\sin(-t) = -\sin t$ by using the graph of the sine function.

15. Verify that $\cos(-t) = \cos t$ by using the graph of the cosine function.

16. Determine the domain and range of the functions in Exercises 1 through 10 by examining the graph of each function.

5.4
**VALUES OF
SINE, COSINE,
AND TANGENT**

Although we now know the shape of the curves $y = \sin t$ and $y = \cos t$, we still don't know the precise value of $\sin t$ and $\cos t$ for an arbitrary value of t. In Table V of the Tables Appendix, you will find the values for $\sin t$ and $\cos t$ for $0 \le t \le 1.57$ in increments of 0.01, which corresponds (approximately) to $0 \le t \le \pi/2$. This raises some important questions.

(1) Since the period of sine and cosine is 2π, we do not need a table of values beyond 2π. But why does the table terminate at $\pi/2$? How can $\sin t$ be found if $\pi/2 < t < 2\pi$?

(2) How were the values determined so precisely? How can $\sin t$ and $\cos t$ be found and listed in the table for arbitrary values of t?

We will answer the first set of questions by showing how to use Table V to find $\sin t$ and $\cos t$ if $\pi/2 < t < 2\pi$. In Figure 24 we illustrate the cases in which

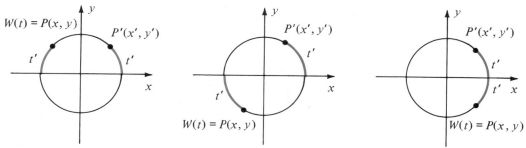

FIGURE 24

$W(t) = P(x, y)$ lies in quadrants II, III, or IV. We define the **reference number** t' associated with t as the shortest arc of the unit circle between $W(t)$ and the x-axis. Clearly, if $W(t)$ is not on a coordinate axis, then the reference number t' is less than $\pi/2$, that is, $0 < t' < \pi/2$. If we locate $W(t')$ by using the wrapping function, we arrive at a point $P'(x', y')$ in quadrant I. By the symmetries of the unit circle, in all three cases we have

$$x' = |x| \quad \text{and} \quad y' = |y|$$

Since $\sin t = y$ and $\cos t = x$, we can determine $\sin t$ and $\cos t$ by first finding $\sin t' = y'$ and $\cos t' = x'$ and then adding the proper sign. The procedure we have outlined is known as the **Reference Number Rule** for finding the value of a trigonometric function of a number t, $\pi/2 < t < 2\pi$. We now list the steps involved in this procedure and illustrate with an example.

Reference Number Rule	
Procedure	**Example**
	Find $\cos 2.55$
Step 1. Find the reference number t' associated with t. (a) If $\pi/2 < t < \pi$, then $W(t)$ is in quadrant II and $t' = \pi - t$. (b) If $\pi < t < 3\pi/2$, then $W(t)$ is in quadrant III and $t' = t - \pi$. (c) If $3\pi/2 < t < 2\pi$, then $W(t)$ is in quadrant IV and $t' = 2\pi - t$.	*Step 1.* Since $\pi/2 \approx 1.57$ and $\pi \approx 3.14$, $W(2.55)$ is in quadrant II. Thus, $$\begin{aligned} t' &= \pi - t \\ &= 3.14 - 2.55 \\ &= 0.59 \end{aligned}$$
Step 2. Obtain the (approximate) value of the required trigonometric function for the reference number t' from Table V in the Tables Appendix, or from a calculator.	*Step 2.* $$\cos 0.59 = 0.8309$$
Step 3. Append the appropriate sign according to the quadrant in which $W(t)$ lies and the trigonometric function being sought.	*Step 3.* Since $\cos t$ is negative in quadrant II, we have $$\cos 2.55 = -\cos 0.59 = -0.8309$$

EXAMPLE 1
Find sin 3.62 using Table V in the Tables Appendix.

SOLUTION
You may verify that $\pi < 3.62 < 3\pi/2$. Since $W(3.62)$ lies in the third quadrant, the reference number is $3.62 - \pi = 3.62 - 3.14 = 0.48$. From Table V we have sin $0.48 = 0.4618$. Recalling that sine is negative in the third quadrant, we must have sin $3.62 = -0.4618$.

PROGRESS CHECK
Find tan 5.96 using $\pi \approx 3.14$ and Table V.

ANSWER
-0.3314

Our second set of questions involves the calculation of tables of values of the trigonometric functions for arbitrary values of the variable t, $0 \le t < \pi/2$. The methods used are derived in calculus and result in polynomials that provide approximations to the desired function. The *approximations*

$$\sin t \approx t - \frac{t^3}{6} + \frac{t^5}{120} - \frac{t^7}{5040}$$

and

$$\cos t \approx 1 - \frac{t^2}{2} + \frac{t^4}{24} - \frac{t^6}{720}$$

will yield values for sine and cosine accurate to three decimal places.

CALCULATORS

There is a class of inexpensive "scientific calculators" that provide sin t, cos t, and tan t as "built-in functions"; that is, they have keys labeled *sin*, *cos*, and *tan*. By now you may realize that they, too, use polynomials or other means to approximate the trigonometric functions. You have probably noticed that the calculator appears to hesitate and take some time in calculating sin t; it is going through the steps of evaluating a polynomial or other function for the specific value of t that you entered.

The procedure for using hand-held calculators is different for each manufacturer. The first step is to set a selector switch for radians; this switch is usually marked *RAD/DEG* or *DRG*. The meaning of the various settings will be clear when we study angular measure in the next chapter.

In general, the value of t must be entered and the appropriate key depressed. Some calculators require that you determine the reference number if $t > \pi/2$; others perform this task for you.

EXAMPLE 2

Using the instruction booklet for your particular calculator, verify the following approximations. (Use $\pi \approx 3.14$ if needed.)

(a) $\sin 5.763 \approx -0.4970$ (b) $\tan 5.11 \approx -2.3811$

(c) $\cos 6.83 \approx 0.8542$

EXERCISE SET 5.4

Use Table V in the Tables Appendix to find each of the following. (Use $\pi \approx 3.14$.)

1. $\cos 1.12$	2. $\sin 0.48$	3. $\tan(-1.39)$	4. $\sin 4.86$
5. $\tan 3.44$	6. $\cos(-4.79)$	7. $\sin(-5.28)$	8. $\tan 6.05$
9. $\cos(-2.91)$	10. $\sin 2.43$	11. $\tan(-3.27)$	12. $\cos 1.72$

Use a calculator and the polynomial approximations

$$\sin t \approx t - \frac{t^3}{6} + \frac{t^5}{120} - \frac{t^7}{5040}$$

and

$$\cos t \approx 1 - \frac{t^2}{2} + \frac{t^4}{24} - \frac{t^6}{720}$$

to find each of the following.

13. $\sin 0.80$	14. $\cos 1.10$	15. $\sin(-0.20)$	16. $\cos(-0.75)$
17. $\tan 0.1$	18. $\tan(-1.2)$		

Use a calculator to find the following.

19. $\tan 0.97$	20. $\sin 1.32$	21. $\cos 0.15$	22. $\sin 1.84$
23. $\tan 2.77$	24. $\sin(-0.65)$	25. $\cos(-5.61)$	26. $\cos 7.08$
27. $\sin(-8.94)$	28. $\tan(-6.67)$	29. $\sin 2.62$	30. $\cos(-3.11)$

31. Using the polynomial approximation for $\sin t$, show that sine is an odd function, that is, $\sin(-t) = -\sin t$.

32. Using the polynomial approximation for $\cos t$, show that cosine is an even function, that is, $\cos(-t) = \cos t$.

5.5

GRAPHS: AMPLITUDE, PERIOD, AND PHASE SHIFT

Our objective in this section is to sketch the graph of $f(x) = A \sin(Bx + C)$, where A, B, and C are real numbers and $B > 0$. Note that we are now using the familiar symbol "x" to indicate the independent variable, rather than the symbol "t" used in earlier sections of this chapter. Of course, any symbol can be used to denote a variable; however, the symbol "x" used here is not to be confused with the x-coordinate of the point $W(t) = P(x, y)$. The results that we obtain throughout this section will also apply to the form $A \cos(Bx + C)$.

AMPLITUDE

Since the sine function has a maximum value of $+1$ and a minimum value of -1, it is clear that the function $f(x) = A \sin x$ has a maximum value of $|A|$ and a minimum value of $-|A|$. If we define the **amplitude** of a periodic function as half the difference of the maximum and minimum values, we see that the amplitude of $f(x) = A \sin x$ is $[|A| - (-|A|)]/2 = |A|$.

The amplitude of $f(x) = A \sin x$ is $|A|$.

The multiplier A acts as a vertical "stretch" factor when $|A| > 1$, and as a vertical "shrinkage" factor when $|A| < 1$. These remarks hold for both $y = A \sin x$ and $y = A \cos x$. Here are some examples.

EXAMPLE 1
Sketch the graphs of $y = 2 \sin x$ and $y = \frac{1}{2} \sin x$ on the same coordinate axes.

SOLUTION
The graph of $y = 2 \sin x$ has an amplitude of 2; the maximum value of y is $+2$ and the minimum is -2. Similarly, the amplitude of $y = \frac{1}{2} \sin x$ is $\frac{1}{2}$ and the graph has a maximum value of $+\frac{1}{2}$ and a minimum of $-\frac{1}{2}$ (Figure 25).

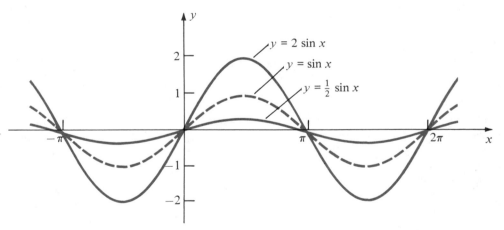

FIGURE 25

EXAMPLE 2
Sketch the graph of $f(x) = -3 \cos x$.

SOLUTION
The amplitude is 3, and $y = -3 \cos x$ has maximum and minimum values of $+3$ and -3, respectively. Since $A = -3$, each y-coordinate will be that of $\cos x$ multiplied by -3.

The graph of $y = -3 \cos x$ shown in Figure 26 is said to be a **reflection** about the x-axis of the graph of $y = 3 \cos x$.

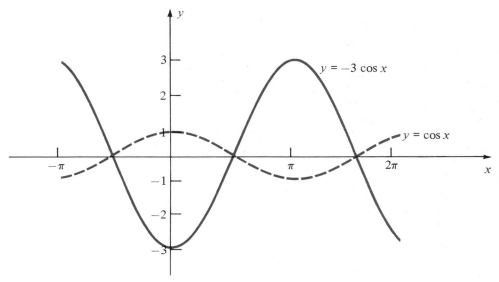

FIGURE 26

PERIOD

We now seek to determine the period of the function $f(x) = \sin Bx$. Since $y = \sin x$ has period 2π, the sine function completes one cycle or wave as x varies from 0 to 2π. Then $\sin Bx$ will complete a cycle as Bx varies from 0 to 2π, which leads to the equations

$$Bx = 0 \qquad Bx = 2\pi$$

$$x = 0 \qquad x = \frac{2\pi}{B}$$

Thus, $f(x) = \sin Bx$ completes a cycle or wave as x varies from 0 to $2\pi/B$. We conclude the following:

The period of $f(x) = \sin Bx$ is $\dfrac{2\pi}{B}$.

The multiplier B acts as a horizontal "stretch" factor if $0 < B < 1$ and as a horizontal "shrinkage" factor if $B > 1$.

EXAMPLE 3
Sketch the graph of $f(x) = \sin 2x$.

SOLUTION

Since $B = 2$, the period is $2\pi/2 = \pi$. The graph will complete a cycle every π units (Figure 27).

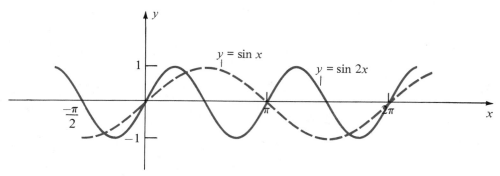

FIGURE 27

EXAMPLE 4

Sketch the graph of $f(x) = 2 \cos \frac{1}{2}x$.

SOLUTION

Since $B = \frac{1}{2}$, the period is $2\pi/\frac{1}{2} = 4\pi$. The graph will complete a cycle every 4π units. Note that the amplitude is 2, which provides us with maximum and minimum values of 2 and -2, respectively. The graph is shown in Figure 28.

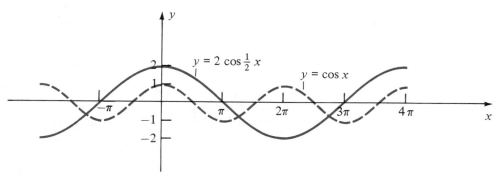

FIGURE 28

PHASE SHIFT

Let's examine the behavior of the function $f(x) = A \sin(Bx + C)$. Since $y = \sin x$ completes a cycle as x varies from 0 to 2π, the function f will complete a cycle as $Bx + C$ varies from 0 to 2π. Solving the equations

$$Bx + C = 0 \qquad Bx + C = 2\pi$$

PREDATOR-PREY INTERACTION

In the natural world we frequently find that two plant or animal species interact in their environment in such a manner that one species (the prey) serves as the primary food supply for the second species (the predator). Examples of such interaction are the relationships between trees (prey) and insects (predators) and between rabbits (prey) and lynxes (predators). As the population of the prey increases, the additional food supply results in an increase in the population of the predators. More predators consume more food, so the population of the prey will decrease, which, in turn, will lead to a decrease in the population of the predators. The reduction in the predator population results in an increase in the number of prey and the cycle will start all over again.

The accompanying figure, adapted from *Mathematics: Ideas and Applications,* by Daniel D. Benice, Academic Press, 1978 (used with permission), shows the interaction between lynx and rabbit populations. Both curves demonstrate periodic behavior and can be described by trigonometric functions.

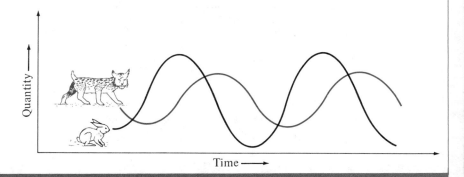

we have

$$x = -\frac{C}{B} \qquad\qquad x = \frac{2\pi - C}{B} = \frac{2\pi}{B} - \frac{C}{B}$$

The number $-C/B$ is called the **phase shift** and indicates that the graph of the function is shifted right $-C/B$ units if $-C/B > 0$ and is shifted left if $-C/B$ is negative.

The phase shift of
$$f(x) = A\,\sin(Bx + C)$$
is $-C/B$.

Note that the amplitude of f is $|A|$ and the period is $2\pi/B$; that is, the introduction of a phase shift has not altered our earlier results.

EXAMPLE 5
Sketch the graph of $f(x) = 3 \sin(2x - \pi)$.

SOLUTION
We have $A = 3$, $B = 2$, and $C = -\pi$. Then

$$\text{amplitude} = |A| = 3$$

$$\text{period} = \frac{2\pi}{B} = \frac{2\pi}{2} = \pi$$

$$\text{phase shift} = -\frac{C}{B} = -\left(-\frac{\pi}{2}\right) = \frac{\pi}{2}$$

Alternatively, solve the equation $2x - \pi = 0$ to obtain $x = \pi/2$ as the phase shift to the right.

The graphs of $y = 3 \sin 2x$ and $y = 3 \sin(2x - \pi)$ are both shown in Figure 29. To clearly demonstrate the effect of the phase shift, $y = 3 \sin(2x - \pi)$ is graphed for $\pi/2 \le x \le 5\pi/2$.

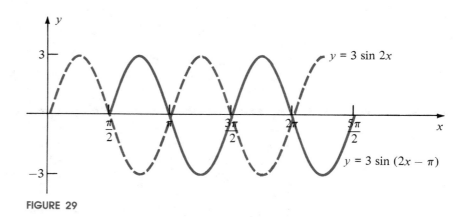

FIGURE 29

PROGRESS CHECK
If $f(x) = 2 \cos(2x + \pi/2)$, find the amplitude, period, and phase shift of f. Sketch the graph of the function.

ANSWER

$$\text{amplitude} = 2 \quad \text{period} = \pi \quad \text{phase shift} = -\frac{\pi}{4} \left(\text{or shift left } \frac{\pi}{4}\right) \text{ (Figure 30)}$$

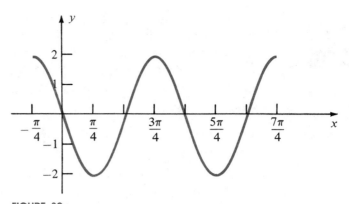

FIGURE 30

EXERCISE SET 5.5

Determine the amplitude and period and sketch the graph of each of the following functions.

1. $f(x) = 3 \sin x$

2. $f(x) = \dfrac{1}{4} \cos x$

3. $f(x) = \cos 4x$

4. $f(x) = \sin \dfrac{x}{4}$

5. $f(x) = -2 \sin 4x$

6. $f(x) = -\cos \dfrac{x}{4}$

7. $f(x) = 2 \cos \dfrac{x}{3}$

8. $f(x) = 4 \sin 4x$

9. $f(x) = \dfrac{1}{4} \sin \dfrac{x}{4}$

10. $f(x) = \dfrac{1}{2} \cos \dfrac{x}{4}$

11. $f(x) = -3 \cos 3x$

12. $f(x) = -2 \sin 3x$

For each given function, determine the amplitude, period, and phase shift. Sketch the graph of the function.

13. $f(x) = 2 \sin(x - \pi)$

14. $f(x) = \dfrac{1}{2} \cos\left(x + \dfrac{\pi}{2}\right)$

15. $f(x) = 3 \cos(2x - \pi)$

16. $f(x) = 4 \sin\left(x + \dfrac{\pi}{4}\right)$

17. $f(x) = \dfrac{1}{3} \sin\left(3x + \dfrac{3\pi}{4}\right)$

18. $f(x) = 2 \cos\left(2x + \dfrac{\pi}{2}\right)$

19. $f(x) = 2 \cos\left(\dfrac{x}{4} - \pi\right)$

20. $f(x) = 6 \sin\left(\dfrac{x}{2} + \dfrac{\pi}{2}\right)$

Use the identities $\sin(-t) = -\sin t$ and $\cos(-t) = \cos t$ to rewrite each equation as an equivalent equation with $B > 0$.

21. $y = -2 \sin(-2x + \pi)$

22. $y = 4 \cos\left(-\dfrac{x}{2} + \dfrac{\pi}{2}\right)$

23. $y = 3 \cos\left(-\dfrac{x}{3} + \dfrac{2\pi}{3}\right)$

24. $y = -5 \sin(-2x - \pi)$

5.6
SECANT, COSECANT, AND COTANGENT

We stated earlier in this chapter that there are six trigonometric or circular functions and that the remaining three functions are reciprocals of sine, cosine, and tangent. These functions are called the **secant, cosecant,** and **cotangent** and are written as **sec, csc,** and **cot,** respectively. We now formally define these functions.

Definition of sec t, csc t, and cot t

$$\sec t = \frac{1}{\cos t}, \quad \cos t \neq 0$$

$$\csc t = \frac{1}{\sin t}, \quad \sin t \neq 0$$

$$\cot t = \frac{1}{\tan t}, \quad \tan t \neq 0$$

By using these definitions, we can apply the results that we have obtained for sine, cosine, and tangent to these new functions.

EXAMPLE 1
Find sec $\pi/3$.

SOLUTION
Since $\cos \pi/3 = \frac{1}{2}$ (from Table 1 in Section 5.2) we see that

$$\sec \frac{\pi}{3} = \frac{1}{\cos \dfrac{\pi}{3}} = \frac{1}{\frac{1}{2}} = 2$$

EXAMPLE 2
Find the real number t, $0 \leq t \leq \pi/2$, such that $\cot t = \sqrt{3}$.

SOLUTION
We seek the real number t such that

$$\tan t = \frac{1}{\cot t} = \frac{1}{\sqrt{3}} = \frac{\sqrt{3}}{3}$$

Thus, $t = \pi/6$ since (from Table 1 in Section 5.2) $\tan \pi/6 = \sqrt{3}/3$.

We know that a real number and its reciprocal have the same sign; that is, if $x > 0$, then $1/x > 0$ and if $x < 0$, then $1/x < 0$. From this, we can immediately extend our conclusions (see Figure 14b) concerning the signs of the trigonometric functions in each quadrant (Figure 31). You don't have to memorize these; simply associate each function with its reciprocal.

EXAMPLE 3
Find the quadrant in which $W(t)$ lies if $\sin t > 0$ and $\sec t < 0$.

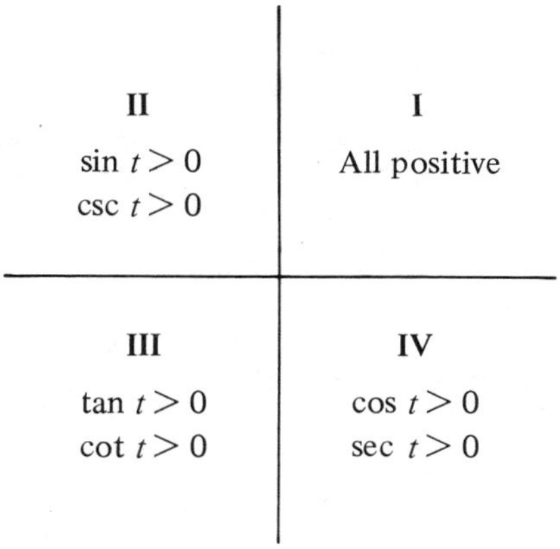

II	**I**
$\sin t > 0$	All positive
$\csc t > 0$	
III	**IV**
$\tan t > 0$	$\cos t > 0$
$\cot t > 0$	$\sec t > 0$

FIGURE 31

SOLUTION
If $\sec t < 0$, then $\cos t < 0$. We know that sine is positive in quadrants I and II, cosine is negative in quadrants II and III (Figure 14). Both conditions are satisfied in quadrant II.

PROGRESS CHECK
Find the quadrant in which $W(t)$ lies if $\tan t < 0$ and $\csc t < 0$.

ANSWER
quadrant IV

EXAMPLE 4
Find t if $\sin t = \sqrt{3}/2$ and $\sec t < 0$.

SOLUTION
Since $\sec t < 0$, we have $\cos t < 0$. Then t must lie in quadrant II, since sine is positive and cosine is negative only in quadrant II (Figure 14). Finally, we know (from Table 1 in Section 5.2) that $\sin \pi/3 = \sqrt{3}/2$. Thus, $\pi/3$ is the reference number of t; that is,

$$\pi - t = \frac{\pi}{3}$$

$$t = \frac{2\pi}{3}$$

PROGRESS CHECK
Find t if $\cos t = -\frac{1}{2}$ and $\cot t > 0$.

ANSWER
$4\pi/3$

We can also employ the definition of cosecant to aid in sketching the graph of the function. Since csc $t = 1/\sin t$, we compute the reciprocal of the y-coordinate of $\sin t$ at a point to determine the y-coordinate of csc t at that point. Of course, we cannot form a reciprocal when $\sin t = 0$, that is, when $t = n\pi$, where n is an integer. The situation at these values of t is analogous to that of the tangent function when $t = \pi/2 + n\pi$. We conclude that the graph of csc t has vertical asymptotes when $t = n\pi$, for all integer values of n. In Figure 32 we have sketched the graph of the sine function with dashed lines, to aid in sketching the reciprocal values of the y-coordinates for the cosecant function.

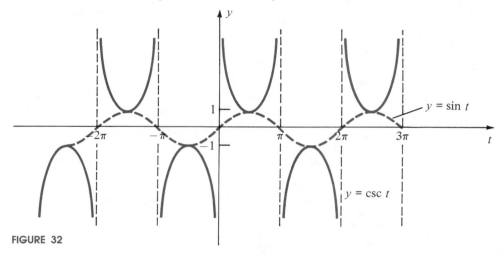

FIGURE 32

A similar approach yields the graphs of sec t and cot t shown in Figures 33 and 34.

Table 5 summarizes the significant properties of the trigonometric functions.

TABLE 5

	Positive in Quadrant	$-t$	Period	Domain	Range
sin	I, II	$-\sin t$	2π	all real numbers	$[-1, 1]$
cos	I, IV	$\cos t$	2π	all real numbers	$[-1, 1]$
tan	I, III	$-\tan t$	π	$t \neq \dfrac{\pi}{2} + n\pi$	$(-\infty, \infty)$
csc	I, II	$-\csc t$	2π	$t \neq n\pi$	$(-\infty, -1], [1, \infty)$
sec	I, IV	$\sec t$	2π	$t \neq \dfrac{\pi}{2} + n\pi$	$(-\infty, -1], [1, \infty)$
cot	I, III	$-\cot t$	π	$t \neq n\pi$	$(-\infty, \infty)$

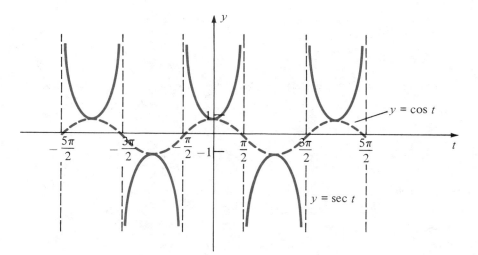

FIGURE 33

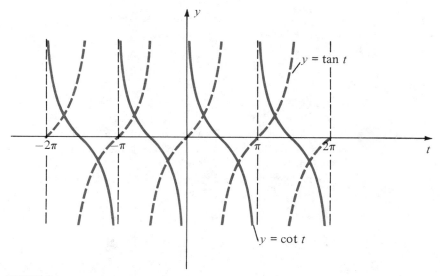

FIGURE 34

EXERCISE SET 5.6

Use the definitions of secant, cosecant, and cotangent to determine sec t, csc t, and cot t for each of the following values of t.

1. $\dfrac{\pi}{3}$

2. $\dfrac{\pi}{6}$

3. $\dfrac{\pi}{4}$

4. $\dfrac{\pi}{2}$

5. $\dfrac{5\pi}{6}$ 6. $\dfrac{4\pi}{3}$ 7. $\dfrac{3\pi}{2}$ 8. $\dfrac{7\pi}{4}$

9. $\dfrac{3\pi}{4}$ 10. $\dfrac{11\pi}{6}$ 11. $\dfrac{5\pi}{4}$ 12. $\dfrac{7\pi}{6}$

Determine the value(s) of t, $0 \le t \le 2\pi$, that satisfy each of the following.

13. $\sec t = 1$ 14. $\sec t = -1$ 15. $\csc t = -2$ 16. $\csc t = 0$

17. $\cot t = 1$ 18. $\cot t = \sqrt{3}$ 19. $\cot t = -1$ 20. $\cot t = \dfrac{\sqrt{3}}{3}$

21. $\sec t = \sqrt{2}$ 22. $\csc t = -\sqrt{2}$ 23. $\cot t = -\sqrt{3}$ 24. $\csc t = 2\dfrac{\sqrt{3}}{3}$

Find the quadrant in which $W(t)$ lies if the following conditions hold.

25. $\sec t < 0$, $\sin t < 0$ 26. $\tan t < 0$, $\sec t < 0$ 27. $\csc t > 0$, $\sec t < 0$ 28. $\sin t < 0$, $\cot t > 0$

29. $\sec t < 0$, $\cot t > 0$ 30. $\cot t < 0$, $\sin t > 0$ 31. $\sec t < 0$, $\csc t < 0$ 32. $\csc t < 0$, $\cot t > 0$

Determine the value of t, $0 \le t \le 2\pi$, that satisfies each of the following.

33. $\sin t = 1/2$, $\sec t < 0$ 34. $\tan t = \sqrt{3}$, $\csc t < 0$

35. $\sec t = -2$, $\csc t > 0$ 36. $\csc t = -2$, $\cot t > 0$

37. $\csc t = -\sqrt{2}$, $\sec t < 0$ 38. $\sec t = \sqrt{2}$, $\cot t > 0$

39. $\cot t = -1$, $\sec t < 0$ 40. $\cot t = \sqrt{3}$, $\csc t < 0$

Use Table V in the Tables Appendix to find each of the following. (Assume $\pi \approx 3.14$ to find the reference number.)

41. $\cot 3.37$ 42. $\sec 0.48$ 43. $\csc(-4.68)$ 44. $\csc 2.48$

45. $\sec 1.26$ 46. $\cot(-1.82)$

5.7
THE INVERSE TRIGONOMETRIC FUNCTIONS

Inverse functions were introduced in Section 2.8 and were used to define logarithmic functions in Section 4.2. We have seen that if f is a one-to-one function whose domain is the set X and whose range is the set Y, then the inverse function f^{-1} reverses the correspondence; that is,

$$f^{-1}(y) = x$$

if and only if

$$f(x) = y \quad \text{for all } x \in X$$

Using this definition, we saw that the following identities characterize inverse functions.

$$f^{-1}[f(x)] = x \quad \text{for all } x \text{ in } X$$
$$f[f^{-1}(y)] = y \quad \text{for all } y \text{ in } Y$$

If we attempt to find an inverse of the sine function, we have an immediate problem. Since sine is a periodic function, it is not a one-to-one function and has no inverse. However, we can resolve this problem by defining a function that agrees with the sine function but over a restricted domain. That is, we would like to find an interval such that $y = \sin x$ is one-to-one and y assumes all values between -1 and $+1$ over this interval. If we define the function f by

$$f(x) = \sin x, \quad -\frac{\pi}{2} \le x \le \frac{\pi}{2}$$

then f takes on the same values as the sine function over the interval $[-\pi/2, \pi/2]$ and assumes all real values in the interval $[-1, 1]$. The graph of $\sin x$ over the interval $[-\pi/2, \pi/2]$ shows that f is an increasing function and is therefore one-to-one. Consequently, f has an inverse, and we are led to the following definition.

The inverse sine function, denoted by **arcsin** or **sin^{-1}**, is defined by

$$\sin^{-1} y = x \quad \text{if and only if} \quad \sin x = y$$

where $-\dfrac{\pi}{2} \le x \le \dfrac{\pi}{2}$.

Note that $-1 \le y \le 1$, so the domain of the inverse sine function is the set of all real numbers in the interval $[-1, 1]$.

WARNING   When we defined $\sin^n t = (\sin t)^n$ we said that this definition does not hold when $n = -1$, allowing us to reserve the notation $\sin^{-1}$ for the inverse sine function. Therefore, $\sin^{-1} y$ is not to be confused with

$$(\sin y)^{-1} = \frac{1}{\sin y}$$

The notations arcsin and $\sin^{-1}$ are both in common use. We will therefore employ both notations to accustom you to their use. Note that if $x = \arcsin y$, then $\sin x = y$. On a unit circle, $W(x)$ determines an *arc whose sine is y,* which is the origin of the notation arcsin y.

We would like to sketch the graph of $y = \sin^{-1} x$. (Since x and y are simply symbols for variables, we have reverted to the usual practice of letting x be the independent variable.) The graph, of course, is the same as that of $\sin y = x$, with the restriction that $-\pi/2 \le y \le \pi/2$. We form a table of values and sketch the graph in Figure 35.

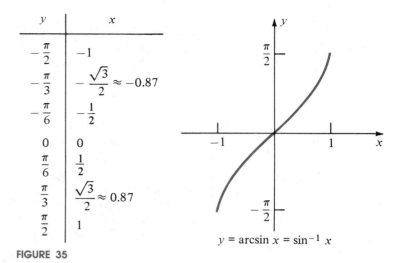

y	x
$-\dfrac{\pi}{2}$	-1
$-\dfrac{\pi}{3}$	$-\dfrac{\sqrt{3}}{2} \approx -0.87$
$-\dfrac{\pi}{6}$	$-\dfrac{1}{2}$
0	0
$\dfrac{\pi}{6}$	$\dfrac{1}{2}$
$\dfrac{\pi}{3}$	$\dfrac{\sqrt{3}}{2} \approx 0.87$
$\dfrac{\pi}{2}$	1

$y = \arcsin x = \sin^{-1} x$

FIGURE 35

EXAMPLE 1

Find   (a) $\arcsin \frac{1}{2}$ (b) $\arcsin(-1)$.

SOLUTION

(a)   If $y = \arcsin \frac{1}{2}$, then $\sin y = \frac{1}{2}$ where y is restricted to the interval $[-\pi/2, \ \pi/2]$. Thus, $y = \pi/6$ is the *only* correct answer.

(b)   If $y = \arcsin(-1)$, then $\sin y = -1$ where $-\pi/2 \leq y \leq \pi/2$. Thus, $-\pi/2$ is the *only* correct answer.

EXAMPLE 2

Evaluate $\sin^{-1}\left(\cos \dfrac{\pi}{4}\right)$.

SOLUTION

Since $\cos \dfrac{\pi}{4} = \dfrac{\sqrt{2}}{2}$, we have

$$\sin^{-1}\left(\cos \frac{\pi}{4}\right) = \sin^{-1}\left(\frac{\sqrt{2}}{2}\right)$$

We let

$$y = \sin^{-1}\left(\frac{\sqrt{2}}{2}\right)$$

Then

$$\sin y = \frac{\sqrt{2}}{2} \quad \text{where} \quad -\frac{\pi}{2} \leq y \leq \frac{\pi}{2}$$

$$y = \frac{\pi}{4}$$

which is the *only* solution.

PROGRESS CHECK

Find   (a) $\sin^{-1}\left(-\dfrac{\sqrt{3}}{2}\right)$ (b) $\arcsin\left(\tan\dfrac{5\pi}{4}\right)$.

ANSWERS

(a) $-\dfrac{\pi}{3}$ (b) $\dfrac{\pi}{2}$

We may use a similar approach to define the inverse cosine function. If we define the function f by

$$f(x) = \cos x, \quad 0 \le x \le \pi$$

then f agrees with the cosine function over the interval $[0, \pi]$, assumes all real values in the interval $[-1, 1]$, and is a decreasing function. Consequently, f is a one-to-one function and has an inverse.

> The inverse cosine function, denoted by **arccos** or **cos^{-1}**, is defined by
> $$\cos^{-1} y = x \quad \text{if and only if} \quad \cos x = y$$
> where $0 \le x \le \pi$.

Since $-1 \le y \le 1$, the domain of the inverse cosine function is the set of all real numbers in the interval $[-1, 1]$.

To sketch the graph of $y = \cos^{-1} x$ we sketch the graph of $\cos y = x$ as in Figure 36.

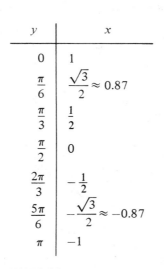

y	x
0	1
$\dfrac{\pi}{6}$	$\dfrac{\sqrt{3}}{2} \approx 0.87$
$\dfrac{\pi}{3}$	$\dfrac{1}{2}$
$\dfrac{\pi}{2}$	0
$\dfrac{2\pi}{3}$	$-\dfrac{1}{2}$
$\dfrac{5\pi}{6}$	$-\dfrac{\sqrt{3}}{2} \approx -0.87$
π	-1

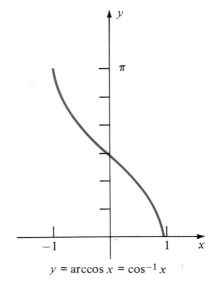

$y = \arccos x = \cos^{-1} x$

FIGURE 36

EXAMPLE 3

Find (a) $\cos^{-1}\left(-\dfrac{1}{2}\right)$ (b) $\arccos\left(\sin\dfrac{\pi}{2}\right)$.

SOLUTION

(a) If $y = \cos^{-1}\left(-\dfrac{1}{2}\right)$, then $\cos y = -\dfrac{1}{2}$ where y is restricted to the interval $[0, \pi]$. Consequently, $y = \dfrac{2\pi}{3}$ is the *only* correct answer.

(b) Since $\sin\dfrac{\pi}{2} = 1$, we let $y = \arccos(1)$. Then $\cos y = 1$ where $0 \le y \le \pi$. Therefore, $y = 0$ is the *only* correct answer.

If we restrict the tangent function to the interval $(-\pi/2, \pi/2)$, we can define the inverse tangent function.

The inverse tangent function, denoted by **arctan** or $\mathbf{tan^{-1}}$, is defined by

$$\tan^{-1} y = x \quad \text{if and only if} \quad \tan x = y$$

where $-\dfrac{\pi}{2} < x < \dfrac{\pi}{2}$.

Note that the domain of the inverse tangent function is the set of all real numbers. Proceeding as before, we sketch the graph of $y = \tan^{-1} x$ in Figure 37.

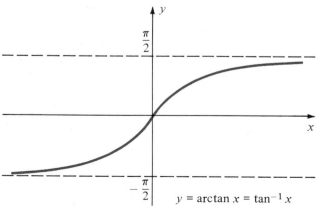

$y = \arctan x = \tan^{-1} x$

FIGURE 37

EXAMPLE 4

Find (a) $\tan^{-1}(\sqrt{3})$ (b) $\tan^{-1}(4.256)$.

SOLUTION

(a) If $y = \tan^{-1}(\sqrt{3})$, then $\tan y = \sqrt{3}$. Since $-\pi/2 < y < \pi/2$, we must have $y = \pi/3$.

(b) If $y = \tan^{-1}(4.256)$, then $\tan y = 4.256$ where $-\pi/2 < y < \pi/2$. Using Table V in the Tables Appendix or a calculator, we find $y = 1.34$.

EXAMPLE 5

Find $\cos\left(\arctan \dfrac{4}{3}\right)$ without using tables or a calculator.

SOLUTION

If we let $x = \arctan \dfrac{4}{3}$, then $\tan x = \dfrac{4}{3}$ and $0 \le x < \pi/2$. Using trigonometric identities,

$$\tan x = \frac{\sin x}{\cos x} = \frac{4}{3}$$

$$3 \sin x = 4 \cos x \qquad \text{Clearing fractions}$$

$$9 \sin^2 x = 16 \cos^2 x \qquad \text{Squaring both sides}$$

$$9(1 - \cos^2 x) = 16 \cos^2 x \qquad \sin^2 x = 1 - \cos^2 x$$

$$\cos^2 x = \frac{9}{25}$$

$$\cos x = \pm\frac{3}{5}$$

Since $x \in [0, \pi/2]$, we conclude that $\cos x = 3/5$.

PROGRESS CHECK

Without using tables or a calculator, find $\cot\left(\sin^{-1} -\dfrac{5}{13}\right)$.

ANSWER

$-\dfrac{12}{5}$

EXAMPLE 6

Find the solutions of the equation $5 \cos^2 x - 3 = 0$ that are in the interval $[0, \pi]$.

SOLUTION

We treat the equation as a quadratic in $\cos x$. Then

$$5 \cos^2 x = 3$$

$$\cos x = \pm \sqrt{\frac{3}{5}} = \pm \frac{\sqrt{15}}{5}$$

We may then write

$$x = \arccos\left(\frac{\sqrt{15}}{5}\right) \quad \text{or} \quad x = \arccos\left(-\frac{\sqrt{15}}{5}\right)$$

These are exact expressions for the solutions. Numerical approximations can be obtained using Table V in the Tables Appendix or a calculator. The student is urged to verify that

$$x \approx 0.6847 \quad \text{and} \quad x \approx 2.4568$$

are appropriate solutions of the original equation.

PROGRESS CHECK

Find the solutions of the equation $2 \sin^2 x + 2 \sin x - 1 = 0$ that are in the interval $[-\pi/2, \pi/2]$.

ANSWER

$\arcsin(-\frac{1}{2} \pm \frac{1}{2}\sqrt{3})$

WARNING It is important to remember that the range of each of the inverse trigonometric functions is a subset of the domain of the corresponding trigonometric function. Given the equation

$$t = \sin^{-1}\left(-\frac{1}{2}\right)$$

students often write $t = 7\pi/6$, which is incorrect since t must lie in the interval $[-\pi/2, \pi/2]$. The only correct answer is $-\pi/6$.

EXERCISE SET 5.7

In Exercises 1–18 evaluate the given expression.

1. $\sin^{-1}\left(-\frac{1}{2}\right)$

2. $\arccos\left(\frac{\sqrt{3}}{2}\right)$

3. $\arctan \sqrt{3}$

4. $\tan^{-1} 0$

5. $\arcsin\left(-\frac{\sqrt{2}}{2}\right)$

6. $\cos^{-1}(-1)$

7. $\arccos\left(-\frac{\sqrt{3}}{2}\right)$

8. $\tan^{-1}\left(\frac{\sqrt{3}}{3}\right)$

9. $\sin^{-1}(-1)$

10. $\arctan 1$

11. $\cos^{-1} 0$

12. $\sin^{-1}\left(-\frac{\sqrt{3}}{2}\right)$

13. $\cos^{-1} 1$

14. $\arcsin\left(\dfrac{\sqrt{2}}{2}\right)$

15. $\arctan(-1)$

16. $\sin^{-1} 0$

17. $\cos^{-1}\left(-\dfrac{1}{2}\right)$

18. $\arcsin\left(\dfrac{1}{2}\right)$

In Exercises 19–24 use Table V in the Tables Appendix to approximate the given expression.

19. $\sin^{-1}(0.3709)$

20. $\arctan(1.398)$

21. $\cos^{-1}(-0.7648)$

22. $\tan^{-1}(-3.010)$

23. $\arcsin(0.9636)$

24. $\arccos(-0.921)$

In Exercises 25–32 evaluate the given expression.

25. $\sin(\arctan 1)$

26. $\cos\left(\arcsin -\dfrac{1}{2}\right)$

27. $\tan^{-1}\left(\cos \dfrac{\pi}{2}\right)$

28. $\sin^{-1}(\sin 0.62)$

29. $\cos^{-1}\left(\sin \dfrac{9\pi}{4}\right)$

30. $\tan(\sin^{-1} 0)$

31. $\cos^{-1}\left(\cos \dfrac{2\pi}{3}\right)$

32. $\sin^{-1}\left(\cos \dfrac{\pi}{6}\right)$

In Exercises 33–38 use the inverse trigonometric functions to express the solutions of the given equation exactly.

33. $7 \sin^2 x - 1 = 0$, $x \epsilon [-\pi/2, \pi/2]$

34. $6 \cos^2 y - 5 = 0$, $y \epsilon [0, \pi]$

35. $12 \cos^2 x - \cos x - 1 = 0$, $x \epsilon [0, \pi]$

36. $2 \tan^2 t + 4 \tan t - 3 = 0$, $t \epsilon [-\pi/2, \pi/2]$

37. $9 \sin^2 t - 12 \sin t + 4 = 0$, $t \epsilon [-\pi/2, \pi/2]$

38. $3 \cos^2 x - 7 \cos x - 6 = 0$, $x \epsilon [0, \pi]$

In Exercises 39 and 40 provide a value for x to show that the equation is not an identity.

39. $\sin^{-1} x = \dfrac{1}{\sin x}$

40. $(\sin^{-1} x)^2 + (\cos^{-1} x)^2 = 1$

TERMS AND SYMBOLS

unit circle (p. 213)
wrapping function (p. 214)
$W(t)$ (p. 214)
trigonometric functions
  (p. 224)
circular functions (p. 224)
sine (sin) (p. 224)
cosine (cos) (p. 224)

tangent (tan) (p. 224)
odd function (p. 228)
even function (p. 228)
identity (p. 228)
periodic function (p. 231)
period (p. 231)
approaches positive
  infinity (p. 233)

approaches negative
  infinity (p. 233)
reference number (p. 237)
Reference Number Rule
  (p. 237)
amplitude (p. 240)
reflection (p. 241)
phase shift (p. 243)

secant (sec) (p. 246)
cosecant (csc) (p. 246)
cotangent (cot) (p. 246)
arcsin (sin^{-1}) (p. 251)
arccos (cos^{-1}) (p. 253)
arctan (tan^{-1}) (p. 254)

KEY IDEAS FOR REVIEW

☐ The wrapping function W establishes a correspondence between the set of all real numbers and the set of all points on the unit circle. Since

$$W(t) = W(t + 2\pi n), \quad n \text{ an integer,}$$

the wrapping function is periodic and consequently is not one-to-one.

☐ In general, it is difficult to determine the rectangular coordinates (x, y) corresponding to the point $W(t)$ on the unit circle. When $t = 0$, $\pi/2$, π, or $3\pi/2$, then $W(t)$ lies

on a coordinate axis and the rectangular coordinates are obvious. When $t = \pi/6$, $\pi/4$, or $\pi/3$, we can employ geometric arguments to show that

$$W\left(\frac{\pi}{6}\right) = \left(\frac{\sqrt{3}}{2}, \frac{1}{2}\right)$$

$$W\left(\frac{\pi}{4}\right) = \left(\frac{\sqrt{2}}{2}, \frac{\sqrt{2}}{2}\right)$$

$$W\left(\frac{\pi}{3}\right) = \left(\frac{1}{2}, \frac{\sqrt{3}}{2}\right)$$

☐ The circular or trigonometric functions sine, cosine, and tangent are defined in terms of the rectangular coordinates of a point on the unit circle determined by the wrapping function W. If t is a real number and $W(t) = (x, y)$, then

$$\sin t = y$$

$$\cos t = x$$

$$\tan t = \frac{y}{x}, \quad x \neq 0$$

☐ The signs of the trigonometric functions in each of the quadrants follow from the definitions and are displayed in Figure 31.

☐ Sine is an odd function and cosine is an even function. That is,

$$\sin(-t) = -\sin t$$

$$\cos(-t) = \cos t$$

☐ The trigonometric functions are all periodic. The period of the sine and cosine functions is 2π; the period of the tangent function is π.

☐ Standard tables of the values of the trigonometric functions (see Table V in the Tables Appendix) display the independent variable t from 0 to $\pi/2$. If it is desired to find $\sin t'$ where $\pi/2 < t' < 2\pi$, the Reference Number Rule is used to determine a real number t, $0 \leq t \leq \pi/2$, such that $|\sin t| = |\sin t'|$. The appropriate sign is then added depending on the quadrant of $W(t')$.

☐ To sketch the graph of $f(x) = A \sin(Bx + C)$, note that
(i) the amplitude is $|A|$;
(ii) the period is $2\pi/B$;
(iii) the phase shift is $-C/B$.
The same observations hold for $f(x) = A \cos(Bx + C)$.

☐ The secant, cosecant, and cotangent functions are defined as the reciprocals of cosine, sine, and tangent, respectively.

☐ To define the inverse trigonometric functions, it is necessary to restrict the domain of the trigonometric functions to ensure that the result is a one-to-one function.

REVIEW EXERCISES

Solutions to exercises whose numbers are in color are in the Solutions section in the back of the book.

5.1 In Exercises 1–5 replace each given real number t by t', $0 \leq t' < 2\pi$, so that $W(t') = W(t)$.

1. $\dfrac{9\pi}{2}$

2. $-\dfrac{15\pi}{2}$

3. -6π

4. $\dfrac{23\pi}{3}$

5. $\dfrac{16\pi}{5}$

In Exercises 6–10 find the rectangular coordinates of the given point.

6. $W\left(\dfrac{7\pi}{6}\right)$

7. $W\left(-\dfrac{8\pi}{3}\right)$

8. $W\left(\dfrac{5\pi}{6}\right)$

9. $W\left(-\dfrac{7\pi}{4}\right)$

10. $W\left(\dfrac{11\pi}{6}\right)$

In Exercises 11–15, $W(t) = (4/5, -3/5)$. Use the symmetries of the circle to find the rectangular coordinates of the given point.

11. $W(t - \pi)$

12. $W\left(t + \dfrac{\pi}{2}\right)$

13. $W(-t)$

14. $W\left(t - \dfrac{\pi}{2}\right)$

15. $W(-t - \pi)$

5.2 In Exercises 16–19 determine the value of the indicated trigonometric function, without the use of tables or a calculator.

16. $\sin \dfrac{2\pi}{3}$

17. $\sec\left(-\dfrac{5\pi}{4}\right)$

18. $\tan \dfrac{5\pi}{6}$

19. $\csc\left(-\dfrac{\pi}{6}\right)$

In Exercises 20–23 find a value of $t \in [0, 2\pi]$ satisfying the given conditions.

20. $\sin t = -\dfrac{\sqrt{2}}{2}$, $W(t)$ in quadrant III

21. $\cos t = \dfrac{\sqrt{3}}{2}$, $W(t)$ in quadrant IV

22. $\cot t = \dfrac{\sqrt{3}}{3}$, $W(t)$ in quadrant I

23. $\sec t = -2$, $W(t)$ in quadrant II

In Exercises 24–27 use the trigonometric identities

$$\sin^2 t + \cos^2 t = 1 \quad \tan t = \frac{\sin t}{\cos t}$$

to find the indicated value under the given conditions.

24. $\cos t = \dfrac{3}{5}$ and $W(t)$ is in quadrant IV; find $\cot t$.

25. $\sin t = -\dfrac{4}{5}$ and $\tan t > 0$; find $\sec t$.

26. $\sin t = \dfrac{12}{13}$ and $\cos t < 0$; find $\tan t$.

27. $\cos t = -\dfrac{5}{13}$ and $\tan t < 0$; find $\csc t$.

In Exercises 28 and 29 use the trigonometric identities to transform the first expression into the second.

28. $(\sin t)(\sec t)$, $\tan t$

29. $\dfrac{\sin t}{\cos^2 t}$, $(\tan t)(\sec t)$

5.4 In Exercises 30 and 31 use Table V in the Tables Appendix to evaluate the given expression. (Assume $\pi \approx 3.14$.)

30. $\cos 3.71 - \sin 1.44$

31. $\tan(-2.74)$

In Exercises 32 and 33 sketch the graph of the given function.

5.3 32. $f(x) = 1 - \sin x$

33. $f(x) = 2 \sin\left(\dfrac{x}{2} + \pi\right)$

5.5 In Exercises 34–36 determine the amplitude, period, and phase shift of each given function.

34. $f(x) = -\cos(2x - \pi)$

35. $f(x) = 4 \sin\left(-x + \dfrac{\pi}{2}\right)$

36. $f(x) = -2 \sin\left(\dfrac{x}{3} + \dfrac{\pi}{3}\right)$

5.7 In Exercises 37–39, evaluate the given expression.

37. $\arcsin\left(-\dfrac{1}{2}\right)$

38. $\tan(\cos^{-1} 1)$

39. $\tan(\tan^{-1} 5)$

40. Use the inverse cosine function to express the exact solutions of the equation

$$5 \cos^2 x - 4 = 0$$

PROGRESS TEST 5A

In Problems 1–3 replace the given real number t by t', $0 \le t' < 2\pi$, so that $W(t') = W(t)$.

1. $\dfrac{19\pi}{3}$

2. -22π

3. $\dfrac{17\pi}{4}$

In Problems 4 and 5 find the rectangular coordinates of the given point.

4. $W\left(\dfrac{29\pi}{6}\right)$

5. $W\left(-\dfrac{\pi}{3}\right)$

In Problems 6–8, $W(t) = (-5/13, 12/13)$. Use the symmetries of the circle to find the rectangular coordinates of the given point.

6. $W(t + \pi)$

7. $W\left(t - \dfrac{\pi}{2}\right)$

8. $W(-t)$

In Problems 9 and 10 determine the value of the indicated trigonometric function without the use of the tables or a calculator.

9. $\cos\left(\dfrac{7\pi}{3}\right)$

10. $\csc\left(-\dfrac{2\pi}{3}\right)$

In Problems 11 and 12 find a value of $t \in [0, 2\pi]$ satisfying the given conditions.

11. $\tan t = 1$, $W(t)$ in quadrant III

12. $\sec t = \sqrt{2}$, $W(t)$ in quadrant IV

In Problems 13 and 14 use the trigonometric identities

$$\sin^2 t + \cos^2 t = 1, \quad \tan t = \dfrac{\sin t}{\cos t}$$

to find the indicated value under the given conditions.

13. $\cos t = -\dfrac{12}{13}$ and $\tan t > 0$; find $\sin t$.

14. $\sin t = \dfrac{3}{5}$ and $W(t)$ is in quadrant II; find $\sec t$.

15. Use the trigonometric identities given for Problems 13 and 14 to transform

$$1 - \tan x \quad \text{to} \quad \dfrac{\cos x - \sin x}{\cos x}$$

In Problems 16 and 17 use Table V in the Tables Appendix to evaluate the given expression. (Assume $\pi \approx 3.14$.)

16. $\tan(-3.68)$

17. $\cos 1.15 - \sin 0.72$

18. Sketch the graph of the function f defined by $f(x) = x + \cos x$.

In Problems 19 and 20 determine the amplitude, period, and phase shift of each given function.

19. $f(x) = -2 \cos(\pi - x)$

20. $f(x) = 2 \sin\left(\dfrac{x}{2} - \dfrac{\pi}{2}\right)$

In Problems 21 and 22 evaluate the given expression.

21. $\tan^{-1}(-\sqrt{3})$

22. $\cos\left(\sin^{-1} \dfrac{\sqrt{3}}{2}\right)$

23. Use the inverse tangent function to express the exact solutions of the equation

$$6 \tan^2 x - 13 \tan x + 6 = 0$$

PROGRESS TEST 5B

In Problems 1–3 replace the given real number t by t', $0 \le t' < 2\pi$, so that $W(t') = W(t)$.

1. -14π

2. $\dfrac{51\pi}{5}$

3. $-\dfrac{19\pi}{6}$

In Problems 4 and 5 find the rectangular coordinates of the given point.

4. $W\left(\dfrac{23\pi}{6}\right)$

5. $W\left(-\dfrac{3\pi}{4}\right)$

In Problems 6–8 $W(t) = \left(-\dfrac{4}{5}, -\dfrac{3}{5}\right)$. Use the symmetries of the circle to find the rectangular coordinates of the given point.

6. $W(-t)$

7. $W\left(t + \dfrac{\pi}{2}\right)$

8. $W(-t + \pi)$

In Problems 9 and 10 determine the value of the indicated trigonometric function without the use of the tables or a calculator.

9. $\tan\left(\dfrac{7\pi}{4}\right)$

10. $\sin\left(-\dfrac{3\pi}{2}\right)$

In Problems 11 and 12 find a value of $t \in [0, 2\pi]$ satisfying the given conditions.

11. $\sin t = \dfrac{\sqrt{3}}{2}$, $W(t)$ in quadrant I

12. $\sec t = -2$, $W(t)$ in quadrant II

In Problems 13 and 14 use the trigonometric identities

$$\sin^2 t + \cos^2 t = 1 \quad \tan t = \dfrac{\sin t}{\cos t}$$

to find the indicated value under the given conditions.

13. $\sin t = -\dfrac{5}{13}$ and $\tan t < 0$; find $\tan t$.

14. $\cos t = \dfrac{3}{5}$ and $\cot t < 0$; find $\cot t$.

15. Use the trigonometric identities of Problems 13 and 14 to transform $\sec^2 t \cot t$ to $\csc t$.

In Problems 16 and 17 use Table V in the Tables Appendix to evaluate the given expression. (Assume $\pi \approx 3.14$.)

16. $\sin(2.45)$

17. $\tan(-1.25) + \cos 1.67$

18. Sketch the graph of the function f defined by

$$f(x) = \sin x + \sin \dfrac{x}{2}$$

In Problems 19 and 20 determine the amplitude, period, and phase shift of each given function.

19. $f(x) = 4\sin(3x - \pi)$

20. $f(x) = -\dfrac{1}{2}\cos\left(2x + \dfrac{\pi}{2}\right)$

In Problems 21 and 22 evaluate the given expression.

21. $\sin^{-1}\left(\cos\dfrac{\pi}{3}\right)$

22. $\tan\left(\cos^{-1}\dfrac{\sqrt{2}}{2}\right)$

23. Use the inverse sine function to express the exact solutions of the equation $25\sin^2 x - 2\sin x - 3 = 0$.

CHAPTER
6

ANGLES
AND TRIANGLES

In the previous chapter we discussed trigonometry in terms of functions of real numbers. This approach has the advantage of illustrating the centrality of the function concept in much of modern mathematical thinking.

We now turn to the more traditional approach to trigonometry, which revolves about the measurement of triangles. We will show that it is possible to define the trigonometric functions as functions of angles rather than as functions of real numbers, and that the two definitions are related in a simple manner. Our attention will then turn to the right triangle, and we will have our first opportunity to explore a wide variety of applications that clearly demonstrate the usefulness of trigonometry in such fields as surveying and navigation.

We will conclude by examining the law of sines and the law of cosines, two important rules that can be employed when dealing with an oblique triangle, that is, a triangle that does not contain a right angle.

6.1
ANGLES AND THEIR
MEASUREMENT

DEFINITION OF
AN ANGLE

In the study of trigonometry we view an angle as the result of a ray or half-line that rotates about its endpoint. When the ray coincides with the positive x-axis with its endpoint at the origin, the angle generated is said to be in **standard position.** In Figure 1 the x-axis, called the **initial side,** rotates in a counterclockwise direction until it coincides with the **terminal side,** forming the angle α. In this case we say that α is a **positive angle.** In Figure 2 the ray has been rotated in a clockwise direction to form the angle β. In this case, we say that β is a **negative angle.** If the terminal side coincides with a coordinate axis, the angle is called a **quadrantal angle;** otherwise, the angle is said to lie in the same quadrant as its terminal side.

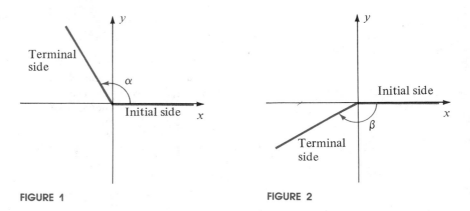

FIGURE 1   FIGURE 2

ANGULAR MEASUREMENT:
DEGREES AND RADIANS

There are two commonly used units for measuring angles. An angle is said to have a measure of one **degree** (written $1°$) if the angle is formed by rotating the initial side $\frac{1}{360}$ of a complete rotation in a counterclockwise direction. It follows that an angle obtained by a complete rotation of the initial side has a measure of $360°$, and an angle obtained by one-fourth of a complete rotation has a measure of $\frac{1}{4}(360°) = 90°$. One degree is subdivided into 60 **minutes** (written $60'$), and one minute is subdivided into 60 **seconds** (written $60''$). For example, the notation $14°24'18''$ is read *14 degrees, 24 minutes, and 18 seconds.* An angle between $0°$ and $90°$ lies in the first quadrant and is called an **acute angle;** an angle between $90°$ and $180°$ lies in the second quadrant and is called an **obtuse angle.** An angle measuring $90°$ is a quadrantal angle and is called a **right angle;** angles measuring $180°$ and $270°$ are also quadrantal angles.

To define the second unit of angular measurement, consider an angle θ in standard position, and let P be the point of intersection of the terminal side of the angle with the unit circle (Figure 3). We may think of P as the point assigned by the wrapping function for some real number t; that is, $P = W(t)$. We then say that θ is an **angle of t radians,** by which we mean that the measure of the angle θ is

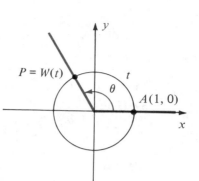

FIGURE 3

defined as the measure of the arc that θ intercepts on the unit circle. We then write $\theta = t$ radians or simply $\theta = t$. The radian measure of angles of 1 radian, -2 radians, and $3\pi/4$ radians is shown in Figures 4a, 4b, and 4c, respectively.

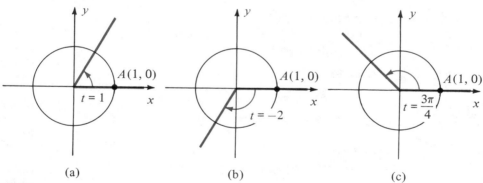

(a)   (b)   (c)

FIGURE 4

An angle of 360° traces a complete revolution in the counterclockwise direction, which corresponds to an arc of length 2π on the unit circle. Thus, 2π radians = 360° or

$$\pi \text{ radians} = 180°$$

This relationship enables us to transform angular measure from radians to degrees and vice versa. The simplest way to handle such conversions for any angle θ is by establishing a proportion.

$$\frac{\text{radian measure of angle } \theta}{\pi \text{ radians}} = \frac{\text{degree measure of angle } \theta}{180°}$$

EXAMPLE 1
Convert 150° to radian measure.

SOLUTION
With θ representing the radian measure of the angle, we establish the proportion

$$\frac{\theta}{\pi} = \frac{150}{180}$$

Solving, we have

$$\theta = \frac{150\pi}{180} = \frac{5\pi}{6}$$

Thus, $\theta = 5\pi/6$ radians.

PROGRESS CHECK
Convert the following from degree to radian measure.
(a) $-210°$ (b) $390°$

ANSWERS
(a) $-\dfrac{7\pi}{6}$ radians   (b) $\dfrac{13\pi}{6}$ radians

EXAMPLE 2
Convert $2\pi/3$ radians to degree measure.

SOLUTION
With θ denoting the degree measure of the angle, we establish the proportion

$$\frac{\frac{2\pi}{3}}{\pi} = \frac{\theta}{180}$$

Solving, we have

$$\theta = \frac{2}{3}(180) = 120$$

Thus, $\theta = 120°$.

PROGRESS CHECK

Convert the following from radian measure to degrees.

(a) $\dfrac{9\pi}{2}$ radians   (b) $-\dfrac{4\pi}{3}$ radians

ANSWERS

(a) 810°   (b) −240°

THE REFERENCE ANGLE

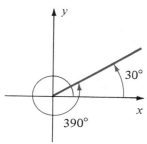

FIGURE 5

Since a complete revolution about a circle returns to the initial position, different angles in standard position may have the same terminal side. For instance, angles of 30° and 390° in standard position have the same terminal side (Figure 5). Such angles are said to be **coterminal.** For an angle in standard position that is not a quadrantal angle, it is convenient to define an acute angle that is called the **reference angle.** The reference angle θ' associated with the angle θ is the acute angle formed by the terminal side of θ and the x-axis. If θ lies in quadrant I, it is an acute angle and $\theta' = \theta$. The other cases are illustrated in Figure 6.

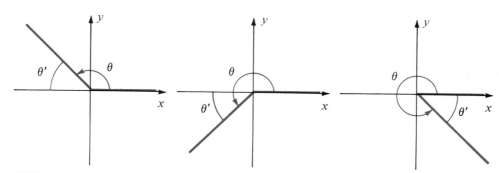

FIGURE 6

You have probably observed that the reference angle is analogous to the reference number. The reference number permits us to restrict our tables of values of the circular functions to real numbers in the interval $[0, \pi/2]$; the reference angle, when measured in radians, is also in the interval $[0, \pi/2]$ and, when measured in degrees, is in the interval $[0°, 90°]$. We will make use of the reference angle in the following section when we deal with tables of values of the trigonometric functions.

EXAMPLE 3
Find the reference angle θ' if

(a) $\theta = 240°$ (b) $\theta = \dfrac{5\pi}{3}$ radians

SOLUTION
(a)   Since $\theta = 240°$ lies in the third quadrant, the reference angle is

$$\theta' = 240° - 180° = 60°$$

(b)   Since $\theta = 5\pi/3$ radians lies in the fourth quadrant, the reference angle is

$$\theta' = 2\pi - \dfrac{5\pi}{3} = \dfrac{\pi}{3}$$

PROGRESS CHECK
Find the reference angle θ' if

(a) $\theta = 160°$ (b) $\theta = \dfrac{4\pi}{3}$ radians

ANSWERS

(a) $20°$ (b) $\dfrac{\pi}{3}$ radians

THE CENTRAL ANGLE FORMULA

The radian measure of an angle can be found by using a circle other than a unit circle. In Figure 7 the central angle θ subtends an arc of length t on the unit circle

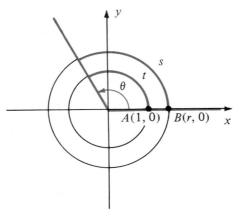

FIGURE 7

and an arc of length s on a circle of radius r. By definition, $\theta = t$. Since the ratio of the arcs is the same as the ratio of the radii, we have

$$\frac{t}{s} = \frac{1}{r}$$

or

$$t = \frac{s}{r}$$

Since $\theta = t$, we have this useful result.

If a central angle θ subtends an arc of length s on a circle of radius r, then

$$\theta = \frac{s}{r}$$

Note that when $s = r$, $\theta = 1$ radian, which tells us that an arc of 1 radian has length equal to the radius of the circle. Also, when $\theta = 2\pi$ we see that $s = 2\pi r$ corresponds to the circumference of the circle.

EXAMPLE 4

A central angle θ subtends an arc of length 12 inches on a circle whose radius is 6 inches. Find the radian measure of the central angle.

SOLUTION

We have $s = 12$ and $r = 6$, so that

$$\theta = \frac{s}{r} = \frac{12}{6} = 2 \text{ radians}$$

WARNING The formula

$$\theta = \frac{s}{r}$$

can only be applied if the angle θ is in radian measure.

EXAMPLE 5

A designer has to place the word ALMONDS on a can using equally spaced letters (see Figure 8a). For good visibility, the letters must cover a sector of the circle having a 90° central angle. If the base of the can is a circle of radius 2 inches (see Figure 8b), what is the maximum width of each letter?

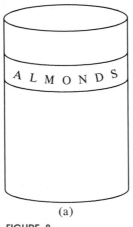

(a)

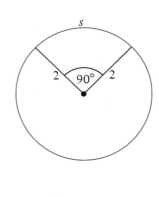

(b)

FIGURE 8

SOLUTION

Since $\theta = 90° = \pi/2$ radians, the arc has length

$$s = r\theta = \frac{\pi}{2}(2) = \pi$$

Each of the seven letters can occupy 1/7 of this arc, or $\pi/7$ inches.

EXERCISE SET 6.1

If the angle θ is in standard position, determine the quadrant in which the angle lies.

1. $\theta = 313°$ 2. $\theta = 182°$ 3. $\theta = 14°$ 4. $\theta = 227°$

5. $\theta = 141°$ 6. $\theta = -167°$ 7. $\theta = -345°$ 8. $\theta = 555°$

9. $\theta = 618°$ 10. $\theta = -428°$ 11. $\theta = -195°$ 12. $\theta = 730°$

13. $\theta = \dfrac{7\pi}{8}$ 14. $\theta = \dfrac{-3\pi}{5}$ 15. $\theta = \dfrac{-8\pi}{3}$ 16. $\theta = \dfrac{3\pi}{8}$

17. $\theta = \dfrac{13\pi}{3}$ 18. $\theta = \dfrac{9\pi}{5}$

Convert from degree measure to radian measure.

19. $30°$ 20. $200°$ 21. $-150°$ 22. $-330°$

23. $75°$ 24. $570°$ 25. $-450°$ 26. $-570°$

27. $135°$ 28. $405°$ 29. $120°$ 30. $90°$

31. $45.22°$ 32. $196.54°$

Convert from radian measure to degree measure.

33. $\dfrac{\pi}{4}$ 34. $\dfrac{\pi}{3}$ 35. $\dfrac{3\pi}{2}$ 36. $\dfrac{5\pi}{6}$

37. $\dfrac{-\pi}{2}$

38. $\dfrac{-7\pi}{12}$

39. $\dfrac{4\pi}{3}$

40. 3π

41. $\dfrac{5\pi}{2}$

42. -5π

43. $\dfrac{-5\pi}{3}$

44. $\dfrac{9\pi}{2}$

 45. 1.72

 46. 24.98

For each pair of angles, write T if they are coterminal and F if they are not coterminal.

47. 30°, 390°

48. 50°, −310°

49. 45°, −45°

50. 120°, $\dfrac{14\pi}{3}$

51. $\dfrac{\pi}{2}, \dfrac{7\pi}{2}$

52. −60°, 760°

For each given angle in Exercises 53–64, find the reference angle.

53. 130°

54. $\dfrac{5\pi}{6}$

55. −20°

56. 25°

57. −455°

58. 700°

59. $\dfrac{12\pi}{5}$

60. $\dfrac{5\pi}{4}$

61. 72°

62. $\dfrac{-2\pi}{3}$

63. $\dfrac{9\pi}{4}$

64. $\dfrac{5\pi}{3}$

65. If a central angle θ subtends an arc of length 4 centimeters on a circle of radius 7 centimeters, find the approximate measure of θ in radians and in degrees.

66. Find the length of arc subtended by a central angle of $\pi/5$ radians on a circle of radius 6 inches.

67. Find the radius of a circle if a central angle of $2\pi/3$ radians subtends an arc of 4 meters.

68. In a circle of radius 150 centimeters, what is the length of arc subtended by a central angle of 45°?

69. A subcompact car uses a tire whose radius is 13 inches. How far has the car moved when the tire completes one rotation? How many rotations are completed when the tire has traveled one mile? (Assume $\pi \approx 3.14$.)

70. A builder intends placing 7 equally spaced homes on a semicircular plot as shown in the accompanying figure. If the circle has a diameter of 400 feet, what is the distance between any two adjacent homes? (Use $\pi \approx 3.14$.)

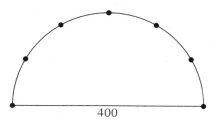

400

71. How many ribs are there in an umbrella if the length of each rib is 1.5 feet and the arc between two adjacent ribs measures $3\pi/10$ feet?

72. A microcomputer accepts both $5\frac{1}{4}$-inch and 8-inch floppy disks. If both disks are divided into 8 sectors, find the ratio of the arc length of a sector of the 8-inch disk to a sector of the $5\frac{1}{4}$-inch disk.

6.2
TRIGONOMETRIC
FUNCTIONS
OF ANGLES

DEFINITION

In Chapter 5 we dealt with the trigonometric functions as functions of real numbers, in which case they are also called the circular functions; we will now define the trigonometric functions as functions of angles. It will then be clear that the results of Chapter 5 will hold for angular measure.

We have previously defined the measure of angle θ as the measure of the arc of length t on the unit circle; that is, $\theta = t$. We now extend this definition to the trigonometric functions of θ.

Definition of sin θ,	$\sin \theta = \sin t$
cos θ, and tan θ	$\cos \theta = \cos t$
	$\tan \theta = \tan t$

The remaining trigonometric functions are again defined as reciprocals; that is, csc $\theta = 1/\sin \theta$, and so on.

To avoid confusion, we will continue to use small Greek letters to denote angles. Further, we will also use the degree symbol to indicate angular measure in degrees when dealing with numeric values. Thus, sin 2° and sin 2 are distinct quantities: sin 2° denotes degree measure and sin 2 denotes radian measure. Note that sin 2 may be interpreted as the value of the sine function either for the real number 2 or for the angle of 2 radians. Since sin $\theta = \sin t$, either interpretation is acceptable and will yield the same value.

Since we can convert from degree measure to radian measure, we can evaluate the trigonometric functions of angles.

EXAMPLE 1
Find sin 30°.

SOLUTION
Converting to radians, $30° = \pi/6$ radians. Then

$$\sin 30° = \sin \frac{\pi}{6} = \frac{1}{2}$$

THE REFERENCE ANGLE RULE

Table V in the Tables Appendix can be used if we first convert from degree measure to radian measure. This cumbersome process can be avoided by performing the conversion for every angle between 0° and 90° at intervals of 10′ and evaluating the trigonometric functions at each of these values. The results are displayed in Table VI in the Tables Appendix, which gives values of the trigonometric functions for angles in degree measure in the interval [0°, 90°]. For example, we can use this table to obtain tan 27°30′ = 0.5206. (When using a calculator, you will first have to convert minutes to fractions of a degree, that is, tan 27°30′ = tan 27.5°. You may then enter 27.5 in the degree mode and press TAN to find the answer.)

In Section 5.4 we showed that the values of the trigonometric functions of any real number can be determined if the values are known in the interval [0, $\pi/2$]. Since $\pi/2$ radians = 90°, we see that Table VI in the Tables Appendix can stop at 90°. For an angle θ such that $90° < \theta < 360°$, we use the **Reference Angle Rule** to find the value of a trigonometric function of θ. We now list the steps involved in this rule and illustrate with an example.

REFERENCE ANGLE RULE	
Procedure	**Example** **Find cos 160°**
Step 1. Find the reference angle associated with θ. (a) If $90° < \theta < 180°$, then θ is in quadrant II, and $\theta' = 180° - \theta$. (b) If $180° < \theta < 270°$, then θ is in quadrant III, and $\theta' = \theta - 180°$. (c) If $270° < \theta < 360°$, then θ is in quadrant IV, and $\theta' = 360° - \theta$.	*Step 1.* Since the terminal side of an angle of $160°$ is in quadrant II, we have $$\begin{aligned} \theta' &= 180° - \theta \\ &= 180° - 160° \\ &= 20° \end{aligned}$$
Step 2. Obtain the value of the required trigonometric function from Table VI in the Tables Appendix.	*Step 2.* $$\cos 20° = 0.9397$$
Step 3. Add the appropriate sign according to the quadrant in which θ lies and the desired trigonometric function.	*Step 3.* Since $\cos \theta$ is negative in quadrant II, we have $$\cos 160° = -\cos 20° = -0.9397$$

EXAMPLE 2
Find tan 332°.

SOLUTION
Since the angle is in quadrant IV, the reference angle θ' is given by

$$\theta' = 360° - 332° = 28°$$

From Table VI in the Tables Appendix, $\tan 28° = 0.5317$. Since tangent is negative in quadrant IV, $\tan 332° = -\tan 28° = -0.5317$.

EXAMPLE 3
Find sin 611°20′.

SOLUTION
Whenever the angle exceeds 360° in absolute value, we must first convert to an angle in the interval $[0, 360°)$ which has the same functional values. To do this, we simply add or subtract multiples of 360°. Thus,

$$\begin{aligned} \sin 611°20' &= \sin (611°20' - 360°) \\ &= \sin 251°20' \end{aligned}$$

The reference angle θ' for this third-quadrant angle is given by

$$\theta' = 251°20' - 180° = 71°20'$$

From Table VI in the Tables Appendix, sin 71°20′ = 0.9474. Since sine is negative in quadrant III, we have

$$\sin 611°20′ = -\sin 71°20′ = -0.9474$$

EXERCISE SET 6.2

Without using tables, find the values of the six trigonometric functions at the following values of an angle θ.

1. 135°   2. 300°   3. −30°   4. −300°
5. 315°   6. 30°   7. 270°   8. −180°

Using Table VI in the Tables Appendix, determine each of the following.

9. cos 42°40′   10. tan 65°50′   11. sin 14°30′   12. sec 17°40′
13. cot 25°10′   14. csc 26°50′   15. tan 272°20′   16. cos 144°40′
17. sin 246°10′   18. cos(−192°50′) 19. sin(−344°20′) 20. tan(−143°30′)
21. sin 554°   22. cos(−684°)

23. Use a calculator to verify the results of Exercises 9–22.

6.3
RIGHT TRIANGLE
TRIGONOMETRY

We are now prepared to show that the trigonometric functions of an angle are related to the ratios of the sides of a right triangle. In Figure 9a we display a right triangle with sides a and b, hypotenuse r, and an acute angle θ. We can place this triangle on a Cartesian coordinate system with θ in standard position (Figure 9b). We can draw a unit circle and let $N(x, y)$ denote the point of intersection of the circle and the hypotenuse OP. If we drop the perpendicular NM as indicated, we see that the triangles OMN and OQP are similar. The corresponding sides must then be proportional so that

$$\frac{\overline{MN}}{\overline{ON}} = \frac{\overline{QP}}{\overline{OP}} \quad \text{and} \quad \frac{\overline{OM}}{1} = \frac{\overline{OQ}}{\overline{OP}}$$

Since $\overline{OP} = r$, we obtain by substitution

$$\frac{y}{1} = \frac{b}{r} \quad \text{and} \quad \frac{x}{1} = \frac{a}{r}$$

By definition, $\sin \theta = y$ and $\cos \theta = x$. Substituting, we have

$$\sin \theta = y = \frac{b}{r}$$

$$\cos \theta = x = \frac{a}{r}$$

$$\tan \theta = \frac{\sin \theta}{\cos \theta} = \frac{b}{a}$$

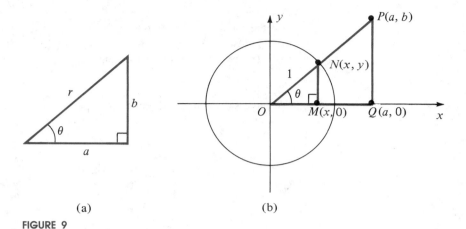

(a)   (b)

FIGURE 9

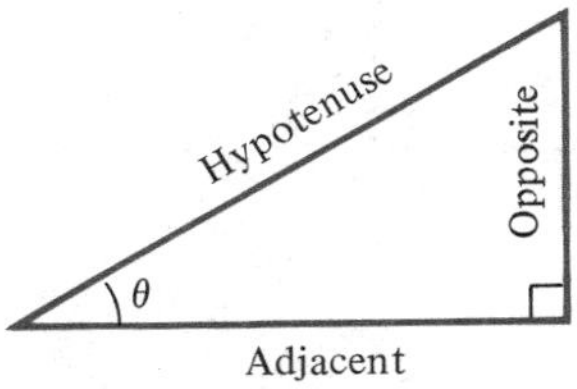

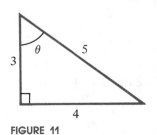

FIGURE 10

If we denote the sides a and b of the right triangle in Figure 9a as the adjacent and opposite sides relative to the angle θ (see Figure 10), then this last result expresses the trigonometric functions as ratios of the lengths of the sides of the right triangle.

$$\sin \theta = \frac{\text{side opposite } \theta}{\text{hypotenuse}} \qquad \csc \theta = \frac{\text{hypotenuse}}{\text{side opposite } \theta}$$

$$\cos \theta = \frac{\text{side adjacent to } \theta}{\text{hypotenuse}} \qquad \sec \theta = \frac{\text{hypotenuse}}{\text{side adjacent to } \theta}$$

$$\tan \theta = \frac{\text{side opposite } \theta}{\text{side adjacent to } \theta} \qquad \cot \theta = \frac{\text{side adjacent to } \theta}{\text{side opposite } \theta}$$

EXAMPLE 1

Find the values of the trigonometric functions of the angle θ in Figure 11.

SOLUTION

$$\sin \theta = \frac{4}{5} \qquad \csc \theta = \frac{5}{4}$$

$$\cos \theta = \frac{3}{5} \qquad \sec \theta = \frac{5}{3}$$

$$\tan \theta = \frac{4}{3} \qquad \cot \theta = \frac{3}{4}$$

FIGURE 11

EXAMPLE 2

Use the values of the trigonometric functions to find the following.

(a)   The sides of a 30°–60°–90° right triangle whose hypotenuse is of length 2.

(b)   The hypotenuse of an isosceles right triangle whose sides are of length 1.

SOLUTION

We have sketched the triangles in Figure 12.

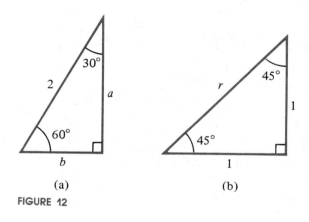

(a)

(b)

FIGURE 12

(a)   Since $\cos 60° = \tfrac{1}{2}$, we have

$$\cos 60° = \frac{1}{2} = \frac{b}{2} \quad \text{or} \quad b = 1$$

Similarly, we can establish that

$$\sin 60° = \frac{\sqrt{3}}{2} = \frac{a}{2} \quad \text{or} \quad a = \sqrt{3}$$

The student is urged to verify these results using the 30° angle and to verify that these values of a and b satisfy the Pythagorean theorem.

(b)   We know that $\sin 45° = \sqrt{2}/2$, so

$$\sin 45° = \frac{\sqrt{2}}{2} = \frac{1}{r} \quad \text{or} \quad r = \frac{2}{\sqrt{2}} = \sqrt{2}$$

Of course, we could have obtained r directly by using the Pythagorean theorem.

EXAMPLE 3

Find $\sec \theta$ if the point $P(-5, -12)$ lies on the terminal side of θ.

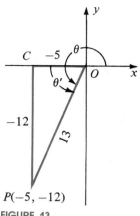

FIGURE 13

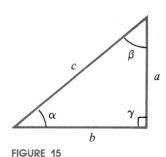

SOLUTION

(See Figure 13.) We construct a perpendicular from P to the x-axis to form right triangle PCO and use the Pythagorean theorem to find $\overline{OP} = 13$. Then

$$\sec \theta' = \frac{\text{hypotenuse}}{\text{adjacent}} = \frac{13}{5}$$

Since θ is in the third quadrant, by the Reference Angle Rule,

$$\sec \theta = -\sec \theta' = -\frac{13}{5}$$

EXAMPLE 4

Find $\cos(\arctan \frac{4}{3})$ without using tables or a calculator.

SOLUTION

We have repeated Example 5 of Section 5.7 to illustrate an alternative, simpler approach using right triangle trigonometry. We let $\theta = \arctan \frac{4}{3}$ so that $\tan \theta = \frac{4}{3}$ and $0 \le \theta \le \pi/2$. The angle θ in Figure 14 satisfies these conditions. Then we see that

$$\cos\left(\arctan \frac{4}{3}\right) = \cos \theta = \frac{3}{5}$$

FIGURE 14

SOLVING A TRIANGLE

The expression "to solve a triangle" is used to indicate that we seek all parts of the triangle, that is, the length of each side and the measure of each angle. For any right triangle, given any two sides, or given one side and an acute angle, it is always possible to solve the triangle. We will standardize the notation as shown in Figure 15 so that (a) the acute angles are labeled α and β, the right angle is labeled γ, and (b) the sides opposite angles α, β, and γ are labeled a, b, and c, respectively.

EXAMPLE 5

In triangle ABC, $\gamma = 90°$, $\beta = 27°$, and $b = 8.6$. Find approximate values for the remaining parts of the triangle.

SOLUTION

We begin by labeling a right triangle as in Figure 16. Since the sum of the angles of a triangle is 180°, we see that $\alpha = 63°$. Using the trigonometric functions of

FIGURE 15

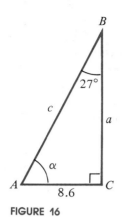

FIGURE 16

angle β we have

$$\sin 27° = \frac{8.6}{c} \quad \text{and} \quad \tan 27° = \frac{8.6}{a}$$

Solving for a and c yields

$$c = \frac{8.6}{\sin 27°} \qquad a = \frac{8.6}{\tan 27°}$$

From Table VI in the Tables Appendix, $\sin 27° = 0.4540$ and $\tan 27° = 0.5095$, so

$$c = 8.6/0.4540 \approx 18.9$$
$$a = 8.6/0.5095 \approx 16.9$$

PROGRESS CHECK

In triangle ABC, $\gamma = 90°$, $\alpha = 64°$, and $b = 24.7$. Solve the triangle.

ANSWERS

$\beta = 26° \qquad a = 50.6 \qquad c = 56.3$

EXAMPLE 6

In triangle ABC, $\gamma = 90°$, $a = 22.5$, and $b = 12.8$. Find approximate values for the remaining parts of the triangle.

SOLUTION

Figure 17 displays the parts of the triangle. Using angle β we have

$$\tan \beta = \frac{12.8}{22.5} = 0.5689$$

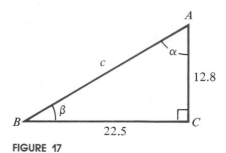

FIGURE 17

Using Table VI in the Tables Appendix, the closest entry yields $\beta \approx 29°40'$. Since the sum of the angles is $180°$, we must have $\alpha \approx 60°20'$. Alternatively,

$$\tan \alpha = \frac{22.5}{12.8} = 1.7578$$

also yields $\alpha \approx 60°20'$ by use of Table VI.

Finally, c can be found by the Pythagorean theorem or by trigonometry.

$$\sin \beta = \sin 29°40' = \frac{12.8}{c}$$

$$c = \frac{12.8}{0.4950} \approx 25.9$$

PROGRESS CHECK

In triangle ABC, $\gamma = 90°$, $a = 17.4$, and $b = 38.2$. Solve the triangle.

ANSWERS

$\alpha = 24°30' \qquad \beta = 65°30' \qquad c = 42$

APPLICATIONS

Many applied problems involve right triangles. We are now prepared to use our ability in solving triangles to tackle a variety of interesting problems.

EXAMPLE 7

A ladder leaning against a building makes an angle of $35°$ with the ground. If the bottom of the ladder is 5 meters from the building, how long is the ladder? To what height does it rise along the building?

SOLUTION

In Figure 18 we seek the length d of the ladder and the height h along the building. Using right triangle trigonometry,

$$\cos 35° = \frac{5}{d} \qquad \text{and} \quad \tan 35° = \frac{h}{5}$$

$$d = \frac{5}{\cos 35°} \qquad\qquad h = 5 \tan 35°$$

$$d = \frac{5}{0.8192} \qquad\qquad h = 5(0.7002)$$

$$d \approx 6.1 \text{ meters} \quad \text{and} \qquad h \approx 3.5 \text{ meters}$$

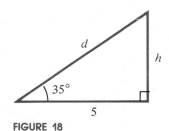

FIGURE 18

PROGRESS CHECK

The string of a kite makes an angle of 32°30′ with the ground. If 125 meters of string have been let out, how high is the kite?

ANSWER

67 meters

There are two technical terms that will occur frequently in our word problems. The **angle of elevation** is the angle between the horizontal and the line of sight. In Figure 19a, θ is the angle of elevation of the top T of a tree from a point x meters from the base of the tree.

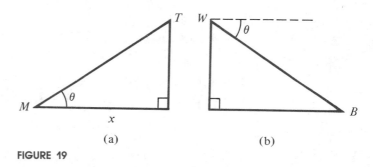

(a)   (b)

FIGURE 19

The **angle of depression** is the angle between the horizontal and the line of sight when looking down. In Figure 19b, θ is the angle of depression of a boat B as seen from a watchtower W.

EXAMPLE 8

A vendor of balloons inadvertently releases a balloon, which rises straight up. A child standing 50 feet from the vendor watches the balloon rise. When the angle of elevation of the balloon reaches 44°, how high is the balloon?

SOLUTION

We seek the height h in Figure 20. Thus,

$$\tan 44° = \frac{h}{50}$$

$$h = 50 \tan 44°$$

$$h = 50(0.9657) \approx 48$$

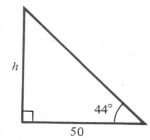

FIGURE 20

The balloon has risen approximately 48 feet.

EXAMPLE 9

A forest ranger is in a tower 65 feet above the ground. If the ranger spots a fire at an angle of depression of 6°40′, how far is the fire from the base of the tower (assuming level terrain)?

SOLUTION

We need to find the distance d in Figure 21. Since $\theta + 6°40′ = 90°$, $\theta = 83°20′$.

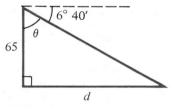

FIGURE 21

Then

$$\tan \theta = \frac{d}{65}$$

$$d = 65 \tan 83°20′$$

$$d = 65(8.5555) \approx 556$$

The fire is approximately 556 feet from the base of the tower.

EXAMPLE 10

A mathematics professor walks toward the university clock tower on the way to her office, and decides to find the height of the clock above ground. She determines the angle of elevation to be 30° and, after proceeding an additional 60 feet toward the base of the tower, finds the angle of elevation to be 40°. What is the height of the clock tower?

SOLUTION

This problem is somewhat more sophisticated since it involves more than one right triangle. In Figure 22 we seek to determine h. From triangle ACD,

$$\tan 30° = \frac{h}{d + 60} \quad \text{or} \quad h = (d + 60)(\tan 30°)$$

and from triangle ACB,

$$\tan 40° = \frac{h}{d} \quad \text{or} \quad h = d \tan 40°$$

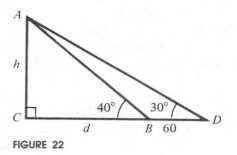

FIGURE 22

Equating the two expressions for h yields

$$(d + 60)(\tan 30°) = d \tan 40°$$

$$60 \tan 30° = d(\tan 40° - \tan 30°)$$

$$d = \frac{60 \tan 30°}{\tan 40° - \tan 30°} \approx 132 \text{ feet}$$

$$h = d \tan 40° = (132)(0.8391) \approx 110.8$$

The height of the clock tower is 110.8 feet.

In navigation and surveying, directions are often given by **bearings,** which specify an acute angle and its direction from the north–south line. In Figure 23a the bearing of point B from point A is N 40° E, that is, 40° east of north; in Figure 23b the bearing of point B from point A is S 60° W; and in Figure 23c it is S 20° E.

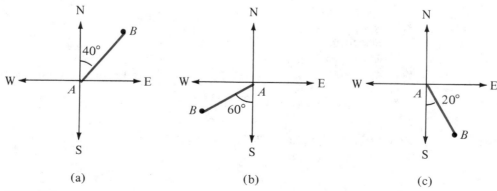

(a)   (b)   (c)

FIGURE 23

EXAMPLE 11

A ship leaves port at 10 A.M. and heads due east at a rate of 22 miles per hour. At 11 A.M. the course is changed to S 52° E. Find the distance and bearing of the ship from the dock at noon.

SOLUTION

The situation is depicted in Figure 24. We find angle $\beta = 38°$. From right triangle BCE,

$$\cos \beta = \frac{e}{22} \quad \text{or} \quad e = 22 \cos 38° \approx 17.3 \text{ miles}$$

$$\sin \beta = \frac{b}{22} \quad \text{or} \quad b = 22 \sin 38° \approx 13.5 \text{ miles}$$

We now know two sides of right triangle ACE, namely

$$\overline{AC} = 22 + e \approx 22 + 17.3 \approx 39.3$$
$$\overline{CE} = b \approx 13.5$$

We can now solve triangle ACE to obtain

$$\tan \alpha \approx \frac{13.5}{39.3} \quad \text{or} \quad \alpha \approx 19°$$

From triangle ACE,

$$\sin \alpha = \frac{b}{d} \approx \frac{13.5}{d}$$

$$d = \frac{13.5}{\sin 19°} \approx 41.5 \text{ miles}$$

The ship is 41.5 miles from port at a bearing of S 71° E.

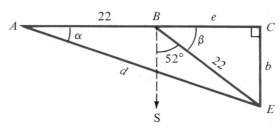

FIGURE 24

EXERCISE SET 6.3

Find the values of the trigonometric functions of the angle θ in each of the following right triangles.

1.
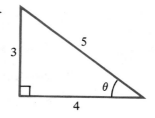

2.
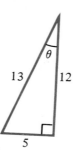

3.
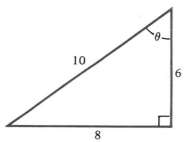

4.
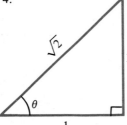

5.
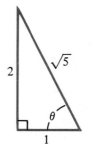

6.
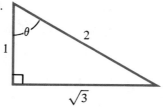

7.
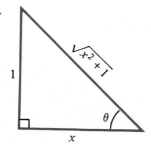

8.
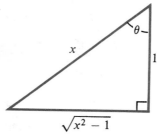

Find the values of the trigonometric functions of the angle θ if the point P lies on the terminal side of θ.

9. $P(-5, 12)$ 10. $P(3, -4)$ 11. $P(-1, -1)$ 12. $P(1, 2)$

13. $P(-8, 6)$ 14. $P(12, 5)$ 15. $P(12, -5)$ 16. $P(-1, \sqrt{3})$

17. $P(-12, -5)$ 18. $P(-3, 4)$ 19. $P(-2, 1)$ 20. $P(-2, -1)$

In each of the following right triangles, express the length h as a trigonometric function of the angle θ.

21.

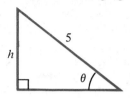

22.

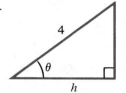

23.

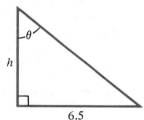

24.

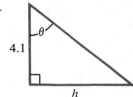

25.

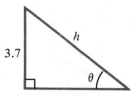

26.

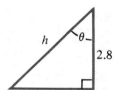

In triangle ABC, $\gamma = 90°$. Find the required parts of the triangle in each of the following.

27. $a = 12$, $b = 16$; find α.   28. $a = 5$, $b = 15$; find β.

29. $b = 40$, $\beta = 40°$; find c.   30. $a = 22$, $\alpha = 36°$; find b.

31. $a = 75$, $\beta = 22°$; find b.   32. $b = 60$, $\alpha = 53°$; find c.

33. $a = 25$, $\beta = 42°30'$; find c.   34. $b = 50$, $\alpha = 36°20'$; find a.

Evaluate the given expression without using tables or a calculator.

35. $\tan\left(\sin^{-1} -\dfrac{5}{13}\right)$

36. $\sin\left(\arctan -\dfrac{12}{5}\right)$

37. $\cos\left(\sin^{-1} \dfrac{4}{5}\right)$

38. $\cos\left(\arcsin -\dfrac{2}{3}\right)$

39. $\tan\left(\cos^{-1} -\dfrac{3}{5}\right)$

40. A ladder 7 meters in length leans against a vertical wall. If the ladder makes an angle of 65° with the ground, find the height the ladder reaches above the ground.

41. A ladder 20 feet in length touches a wall at a point 16 feet above the ground. Find the angle the ladder makes with the ground.

42. A monument is 550 feet high. What is the length of the shadow cast by the monument when the sun is 64° above the horizon?

43. Find the angle of elevation of the sun when a tower 45 meters in height casts a horizontal shadow 25 meters in length.

44. A technician positioned on an oil-drilling rig 120 feet above the water spots a boat at an angle of depression of 16°. How far is the boat from the rig?

45. A mountainside hotel is located 8000 feet above sea level. From the hotel, a trail leads farther up the mountain to an inn at an elevation of 10,400 feet. If the trail has an angle of inclination of 18° (that is, the angle of elevation of the inn from the hotel is 18°), find the distance along the trail from the hotel to the inn.

46. A hill is known to be 200 meters high. A surveyor standing on the ground finds the angle of elevation of the top of the hill to be 42°50′. Find the distance from the surveyor to a point directly below the top of the hill. (Ignore the height of the surveyor.)

47. An observer is 425 meters from a launching pad when a rocket is launched vertically. If the angle of elevation of the rocket at its apogee (highest point) is 66°20′, how high does the rocket rise?

48. An airplane pilot wants to climb from an altitude of 6000 feet to an altitude of 16,000 feet. If the plane climbs at an angle of 9° with a constant speed of 22,000 feet per minute, how long will it take to reach the increased altitude?

49. A rectangle is 16 inches long and 13 inches wide. Find the measures of the angles formed by a diagonal with the sides.

50. The sides of an isosceles triangle are 15, 15, and 26 centimeters. Find the measures of the angles of the triangle. (*Hint:* The altitude of an isosceles triangle bisects the base.)

51. The side of a regular pentagon is 22 centimeters. Find the radius of the circle circumscribed about the pentagon. (*Hint:* The radii from the center of the circumscribed circle to any two adjacent vertices of the regular pentagon form an isosceles triangle. The altitude of an isosceles triangle bisects the base.)

52. To determine the width of a river, markers are placed at each side of the river in line with the base of a tower that rises 23.4 meters above the ground. From the top of the tower, the angles of depression of the markers are 58°20′ and 11°40′. Find the width of the river.

53. The angle of elevation of the top of building B from the base of building A is 29°. From the top of building A, the angle of depression of the base of building B is 15°. If building B is 110 feet high, find the height of building A.

54. A ship leaves port at 2 P.M. and heads due east at a rate of 40 kilometers per hour. At 4 P.M. the course is changed to N 32° E. Find the distance and bearing of the ship from the dock at 6 P.M.

55. An attendant in a lighthouse receives a request for aid from a stalled craft located 15 miles due east of the lighthouse. The attendant contacts a second boat located 14 miles from the lighthouse at a bearing of N 23° W. What is the distance of the rescue ship from the stalled craft?

6.4
LAW OF COSINES

In Section 6.3 we studied the trigonometry of a right triangle. In this and the next section we will examine an **oblique triangle,** a triangle that does not contain a right angle.

We can always solve an oblique triangle by dropping a perpendicular as in Figures 25a and 25b and treating the resulting right triangles *ADC* and *BDC*. It is, however, worthwhile to perform the analysis in a general way. This yields two results, known as the *law of sines* and the *law of cosines*. We shall now state

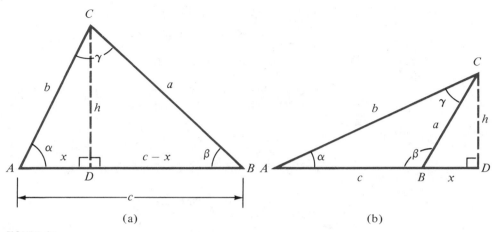

FIGURE 25

and prove the law of cosines. We maintain the notation of the last section; thus, the angles of triangle ABC are denoted by α, β, and γ, with opposite sides a, b, and c, respectively.

The Law of Cosines	In triangle ABC,

$$a^2 = b^2 + c^2 - 2bc \cos \alpha \qquad (1)$$
$$b^2 = a^2 + c^2 - 2ac \cos \beta \qquad (2)$$
$$c^2 = a^2 + b^2 - 2ab \cos \gamma \qquad (3)$$

The student is urged to note the *pattern* of the three forms of the law of cosines as an aid in their memorization.

To prove the law of cosines, we deal with the cases shown in Figure 25.

Case 1. The angles of triangle ABC are all acute (Figure 25a). We construct the perpendicular CD to side AB. Applying the Pythagorean theorem to right triangle BDC, we have

$$\begin{aligned} a^2 &= h^2 + (c - x)^2 \\ &= h^2 + c^2 - 2cx + x^2 \\ &= (h^2 + x^2) + c^2 - 2cx \\ &= b^2 + c^2 - 2cx \end{aligned} \qquad (4)$$

The last step results from application of the Pythagorean theorem to right triangle ADC. Also,

$$\cos \alpha = \frac{x}{b} \quad \text{or} \quad x = b \cos \alpha$$

which we then substitute in Equation (4) to yield

$$a^2 = b^2 + c^2 - 2bc \cos \alpha$$

This establishes the desired result of Equation (1).

Case 2. Triangle *ABC* has an obtuse angle β (Figure 25b). We construct the perpendicular *CD* to side *AB*. The Pythagorean theorem can be applied to right triangle *BDC* to give

$$a^2 = h^2 + x^2 \tag{5}$$

Next, we use the trigonometry of the right triangle *ADC* to obtain

$$\sin \alpha = \frac{h}{b} \quad \text{or} \quad h = b \sin \alpha$$

$$\cos \alpha = \frac{c + x}{b} \quad \text{or} \quad x = b \cos \alpha - c$$

Substituting for *h* and *x* in Equation (5) we have

$$\begin{aligned}
a^2 &= b^2 \sin^2 \alpha + (b \cos \alpha - c)^2 \\
&= b^2 \sin^2 \alpha + b^2 \cos^2 \alpha - 2bc \cos \alpha + c^2 \\
&= b^2(\sin^2 \alpha + \cos^2 \alpha) - 2bc \cos \alpha + c^2 \\
&= b^2 + c^2 - 2bc \cos \alpha \quad (\text{Since } \sin^2 \alpha + \cos^2 \alpha = 1)
\end{aligned}$$

Once again, this is the desired result of Equation (1).

We have thus established the first form of the law of cosines for both cases. A similar argument can be used to establish the other two forms, given in Equations (2) and (3).

Examination of the law of cosines shows that it can be used in the following circumstances.

Applying the Law of Cosines

The law of cosines may be used when
(a) three sides of a triangle are known (SSS), or
(b) two sides of a triangle are known and the measure of the angle formed by those sides is known (SAS).

EXAMPLE 1

Highway engineers who are to dig a tunnel through a small mountain wish to determine the length of the tunnel. Points *A* and *B* are chosen as the endpoints of the tunnel. Then a point *C* is selected from which the distances to *A* and *B* are found to be 190 feet and 230 feet, respectively. If angle *ACB* measures 48°, find the approximate length of the tunnel.

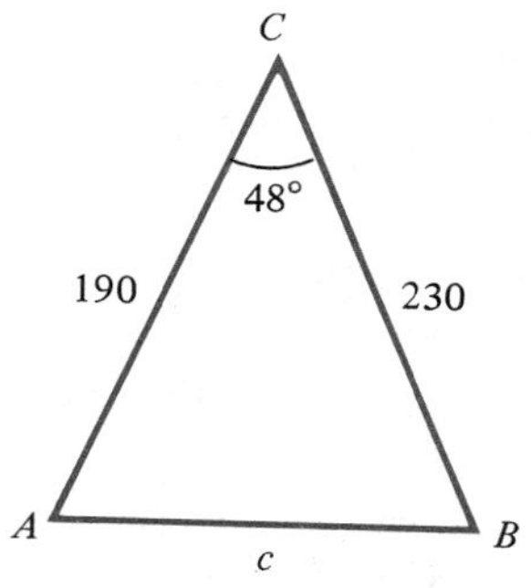

FIGURE 26

SOLUTION

The known information is displayed in Figure 26. Applying the law of cosines,

$$c^2 = a^2 + b^2 - 2ab \cos \gamma$$
$$c^2 = 230^2 + 190^2 - 2(230)(190) \cos 48°$$
$$c^2 \approx 30{,}518$$
$$c \approx 175 \text{ feet}$$

EXAMPLE 2

Find the approximate measure of the angles of triangle ABC if $a = 150$, $b = 100$, and $c = 75$.

SOLUTION

Substituting in the equation

$$a^2 = b^2 + c^2 - 2bc \cos \alpha$$
$$150^2 = 100^2 + 75^2 - 2(100)(75) \cos \alpha$$
$$22{,}500 = 10{,}000 + 5625 - 15{,}000 \cos \alpha$$
$$\cos \alpha = -0.4583$$

Since $\cos \alpha$ is negative, angle α must lie in the second quadrant and is an obtuse angle. Using Table VI in the Tables Appendix, or a calculator,

$$\alpha \approx 180° - 62°40' \approx 117°20'$$

Similarly,

$$b^2 = a^2 + c^2 - 2ac \cos \beta$$
$$100^2 = 150^2 + 75^2 - 2(150)(75) \cos \beta$$
$$10{,}000 = 22{,}500 + 5625 - 22{,}500 \cos \beta$$
$$\cos \beta = 0.8056$$
$$\beta \approx 36°20'$$

Finally, we may easily determine γ since the sum of the angles of a triangle is 180°.

$$\gamma \approx 180° - (117°20' + 36°20') \approx 26°20'$$

The student should verify this result by substituting in the equation

$$c^2 = a^2 + b^2 - 2ab \cos \gamma$$

EXERCISE SET 6.4

In Exercises 1–10 use the law of cosines to approximate the required part of triangle ABC.

1. $a = 10$, $b = 15$, $c = 21$; find β.
2. $a = 5$, $b = 12$, $c = 15$; find γ.

3. $a = 25$, $c = 30$, $\beta = 28°30'$; find b.

5. $a = 10$, $b = 12$, $\gamma = 108°$; find c.

7. $b = 6$, $a = 7$, $\gamma = 68°$; find α.

9. $a = 9$, $b = 12$, $c = 15$; find γ.

4. $b = 20$, $c = 13$, $\alpha = 19°10'$; find a.

6. $a = 30$, $c = 40$, $\beta = 122°$; find b.

8. $a = 6$, $b = 15$, $c = 16$; find β.

10. $a = 11$, $c = 15$, $\beta = 33°$; find γ.

11. The sides of a parallelogram measure 25 centimeters and 40 centimeters, and the longer diagonal measures 50 centimeters. Find the approximate measure of the smaller angle of the parallelogram.

12. The sides of a parallelogram measure 40 inches and 70 inches, and one of the angles is 108°. Find the approximate length of each diagonal of the parallelogram.

13. A ship leaves port at 9 A.M. and travels due west at a rate of 15 miles per hour. At 11 A.M. the ship changes direction to S 32° W. What is the distance and bearing of the ship from port at 1 P.M.?

14. A ship leaves from port A intending to travel direct to port B, a distance of 25 kilometers. After traveling 12 kilometers the captain finds that his course has been in error by 10°. How far is the ship from port B?

15. Two trains leave Pennsylvania Station in New York City at 2 P.M. and travel in directions that differ by 55°. If the trains travel at constant rates of 50 miles per hour and 80 miles per hour, respectively, what is the distance between them at 2:30 P.M.?

16. Hurricane David has left a telephone pole in a nonvertical position. Workmen place a 30-foot ladder at a point 10 feet from the base of the pole. If the ladder touches the pole at a point 26 feet up the pole, find the angle the pole makes with the ground.

17. Find the approximate perimeter of triangle ABC if $a = 20$, $b = 30$, and $\gamma = 37°$.

18. A hill makes an angle of 10° with the horizontal. An antenna 50 feet in height is erected at the top of the hill and a guy wire is run to a point 30 feet from the base of the antenna. What is the length of the guy wire?

19. Prove that if ABC is a right triangle, the law of cosines reduces to the Pythagorean theorem.

20. Prove the following in triangle ABC.
    (a) $a^2 + b^2 + c^2$
    $$= 2(bc \cos \alpha + ac \cos \beta + ab \cos \gamma)$$
    (b) $\dfrac{\cos \alpha}{a} + \dfrac{\cos \beta}{b} + \dfrac{\cos \gamma}{c} = \dfrac{a^2 + b^2 + c^2}{2abc}$

21. Prove that if
    $$\frac{\cos \beta}{a} = \frac{\cos \alpha}{b}$$
    triangle ABC is either a right triangle or an isosceles triangle.

6.5
LAW OF SINES

In the last section we applied the law of cosines to an oblique triangle. That law derives its name from the appearance of the cosine function in its statement.

We will now state and prove the law of sines, which also applies to an oblique triangle. Not surprisingly, the law of sines involves the sine function. Once again, we denote the angles of triangle ABC by α, β, and γ, with opposite sides a, b, and c, respectively.

The Law of Sines   In triangle ABC,

$$\frac{a}{\sin \alpha} = \frac{b}{\sin \beta} = \frac{c}{\sin \gamma}$$

The two cases are illustrated in Figure 27.

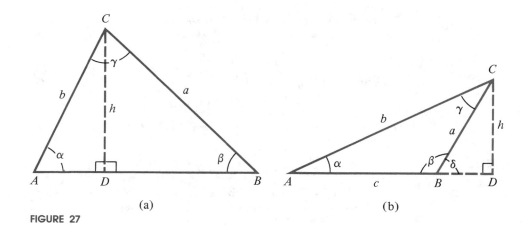

FIGURE 27

(a)   (b)

Case 1. The angles of triangle *ABC* are all acute (Figure 27a). We construct the perpendicular *CD* to side *AB*. Then triangles *ADC* and *BDC* are both right triangles, and we can apply trigonometry of a right triangle to obtain

$$\sin \alpha = \frac{h}{b} \quad \text{or} \quad h = b \sin \alpha$$

$$\sin \beta = \frac{h}{a} \quad \text{or} \quad h = a \sin \beta$$

Equating the expressions for *h* yields

$$b \sin \alpha = a \sin \beta$$

which can be written in the convenient form

$$\frac{a}{\sin \alpha} = \frac{b}{\sin \beta}$$

Case 2. Triangle *ABC* has an obtuse angle β (Figure 27b). We construct the perpendicular *CD* to side *AB*. Applying right triangle trigonometry to triangles *ADC* and *BDC*, and noting that $\delta = 180° - \beta$, we obtain

$$\sin \alpha = \frac{h}{b} \quad \text{or} \quad h = b \sin \alpha$$

$$\sin \delta = \sin(180° - \beta) = \frac{h}{a} \quad \text{or} \quad h = a \sin(180° - \beta)$$

Equating the expressions for *h* yields

$$b \sin \alpha = a \sin(180° - \beta)$$

Since sine is positive in both the first and second quadrants, the Reference Angle Rule tells us that

$$\sin(180° - \beta) = \sin \beta$$

Substituting, we again obtain

$$b \sin \alpha = a \sin \beta$$

or

$$\frac{a}{\sin \alpha} = \frac{b}{\sin \beta}$$

To complete the proof of the law of sines we need only drop a perpendicular from A to BC and use a similar argument to show that

$$\frac{b}{\sin \beta} = \frac{c}{\sin \gamma}$$

The law of sines then follows from the transitive property of equality. The law of sines can be used in the following circumstances.

Applying the Law of Sines

The law of sines may be used when the known parts of a triangle are
(a) one side and two angles (SAA), or
(b) two sides and an angle opposite one of these sides (SSA).

Remember that if two angles of a triangle are known, we can immediately determine the third angle. Here is an example.

EXAMPLE 1

In triangle ABC, $\alpha = 38°$, $\beta = 64°$, and $c = 24$. Find approximate values for the remaining parts of the triangle.

SOLUTION

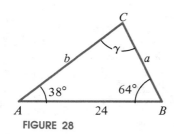

FIGURE 28

(See Figure 28.) Since α and β are known,

$$\gamma = 180° - (\alpha + \beta) = 180° - (38° + 64°) = 78°$$

Applying the law of sines,

$$\frac{a}{\sin \alpha} = \frac{c}{\sin \gamma}$$

$$\frac{a}{\sin 38°} = \frac{24}{\sin 78°}$$

$$a = \frac{24 \sin 38°}{\sin 78°} = \frac{24(0.6157)}{(0.9781)} \approx 15.1$$

Similarly, from

$$\frac{b}{\sin \beta} = \frac{c}{\sin \gamma}$$

we obtain

$$b \approx 22.1$$

When the given parts of a triangle are two sides and an angle opposite one of them, the situation is not straightforward since a *unique* triangle is not always determined. In Figure 29 we have constructed angle α and side b and then used a compass to construct a side of length a with an endpoint at C. In Figure 29a no triangle exists satisfying the given conditions; Figure 29b shows that we may obtain a right triangle; Figure 29c illustrates the possibility that two triangles will satisfy the given conditions; Figure 29d shows that precisely one acute triangle may be possible.

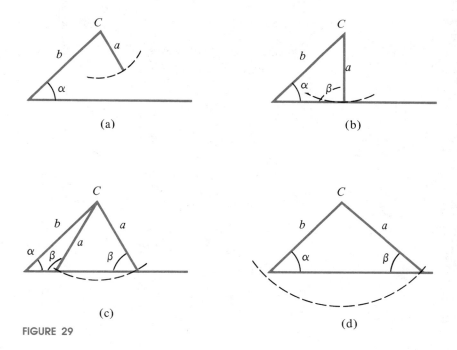

(a)

(b)

(c)

(d)

FIGURE 29

In Exercise 23 you will be asked to prove a number of inequalities that determine which of the four cases applies to a given set of conditions. In practice, we prefer to have you go ahead with the law of sines and let the results lead you to the appropriate answer.

Assume that sides a and b and angle α of triangle ABC are known and that we use the law of sines to determine angle β. These are the results that correspond to the possibilities of Figure 29.

(a) $\sin \beta > 1$. Since $|\sin \theta| \leq 1$ for all θ, there is no angle β satisfying the given conditions. This corresponds to the illustration in Figure 29a.

(b) $\sin \beta = 1$. Then $\beta = 90°$ and the given parts determine a unique right triangle (Figure 29b).

(c) $0 < \sin \beta < 1$. There are two possible choices for β, which is why this is called the **ambiguous case**. Since the sine function is positive in quadrants I and II, one choice will be an acute angle and one will be an obtuse angle (Figure 29c).

(d) $0 < \sin \beta < 1$. There are two possible choices for β but the obtuse angle does not form a triangle (Figure 29d). This case is signaled by $\alpha + \beta$ exceeding 180°.

Here are several illustrations of the law of sines when two sides and an angle opposite one of these sides are known.

EXAMPLE 2

In triangle ABC, $\alpha = 60°$, $a = 5$, and $b = 7$. Find angle β.

SOLUTION

Using the law of sines,

$$\frac{a}{\sin \alpha} = \frac{b}{\sin \beta}$$

$$\sin \beta = \frac{b \sin \alpha}{a} = \frac{7 \sin 60°}{5} \approx 1.2$$

Since the sine function has a maximum value of 1, there is no angle β such that $\sin \beta = 1.2$. Hence, there is no triangle with the given parts. This example corresponds to Figure 29a.

EXAMPLE 3

In triangle ABC, $a = 5$, $b = 8$, and $\alpha = 22°$. Find the remaining angles of the triangle.

SOLUTION

Using the law of sines,

$$\frac{a}{\sin \alpha} = \frac{b}{\sin \beta}$$

$$\sin \beta = \frac{b \sin \alpha}{a} = \frac{8 \sin 22°}{5} \approx 0.5993$$

Using tables or a calculator, we find that $\beta \approx 36°50'$. Thus, the angles are $\alpha = 22°$, $\beta = 36°50'$, and $\gamma = 121°10'$.

However, the angle $\beta = 180° - 36°50' = 143°10'$ also satisfies the requirement that $\sin \beta = 0.5993$. Therefore, another satisfactory triangle has angles $\alpha = 22°$, $\beta = 143°10'$, and $\gamma = 14°50'$.

This is an example of the ambiguous case, and corresponds to Figure 29c.

EXAMPLE 4

In triangle ABC, $a = 9$, $b = 6$, and $\alpha = 35°$. Find angles β and γ.

SOLUTION

We again apply the law of sines.

$$\frac{a}{\sin \alpha} = \frac{b}{\sin \beta}$$

$$\sin \beta = \frac{b \sin \alpha}{a} = \frac{6 \sin 35°}{9} \approx 0.3824$$

Using tables or a calculator yields $\beta \approx 22°30'$. A triangle satisfying the given conditions has $\alpha = 35°$, $\beta = 22°30'$, and $\gamma = 122°30'$.

The angle $\beta = 180° - 22°30' = 157°30'$ also satisfies the requirement that $\sin \beta = 0.3824$. But this "solution" must be rejected since $\alpha + \beta > 180°$.

This example corresponds to Figure 29d.

EXERCISE SET 6.5

In Exercises 1–12 use the law of sines to approximate the required part(s) of triangle ABC. Give both solutions if more than one triangle satisfies the given conditions.

1. $\alpha = 25°$, $\beta = 82°$, $a = 12.4$; find b.

2. $\alpha = 74°$, $\gamma = 36°$, $c = 6.8$; find a.

3. $\beta = 23°$, $\gamma = 47°$, $a = 9.3$; find c.

4. $\alpha = 46°$, $\beta = 88°$, $c = 10.5$; find b.

5. $\alpha = 42°20'$, $\gamma = 78°40'$, $b = 20$; find a.

6. $\beta = 16°30'$, $\gamma = 84°40'$, $a = 15$; find c.

7. $\alpha = 65°$, $a = 25$, $b = 30$; find β.

8. $\beta = 32°$, $b = 20$, $c = 14$; find α and γ.

9. $\gamma = 30°$, $a = 12.6$, $c = 6.3$; find b.

10. $\beta = 64°$, $a = 10$, $b = 8$; find c.

11. $\gamma = 45°$, $b = 7$, $c = 6$; find a.

12. $\alpha = 64°$, $a = 11$, $b = 12$; find β and γ.

13. Points A and B are chosen on opposite sides of a rock quarry. A point C is 160 meters from B, and the measures of angles BAC and ABC are found to be 95° and 47°, respectively. Find the width of the quarry.

14. A tunnel is to be dug between points A and B on opposite sides of a hill. A point C is chosen that is 150 meters from A and 180 meters from B. If angle ABC measures 54°, find the length of the tunnel.

15. A ski lift 750 meters in length rises to the top of a mountain at an angle of inclination of 40°. A second lift is to be built whose base is in the same horizontal plane as the initial lift. If the angle of elevation of the second lift is 45°, what is the length of the second lift?

16. A tree leans away from the sun at an angle of 9° from the vertical. The tree casts a shadow 20 meters in length when the angle of elevation of the sun is 62°. Find the height of the tree.

17. A ship is sailing due north at a rate of 22 miles per hour. At 2 P.M. a lighthouse is seen at a bearing of N 15° W. At 4 P.M., the bearing of the same lighthouse is S 65° W. Find the distance of the ship from the lighthouse at 2 P.M.

18. A plane leaves airport A and flies at a bearing of N 32° E. A few moments later, the plane is spotted from airport B at a bearing of N 56° W. If airport B lies 15 miles due east of airport A, find the distance of the plane from airport B at the moment it is spotted.

19. A guy wire attached to the top of a vertical pole has an angle of inclination of 65° with the ground. From a point 10 meters farther from the pole, the angle of elevation of the top of the pole is 45°. Find the height of the pole.

20. At 5 P.M. a sailor on board a ship sailing at a rate of 18 miles per hour spots an island due east of the ship. The ship maintains a bearing of N 26° E. At 6 P.M. the sailor finds the bearing of the island to be S 37° E. Find the distance of the island from the ship at 6 P.M.

21. The short side of a parallelogram and the shorter diagonal measure 80 centimeters and 100 centimeters, respectively. If the angle between the longer side and the shorter diagonal is 43°, find the length of the longer side.

22. An archaeological mound is discovered in a jungle in Central America. To determine the height of the mound, a point A is chosen from which the angle of elevation of the top of the mound is found to be 31°. A second point B is chosen on a line with A and the base of the mound, 30 meters closer to the base of the mound. If the angle of elevation of the top of the mound from point B is 39°, find the height of the mound.

23. In a triangle, sides of length a and b and an angle α are given. Prove the following.
    (a) If $b \sin \alpha > a$, there is no triangle with the given parts.
    (b) If $b \sin \alpha = a$, the parts determine a right triangle.
    (c) If $b \sin \alpha < a < b$, there are two triangles with the given parts.
    (d) If $b \leq a$, there is one acute triangle with the given parts.

TERMS AND SYMBOLS

standard position of an angle (p. 264)
initial side of an angle (p. 264)
terminal side of an angle (p. 264)
positive angle (p. 264)

negative angle (p. 264)
quadrantal angle (p. 264)
degree (p. 264)
minute (p. 264)
second (p. 264)
acute angle (p. 264)
obtuse angle (p. 264)

right angle (p. 264)
radian (p. 264)
coterminal (p. 267)
reference angle (p. 267)
Reference Angle Rule (p. 272)
angle of elevation (p. 280)

angle of depression (p. 280)
bearing (p. 282)
oblique triangle (p. 286)
law of cosines (p. 287)
law of sines (p. 290)
ambiguous case of the law of sines (p. 294)

KEY IDEAS FOR REVIEW

☐ An angle may be measured in either degrees or in radians. The two forms of measure are related by the equation π radians $= 180°$.

☐ A trigonometric function of an angle is the same as the trigonometric function of the arc on the unit circle that the angle intercepts.

☐ The Reference Angle Rule is analogous to the Reference Number Rule. It enables us to find the value of a trigonometric function of an angle greater than 90° by using Table VI in the Tables Appendix, which gives the values for angles between 0° and 90°.

☐ Right triangle trigonometry relates a trigonometric function of an angle θ of a right triangle to the ratio of the lengths of two of its sides as follows:

$$\sin \theta = \frac{\text{side opposite } \theta}{\text{hypotenuse}}$$

$$\cos \theta = \frac{\text{side adjacent to } \theta}{\text{hypotenuse}}$$

$$\tan \theta = \frac{\text{side opposite to } \theta}{\text{side adjacent to } \theta}$$

☐ Right triangle trigonometry can be used to solve a wide variety of applied problems.

☐ The law of cosines and the law of sines are useful in solving problems that involve an oblique triangle. The derivation of these laws is accomplished by using right triangle trigonometry.

REVIEW EXERCISES

Solutions to exercises whose numbers are in color are in the Solutions section in the back of the book.

6.1 In Exercises 1–4 convert from degree measure to radian measure or from radian measure to degree measure.

1. $-60°$

2. $\dfrac{3\pi}{2}$

3. $-\dfrac{5\pi}{12}$

4. $45°$

In Exercises 5–7 determine if the pair of angles are coterminal.

5. $100°, \dfrac{5\pi}{9}$

6. $\dfrac{4\pi}{3}, \ 480°$

7. $\dfrac{5\pi}{4}, \ -135°$

In Exercises 8–10 find the reference angle of the given angle.

8. $310°$

9. $-185°$

10. $405°$

11. If a central angle θ subtends an arc of length 14 centimeters on a circle whose radius is 10 centimeters, find the radian measure of θ.

12. A central angle of $2\pi/3$ radians subtends an arc of length $5\pi/2$ centimeters. Find the radius of the circle.

6.3 In Exercises 13–15 express the required trigonometric function as a ratio of the given parts of the right triangle ABC with $\gamma = 90°$.

13. $a = 5, b = 12$; find $\sin \alpha$.

14. $a = 3, c = 5$; find $\tan \beta$.

15. $a = 4, b = 7$; find $\sec \alpha$.

In Exercises 16–19 the point P lies on the terminal side of the angle θ. Find the value of the required trigonometric function without using tables or a calculator.

16. $P(-\sqrt{3}, 1)$; $\csc \theta$

17. $P(\sqrt{2}, -\sqrt{2})$; $\cot \theta$

18. $P(-1, -\sqrt{3})$; $\cos \theta$

19. $P(\sqrt{2}, \sqrt{2})$; $\sin \theta$

In Exercises 20–23 find the required part of triangle ABC with $\gamma = 90°$. Use Table VI in the Tables Appendix, or a calculator.

20. $a = 50, b = 60$; find α.

21. $a = 40, \beta = 20°$; find b.

22. $a = 20, \alpha = 52°$; find c.

23. $b = 15, \alpha = 25°$; find c.

24. A ladder 6 meters in length leans against a vertical wall. If the ladder makes an angle of $65°$ with the ground, find the height that the ladder reaches above the ground.

25. Find the angle of elevation of the sun when a tree 25 meters in height casts a horizontal shadow 10 meters in length.

26. A rectangle is 22 centimeters long and 16 centimeters wide. Find the measure of the smaller angle formed by the diagonal with a side.

6.4–6.5 In Exercises 27 and 28 use the law of cosines or the law of sines to approximate the required part of triangle ABC.

27. $a = 12, b = 7, c = 15$; find α.

28. $a = 20, b = 15, \alpha = 55°$; find β.

PROGRESS TEST 6A

In Problems 1–3 convert from degree measure to radian measure or from radian measure to degree measure.

1. $\dfrac{5\pi}{3}$

2. $-200°$

3. $75°$

In Problems 4 and 5 find an angle θ, $0 \le \theta < 360°$, that is coterminal with the given angle.

4. $-25°$

5. $\dfrac{17\pi}{4}$

In Problems 6 and 7 find the reference angle of the given angle.

6. $160°$

7. $\dfrac{7\pi}{4}$

8. If a central angle θ subtends an arc of length 12 inches on a circle whose radius is 15 inches, find the radian measure of θ.

In Problems 9 and 10 ABC is a right triangle with $\gamma = 90°$. Express the required trigonometric function as a ratio of the given parts of the triangle.

9. $a = 7$, $b = 5$; tan α

10. $b = 5$, $c = 15$; sec α

In Problems 11–13 the point P lies on the terminal side of the angle θ. Find the value of the required trigonometric function without using tables or a calculator.

11. $P(-\sqrt{2}, \sqrt{2})$; cot θ

12. $P(0, -5)$; sin θ

13. $P(2, 2\sqrt{3})$; sec θ

In Problems 14–16 use Table VI in the Tables Appendix, or a calculator, to find the required part of triangle ABC with $\gamma = 90°$.

14. $a = 25$, $c = 30$; find α.

15. $b = 20$, $\alpha = 32°$; find c.

16. $a = 15$, $b = 20$; find β.

17. From the top of a hill 100 meters in height, the angle of depression of the entrance to a castle is $36°$. Find the distance of the castle from the base of the hill.

PROGRESS TEST 6B

In Problems 1–3 convert from degree measure to radian measure or from radian measure to degree measure.

1. $-135°$

2. $\dfrac{3\pi}{4}$

3. $-\dfrac{5\pi}{6}$

In Problems 4 and 5 find an angle θ, $0 \le \theta < 360°$, that is coterminal with the given angle.

4. $430°$

5. $-\dfrac{2\pi}{3}$

In Problems 6 and 7 find the reference angle of the given angle.

6. $\dfrac{9\pi}{16}$

7. $345°$

8. A central angle of $100°$ subtends an arc of length $7\pi/3$ centimeters. Find the radius of the circle.

In Problems 9 and 10 ABC is a right triangle with $\gamma = 90°$. Express the required trigonometric function as a ratio of the given parts of the triangle.

9. $a = 6$, $b = 8$; csc β

10. $a = 7$, $b = 6$; cot α

In Problems 11–13 the point P lies on the terminal side of the angle θ. Find the value of the required trigonometric function without using tables or a calculator.

11. $P(-3, 0)$; csc θ

12. $P(2, 2\sqrt{3})$; csc θ

13. $P(-\sqrt{2}, -\sqrt{2})$; tan θ

In Problems 14–16 use Table VI in the Tables Appendix, or a calculator, to find the required part of triangle ABC with $\gamma = 90°$.

14. $a = 5$, $\beta = 61°$; find c.

15. $b = 6$, $c = 15$; find α.

16. $a = 7$, $b = 10$; find c.

17. A surveyor finds the angle of elevation of the top of a tree to be $42°$. If the surveyor is 75 feet from the base of the tree, find the height of the tree.

CHAPTER 7

ANALYTIC TRIGONOMETRY

Much of the language and terminology of algebra carries over to trigonometry. For example, we have seen that algebraic expressions involve variables, constants, and algebraic operations. **Trigonometric expressions** involve these same elements but also permit trigonometric functions of variables and constants. They also allow algebraic operations upon these trigonometric functions. Thus,

$$x + \sin x \qquad \sin x + \tan x \qquad \frac{1 - \cos x}{\sec^2 x}$$

are all examples of trigonometric expressions.

The distinction between an identity and an equation also carries over to trigonometry. Thus, a **trigonometric identity** is true for all real values in the domain of the variable, but a **trigonometric equation** is true only for certain values called **solutions.** (Note that the solutions of a trigonometric equation may be expressed as real numbers or as angles.) As usual, the set of all solutions of a trigonometric equation is called the **solution set.**

7.1
TRIGONOMETRIC
IDENTITIES

FUNDAMENTAL
IDENTITIES

In Section 5.2 we established the identity

$$\sin^2 t + \cos^2 t = 1 \tag{1}$$

If $\cos t \neq 0$, we may divide both sides of Equation (1) by $\cos^2 t$ to obtain

$$\frac{\sin^2 t}{\cos^2 t} + \frac{\cos^2 t}{\cos^2 t} = \frac{1}{\cos^2 t}$$

or

$$\tan^2 t + 1 = \sec^2 t \tag{2}$$

Similarly, if $\sin t \neq 0$, dividing Equation (1) by $\sin^2 t$ yields

$$\frac{\sin^2 t}{\sin^2 t} + \frac{\cos^2 t}{\sin^2 t} = \frac{1}{\sin^2 t}$$

or

$$\cot^2 t + 1 = \csc^2 t \tag{3}$$

Observe that $\tan t$ and $\cot t$ are undefined for exactly those values of t for which $\cos t$ and $\sin t$ are 0, respectively. It follows that the identities (2) and (3) are true for all values of t for which the trigonometric expressions are defined.

The two identities that we have just established, together with the identities discussed in Sections 5.2 and 5.6, are called the **fundamental identities.** Since we will use these eight identities throughout this chapter, it is essential that you know and recognize them in their various forms. Here is a list for review.

Fundamental Identity	Alternate Form(s)
$\tan t = \dfrac{\sin t}{\cos t}$	
$\cot t = \dfrac{\cos t}{\sin t}$	
$\csc t = \dfrac{1}{\sin t}$	$\sin t = \dfrac{1}{\csc t}$
$\sec t = \dfrac{1}{\cos t}$	$\cos t = \dfrac{1}{\sec t}$

$$\cot t = \frac{1}{\tan t}$$

$$\sin^2 t + \cos^2 t = 1$$

$$\tan^2 t + 1 = \sec^2 t$$

$$\cot^2 t + 1 = \csc^2 t$$

$$\tan t = \frac{1}{\cot t}$$

$$\sin^2 t = 1 - \cos^2 t$$

$$\cos^2 t = 1 - \sin^2 t$$

$$\tan^2 t = \sec^2 t - 1$$

$$\cot^2 t = \csc^2 t - 1$$

In Section 5.2 we saw that trigonometric identities can be used to simplify a trigonometric expression. Here is another example, in which we use the identities developed in this section.

EXAMPLE 1
Simplify the expression $\sin^2 x + \sin^2 x \tan^2 x$.

SOLUTION
We begin by noting that $\sin^2 x$ appears in both terms, which suggests that we factor.

$$
\begin{aligned}
\sin^2 x + \sin^2 x \tan^2 x &= \sin^2 x(1 + \tan^2 x) && \text{Factoring} \\
&= \sin^2 x \sec^2 x && 1 + \tan^2 x = \sec^2 x \\
&= \frac{\sin^2 x}{\cos^2 x} && \sec x = \frac{1}{\cos x} \\
&= \tan^2 x && \frac{\sin x}{\cos x} = \tan x
\end{aligned}
$$

PROGRESS CHECK
Simplify the expression $\dfrac{\csc \theta}{1 + \cot^2 \theta}$.

ANSWER
$\sin \theta$

TRIGONOMETRIC IDENTITIES

The fundamental identities can be employed to prove or, more properly, to verify various trigonometric identities. The principal reasons for including this topic are (a) to improve your skills in recognizing and using the fundamental identities, and (b) to sharpen your reasoning processes. There are also times in calculus and applied mathematics when simplification of a trigonometric expression may enable us to see a relationship that would otherwise be obscured. Finally, in computer applications it is much more efficient to evaluate a simple trigonometric expression than an involved one.

The preferred method of verifying an identity is to transform one side of the equation into the other. We will use this method whenever practical. Unfortunately, we cannot outline a rigid set of steps that will "work" to transform one side into the other; in fact, there are often many ways to tackle a given identity. We will provide a number of examples that demonstrate some of the techniques that can be employed. If you should make a false start and find yourself trying something that doesn't appear to be working, start again and try another approach. With practice your skills will improve.

EXAMPLE 2

Verify the identity $\cos x \tan x \csc x = 1$.

SOLUTION

It is often helpful to write all of the trigonometric functions in terms of sine and cosine. The student should supply a reason for each step.

$$\cos x \tan x \csc x = \cos x \, \frac{\sin x}{\cos x} \, \frac{1}{\sin x}$$

$$= 1$$

PROGRESS CHECK

Prove the identity $\sin x \sec x = \tan x$.

EXAMPLE 3

Verify the identity $\dfrac{1}{1 - \sin x} + \dfrac{1}{1 + \sin x} = 2 \sec^2 x$.

SOLUTION

Another useful technique is to begin with the more complicated expression and complete the indicated operations. We will begin with the left-hand side and will combine the fractions.

$$\frac{1}{1 - \sin x} + \frac{1}{1 + \sin x} = \frac{1 + \sin x + 1 - \sin x}{(1 - \sin x)(1 + \sin x)}$$

$$= \frac{2}{1 - \sin^2 x} = \frac{2}{\cos^2 x}$$

$$= 2 \sec^2 x$$

PROGRESS CHECK

Verify the identity $\cos x + \tan x \sin x = \sec x$.

EXAMPLE 4

Verify the identity $\sin \alpha - \sin^2 \alpha = \dfrac{1 - \sin \alpha}{\csc \alpha}$.

SOLUTION

Factoring will sometimes help to simplify an expression. The student should supply a reason for each step.

$$\sin \alpha - \sin^2 \alpha = \sin \alpha(1 - \sin \alpha)$$

$$= \frac{1 - \sin \alpha}{\csc \alpha}$$

PROGRESS CHECK

Verify the identity $\dfrac{\sin^2 y - 1}{1 - \sin y} = -1 - \sin y.$

EXAMPLE 5

Verify the identity $\dfrac{\cos \theta}{1 - \sin \theta} = \sec \theta + \tan \theta.$

SOLUTION

Multiplying the numerator and denominator of a rational expression by the same quantity is a useful technique. Of course, this quantity should be selected carefully. In this example, multiplying the denominator $1 - \sin \theta$ by $1 + \sin \theta$ will produce $1 - \sin^2 \theta = \cos^2 \theta$. (Similarly, should $\sec x - 1$ appear in a denominator, you might try multiplying by $\sec x + 1$ to obtain $\sec^2 x - 1 = \tan^2 x$.)

The student should supply a reason for each of the following steps.

$$\frac{\cos \theta}{1 - \sin \theta} = \frac{\cos \theta}{1 - \sin \theta} \cdot \frac{1 + \sin \theta}{1 + \sin \theta}$$

$$= \frac{\cos \theta(1 + \sin \theta)}{1 - \sin^2 \theta}$$

$$= \frac{\cos \theta(1 + \sin \theta)}{\cos^2 \theta}$$

$$= \frac{1 + \sin \theta}{\cos \theta}$$

$$= \frac{1}{\cos \theta} + \frac{\sin \theta}{\cos \theta}$$

$$= \sec \theta + \tan \theta$$

PROGRESS CHECK

Verify the identity $\dfrac{1 + \cos t}{\sin t} + \dfrac{\sin t}{1 + \cos t} = 2 \csc t.$

We said earlier that the preferred way of verifying an identity is to transform one side of the equation into the other. At times, both sides may involve complicated expressions and this approach may not be practical. We can then try to transform each side of the equation into the same expression, being careful to use only procedures that are reversible. Here is an example.

EXAMPLE 6

Verify the identity $\dfrac{\cot u - \tan u}{\sin u \cos u} = \csc^2 u - \sec^2 u.$

SOLUTION

Beginning with the left-hand side we have

$$\frac{\cot u - \tan u}{\sin u \cos u} = \frac{\dfrac{\cos u}{\sin u} - \dfrac{\sin u}{\cos u}}{\sin u \cos u}$$

$$= \frac{\cos^2 u - \sin^2 u}{\sin^2 u \cos^2 u}$$

We then transform the right-hand side of the equation by writing all trigonometric functions in terms of sine and cosine.

$$\csc^2 u - \sec^2 u = \frac{1}{\sin^2 u} - \frac{1}{\cos^2 u}$$

$$= \frac{\cos^2 u - \sin^2 u}{\sin^2 u \cos^2 u}$$

We have successfully transformed both sides of the equation into the same expression. Since all the steps are reversible, we have verified the identity.

PROGRESS CHECK

Verify the identity $\dfrac{\sin x + \cos x}{\tan^2 x - 1} = \dfrac{\cos^2 x}{\sin x - \cos x}.$

EXERCISE SET 7.1

Verify each of the following identities.

1. $\csc \gamma - \cos \gamma \cot \gamma = \sin \gamma$

2. $\cot x \sec x = \csc x$

3. $\sec v + \tan v = \dfrac{1 + \sin v}{\cos v}$

4. $\cos \theta + \tan \theta \sin \theta = \sec \theta$

5. $\sin \alpha \sec \alpha = \tan \alpha$

6. $\sec \beta - \cos \beta = \sin \beta \tan \beta$

7. $3 - \sec^2 x = 2 - \tan^2 x$

8. $1 - 2 \sin^2 t = 2 \cos^2 t - 1$

9. $\dfrac{\sec^2 y}{\tan y} = \tan y + \cot y$

10. $\dfrac{\sin x + \cos x}{\cos x} = 1 + \tan x$

11. $\dfrac{\sin u}{\csc u} + \dfrac{\cos u}{\sec u} = 1$

12. $\dfrac{\tan^2 \alpha}{1 + \sec \alpha} = \sec \alpha - 1$

13. $\dfrac{\sec^2 \theta - 1}{\sec^2 \theta} = \sin^2 \theta$

14. $\sin^4 x + 2 \sin^2 x \cos^2 x + \cos^4 x = 1$

15. $\cos \gamma + \cos \gamma \tan^2 \gamma = \sec \gamma$

16. $\dfrac{1}{\tan u + \cot u} = \cos u \sin u$

17. $\dfrac{\sec w \sin w}{\tan w + \cot w} = \sin^2 w$

18. $(1 - \cos^2 \beta)(1 + \cot^2 \beta) = 1$

19. $(\sin \alpha + \cos \alpha)^2 + (\sin \alpha - \cos \alpha)^2 = 2$

20. $\dfrac{1 + \tan^2 u}{\csc^2 u} = \tan^2 u$

21. $\sec^2 v + \cos^2 v = \dfrac{\sec^4 v + 1}{\sec^2 v}$

22. $\sin^2 \theta - \tan^2 \theta = -\tan^2 \theta \sin^2 \theta$

23. $\dfrac{\sin^2 \alpha}{1 + \cos \alpha} = 1 - \cos \alpha$

24. $\cot x \sin^2 x = \cos x (1 - \sin x)$

25. $\dfrac{\cos t}{1 + \sin t} = \dfrac{1 - \sin t}{\cos t}$

26. $\dfrac{\sin \beta}{1 + \cos \beta} + \dfrac{1 + \cos \beta}{\sin \beta} = 2 \csc \beta$

27. $\csc^2 \theta - \dfrac{\cos^2 \theta}{\sin^2 \theta} = 1$

28. $\dfrac{\cos^2 u}{1 - \sin u} = 1 + \sin u$

29. $\dfrac{\cot y}{1 + \cot^2 y} = \sin y \cos y$

30. $\dfrac{1 + \tan^2 x}{\tan^2 x} = \csc^2 x$

31. $\cos(-t) \csc(-t) = -\cot t$

32. $\sin(-\theta) \sec(-\theta) = -\tan \theta$

33. $\dfrac{\sec x + \csc x}{1 + \tan x} = \csc x$

34. $\dfrac{\sec u}{\sec u - 1} = \dfrac{1}{1 - \cos u}$

35. $\dfrac{1 + \tan x}{1 + \cot x} = \dfrac{\sec x}{\csc x}$

36. $(\tan u + \sec u)^2 = \dfrac{1 + \sin u}{1 - \sin u}$

37. $\dfrac{1 - \sin t}{1 + \sin t} = (\sec t - \tan t)^2$

38. $2 \csc^2 \theta - \csc^4 \theta = 1 - \cot^4 \theta$

39. $\dfrac{\sin^2 w}{\cos^4 w + \cos^2 w \sin^2 w} = \tan^2 w$

40. $\dfrac{\sin z + \tan z}{1 + \cos z} = \tan z$

41. $\dfrac{\sec \gamma - \csc \gamma}{\sec \gamma + \csc \gamma} = \dfrac{\tan \gamma - 1}{\tan \gamma + 1}$

42. $\dfrac{\cot x - 1}{1 - \tan x} = \dfrac{\csc x}{\sec x}$

43. $\dfrac{\tan \gamma - \sin \gamma}{\tan \gamma} = \dfrac{\sin^2 \gamma}{1 + \cos \gamma}$

44. $\cos^4 u - \sin^4 u = \cos^2 u - \sin^2 u$

45. $\dfrac{\csc x}{1 + \csc x} - \dfrac{\csc x}{1 - \csc x} = 2 \sec^2 x$

46. $\sin^3 \theta + \cos^3 \theta = (1 - \sin \theta \cos \theta)(\sin \theta + \cos \theta)$

Show that each of the following equations is not an identity by finding a value of the variable for which the equation is not true.

47. $\sin x = \sqrt{1 - \cos^2 x}$

48. $\tan x = \sqrt{\sec^2 x - 1}$

49. $(\sin t + \cos t)^2 = \sin^2 t + \cos^2 t$

50. $\sin \theta + \cos \theta = \sec \theta + \csc \theta$

51. $\sqrt{\cos^2 x} = \cos x$

52. $\sqrt{\cot^2 x} = \cot x$

7.2
THE ADDITION
FORMULAS

The identities that we verified in the examples and exercises of Section 7.1 were themselves of no special significance; we were primarily interested in having you practice manipulation with the fundamental identities. There are, however, many trigonometric identities that are indeed of importance; these identities are called **trigonometric formulas.** Such formulas are used so frequently that it is probably best for you to memorize them. We will develop these formulas in a logical sequence so that you will be able to derive them yourself should you wish to verify that your memorization is correct.

Our first objective is to develop the **addition formula** for $\cos(s + t)$ where s and t are any real numbers. It happens that it is easier to begin with $\cos(s - t)$, which demonstrates that the mathematician may at times have to take a circuitous route to establish a result!

For convenience, we assume that s, t, and $s - t$ are all positive and less than 2π. We let $P = W(s)$, $Q = W(t)$, $R = W(s - t)$ be the points on the unit circle determined by the wrapping function W (see Figure 1). Then $\widehat{AP} = s$, $\widehat{AQ} = t$, $\widehat{AR} = s - t$, and by the definitions of sine and cosine, the coordinates of the points can be written as

$$P(\cos s, \sin s) \quad Q(\cos t, \sin t) \quad R(\cos(s - t), \sin(s - t))$$

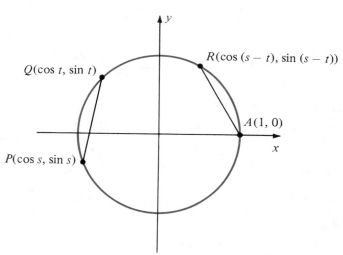

FIGURE 1

Since the arcs $\widehat{QP}$ and $\widehat{AR}$ are both of length $s - t$, the chords QP and AR are also of equal length. By the distance formula, we have

$$\overline{AR} = \overline{QP}$$
$$\sqrt{[\cos(s - t) - 1]^2 + [\sin(s - t)]^2} = \sqrt{(\cos s - \cos t)^2 + (\sin s - \sin t)^2}$$

Squaring both sides and rearranging terms, we have

$$\sin^2(s - t) + \cos^2(s - t) - 2\cos(s - t) + 1$$
$$= \sin^2 s + \cos^2 s + \sin^2 t + \cos^2 t - 2\cos s \cos t - 2\sin s \sin t$$

Since each of the expressions $\sin^2(s - t) + \cos^2(s - t)$, $\sin^2 s + \cos^2 s$, and $\sin^2 t + \cos^2 t$ equals 1, we have

$$2 - 2\cos(s - t) = 2 - 2\cos s \cos t - 2\sin s \sin t$$

Solving for $\cos(s - t)$ yields the formula

$$\cos(s - t) = \cos s \cos t + \sin s \sin t \qquad (1)$$

Now it's easy to obtain the addition formula for $\cos(s + t)$. By writing

$$s + t = s - (-t)$$

we have

$$\cos(s + t) = \cos(s - (-t))$$
$$= \cos s \cos(-t) + \sin s \sin(-t)$$

Since $\cos(-t) = \cos t$ and $\sin(-t) = -\sin t$,

$$\cos(s + t) = \cos s \cos t - \sin s \sin t \qquad (2)$$

EXAMPLE 1
Find $\cos 15°$ without the use of tables or a calculator.

SOLUTION
Since $15° = 45° - 30°$, we may use the formula for $\cos(s - t)$ to obtain

$$\cos 15° = \cos(45° - 30°)$$
$$= \cos 45° \cos 30° + \sin 45° \sin 30°$$
$$= \frac{\sqrt{2}}{2} \cdot \frac{\sqrt{3}}{2} + \frac{\sqrt{2}}{2} \cdot \frac{1}{2}$$
$$= \frac{\sqrt{6} + \sqrt{2}}{4}$$

PROGRESS CHECK
Solve Example 1 using the identity $15° = 60° - 45°$.

EXAMPLE 2
Find the exact value of $\cos(5\pi/12)$.

SOLUTION
We note that $5\pi/12 = 2\pi/12 + 3\pi/12 = \pi/6 + \pi/4$. Then

$$\cos\left(\frac{5\pi}{12}\right) = \cos\left(\frac{\pi}{6} + \frac{\pi}{4}\right)$$

$$= \cos\frac{\pi}{6}\cos\frac{\pi}{4} - \sin\frac{\pi}{6}\sin\frac{\pi}{4}$$

$$= \frac{\sqrt{3}}{2} \cdot \frac{\sqrt{2}}{2} - \frac{1}{2} \cdot \frac{\sqrt{2}}{2}$$

$$= \frac{\sqrt{6} - \sqrt{2}}{4}$$

PROGRESS CHECK
Solve Example 2 using the identity $5\pi/12 = 9\pi/12 - 4\pi/12$.

Before tackling $\sin(s + t)$, we first establish the following important functional relationships.

$$\cos\left(\frac{\pi}{2} - t\right) = \sin t \qquad (3)$$

$$\sin\left(\frac{\pi}{2} - t\right) = \cos t \qquad (4)$$

$$\tan\left(\frac{\pi}{2} - t\right) = \cot t \qquad (5)$$

Using the difference formula for cosine we have

$$\cos\left(\frac{\pi}{2} - t\right) = \cos\frac{\pi}{2}\cos t + \sin\frac{\pi}{2}\sin t$$

$$= 0 \cdot \cos t + 1 \cdot \sin t$$

$$= \sin t$$

which establishes Equation (3). Replacing t with $\frac{\pi}{2} - t$ in this identity yields

$$\cos\left[\frac{\pi}{2} - \left(\frac{\pi}{2} - t\right)\right] = \sin\left(\frac{\pi}{2} - t\right)$$

$$\cos t = \sin\left(\frac{\pi}{2} - t\right)$$

which establishes Equation (4). The third identity follows from the definition of tangent and from Equations (3) and (4):

$$\tan\left(\frac{\pi}{2} - t\right) = \frac{\sin\left(\frac{\pi}{2} - t\right)}{\cos\left(\frac{\pi}{2} - t\right)} = \frac{\cos t}{\sin t} = \cot t$$

Functions satisfying the properties of the identities (3) and (4) are called **cofunctions.** Thus, sine and cosine are cofunctions. So, too, the tangent and cotangent functions are cofunctions, as are secant and cosecant. This is the origin of the prefix *co* in *co*sine, *co*secant, and *co*tangent.

EXAMPLE 3

Use trigonometry of the right triangle to show that sine and cosine are cofunctions.

SOLUTION

In right triangle ABC, angle $\gamma = 90°$ (Figure 2). Then $\sin \alpha = a/c = \cos \beta$. But angles α and β are complementary; that is, $\alpha + \beta = 90°$. Thus $\sin \alpha = \cos(90° - \alpha)$ and $\cos \beta = \sin(90° - \beta)$, which establishes that they are cofunctions.

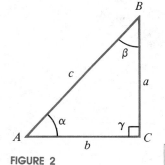

FIGURE 2

We are now prepared to prove the following.

$$\sin(s + t) = \sin s \cos t + \cos s \sin t \qquad (6)$$
$$\sin(s - t) = \sin s \cos t - \cos s \sin t \qquad (7)$$

We supply the steps for a proof of Equation (6); the student should supply a reason for each step.

$$\sin(s + t) = \cos\left[\frac{\pi}{2} - (s + t)\right]$$

$$= \cos\left[\left(\frac{\pi}{2} - s\right) - t\right]$$

$$= \cos\left(\frac{\pi}{2} - s\right)\cos t + \sin\left(\frac{\pi}{2} - s\right)\sin t$$

$$= \sin s \cos t + \cos s \sin t$$

The student should now prove Equation (7) by using
$$\sin(s - t) = \sin[s + (-t)]$$
We conclude with the addition formulas for the tangent function.

$$\tan(s + t) = \frac{\tan s + \tan t}{1 - \tan s \tan t} \qquad (8)$$

$$\tan(s - t) = \frac{\tan s - \tan t}{1 + \tan s \tan t} \qquad (9)$$

Again, we supply the steps for a proof of Equation (8) and will let the student supply a reason for each step.

$$\tan(s + t) = \frac{\sin(s + t)}{\cos(s + t)}$$

$$= \frac{\sin s \cos t + \cos s \sin t}{\cos s \cos t - \sin s \sin t}$$

$$= \frac{\left(\dfrac{\sin s}{\cos s} \cdot \dfrac{\cos t}{\cos t}\right) + \left(\dfrac{\cos s}{\cos s} \cdot \dfrac{\sin t}{\cos t}\right)}{\left(\dfrac{\cos s}{\cos s} \cdot \dfrac{\cos t}{\cos t}\right) - \left(\dfrac{\sin s}{\cos s} \cdot \dfrac{\sin t}{\cos t}\right)}$$

$$= \frac{\tan s + \tan t}{1 - \tan s \tan t}$$

The student should now prove Equation (9) by using

$$\tan(s - t) = \tan[s + (-t)]$$

EXAMPLE 4
Show that $\sin(x + 2\pi) = \sin x$.

SOLUTION
Using the addition formula,

$$\sin(x + 2\pi) = \sin x \cos 2\pi + \cos x \sin 2\pi$$
$$= \sin x \cdot 1 + \cos x \cdot 0$$
$$= \sin x$$

PROGRESS CHECK
Show that $\tan(x + \pi) = \tan x$.

EXAMPLE 5
Given $\sin \alpha = -\frac{4}{5}$, with α an angle in quadrant III, and $\cos \beta = -\frac{5}{13}$, with β an angle in quadrant II, use the addition formula to find $\sin(\alpha + \beta)$ and the quadrant in which $\alpha + \beta$ lies.

COMPUTING SINE AND COSINE

```
10  LET S1 = 0.01745
20  LET C1 = 0.99985
30  PRINT
    ''DEGREES'',
    ''SIN'',''COS''
40  PRINT ''1'',
    S1, C1
50  LET S2 = S1
60  LET C2 = C1
70  FOR I = 2 TO 90
80    LET S3 = S2
90    LET S2 =
      (S1 * C2) +
      (C1 * S2)
100   LET C2 =
      (C1 * C2) -
      (S1 * S3)
110   PRINT I, S2,
      C2
120 NEXT I
130 END
```

We can make use of the trigonometric formulas to generate a table of sine and cosine values. Suppose we have determined that

$$\sin 1° = 0.01745 \quad \cos 1° = 0.99985 \tag{1}$$

We can then write

$$\sin(1° + \alpha) = \sin 1° \cos \alpha + \cos 1° \sin \alpha$$
$$\cos(1° + \alpha) = \cos 1° \cos \alpha - \sin 1° \sin \alpha$$

Substituting for $\sin 1°$ and $\cos 1°$ from Equations (1),

$$\sin(1° + \alpha) = 0.01745 \cos \alpha + 0.99985 \sin \alpha \tag{2}$$
$$\cos(1° + \alpha) = 0.99985 \cos \alpha - 0.01745 \sin \alpha \tag{3}$$

Now, if we let $\alpha = 1°$, Equations (2) and (3) can be used to calculate $\sin 2°$ and $\cos 2°$. We can then repeat the process with $\alpha = 2°$ to calculate $\sin 3°$ and $\cos 3°$, and so on. Since this is an iterative procedure well suited for a computer, we are providing a program in BASIC that will calculate sine and cosine values from 2° to 90° in increments of 1°. Although this is a neat illustration of the use of trigonometric formulas, the method of polynomial approximation used in Section 5.4 is of greater practical value.

SOLUTION

The addition formula

$$\sin(\alpha + \beta) = \sin \alpha \cos \beta + \cos \alpha \sin \beta$$

requires that we know $\sin \alpha$, $\cos \alpha$, $\sin \beta$, and $\cos \beta$. Using the fundamental identity $\sin^2 \alpha + \cos^2 \alpha = 1$, we have

$$\cos^2 \alpha = 1 - \sin^2 \alpha = 1 - \frac{16}{25} = \frac{9}{25}$$

Taking the square root of both sides, we must have $\cos \alpha = -\frac{3}{5}$ since α is in quadrant III. Similarly,

$$\sin^2 \beta = 1 - \cos^2 \beta = 1 - \frac{25}{169} = \frac{144}{169}$$

Taking the square root of both sides, we must have $\sin \beta = \frac{12}{13}$ since β is in quadrant II. Thus,

$$\sin(\alpha + \beta) = \left(-\frac{4}{5}\right)\left(-\frac{5}{13}\right) + \left(-\frac{3}{5}\right)\left(\frac{12}{13}\right)$$

$$= \frac{20}{65} - \frac{36}{65} = -\frac{16}{65}$$

Since $\sin(\alpha + \beta)$ is negative, $\alpha + \beta$ lies in either quadrant III or quadrant IV. However, the sum of an angle that lies in quadrant III and an angle that lies in quadrant II cannot lie in quadrant III. Thus, $\alpha + \beta$ lies in quadrant IV.

PROGRESS CHECK
Given $\cos \alpha = -\frac{4}{5}$, with α in quadrant III, and $\cos \beta = \frac{3}{5}$, with β in quadrant I, find $\cos(\alpha - \beta)$ and the quadrant in which $\alpha - \beta$ lies.

ANSWER
$-\frac{24}{25}$; quadrant II

EXERCISE SET 7.2
In Exercises 1–6 show that the given equation is not an identity. (*Hint:* For each equation, find a pair of values of s and t for which the equation is not true.)

1. $\cos(s - t) = \cos s - \cos t$
2. $\sin(s + t) = \sin s + \sin t$
3. $\sin(s - t) = \sin s - \sin t$
4. $\cos(s + t) = \cos s + \cos t$
5. $\tan(s + t) = \tan s + \tan t$
6. $\tan(s - t) = \tan s - \tan t$

In Exercises 7–22 use the addition formulas to find exact values.

7. $\cos\left(\frac{\pi}{6} + \frac{\pi}{4}\right)$
8. $\sin\left(\frac{\pi}{6} - \frac{\pi}{4}\right)$
9. $\sin\left(\frac{\pi}{4} + \frac{\pi}{3}\right)$
10. $\cos\left(\frac{\pi}{3} - \frac{\pi}{4}\right)$
11. $\cos(30° + 180°)$
12. $\tan(60° + 300°)$
13. $\tan(300° - 60°)$
14. $\sin(270° - 45°)$
15. $\sin 11\pi/12$ (*Hint:* $11\pi/12 = \pi/6 + 3\pi/4$)
16. $\tan 7\pi/12$ (*Hint:* $7\pi/12 = \pi/4 + \pi/3$)
17. $\cos 7\pi/12$ (*Hint:* $7\pi/12 = 5\pi/6 - \pi/4$)
18. $\tan 75°$ (*Hint:* $75° = 135° - 60°$)
19. $\sin 7\pi/6$
20. $\cos 5\pi/6$
21. $\tan 15°$
22. $\tan 165°$

In Exercises 23–28 write the given expression in terms of cofunctions of complementary angles.

23. $\sin 47°$
24. $\cos 78°$
25. $\tan \pi/6$
26. $\tan 84°$
27. $\cos \pi/3$
28. $\sin 72°30'$

29. If $\sin t = -3/5$, with $W(t)$ in quadrant III, find $\sin(\pi/2 - t)$.

30. If $\cos t = -5/13$, with $W(t)$ in quadrant II, find $\sin(t - \pi)$.

31. If $\tan \theta = 4/3$ and angle θ lies in quadrant III, find $\tan(\theta + \pi/4)$.

32. If $\sec \theta = 5/3$ and angle θ lies in quadrant I, find $\sin(\theta + \pi/6)$.

33. If $\cos t = 0.4$, with $W(t)$ in quadrant IV, find $\tan(t + \pi)$.

34. If $\sec \alpha = 1.2$ and angle α lies in quadrant IV, find $\tan(\alpha - \pi)$.

35. If $\sin s = 3/5$ and $\cos t = -12/13$, with $W(s)$ in quadrant II and $W(t)$ in quadrant III, find $\sin(s + t)$.

36. If $\sin s = -4/5$ and $\csc t = 13/5$, with $W(s)$ in quadrant IV and $W(t)$ in quadrant II, find $\cos(s - t)$.

37. If $\cos \alpha = 5/13$ and $\tan \beta = -2$, with angle α in quadrant I and angle β in quadrant II, find

$$\tan(\alpha + \beta)$$

38. If $\sec \alpha = 5/3$ and $\cot \beta = 15/8$, with angle α in quadrant IV and angle β in quadrant III, find $\tan(\alpha - \beta)$.

Prove each of the following identities by transforming the left-hand side of the equation into the expression on the right-hand side.

39. $\sin 2\alpha = 2 \sin \alpha \cos \alpha$

40. $\cos 2t = \cos^2 t - \sin^2 t$

41. $\tan 2\alpha = \dfrac{2 \tan \alpha}{1 - \tan^2 \alpha}$

42. $\sin(x + y)\sin(x - y) = \sin^2 x - \sin^2 y$

43. $\cos(x - y) \cos(x + y) = \cos^2 x - \sin^2 x$

44. $\dfrac{\sin(s + t)}{\sin(s - t)} = \dfrac{\tan s + \tan t}{\tan s - \tan t}$

45. $\csc(t + \pi/2) = \sec t$

46. $\tan(\alpha + 90°) = -\cot \alpha$

47. $\tan(x + \pi/4) = \dfrac{1 + \tan x}{1 - \tan x}$

48. $\csc(t - \pi) = -\csc t$

49. $\cot(s - t) = \dfrac{1 + \tan s \tan t}{\tan s - \tan t}$

50. $\cot(u + v) = \dfrac{\cot u \cot v - 1}{\cot u + \cot v}$

51. $\sin(s + t) + \sin(s - t) = 2 \sin s \cos t$

52. $\cos(s + t) + \cos(s - t) = 2 \cos s \cos t$

53. $\dfrac{\sin(x + h) - \sin x}{h} = \sin x\left(\dfrac{\cos h - 1}{h}\right) + \cos x\left(\dfrac{\sin h}{h}\right)$

54. $\dfrac{\cos(x + h) - \cos x}{h} = \cos x\left(\dfrac{\cos h - 1}{h}\right) - \sin x\left(\dfrac{\sin h}{h}\right)$

7.3 DOUBLE- AND HALF-ANGLE FORMULAS

Our initial objective in this section is to derive expressions for $\sin 2t$, $\cos 2t$, and $\tan 2t$ in terms of trigonometric functions of t. We will establish the following **double-angle formulas.**

DOUBLE-ANGLE FORMULAS

$$\sin 2t = 2 \sin t \cos t \qquad (1)$$
$$\cos 2t = \cos^2 t - \sin^2 t \qquad (2)$$
$$\tan 2t = \dfrac{2 \tan t}{1 - \tan^2 t} \qquad (3)$$

Once again, it's best to memorize these formulas. However, the derivations are so straightforward that you can always return to them to verify the results.

To establish Equation (1), we simply rewrite $2t$ as $(t + t)$ and use the addition formula.

$$\sin 2t = \sin(t + t)$$
$$= \sin t \cos t + \cos t \sin t$$
$$= 2 \sin t \cos t$$

We proceed in the same manner to prove Equation (2).

$$\cos 2t = \cos(t + t)$$
$$= \cos t \cos t - \sin t \sin t$$
$$= \cos^2 t - \sin^2 t$$

Using the addition formula for the tangent function yields a proof of Equation (3).

$$\tan 2t = \tan(t + t)$$
$$= \frac{\tan t + \tan t}{1 - \tan t \tan t}$$
$$= \frac{2 \tan t}{1 - \tan^2 t}$$

EXAMPLE 1

If $\cos t = -\frac{3}{5}$ and $W(t)$ is in quadrant II, evaluate $\sin 2t$ and $\cos 2t$. In which quadrant does $W(2t)$ lie?

SOLUTION

We first find $\sin t$ by use of the fundamental identity $\sin^2 t + \cos^2 t = 1$. Thus,

$$\sin^2 t + \frac{9}{25} = 1$$
$$\sin^2 t = \frac{16}{25}$$

Since $W(t)$ is in quadrant II, $\sin t$ must be positive. Therefore,

$$\sin t = \frac{4}{5}$$

Applying the double-angle formulas with $\cos t = -\frac{3}{5}$, $\sin t = \frac{4}{5}$, yields

$$\sin 2t = 2 \sin t \cos t = 2\left(\frac{4}{5}\right)\left(-\frac{3}{5}\right) = -\frac{24}{25}$$

$$\cos 2t = \cos^2 t - \sin^2 t = \frac{9}{25} - \frac{16}{25} = -\frac{7}{25}$$

Since $\sin 2t$ and $\cos 2t$ are both negative, we may conclude that $W(2t)$ lies in quadrant III.

PROGRESS CHECK

If $\sin \theta = 5/13$ and θ is in quadrant I, evaluate $\sin 2\theta$ and $\tan 2\theta$.

ANSWER

$$\sin 2\theta = \frac{120}{169}, \ \tan 2\theta = \frac{120}{119}$$

EXAMPLE 2

Express $\sin 3t$ in terms of $\sin t$ and $\cos t$.

SOLUTION

We write $3t$ as $(2t + t)$. Then

$$\begin{aligned}
\sin 3t &= \sin(2t + t) \\
&= \sin 2t \cos t + \cos 2t \sin t \\
&= 2 \sin t \cos t \cos t + (\cos^2 t - \sin^2 t) \sin t \\
&= 2 \sin t \cos^2 t + \sin t \cos^2 t - \sin^3 t \\
&= 3 \sin t \cos^2 t - \sin^3 t
\end{aligned}$$

PROGRESS CHECK

Express $\cos 3t$ in terms of $\sin t$ and $\cos t$.

ANSWER

$\cos 3t = 4 \cos^3 t - 3 \cos t$

If we begin with the formula for $\cos 2t$ and use the fundamental identity $\cos^2 t = 1 - \sin^2 t$, we obtain

$$\begin{aligned}
\cos 2t &= \cos^2 t - \sin^2 t \\
&= (1 - \sin^2 t) - \sin^2 t \\
&= 1 - 2 \sin^2 t
\end{aligned}$$

Similarly, replacing $\sin^2 t$ by $1 - \cos^2 t$ yields

$$\begin{aligned}\cos 2t &= \cos^2 t - \sin^2 t \\ &= \cos^2 t - (1 - \cos^2 t) \\ &= 2 \cos^2 t - 1\end{aligned}$$

We then have two additional formulas for $\cos 2t$.

$$\cos 2t = 1 - 2 \sin^2 t \qquad (4)$$
$$\cos 2t = 2 \cos^2 t - 1 \qquad (5)$$

EXAMPLE 3

Verify the identity $\dfrac{1 - \cos 2\alpha}{2 \sin \alpha \cos \alpha} = \tan \alpha$.

SOLUTION

Substituting $\cos 2\alpha = 1 - 2 \sin^2 \alpha$, we have

$$\begin{aligned}\frac{1 - \cos 2\alpha}{2 \sin \alpha \cos \alpha} &= \frac{1 - (1 - 2 \sin^2 \alpha)}{2 \sin \alpha \cos \alpha} \\ &= \frac{2 \sin^2 \alpha}{2 \sin \alpha \cos \alpha} \\ &= \frac{\sin \alpha}{\cos \alpha} \\ &= \tan \alpha\end{aligned}$$

PROGRESS CHECK

Verify the identity $\dfrac{1 + \cos 2\theta}{\sin 2\theta} = \cot \theta$.

HALF-ANGLE FORMULAS

If we begin with the alternative forms for $\cos 2t$ given in Equations (4) and (5), we can obtain the following expressions for $\sin^2 t$ and $\cos^2 t$. These expressions are often used in calculus.

$$\sin^2 t = \frac{1 - \cos 2t}{2} \qquad (6)$$

$$\cos^2 t = \frac{1 + \cos 2t}{2} \qquad (7)$$

We will use the identities in Equations (6) and (7) to derive formulas for $\sin t/2$, $\cos t/2$, and $\tan t/2$. Substituting $s = 2t$ in Equations (6) and (7) we obtain

$$\sin^2 \frac{s}{2} = \frac{1 - \cos s}{2}$$

$$\cos^2 \frac{s}{2} = \frac{1 + \cos s}{2}$$

Replacing s with t and solving, we have

$$\sin \frac{t}{2} = \pm\sqrt{\frac{1 - \cos t}{2}} \qquad (8)$$

$$\cos \frac{t}{2} = \pm\sqrt{\frac{1 + \cos t}{2}} \qquad (9)$$

The appropriate sign to use in Equations (8) and (9) depends on the quadrant in which $W(t/2)$ is located. Thus, $\sin t/2$ is positive if $W(t/2)$ lies in quadrant I or II; similarly, we choose the positive root for $\cos t/2$ in Equation (9) if $W(t/2)$ lies in quadrant I or IV.

Using the identity

$$\tan \frac{t}{2} = \frac{\sin \dfrac{t}{2}}{\cos \dfrac{t}{2}}$$

we obtain

$$\tan \frac{t}{2} = \pm\sqrt{\frac{1 - \cos t}{1 + \cos t}} \qquad (10)$$

Formulas (8), (9), and (10) are known as the **half-angle formulas.**

EXAMPLE 4
Find the exact values of $\sin 22.5°$ and $\cos 112.5°$.

SOLUTION

Applying the half-angle formulas with $22.5° = 45°/2$ yields

$$\sin 22.5° = \sin \frac{45°}{2}$$

$$= \sqrt{\frac{1 - \cos 45°}{2}}$$

$$= \frac{\sqrt{1 - \sqrt{2}/2}}{2}$$

$$= \frac{\sqrt{2 - \sqrt{2}}}{2}$$

Note that we choose the positive square root since $22.5°$ is in the first quadrant and the sine function is positive in the first quadrant. Similarly,

$$\cos 112.5° = \cos \frac{225°}{2}$$

$$= -\sqrt{\frac{1 + \cos 225°}{2}}$$

$$= -\sqrt{\frac{1 - \cos 45°}{2}}$$

$$= -\sqrt{\frac{1 - \sqrt{2}/2}{2}}$$

$$= -\frac{\sqrt{2 - \sqrt{2}}}{2}$$

The negative square root was selected since $112.5°$ is in the second quadrant and the cosine function is negative in quadrant II.

PROGRESS CHECK

Use the half-angle formulas to evaluate $\tan \dfrac{\pi}{8}$.

ANSWER

$\sqrt{2} - 1$

**TRIGONOMETRY AND
THE PYTHAGOREAN
THEOREM**

The Pythagorean Theorem can be derived by using trigonometry of the right triangle. In the accompanying figure, ABC is a right triangle, and CD is perpendicular to the hypotenuse AB of length c. Using triangle ABC, you can verify that

$$\sin \alpha = \frac{a}{c} \quad \text{and} \quad \cos \alpha = \frac{b}{c} \tag{1}$$

Now, from right triangle ACD,

$$\overline{AD} = b \cos \alpha \tag{2}$$

Noting that $\beta = 90° - \alpha$ and using right triangle BCD,

$$\overline{BD} = a \cos(90° - \alpha) = a \sin \alpha \tag{3}$$

since $\cos(90° - \alpha) = \sin \alpha$. We can now use Equations (2) and (3) to sum

$$c = \overline{BD} + \overline{AD} = a \sin \alpha + b \cos \alpha$$

and, substituting from Equations (1),

$$c = \frac{a^2}{c} + \frac{b^2}{c}$$

or

$$c^2 = a^2 + b^2$$

This, of course, is a statement of the Pythagorean Theorem.

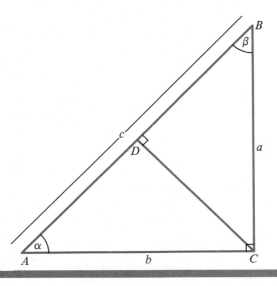

EXAMPLE 5

If $\sin \theta = -\frac{3}{5}$ and θ is in quadrant III, evaluate $\cos \theta/2$.

SOLUTION

We first evaluate $\cos \theta$ by using the identity

$$\cos^2 \theta = 1 - \sin^2 \theta = 1 - \frac{9}{25} = \frac{16}{25}$$

Since θ is in quadrant III, $\cos \theta$ is negative. Thus, $\cos \theta = -\frac{4}{5}$. We can now employ the half-angle formula

$$\cos \frac{\theta}{2} = \pm \sqrt{\frac{1 + \cos \theta}{2}}$$

$$= \pm \sqrt{\frac{1 - \frac{4}{5}}{2}}$$

$$= \pm \frac{\sqrt{10}}{10}$$

Since $180° < \theta < 270°$, we see that $90° < \theta/2 < 135°$. Thus, $\theta/2$ is in quadrant II and $\cos \theta/2$ is negative. We conclude that $\cos \theta/2 = -\sqrt{10}/10$.

PROGRESS CHECK

If $\tan \alpha = \frac{3}{4}$ and α is in quadrant III, evaluate $\tan \alpha/2$.

ANSWER

-3

EXERCISE SET 7.3

Use the given conditions to determine the value of the specified trigonometric function.

1. $\sin u = \frac{2}{3}$ and $W(u)$ is in quadrant II; find $\cos 2u$.

2. $\cos x = -\frac{5}{13}$ and $W(x)$ is in quadrant III; find $\sin 2x$.

3. $\sec \alpha = -2$ and α is in quadrant II; find $\sin 2\alpha$.

4. $\tan \theta = \frac{3}{4}$ and θ is in quadrant I; find $\cos 2\theta$.

5. $\csc t = -\frac{17}{8}$ and $W(t)$ is in quadrant IV; find $\tan 2t$.

6. $\cot \beta = \frac{3}{4}$ and β is in quadrant III; find $\cot 2\beta$.

7. $\sin 2\alpha = -\frac{4}{5}$ and 2α is in quadrant IV; find $\sin 4\alpha$.

8. $\sec 5x = -\frac{13}{12}$ and $W(5x)$ is in quadrant III; find $\tan 10x$.

9. $\cos(\theta/2) = \frac{8}{17}$ and $\theta/2$ is acute; find $\cos \theta$.

10. $\csc(t/4) = -\frac{13}{5}$ and $W(t/4)$ is in quadrant IV; find $\cos(t/2)$.

11. $\sin 42° = 0.67$; find $\cos 84°$.

12. $\cos 77° = 0.22$; find $\cos 154°$.

Use the half-angle formulas to find exact values for each of the following.

13. $\sin 15°$

14. $\cos 75°$

15. $\tan \pi/8$

16. $\sec 5\pi/8$

17. $\csc 165°$

18. $\cot 7\pi/12$

Use the given conditions to determine the exact value of the specified trigonometric function.

19. $\sin \theta = -\frac{4}{5}$ and θ is in quadrant IV; find $\cos \theta/2$.

20. $\cos \theta = \frac{3}{5}$ and θ is in quadrant I; find $\sin \theta/2$.

21. $\sec t = -3$ and $W(t)$ is in quadrant II; find $\sin t/2$.

22. $\tan x = \frac{4}{3}$ and $W(x)$ is in quadrant III; find $\cos x/2$.

23. $\cot \beta = \frac{3}{4}$ and β is in quadrant III; find $\tan \beta/2$.

24. $\csc \alpha = \frac{13}{5}$ and α is in quadrant II; find $\tan \alpha/2$.

25. $\cos 4x = \frac{1}{3}$ and $W(4x)$ is in quadrant IV; find $\cos 2x$.

26. $\sec 6\alpha = -\frac{13}{12}$ and α is in quadrant III; find $\sin 3\alpha$.

Verify the given identities.

27. $\sin 50x = 2 \sin 25x \cos 25x$

28. $(\sin \theta + \cos \theta)^2 = 1 + \sin 2\theta$

29. $\tan 2y = \dfrac{2 \cot y}{\csc^2 y - 2}$

30. $2 \sin^2 2t + \cos 4t = 1$

31. $\sin 4\alpha = 4 \sin \alpha \cos^3 \alpha - 4 \sin^3 \alpha \cos \alpha$

32. $\cos 4\beta = 1 - 8 \sin^2 \beta \cos^2 \beta$

33. $\cos 2u = \dfrac{1 - \tan^2 u}{1 + \tan^2 u}$

34. $\sin 2\theta = \dfrac{2 \tan \theta}{1 + \tan^2 \theta}$

35. $\sin \dfrac{t}{2} \cos \dfrac{t}{2} = \dfrac{\sin t}{2}$

36. $\tan \dfrac{y}{2} = \csc y - \cot y$

37. $\sin \alpha - \cos \alpha \tan \dfrac{\alpha}{2} = \tan \dfrac{\alpha}{2}$

38. $\dfrac{1 - \cos 2\beta}{1 + \cos 2\beta} = \tan^2 \beta$

39. $\cos^4 x - \sin^4 x = \cos 2x$

40. $\dfrac{\sin 2t}{\sin t} - \dfrac{\cos 2t}{\cos t} = \sec t$

41. $\dfrac{2 \tan \alpha}{1 + \tan^2 \alpha} = \sin 2\alpha$

42. $\cos^2 \dfrac{x}{2} = \dfrac{\tan x + \sin x}{2 \tan x}$

43. $\sec 2t = \dfrac{\sec^2 t}{2 - \sec^2 t}$

44. $\cos 2t + \cot 2t = \cot 2t(\sin t + \cos t)^2$

45. $\tan \dfrac{t}{2} = \dfrac{1 - \cos t}{\sin t}$

46. $\tan \dfrac{t}{2} = \dfrac{\sin t}{1 + \cos t}$

7.4
THE PRODUCT-SUM
FORMULAS

The formulas that will be derived in this section are of use in calculus and in other courses in higher mathematics. They are not as important as the formulas that appeared in Sections 7.2 and 7.3 and need not be memorized. Rather, you should be aware of these formulas so that you can look them up when needed.

The following formulas express a product as a sum.

$$\sin s \cos t = \frac{\sin(s + t) + \sin(s - t)}{2} \qquad (1)$$

$$\cos s \sin t = \frac{\sin(s + t) - \sin(s - t)}{2} \qquad (2)$$

$$\cos s \cos t = \frac{\cos(s + t) + \cos(s - t)}{2} \qquad (3)$$

$$\sin s \sin t = \frac{\cos(s - t) - \cos(s + t)}{2} \qquad (4)$$

To prove Equation (1), we begin with the right-hand side of the equation.

$$\frac{\sin(s + t) + \sin(s - t)}{2}$$

$$= \frac{(\sin s \cos t + \cos s \sin t) + (\sin s \cos t - \cos s \sin t)}{2}$$

$$= \frac{2 \sin s \cos t}{2}$$

$$= \sin s \cos t$$

The proofs of Equations (2), (3), and (4) are very similar.

EXAMPLE 1

Express $\sin 4x \cos 3x$ as a sum or a difference.

SOLUTION

Applying Equation (1) we obtain

$$\sin 4x \cos 3x = \frac{\sin(4x + 3x) + \sin(4x - 3x)}{2}$$

$$= \frac{\sin 7x + \sin x}{2}$$

PROGRESS CHECK

Express $\sin 5x \sin 2x$ as a sum or as a difference.

ANSWER

$\frac{1}{2}(\cos 3x - \cos 7x)$

EXAMPLE 2

Evaluate the product $\cos \dfrac{5\pi}{8} \cos \dfrac{3\pi}{8}$ by a product–sum formula.

SOLUTION

Using Equation (3) we have

$$\cos \frac{5\pi}{8} \cos \frac{3\pi}{8} = \frac{1}{2}\left[\cos\left(\frac{5\pi}{8} + \frac{3\pi}{8}\right) + \cos\left(\frac{5\pi}{8} - \frac{3\pi}{8}\right)\right]$$

$$= \frac{1}{2}\left[\cos \pi + \cos \frac{\pi}{4}\right]$$

$$= \frac{1}{2}\left[-1 + \frac{\sqrt{2}}{2}\right]$$

$$= \frac{\sqrt{2} - 2}{4}$$

PROGRESS CHECK

Evaluate $\cos \pi/3 \sin \pi/6$ by a product–sum formula.

ANSWER

$\frac{1}{4}$

The following formulas express a sum as a product.

$$\sin s + \sin t = 2 \sin \frac{s + t}{2} \cos \frac{s - t}{2} \qquad (5)$$

$$\sin s - \sin t = 2 \cos \frac{s + t}{2} \sin \frac{s - t}{2} \qquad (6)$$

$$\cos s + \cos t = 2 \cos \frac{s + t}{2} \cos \frac{s - t}{2} \qquad (7)$$

$$\cos s - \cos t = -2 \sin \frac{s + t}{2} \sin \frac{s - t}{2} \qquad (8)$$

To prove the identity in Equation (5), begin with the right-hand side and apply Equation (1). Then

$$2 \sin \frac{s + t}{2} \cos \frac{s - t}{2} = \frac{1}{2}\left[\sin\left(\frac{s + t}{2} + \frac{s - t}{2}\right) + \sin\left(\frac{s + t}{2} - \frac{s - t}{2}\right)\right]$$

$$= \sin s + \sin t$$

This establishes Equation (5).

EXAMPLE 3

Express $\sin 5x - \sin 3x$ as a product.

SOLUTION

Using Equation (6) we have

$$\sin 5x - \sin 3x = 2 \cos \frac{5x + 3x}{2} \sin \frac{5x - 3x}{2}$$

$$= 2 \cos 4x \sin x$$

PROGRESS CHECK

Express $\cos 6x + \cos 2x$ as a product.

ANSWER

$2 \cos 4x \cos 2x$

EXAMPLE 4

Evaluate $\cos 5\pi/12 - \cos \pi/12$ by using a sum–product formula.

SOLUTION

Using Equation (8), we have

$$\cos \frac{5\pi}{12} - \cos \frac{\pi}{12} = -2 \sin \frac{\pi}{4} \sin \frac{\pi}{6}$$

$$= -2\left(\frac{\sqrt{2}}{2}\right)\frac{1}{2} = -\frac{\sqrt{2}}{2}$$

PROGRESS CHECK

Evaluate $\sin 2\pi/3$ by using a sum–product formula.

ANSWER

$\sqrt{3}/2$

EXERCISE SET 7.4

Express each product as a sum or difference.

1. $2 \sin 5\alpha \cos \alpha$
2. $-3 \cos 6x \sin 2x$
3. $\sin 3x \sin(-2x)$
4. $\cos 7t \cos(-3t)$
5. $-2 \cos 2\theta \cos 5\theta$
6. $\sin \frac{5\theta}{2} \sin \frac{\theta}{2}$
7. $\cos(\alpha + \beta) \cos(\alpha - \beta)$
8. $-\sin 2u \cos 4u$

Evaluate each product by using a product–sum formula.

9. $\cos \frac{7\pi}{8} \sin \frac{5\pi}{8}$
10. $\cos \frac{\pi}{3} \cos \frac{\pi}{6}$
11. $\sin 120° \cos 60°$
12. $\sin \frac{13\pi}{12} \sin \frac{11\pi}{12}$

Express each sum or difference as a product.

13. $\sin 5x + \sin x$

14. $\cos 8t - \cos 2t$

15. $\cos 2\theta + \cos 6\theta$

16. $\sin 5\alpha - \sin 7\alpha$

17. $\sin(\alpha + \beta) + \sin(\alpha - \beta)$

18. $\cos \dfrac{x}{2} - \cos \dfrac{3x}{2}$

19. $\sin 7x - \sin 3x$

20. $\cos 5\theta + \cos 3\theta$

Evaluate each sum by using a sum–product formula.

21. $\cos 75° + \cos 15°$ 22. $\sin \dfrac{5\pi}{12} + \sin \dfrac{\pi}{12}$ 23. $\cos \dfrac{3\pi}{4} - \cos \dfrac{\pi}{4}$ 24. $\sin \dfrac{13\pi}{12} - \sin \dfrac{5\pi}{12}$

Verify the identities in Exercises 25–34.

25. $\sin 40° + \sin 20° = \sin 10°$

26. $\cos 70° - \cos 10° = -\sin 40°$

27. $\dfrac{\sin 5\theta - \sin 3\theta}{\cos 3\theta - \cos 5\theta} = \cot 4\theta$

28. $\dfrac{\cos 5x - \cos x}{\sin 7x + \sin x} = -\tan 3x$

29. $\dfrac{\sin t - \sin s}{\cos t - \cos s} = -\cot \dfrac{s + t}{2}$

30. $\dfrac{\sin s + \sin t}{\cos s + \cos t} = \tan \dfrac{s + t}{2}$

31. $\dfrac{\sin 50° - \sin 10°}{\cos 50° - \cos 10°} = -\sqrt{3}$

32. $2 \sin\!\left(\theta + \dfrac{\pi}{4}\right) \sin\!\left(\theta - \dfrac{\pi}{4}\right) = -\cos 2\theta$

33. $\dfrac{\cot x - \tan x}{\cot x + \tan x} = \cos 2x$

34. $\cos 6t \cos 2t + \sin^2 4x = \cos^2 2x$

35. Express $(\sin ax)(\cos bx)$ as a sum.

36. Express $(\cos ax)(\cos bx)$ as a sum.

7.5
TRIGONOMETRIC
EQUATIONS

Thus far, this chapter has dealt exclusively with trigonometric identities. We now seek to solve trigonometric equations that are not true for all values of the variable but may be true for some values.

We have seen that algebraic equations may have just one or two solutions. The situation is quite different with trigonometric equations since the periodic nature of the trigonometric functions assures us that if there is a solution, there are an infinite number of solutions. To handle this complication, we simply seek all solutions t such that $0 \le t < 2\pi$. Then for every integer value of n, $t + 2\pi n$ is also a solution. The following example illustrates this convenient means for writing the solution set.

EXAMPLE 1
Find all solutions of the equation $\cos t = 0$.

SOLUTION
The only values in the interval $[0, 2\pi)$ for which $\cos t = 0$ are $\pi/2$ and $3\pi/2$. Then every solution is included among those values of t such that

$$t = \dfrac{\pi}{2} + 2\pi n \quad \text{or} \quad t = \dfrac{3\pi}{2} + 2\pi n, \quad n \text{ an integer}$$

Since $\dfrac{3\pi}{2} = \dfrac{\pi}{2} + \pi$, the solution set can be written in the more compact form

$$t = \frac{\pi}{2} + \pi n, \quad n \text{ an integer}$$

Factoring provides the key for solving many trigonometric equations. If we can write the equation in the form $P(x)Q(x) = 0$, we can then find the solutions by setting $P(x) = 0$ and $Q(x) = 0$. Of course, P and Q will themselves generally contain trigonometric functions.

It may also be helpful to think in terms of a substitution of variable. Thus, the equation

$$4 \sin^2 x + 3 \sin x - 1 = 0$$

can be viewed as a quadratic in u

$$4u^2 + 3u - 1 = 0$$

by substituting $u = \sin x$. Here is an example.

EXAMPLE 2
Find all solutions of the equation $2 \cos^2 t - \cos t - 1 = 0$ in the interval $[0, 2\pi)$.

SOLUTION
Factoring the left side of the equation yields

$$(2 \cos t + 1)(\cos t - 1) = 0$$

Setting each factor equal to 0, we have

$$2 \cos t + 1 = 0 \quad \text{or} \quad \cos t - 1 = 0$$

so that

$$\cos t = -\tfrac{1}{2} \quad \text{or} \quad \cos t = 1$$

The solutions of $\cos t = -\frac{1}{2}$ in the interval $[0, 2\pi)$ are $t = 2\pi/3$ and $t = 4\pi/3$; the only solution of $\cos t = 1$ in the interval $[0, 2\pi)$ is $t = 0$. The solutions of the original equation in the interval $[0, 2\pi)$ are

$$t = \frac{2\pi}{3}, \quad t = \frac{4\pi}{3}, \quad \text{and} \quad t = 0$$

PROGRESS CHECK
Find all solutions of the equation $2 \sin^2 t - 3 \sin t + 1 = 0$ in the interval $[0, 2\pi)$.

ANSWER

$$\frac{\pi}{6}, \frac{5\pi}{6}, \frac{\pi}{2}$$

If the solutions of a trigonometric equation are angles, the answer may be given in either radians or degrees.

EXAMPLE 3

Find all solutions of the equation $\tan \theta \cos^2 \theta - \tan \theta = 0$.

SOLUTION

Factoring the left side yields

$$\tan \theta (\cos^2 \theta - 1) = 0$$

Setting each factor equal to 0,

$$\tan \theta = 0 \quad \text{or} \quad \cos^2 \theta = 1$$

so that

$$\tan \theta = 0, \quad \cos \theta = 1, \quad \text{or} \quad \cos \theta = -1$$

These equations yield the following solutions in the interval $[0, 2\pi)$.

$$\tan \theta = 0: \quad \theta = 0 \quad \text{or} \quad \theta = \pi$$
$$\cos \theta = 1: \quad \theta = 0$$
$$\cos \theta = -1: \quad \theta = \pi$$

The solutions of the original equation are

$$\theta = 0 + 2\pi n \quad \text{and} \quad \theta = \pi + 2\pi n, \quad n \text{ an integer}$$

which can be expressed more compactly as

$$\theta = \pi n, \quad n \text{ an integer}$$

In degree measure, the solution is

$$\theta = 180°n, \quad n \text{ an integer}$$

EXAMPLE 4

Find all solutions of the equation $\sin 2\theta - 3 \sin \theta = 0$ in the interval $[0, 2\pi)$.

SOLUTION

Using the identity $\sin 2\theta = 2 \sin \theta \cos \theta$ yields

$$2 \sin \theta \cos \theta - 3 \sin \theta = 0$$
$$\sin \theta (2 \cos \theta - 3) = 0$$
$$\sin \theta = 0 \quad \text{or} \quad 2 \cos \theta - 3 = 0$$
$$\sin \theta = 0 \quad \text{or} \quad \cos \theta = \tfrac{3}{2}$$

The equation $\cos\,\theta = \frac{3}{2}$ has no solutions; the solutions of $\sin\,\theta = 0$ are $\theta = 0$ and $\theta = \pi$. The solutions of the original equation are

$$\theta = 0 \quad \text{and} \quad \theta = \pi$$

or, in degree measure,

$$\theta = 0° \quad \text{and} \quad \theta = 180°$$

PROGRESS CHECK

Find all solutions of the equation $\cos\,2\theta + \cos\,\theta = 0$.

ANSWER

$$\frac{\pi}{3} + 2\pi n, \ \pi + 2\pi n, \ \frac{5\pi}{3} + 2\pi n$$

or

$$60° + 360°n, \ 180° + 360°n, \ 300° + 360°n$$

Equations involving multiple angles can often be solved by using a substitution of variable. The following example shows that we must proceed with caution when seeking solutions in the interval $[0, 2\pi)$.

EXAMPLE 5

Find all solutions of the equation $\cos\,3x = 0$ in the interval $[0, 2\pi)$.

SOLUTION

We are given

$$\cos\,3x = 0, \quad 0 \le x < 2\pi$$

Substituting $t = 3x$, we obtain

$$\cos\,t = 0, \quad 0 \le \frac{t}{3} < 2\pi$$

or

$$\cos\,t = 0, \quad 0 \le t < 6\pi$$

Note that we seek solutions of $\cos\,t = 0$ in the interval $[0, 6\pi)$ rather than $[0, 2\pi)$. The solutions are then

$$t = \frac{\pi}{2} \quad \frac{3\pi}{2} \quad \frac{5\pi}{2} \quad \frac{7\pi}{2} \quad \frac{9\pi}{2} \quad \frac{11\pi}{2}$$

Since $x = t/3$ we obtain

$$x = \frac{\pi}{6} \quad \frac{\pi}{2} \quad \frac{5\pi}{6} \quad \frac{7\pi}{6} \quad \frac{3\pi}{2} \quad \frac{11\pi}{6}$$

It is sometimes possible to treat the trigonometric equation as a quadratic equation after performing a suitable substitution of variable. Here is an example.

EXAMPLE 6

Find all solutions of the equation

$$3 \tan^2 x + \tan x - 1 = 0$$

in the interval $[0, \pi)$.

SOLUTION

The equation doesn't yield to the method of factoring. However, it can be viewed as a quadratic equation in tan x. That is, if we substitute $t = \tan x$ we obtain

$$3t^2 + t - 1 = 0$$

which is a quadratic in t. By the quadratic formula,

$$t = \frac{-1 \pm \sqrt{13}}{6}$$

so that

$$\tan x = \frac{-1 \pm \sqrt{13}}{6}$$

Using Table V in the Tables Appendix, or a calculator, we have

$$x \approx 0.41 \quad \text{and} \quad x \approx 2.49$$

as the solutions of the equation.

EXERCISE SET 7.5

Find all solutions of the given equation in the interval $[0, 2\pi)$. Express the answers in both radian measure and degree measure.

1. $2 \sin \theta - 1 = 0$
2. $2 \cos \theta + 1 = 0$
3. $\cos \alpha + 1 = 0$
4. $\cot \gamma + 1 = 0$
5. $4 \cos^2 \alpha = 3$
6. $\tan^2 \theta = 3$
7. $3 \tan^2 \alpha = 1$
8. $2 \cos^2 \alpha - 1 = 0$
9. $2 \sin^2 \beta = \sin \beta$
10. $\sin \alpha = \cos \alpha$
11. $2 \cos^2 \theta - 3 \cos \theta + 1 = 0$
12. $2 \sin^2 \theta - \sin \theta - 1 = 0$
13. $\sin 5\theta = 1$
14. $\tan 3\beta = -\sqrt{3}$
15. $2 \sin^2 \alpha - 3 \cos \alpha = 0$
16. $\csc 2\theta = 2$
17. $2 \cos^2 \theta - 1 = \sin \theta$
18. $\cos^2 2\alpha = \frac{1}{4}$
19. $\sin^2 \beta + 3 \cos \beta - 3 = 0$
20. $2 \cos^2 \theta \tan \theta - \tan \theta = 0$

Find all the solutions of the given equation.

21. $3 \tan^2 x - 1 = 0$ 22. $2 \sin^2 y - 1 = 0$

23. $3 \cot^2 \theta - 1 = 0$ 24. $1 - 4 \cos^2 t = 0$

25. $\sec 2u - 2 = 0$ 26. $\tan 3x - 1 = 0$

27. $\sin 4x = 0$ 28. $\cos 5t = -1$

29. $4 \cos^2 2t - 3 = 0$ 30. $\csc^2 2x - 2 = 0$

31. $\sin 2t + 2 \cos t = 0$ 32. $\sin 2t + 3 \cos t = 0$

33. $\cos 2t + \sin t = 0$ 34. $2 \cos 2t + 2 \sin t = 0$

35. $\tan^2 x - \tan x = 0$ 36. $\sec^2 x - 3 \sec x + 2 = 0$

37. $2 \sin^2 x + 3 \sin x - 2 = 0$ 38. $2 \cos^2 x - 5 \cos x - 3 = 0$

Find the approximate solutions of the given equations in the interval $[0, 2\pi)$ by using Table V in the Tables Appendix, or a calculator.

39. $5 \sin^2 x - \sin x - 2 = 0$ 40. $\sec^2 y - 5 \sec y + 6 = 0$

41. $3 \tan^2 u + 5 \tan u + 1 = 0$ 42. $\cos^2 t - 2 \sin t + 3 = 0$

TERMS AND SYMBOLS

trigonometric expression (p. 299)	fundamental identities (p. 300)	addition formulas (p. 306)	half-angle formulas (p. 317)
trigonometric identity (p. 299)	trigonometric formulas (p. 306)	cofunctions (p. 309)	product–sum formulas (p. 321)
trigonometric equation (p. 299)		double-angle formulas (p. 313)	

KEY IDEAS FOR REVIEW

☐ A trigonometric identity is true for all real values in the domain of the variable. The fundamental identities are those trigonometric identities that occur so frequently that they must be remembered and recognized.

☐ The fundamental identities can be used to verify other trigonometric identities. The techniques commonly used to verify identities include the following.
- Write all of the trigonometric functions in terms of sine and cosine.
- Factor.
- Complete the indicated operations, especially when this involves the sum of fractional expressions.
- Multiply the numerator and denominator of a fractional expression by the same quantity to produce a simpler product such as $1 - \sin^2 \theta$, $1 - \cos^2 \theta$, or $\sec^2 \theta - 1$.

☐ The most useful of the trigonometric formulas are the following.

Addition Formulas

$$\sin(s + t) = \sin s \cos t + \cos s \sin t$$

$$\cos(s + t) = \cos s \cos t - \sin s \sin t$$

$$\tan(s + t) = \frac{\tan s + \tan t}{1 - \tan s \tan t}$$

Double-Angle Formulas

$$\sin 2t = 2 \sin t \cos t$$

$$\cos 2t = \cos^2 t - \sin^2 t$$

$$\tan 2t = \frac{2 \tan t}{1 - \tan^2 t}$$

Half-Angle Formulas

$$\sin \frac{t}{2} = \pm \sqrt{\frac{1 - \cos t}{2}}$$

$$\cos \frac{t}{2} = \pm \sqrt{\frac{1 + \cos t}{2}}$$

$$\tan \frac{t}{2} = \pm \sqrt{\frac{1 - \cos t}{1 + \cos t}}$$

☐ Since the trigonometric functions are periodic, a trigonometric equation has either no solutions or an infinite number of solutions.

REVIEW EXERCISES

Solutions to exercises whose numbers are in color are in the Solutions section in the back of the book.

7.1 In Exercises 1–3 verify the given identity.

1. $\sin \sigma \sec \sigma + \tan \sigma = 2 \tan \sigma$

2. $\dfrac{\cos^2 x}{1 - \sin x} = 1 + \sin x$

3. $\sin \alpha + \sin \alpha \cot^2 \alpha = \csc \alpha$

7.2 In Exercises 4–7 determine the exact value of the given expression by using the addition formulas.

4. $\sin\left(\dfrac{\pi}{6} + \dfrac{\pi}{4}\right)$

5. $\cos(45° + 90°)$

6. $\tan\left(\dfrac{\pi}{3} + \dfrac{\pi}{4}\right)$

7. $\sin \dfrac{7\pi}{12}$

In Exercises 8–11 write the given expression in terms of cofunctions of complementary angles.

8. $\csc 15°$

9. $\cos 23°$

10. $\sin \dfrac{\pi}{8}$

11. $\tan \dfrac{2\pi}{7}$

12. If $\cos \sigma = -\frac{12}{13}$ and $0 \le \sigma \le 180°$, find $\sin(\pi - \sigma)$.

13. If $\sec \sigma = \frac{10}{8}$ and σ lies in quadrant IV, find $\csc(\sigma + \pi/3)$.

14. If $\sin t = -\frac{3}{5}$ and $W(t)$ is in quadrant III, find $\tan(t + \pi)$.

15. If $\cos \alpha = -\frac{12}{13}$ and $\tan \beta = -\frac{5}{2}$, with angles α and β in quadrant II, find $\tan(\alpha + \beta)$.

16. If $\sin x = \frac{3}{5}$ and $\csc y = \frac{13}{12}$, with $W(x)$ in quadrant II and $W(y)$ in quadrant I, find $\cos(x - y)$.

7.3 17. If $\csc u = -\frac{5}{4}$ and $W(u)$ is in quadrant IV, find $\cos 2u$.

18. If $\tan \sigma = -\frac{3}{4}$ and $0 \le \sigma \le 180°$, find $\sin 2\sigma$.

19. If $\sin 2t = \frac{3}{5}$ and $W(2t)$ is in quadrant I, find $\sin 4t$.

20. If $\sin \sigma = 0.5$ and $\pi/2 \le \sigma \le \pi$, find $\sin 2\sigma$.

21. If $\cos(\sigma/2) = \frac{12}{13}$ and σ is acute, find $\sin \sigma$.

22. If $\sin \alpha = -\frac{3}{5}$ and α is in quadrant III, find $\cos(\alpha/2)$.

23. If $\cot t = -\frac{4}{3}$ and $W(t)$ is in quadrant IV, find $\tan(t/2)$.

24. If $\cos 4x = \frac{2}{3}$ and $W(4x)$ is in quadrant IV, find $\cos 2x$.

25. Find the exact value of $\cos 15°$ by using a half-angle formula.

26. Find the exact value of $\sin \pi/8$ by using a half-angle formula.

27. Find the exact value of $\tan 112.5°$ by using a half-angle formula.

In Exercises 28–30 verify the given identity.

28. $\cos 30x = 1 - 2 \sin^2 15x$

29. $\dfrac{1}{2} \sin 2y = \dfrac{\sin y}{\sec y}$

30. $\tan \dfrac{\alpha}{2} = \dfrac{(1 - \cos \alpha)}{\sin \alpha}$

7.4 31. Express $\sin \dfrac{3\alpha}{2} \sin \dfrac{\alpha}{2}$ as a sum or difference.

32. Express $\cos 3x - \cos x$ as a product.

33. Evaluate $\sin 75° \sin 15°$ by using a product–sum formula.

34. Evaluate $\cos \dfrac{3\pi}{4} + \cos \dfrac{\pi}{4}$ by using a sum–product formula.

7.5 In Exercises 35–37 find all solutions of the given equation in the interval $[0, 2\pi)$. Express the answers in radian measure.

35. $2 \cos^2 \alpha - 1 = 0$

36. $2 \sin \sigma \cos \sigma = 0$

37. $\sin 2t - \sin t = 0$

In Exercises 38–40 find all solutions of the given equation. Express the answers in degree measure.

38. $\cos^2 \alpha - 2 \cos \alpha = 0$

39. $\tan 3x + 1 = 0$

40. $4 \sin^2 2t = 3$

PROGRESS TEST 7A

1. Verify the identity $4 - \tan^2 x = 5 - \sec^2 x$.

In Problems 2 and 3 determine exact values of the given expressions by using the addition formulas.

2. $\cos(270° + 30°)$

3. $\tan\left(\dfrac{\pi}{4} - \dfrac{\pi}{3}\right)$

4. Write $\sin 47°$ in terms of its cofunction.

5. If $\cos \theta = \frac{4}{5}$ and θ lies in quadrant IV, find $\sin(\theta - \pi)$.

6. If $\sin x = -\frac{5}{13}$ and $\tan y = \frac{8}{3}$ with angles x and y in quadrant III, find $\tan(x - y)$.

7. If $\sin v = -\frac{12}{13}$ and $W(v)$ is in quadrant IV, find $\cos 2v$.

8. If $\cos 2\alpha = -\frac{4}{5}$ and 2α is in quadrant II, find $\cos 4\alpha$.

9. If $\csc \alpha = -2$ and α is in quadrant III, find $\cos(\alpha/2)$.

10. Find the exact value of $\tan 15°$ by using a half-angle formula.

11. Verify the identity $\sin \dfrac{x}{4} = 2 \sin \dfrac{x}{8} \cos \dfrac{x}{8}$.

12. Express $\sin 2x + \sin 3x$ as a product.

13. Express $\sin 150° - \sin 30°$ by using a sum–product formula.

14. Find all solutions of the equation $4 \sin^2 \alpha = 3$ in the interval $[0, 2\pi)$. Express the answers in radian measure.

15. Find all solutions of the equation $\sin^2 \theta - \cos^2 \theta = 0$ and express the answers in degree measure.

PROGRESS TEST 7B

1. Verify the identity $\dfrac{\tan u + \cot u}{\sec u \sin u} = \csc^2 u$.

In Problems 2 and 3 determine exact values of the given expression by using the addition formulas.

2. $\csc(180° - 30°)$

3. $\sin \dfrac{7\pi}{12}$

4. Write $\tan 71°$ in terms of its cofunction.

5. If $\sin t = -\frac{5}{13}$ and $W(t)$ lies in quadrant III, find $\sec(t + \pi/4)$.

6. If $\cos \alpha = -0.6$ and $\csc \beta = \frac{5}{4}$ with angles α and β in quadrant II, find $\sin(\alpha - \beta)$.

7. If $\sec \theta = \frac{5}{4}$ and $0 \le \theta \le 180°$, find $\sin 2\theta$.

8. If $\sin \theta/2 = \frac{3}{5}$ and θ is acute, find $\sin 2\theta$.

9. If $\sin 6x = -\frac{12}{13}$ and $W(6x)$ is in quadrant IV, find $\cos 3x$.

10. Find the exact value of $\sin \pi/8$ by using a half-angle formula.

11. Verify the identity $\sec 2t = \dfrac{1 + \tan^2 t}{1 - \tan^2 t}$.

12. Express $\sin \pi/4 \cos \pi/3$ as a sum or difference.

13. Express $\cos 75° \cos 15°$ by using a product–sum formula.

14. Find all solutions of the equation $\tan^2 x + \tan x = 0$ in the interval $[0, 2\pi)$. Express the answers in radian measure.

15. Find all solutions of the equation $2 \sin^2 \alpha - \sin \alpha - 1 = 0$ and express the answers in degree measure.

CHAPTER 8

COMPLEX NUMBERS AND THE FUNDAMENTAL THEOREM OF ALGEBRA

Does a polynomial equation always have a root? The answer to this question depends on the number system under consideration. For example, the equation

$$x^2 + x + 1 = 0$$

has no real roots but does have two complex roots, which we can find by using the quadratic equation. On the other hand, the seemingly simple equation

$$x^4 - 1 = 0$$

has two real roots, ± 1, and two complex roots, $\pm i$.

Questions about the roots of polynomial equations attracted the attention of mathematicians for several hundred years. The examples we have just looked at tell us that our search must involve the set of complex numbers. We will therefore begin this chapter by delving more deeply into this number system and its properties. We will then be in a better position to extract some fundamental relationships between the complex number system and the roots of polynomial equations.

8.1

COMPLEX NUMBERS AND THEIR PROPERTIES

We introduced the complex number system in Section 1.3 and then used this number system in Section 2.6 to provide solutions to quadratic equations. Recall that $a + bi$ is said to be a complex number where a and b are real numbers and the imaginary unit $i = \sqrt{-1}$ has the property that $i^2 = -1$. We defined fundamental operations with complex numbers in the following way.

Equality: $\quad a + bi = c + di$ if $a = c$ and $b = d$

Addition: $\quad (a + bi) + (c + di) = (a + c) + (b + d)i$

Multiplication: $\quad (a + bi)(c + di) = (ac - bd) + (ad + bc)i$

With this background, we can now explore further properties of the complex number system.

The complex number $a - bi$ is called the **complex conjugate** (or simply the **conjugate**) of the complex number $a + bi$. For example, $3 - 2i$ is the conjugate of $3 + 2i$, $4i$ is the conjugate of $-4i$, and 2 is the conjugate of 2. Forming the product $(a + bi)(a - bi)$ we have

$$(a + bi)(a - bi) = a^2 - abi + abi - b^2i^2$$
$$= a^2 + b^2 \qquad \text{Since } i^2 = -1$$

Because a and b are real numbers, $a^2 + b^2$ is also a real number. We can summarize this result as follows.

The product of a complex number and its conjugate is a real number.

$$(a + bi)(a - bi) = a^2 + b^2$$

We can now demonstrate that the quotient of two complex numbers is also a complex number. The quotient

$$\frac{q + ri}{s + ti}$$

can be written in the form $a + bi$ by multiplying both numerator and denominator by $s - ti$, the conjugate of the denominator. We then have

$$\frac{q + ri}{s + ti} = \frac{q + ri}{s + ti} \cdot \frac{s - ti}{s - ti} = \frac{(qs + rt) + (rs - qt)i}{s^2 + t^2}$$

$$= \frac{qs + rt}{s^2 + t^2} + \frac{(rs - qt)}{s^2 + t^2}i$$

which is a complex number of the form $a + bi$. Of course, the reciprocal of the complex number $s + ti$ is the quotient $1/(s + ti)$, which can also be written as a complex number by using the same technique. In summary,

- The quotient of two complex numbers is a complex number.
- The reciprocal of a nonzero complex number is a complex number.

EXAMPLE 1

(a) Write the quotient $\dfrac{-2 + 3i}{3 - 2i}$ in the form $a + bi$.

(b) Write the reciprocal of $2 - 5i$ in the form $a + bi$.

SOLUTION

(a) Multiplying numerator and denominator by the conjugate $3 + 2i$ of the denominator, we have

$$\frac{-2 + 3i}{3 - 2i} = \frac{-2 + 3i}{3 - 2i} \cdot \frac{3 + 2i}{3 + 2i} = \frac{-6 - 4i + 9i + 6i^2}{3^2 + 2^2} = \frac{-6 + 5i + 6(-1)}{9 + 4}$$

$$= \frac{-12 + 5i}{13} = -\frac{12}{13} + \frac{5}{13}i$$

(b) The reciprocal is $1/(2 - 5i)$. Multiplying both numerator and denominator by the conjugate $2 + 5i$, we have

$$\frac{1}{2 - 5i} \cdot \frac{2 + 5i}{2 + 5i} = \frac{2 + 5i}{2^2 + 5^2} = \frac{2 + 5i}{29} = \frac{2}{29} + \frac{5}{29}i$$

Verify that $(2 - 5i)\left(\dfrac{2}{29} + \dfrac{5}{29}i\right) = 1$.

PROGRESS CHECK

Write the following in the form $a + bi$.

(a) $\dfrac{4 - 2i}{5 + 2i}$ (b) $\dfrac{1}{2 - 3i}$ (c) $\dfrac{-3i}{3 + 5i}$

ANSWERS

(a) $\dfrac{16}{29} - \dfrac{18}{29}i$ (b) $\dfrac{2}{13} + \dfrac{3}{13}i$ (c) $-\dfrac{15}{34} - \dfrac{9}{34}i$

If we let $z = a + bi$, it is customary to write the conjugate $a - bi$ as $\overline{z}$. We will have need to use the following properties of complex numbers and their conjugates.

If z and w are complex numbers, then
(1) $\overline{z} = \overline{w}$ if and only if $z = w$
(2) $\overline{z} = z$ if and only if z is a real number
(3) $\overline{z + w} = \overline{z} + \overline{w}$
(4) $\overline{z \cdot w} = \overline{z} \cdot \overline{w}$
(5) $\overline{z^n} = \overline{z}^n$, n a positive integer

To prove Properties (1)–(5), let $z = a + bi$ and $w = c + di$. Properties (1) and (2) follow directly from the definition of equality of complex numbers. To prove Property (3), we note that $z + w = (a + c) + (b + d)i$. Then, by the definition of a complex conjugate,

$$\overline{z + w} = (a + c) - (b + d)i$$
$$= (a - bi) + (c - di)$$
$$= \overline{z} + \overline{w}$$

Properties (4) and (5) can be proved in a similar manner, although a rigorous proof of Property (5) requires the use of mathematical induction, a method we will discuss in a later chapter.

EXAMPLE 2
If $z = 1 + 2i$ and $w = 3 - i$, verify that
(a) $\overline{z + w} = \overline{z} + \overline{w}$ (b) $\overline{z \cdot w} = \overline{z} \cdot \overline{w}$ (c) $\overline{z^2} = \overline{z}^2$

SOLUTION
(a) Adding, we get $z + w = 4 + i$. Therefore $\overline{z + w} = 4 - i$. Also,

$$\overline{z} + \overline{w} = (1 - 2i) + (3 + i) = 4 - i$$

Thus, $\overline{z + w} = \overline{z} + \overline{w}$.
(b) Multiplying, we get $z \cdot w = (1 + 2i)(3 - i) = 5 + 5i$. Therefore $\overline{z \cdot w} = 5 - 5i$. Also,

$$\overline{z} \cdot \overline{w} = (1 - 2i)(3 + i) = 5 - 5i$$

Thus, $\overline{z \cdot w} = \overline{z} \cdot \overline{w}$.
(c) Squaring, we get

$$z^2 = (1 + 2i)(1 + 2i) = -3 + 4i$$

Therefore $\overline{z^2} = -3 - 4i$.

Also,

$$\bar{z}^2 = (1 - 2i)(1 - 2i) = -3 - 4i$$

Thus, $\overline{z^2} = \bar{z}^2$.

PROGRESS CHECK

If $z = 2 + 3i$ and $w = \frac{1}{2} - 2i$, verify that

(a) $\overline{z + w} = \bar{z} + \bar{w}$ (b) $\overline{z \cdot w} = \bar{z} \cdot \bar{w}$

(c) $\overline{z^2} = \bar{z}^2$ (d) $\overline{w^3} = \bar{w}^3$

THE COMPLEX PLANE

We associate the complex number $a + bi$ with the point in the plane whose coordinates are (a, b). Figure 1 illustrates the geometric representation of several complex numbers. Conversely, every point (a, b) in the plane represents a complex number, $a + bi$. When a rectangular coordinate system is used to represent complex numbers, it is called a **complex plane** and the x- and y-axes are called the **real axis** and the **imaginary axis,** respectively.

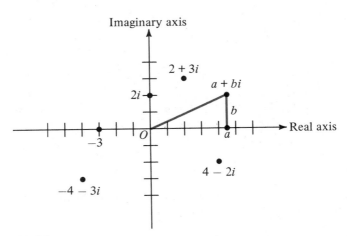

FIGURE 1

We can extend the concept of absolute value to complex numbers in a natural manner. Since $|x|$ represents the distance on a real number line from the origin to a point that corresponds to x, it would be consistent to define the

absolute value $|a + bi|$ as the distance from the origin to the point corresponding to $a + bi$. Applying the distance formula (see Figure 1) we are led to the following definition.

The absolute value of a complex number $a + bi$ is denoted by $|a + bi|$ and is defined by

$$|a + bi| = \sqrt{a^2 + b^2}$$

EXAMPLE 3
Find the absolute value of each of the following complex numbers.
(a) $2 - 3i$ (b) $4i$ (c) -2

SOLUTION
Applying the definition of absolute value,
(a) $|2 - 3i| = \sqrt{4 + 9} = \sqrt{13}$ (b) $|4i| = \sqrt{0 + 16} = 4$
(c) $|-2| = \sqrt{4 + 0} = 2$

EXERCISE SET 8.1
In Exercises 1–6 multiply by the conjugate and simplify.

1. $2 - i$ 2. $3 + i$ 3. $3 + 4i$ 4. $2 - 3i$

5. $-4 - 2i$ 6. $5 + 2i$

In Exercises 7–15 perform the indicated operations and write the answer in the form $a + bi$.

7. $\dfrac{2 + 5i}{1 - 3i}$ 8. $\dfrac{1 + 3i}{2 - 5i}$ 9. $\dfrac{3 - 4i}{3 + 4i}$ 10. $\dfrac{4 - 3i}{4 + 3i}$

11. $\dfrac{3 - 2i}{2 - i}$ 12. $\dfrac{2 - 3i}{3 - i}$ 13. $\dfrac{2 + 5i}{3i}$ 14. $\dfrac{5 - 2i}{-3i}$

15. $\dfrac{4i}{2 + i}$

In Exercises 16–21 find the reciprocal and write the answer in the form $a + bi$.

16. $3 + 2i$ 17. $4 + 3i$ 18. $\frac{1}{2} - i$ 19. $1 - \frac{1}{3}i$

20. $-7i$ 21. $-5i$

22. Prove that the multiplicative inverse of the complex number $a + bi$ (a and b not both 0) is

$$\frac{a}{a^2 + b^2} - \frac{b}{a^2 + b^2} i$$

23. If z and w are complex numbers, prove that

$$\overline{z \cdot w} = \overline{z} \cdot \overline{w}$$

24. If z is a complex number, verify that $\overline{z^2} = \overline{z}^2$ and $\overline{z^3} = \overline{z}^3$.

In Exercises 25–30 find the absolute value of the given complex number.

25. $3 - 2i$

26. $-7 + 6i$

27. $1 + i$

28. $\dfrac{1}{2} + \dfrac{1}{2}i$

29. $-6 - 2i$

30. $3 - i$

8.2
TRIGONOMETRY AND COMPLEX NUMBERS

The representation of a complex number as a point in a coordinate plane can be used to link complex numbers with trigonometry of the right triangle. In Figure 2, $a + bi$ is any nonzero complex number, and we consider the line segment OP to be the terminal side of an angle θ in standard position. Using trigonometry of the right triangle, we see that

$$a = r \cos \theta \quad \text{and} \quad b = r \sin \theta$$

We may then write

$$a + bi = (r \cos \theta) + (r \sin \theta)i$$

or

$$a + bi = r(\cos \theta + i \sin \theta) \tag{1}$$

where $r = \overline{OP} = |a + bi| = \sqrt{a^2 + b^2}$. If $a + bi = 0$, then $r = 0$, and θ may assume any value.

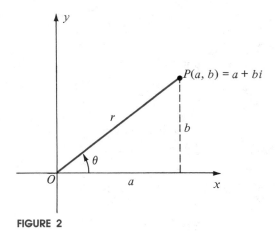

FIGURE 2

Equation (1) is known as the **trigonometric form** or **polar form** of a complex number. Since we have an infinite number of choices for the angle θ, the polar form of a complex number is not unique. We call r the **modulus** and θ the **argument** of the complex number $r(\cos \theta + i \sin \theta)$. If $0 \le \theta < 360°$, then θ is called the **principal argument.**

EXAMPLE 1

Write the complex number $-2 + 2i$ in trigonometric form.

SOLUTION

The geometric representation is shown in Figure 3. The modulus of $-2 + 2i$ is

$$r = |-2 + 2i| = \sqrt{4 + 4} = 2\sqrt{2}$$

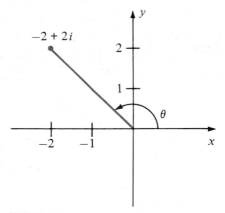

FIGURE 3

The principal argument θ is an angle in the second quadrant such that

$$\tan \theta = \frac{2}{-2} = -1$$

Thus, $\theta = 135°$, and using the trigonometric form of a complex number of Equation (1), we have

$$-2 + 2i = 2\sqrt{2}(\cos 135° + i \sin 135°)$$

PROGRESS CHECK

Write the complex number $1 - \sqrt{3}i$ in trigonometric form.

ANSWER

$2(\cos 300° + i \sin 300°)$

EXAMPLE 2

Write the complex number $2\sqrt{3}(\cos 150° + i \sin 150°)$ in the form $a + bi$.

SOLUTION

We need only substitute $\cos 150° = \dfrac{\sqrt{3}}{2}$ and $\sin 150° = -\dfrac{1}{2}$. Thus,

$$2\sqrt{3}(\cos 150° + i \sin 150°) = 2\sqrt{3}\left(\frac{\sqrt{3}}{2} - \frac{1}{2}i\right)$$
$$= 3 - \sqrt{3}i$$

PROGRESS CHECK

Write the complex number $\sqrt{2}\left(\cos \frac{\pi}{4} + i \sin \frac{\pi}{4}\right)$ in the form $a + bi$.

ANSWER

$1 + i$

Why have we introduced the trigonometric form of a complex number? Because multiplication and division of complex numbers is very simple when this form is used. If $r_1(\cos \theta_1 + i \sin \theta_1)$ and $r_2(\cos \theta_2 + i \sin \theta_2)$ are any two complex numbers, the rules for their multiplication and division are

$$r_1(\cos \theta_1 + i \sin \theta_1) \cdot r_2(\cos \theta_2 + i \sin \theta_2) = \\ r_1 r_2[\cos(\theta_1 + \theta_2) + i \sin(\theta_1 + \theta_2)] \qquad (2)$$

$$\frac{r_1(\cos \theta_1 + i \sin \theta_1)}{r_2(\cos \theta_2 + i \sin \theta_2)} = \frac{r_1}{r_2}[\cos(\theta_1 - \theta_2) + i \sin(\theta_1 - \theta_2)] \qquad (3)$$

Note that the rule for multiplication requires the multiplication of the moduli and addition of the arguments. To prove this we see that

$$r_1(\cos \theta_1 + i \sin \theta_1) \cdot r_2(\cos \theta_2 + i \sin \theta_2)$$
$$= r_1 r_2[(\cos \theta_1 \cos \theta_2 - \sin \theta_1 \sin \theta_2) + i(\sin \theta_1 \cos \theta_2 + \cos \theta_1 \sin \theta_2)]$$
$$= r_1 r_2[\cos(\theta_1 + \theta_2) + i \sin(\theta_1 + \theta_2)]$$

where the last step results from the addition formulas.

The rule for division requires the division of moduli and the subtraction of the arguments. The proof is left as an exercise.

EXAMPLE 3

Find the product of the complex numbers $1 + i$ and $-2i$ (a) by writing the numbers in trigonometric form and (b) by multiplying the numbers algebraically.

SOLUTION

(a)   The trigonometric forms of these complex numbers are

$$1 + i = \sqrt{2}(\cos 45° + i \sin 45°)$$

and

$$-2i = 2(\cos 270° + i \sin 270°)$$

Multiplying, we have

$$\sqrt{2}(\cos 45° + i \sin 45°) \cdot 2(\cos 270° + i \sin 270°)$$
$$= 2\sqrt{2}(\cos 315° + i \sin 315°)$$
$$= 2\sqrt{2}\left(\frac{\sqrt{2}}{2} - i\frac{\sqrt{2}}{2}\right)$$
$$= 2 - 2i$$

(b)   Multiplying algebraically,

$$(1 + i)(-2i) = -2i - 2i^2 = -2i + 2 = 2 - 2i$$

PROGRESS CHECK

Express the complex numbers $1 + \sqrt{3}i$ and $1 - \sqrt{3}i$ in trigonometric form and find their product.

ANSWER

$2(\cos 60° + i \sin 60°)$; $2(\cos 300° + i \sin 300°)$; 4

DE MOIVRE'S THEOREM

Since exponentiation is repeated multiplication, we are led to anticipate a simple result when a complex number in trigonometric form is raised to a power. The theorem that states this result is credited to Abraham De Moivre, a French mathematician. In this theorem $r(\cos \theta + i \sin \theta)$ is a complex number and n is a natural number.

De Moivre's Theorem

$$[r(\cos \theta + i \sin \theta)]^n = r^n(\cos n\theta + i \sin n\theta)$$

We can verify the theorem for some values of n. Thus, by Equation (2),

$$[r(\cos \theta + i \sin \theta)]^2 = r(\cos \theta + i \sin \theta) \cdot r(\cos \theta + i \sin \theta)$$
$$= r^2[\cos(\theta + \theta) + i \sin(\theta + \theta)]$$
$$= r^2(\cos 2\theta + i \sin 2\theta)$$

which is precisely what we obtain by using De Moivre's theorem. If we multiply again by $r(\cos \theta + i \sin \theta)$ and again apply Equation (2), we have

$$[r(\cos \theta + i \sin \theta)]^3 = r^2(\cos 2\theta + i \sin 2\theta) \cdot r(\cos \theta + i \sin \theta)$$
$$= r^3(\cos 3\theta + i \sin 3\theta)$$

Thus, De Moivre's theorem seems "reasonable." A rigorous *proof* requires the application of the method of mathematical induction, which will be discussed in a later chapter.

EXAMPLE 4

Evaluate $(1 - i)^{10}$.

SOLUTION
Writing $1 - i$ in trigonometric form we have

$$1 - i = \sqrt{2}(\cos 315° + i \sin 315°)$$

and

$$(1 - i)^{10} = [\sqrt{2}(\cos 315° + i \sin 315°)]^{10}$$

Applying De Moivre's theorem,

$$(1 - i)^{10} = (\sqrt{2})^{10}[\cos 3150° + i \sin 3150°]$$
$$= 32[\cos 270° + i \sin 270°]$$
$$= 32[0 + i(-1)] = -32i$$

PROGRESS CHECK
Evaluate $(\sqrt{3} + i)^6$.

ANSWER
-64

Recall that a real number a is said to be an nth root of the real number b if $a^n = b$ for a positive integer n. In an analogous manner, we say that the complex number u is an **nth root** of the nonzero complex number z if $u^n = z$. If we express u and z in trigonometric form as

$$u = s(\cos \phi + i \sin \phi) \qquad z = r(\cos \theta + i \sin \theta) \qquad (4)$$

we can then apply De Moivre's theorem to obtain

$$u^n = s^n(\cos n\phi + i \sin n\phi) = r(\cos \theta + i \sin \theta) \qquad (5)$$

Since the two complex numbers u^n and z are equal, they are represented by the same point in the complex plane. Hence, the moduli must be equal, since the modulus is the distance of the point from the origin. Therefore, $s^n = r$ or

$$s = \sqrt[n]{r}$$

Since $z \neq 0$, we know that $r \neq 0$. We may therefore divide Equation (5) by r to obtain

$$\cos n\phi + i \sin n\phi = \cos \theta + i \sin \theta$$

By the definition of equality of complex numbers, we must have

$$\cos n\phi = \cos \theta \quad \sin n\phi = \sin \theta$$

Since both sine and cosine are periodic functions with period 2π, we conclude that

$$n\phi = \theta + 2\pi k$$

or

$$\phi = \frac{\theta + 2\pi k}{n}$$

where k is an integer. Substituting for s and for ϕ in the trigonometric form of u given in Equation (4) yields

---

The *n*th Roots of a Complex Number

The n distinct roots of $r(\cos \theta + i \sin \theta)$ are given by

$$\sqrt[n]{r}\left[\cos\left(\frac{\theta + 2\pi k}{n}\right) + i \sin\left(\frac{\theta + 2\pi k}{n}\right)\right]$$

where $k = 0, 1, 2, \ldots, n - 1$.

Note that when k exceeds $n - 1$, we repeat a previous root. For example, when $k = n$, the angle is

$$\frac{\theta + 2\pi n}{n} = \frac{\theta}{n} + 2\pi = \frac{\theta}{n}$$

which is the same result that is obtained when $k = 0$.

EXAMPLE 5
Find the cube roots of $-8i$.

SOLUTION
In trigonometric form,

$$-8i = 8(\cos 270° + i \sin 270°)$$

We then have $r = 8$, $\theta = 270°$, and $n = 3$.
The cube roots are then

$$\sqrt[3]{8}\left[\cos\left(\frac{270° + 360°k}{3}\right) + i \sin\left(\frac{270° + 360°k}{3}\right)\right]$$

for $k = 0, 1, 2$. Substituting for each value of k we have

$$2(\cos 90° + i \sin 90°) = 2i$$
$$2(\cos 210° + i \sin 210°) = -\sqrt{3} - i$$
$$2(\cos 330° + i \sin 330°) = \sqrt{3} - i$$

When $z = 1$, we call the n distinct nth roots the **nth roots of unity.**

EXAMPLE 6
Find the four fourth roots of unity.

SOLUTION

In trigonometric form,

$$1 = 1(\cos 0° + i \sin 0°)$$

so that $r = 1$, $\theta = 0°$, and $n = 4$. The fourth roots are then given by

$$\sqrt[4]{1}\left[\cos\left(\frac{0° + 360°k}{4}\right) + i \sin\left(\frac{0° + 360°k}{4}\right)\right]$$

for $k = 0, 1, 2, 3$. Substituting these values for k yields

$$\cos 0° + i \sin 0° = 1$$
$$\cos 90° + i \sin 90° = i$$
$$\cos 180° + i \sin 180° = -1$$
$$\cos 270° + i \sin 270° = -i$$

It is easy to verify that each of these answers is indeed a fourth root of unity.

PROGRESS CHECK

Find the two square roots of $\dfrac{\sqrt{3}}{2} - \dfrac{1}{2}i$. Express the answers in trigonometric form.

ANSWER

$\cos 165° + i \sin 165°$, $\cos 345° + i \sin 345°$

EXERCISE SET 8.2

Express the given complex number in trigonometric form.

1. $3 - 3i$ 2. $2 + 2i$ 3. $\sqrt{3} - i$ 4. $-2 - 2\sqrt{3}i$

5. $-1 + i$ 6. $-2i$ 7. -4 8. $3i$

Convert the given complex number from trigonometric form to the algebraic form $a + bi$.

9. $4(\cos 180° + i \sin 180°)$

10. $\dfrac{1}{2}\left(\cos \dfrac{\pi}{2} + i \sin \dfrac{\pi}{2}\right)$

11. $\sqrt{2}(\cos 135° + i \sin 135°)$

12. $2(\cos 120° + i \sin 120°)$

13. $5\left(\cos \dfrac{3\pi}{2} + i \sin \dfrac{3\pi}{2}\right)$

14. $4(\cos 240° + i \sin 240°)$

Find the product of the given complex numbers. Express the answers in trigonometric form.

15. $2(\cos 150° + i \sin 150°) \cdot 3(\cos 210° + i \sin 210°)$

16. $3(\cos 120° + i \sin 120°) \cdot 3(\cos 150° + i \sin 150°)$

17. $2(\cos 10° + i \sin 10°) \cdot (\cos 320° + i \sin 320°)$

18. $3(\cos 230° + i \sin 230°) \cdot 4(\cos 250° + i \sin 250°)$

Express the given complex numbers in trigonometric form, compute the product, and write the answer in the form $a + bi$.

19. $1 - i, 2i$

20. $-\sqrt{3} + i, -2$

21. $-2 + 2\sqrt{3}i, 3 + 3i$

22. $1 - \sqrt{3}i, 1 + \sqrt{3}i$

23. $5, -2 - 2i$

24. $-4i, -3i$

Use De Moivre's theorem to express the given number in the form $a + bi$.

25. $(-2 + 2i)^6$

26. $(\sqrt{3} - i)^{10}$

27. $(1 - i)^9$

28. $(-1 + \sqrt{3}i)^{10}$

29. $(-1 - i)^7$

30. $(-\sqrt{2} + \sqrt{2}i)^6$

Find the indicated roots of the given complex number. Express the answer in the indicated form.

31. The fourth roots of -16; algebraic form $a + bi$.

32. The square roots of -25; trigonometric form.

33. The square roots of $1 - \sqrt{3}i$; trigonometric form.

34. The four fourth roots of unity; algebraic form.

In Exercises 35–38 determine all roots of the given equation.

35. $x^3 + 8 = 0$ 36. $x^3 + 125 = 0$ 37. $x^4 - 16 = 0$ 38. $x^4 + 16 = 0$

39. Prove $\dfrac{r_1(\cos \theta_1 + i \sin \theta_1)}{r_2(\cos \theta_2 + i \sin \theta_2)} = \dfrac{r_1}{r_2}[\cos(\theta_1 - \theta_2) + i \sin (\theta_1 - \theta_2)]$.

8.3
THE FUNDAMENTAL
THEOREM OF ALGEBRA

We began this chapter with the question, Does a polynomial equation always have a root? The answer was supplied by Carl Friedrich Gauss in his doctoral dissertation in 1799. Unfortunately, the proof of this theorem is beyond the scope of this book.

The Fundamental Theorem of Algebra— Part I	Every polynomial $P(x)$ of degree $n \geq 1$ has at least one root among the complex numbers.

Note that the root guaranteed by this theorem may be a real number since the real numbers are a subset of the complex number system.

Gauss, who is considered by many to have been the greatest mathematician of all time, supplied the proof at age 22. The importance of the theorem is reflected in its title. We now see why it was necessary to create the complex numbers and that we need not create any other number system beyond the complex numbers in order to solve polynomial equations.

How many roots does a polynomial of degree n have? The next theorem will bring us closer to an answer.

Linear Factor Theorem	A polynomial $P(x)$ of degree $n \geq 1$ can be written as the product of n linear factors.

$$P(x) = a(x - r_1)(x - r_2) \ldots (x - r_n)$$

Note that a is the leading coefficient of $P(x)$ and that $r_1, r_2, \ldots, r_n$ are, in general, complex numbers.

To prove this theorem, we first note that the Fundamental Theorem of Algebra guarantees us the existence of a root r_1. By the Factor Theorem, $x - r_1$ is a factor and, consequently,

$$P(x) = (x - r_1)Q_1(x) \tag{1}$$

where $Q_1(x)$ is a polynomial of degree $n - 1$. If $n - 1 \geq 1$, then $Q_1(x)$ must have a root r_2. Thus

$$Q_1(x) = (x - r_2)Q_2(x) \tag{2}$$

where $Q_2(x)$ is of degree $n - 2$. Substituting in Equation (1) for $Q_1(x)$ we have

$$P(x) = (x - r_1)(x - r_2)Q_2(x) \tag{3}$$

This process is repeated n times until $Q_n(x) = a$ is of degree 0. Hence,

$$P(x) = a(x - r_1)(x - r_2) \ldots (x - r_n) \tag{4}$$

Since a is the leading coefficient of the polynomial on the right side of Equation (4), it must also be the leading coefficient of $P(x)$.

EXAMPLE 1

Find the polynomial $P(x)$ of degree 3 that has the roots -2, i, and $-i$, and satisfies $P(1) = -3$.

SOLUTION

Since -2, i, and $-i$ are roots of $P(x)$, we may write

$$P(x) = a(x + 2)(x - i)(x + i)$$

To find the constant a, we use the condition $P(1) = -3$.

$$P(1) = -3 = a(1 + 2)(1 - i)(1 + i) = 6a$$

$$a = -\frac{1}{2}$$

so that

$$P(x) = -\frac{1}{2}(x + 2)(x - i)(x + i)$$

Recall that the roots of a polynomial need not be distinct from each other. The polynomial

$$P(x) = x^2 - 2x + 1$$

can be written in the factored form

$$P(x) = (x - 1)(x - 1)$$

which shows that the roots of $P(x)$ are 1 and 1. Since a root is associated with a factor and a factor may be repeated, we may have repeated roots. If the factor $x - r$ appears k times, we say that r is a **root of multiplicity k.**

It is now easy to establish the following, which may be thought of as an alternative form of the Fundamental Theorem of Algebra.

The Fundamental Theorem of Algebra— Part II	If $P(x)$ is a polynomial of degree $n \geq 1$, then $P(x)$ has precisely n roots among the complex numbers when a root of multiplicity k is counted k times.

We may prove this theorem as follows. If we write $P(x)$ in the form of Equation (4), we see that $r_1, r_2, \ldots, r_n$ are roots of the equation $P(x) = 0$ and hence there exist n roots. If there is an additional root r that is distinct from the roots $r_1, r_2, \ldots, r_n$, then $r - r_1, r - r_2, \ldots, r - r_n$ are all different from 0. Substituting r for x in Equation (4) yields

$$P(r) = a(r - r_1)(r - r_2) \ldots (r - r_n) \tag{5}$$

which cannot equal 0, since the product of nonzero numbers cannot equal 0. Thus, $r_1, r_2, \ldots, r_n$ are roots of $P(x)$ and there are no other roots. We conclude that $P(x)$ has precisely n roots.

EXAMPLE 2
Find all roots of the polynomial

$$P(x) = (x - \tfrac{1}{2})^3(x + i)(x - 5)^4$$

SOLUTION
The distinct roots are $\tfrac{1}{2}$, $-i$, and 5. Further, $\tfrac{1}{2}$ is a root of multiplicity 3; $-i$ is a root of multiplicity 1; 5 is a root of multiplicity 4.

EXAMPLE 3
If -1 is a root of multiplicity 2 of $P(x) = x^4 + 4x^3 + 2x^2 - 4x - 3$, find the remaining roots and write $P(x)$ as a product of linear factors.

SOLUTION
Since -1 is a double root of $P(x)$, then $(x + 1)^2$ is a factor of $P(x)$. Therefore,

$$P(x) = (x + 1)^2 Q(x)$$

or

$$P(x) = (x^2 + 2x + 1)Q(x)$$

Using polynomial division, we can divide both sides of the last equation by $x^2 + 2x + 1$ to obtain

$$Q(x) = \frac{x^4 + 4x^3 + 2x^2 - 4x - 3}{x^2 + 2x + 1}$$
$$= x^2 + 2x - 3$$
$$= (x - 1)(x + 3)$$

The roots of the depressed equation $Q(x) = 0$ are 1 and -3, and these are the remaining roots of $P(x)$. By the Linear Factor Theorem,

$$P(x) = (x + 1)^2(x - 1)(x + 3)$$

PROGRESS CHECK
If -2 is a root of multiplicity 2 of $P(x) = x^4 + 4x^3 + 5x^2 + 4x + 4$, write $P(x)$ as a product of linear factors.

ANSWER
$P(x) = (x + 2)(x + 2)(x + i)(x - i)$

We know from the quadratic formula that if a quadratic equation with real coefficients has a complex root $a + bi$, then the conjugate $a - bi$ is the other root. The following theorem extends this result to a polynomial of degree n with real coefficients.

Conjugate Roots Theorem

If $P(x)$ is a polynomial of degree $n \geq 1$ with real coefficients, and if $a + bi$, $b \neq 0$, is a root of $P(x)$, then the complex conjugate $a - bi$ is also a root of $P(x)$.

PROOF OF CONJUGATE ROOTS THEOREM (Optional)

To prove the Conjugate Roots Theorem, we let $z = a + bi$ and make use of the properties of complex conjugates developed earlier in this section. We may write

$$P(x) = a_n x^n + a_{n-1} x^{n-1} + \cdots + a_1 x + a_0 \tag{6}$$

and, since z is a root of $P(x)$,

$$a_n z^n + a_{n-1} z^{n-1} + \cdots + a_n z + a_0 = 0 \tag{7}$$

But if $z = w$, then $\overline{z} = \overline{w}$. Applying this property of complex numbers to both sides of Equation (7), we have

$$\overline{a_n z^n + a_{n-1} z^{n-1} + \cdots + a_1 z + a_0} = \overline{0} = 0 \tag{8}$$

We also know that $\overline{z + w} = \overline{z} + \overline{w}$. Applying this property to the left side of Equation (8) we see that

$$\overline{a_n z^n} + \overline{a_{n-1} z^{n-1}} + \cdots + \overline{a_1 z} + \overline{a_0} = 0 \tag{9}$$

Further, $\overline{z \cdot w} = \overline{z} \cdot \overline{w}$, so we may rewrite Equation (9) as

$$\overline{a_n} \overline{z^n} + \overline{a_{n-1}} \overline{z^{n-1}} + \cdots + \overline{a_1} \overline{z} + \overline{a_0} = 0 \tag{10}$$

Since $a_0, a_1, \ldots, a_n$ are all real numbers, we know that $\overline{a_0} = a_0$, $\overline{a_1} = a_1$, $\ldots, \overline{a_n} = a_n$. Finally, we use the property $\overline{z^n} = \overline{z}^n$ to rewrite Equation (10) as

$$a_n \overline{z}^n + a_{n-1} \overline{z}^{n-1} + \cdots + a_1 \overline{z} + a_0 = 0$$

which establishes that $\overline{z}$ is a root of $P(x)$.

EXAMPLE 4
Find a polynomial $P(x)$ with real coefficients that is of degree 3 and whose roots include -2 and $1 - i$.

SOLUTION
Since $1 - i$ is a root, it follows from the Conjugate Roots Theorem that $1 + i$ is also a root of $P(x)$. By the Factor Theorem, $(x + 2)$, $[x - (1 - i)]$, and $[x - (1 + i)]$ are factors of $P(x)$. Therefore,

$$\begin{aligned}
P(x) &= (x + 2)[x - (1 - i)][x - (1 + i)] \\
&= (x + 2)(x^2 - 2x + 2) \\
&= x^3 - 2x + 4
\end{aligned}$$

PROGRESS CHECK
Find a polynomial $P(x)$ with real coefficients that is of degree 4 and whose roots include i and $-3 + i$.

ANSWER
$P(x) = x^4 + 6x^3 + 11x^2 + 6x + 10$

The following is a corollary of the Conjugate Roots Theorem.

A polynomial $P(x)$ of degree $n \geq 1$ with real coefficients can be written as a product of linear and quadratic factors with real coefficients so that the quadratic factors have no real roots.

By the Linear Factor Theorem, we may write

$$P(x) = a(x - r_1)(x - r_2) \ldots (x - r_n)$$

| A "NATURAL" MATHEMATICIAN | Srinivasa Ramanujan was born in India in 1887. At age 15 he borrowed a copy of Carr's *Synopsis of Pure Mathematics* from a library. Thus began what is possibly the strangest story in the history of mathematics. |

Through the study of this one-volume compendium of theorems, which was already 40 years out-of-date, Ramanujan jumped to the forefront of the mathematicians of his time. With no formal training and little idea of what constituted a mathematical proof, this "natural" mathematician presented to G. H. Hardy of Cambridge a list of 120 theorems, many of remarkable difficulty and insight.

Hardy brought Ramanujan to England in 1913 and acted as his tutor, friend, and coauthor until Ramanujan's return to India, where he died in 1920 at age 33. When visiting Ramanujan in an English hospital, Hardy remarked that the taxicab he had ridden in was No. 1729, a rather dull number.

"No," replied Ramanujan, "it is a very interesting number. It is the smallest number that can be written as the sum of two cubes in two different ways."

Ramanujan was asserting that

$$1729 = a^3 + b^3 \quad \text{and} \quad 1729 = c^3 + d^3$$

where a, b, c, and d are distinct positive integers, and that 1729 is the smallest number that can be written in this manner. By generating a list of cubes less than 1729, you can easily find a, b, c, and d. You might want to devise a computer program to verify Ramanujan's assertion that 1729 is the smallest such number.

where $r_1, r_2, \ldots, r_n$ are the n roots of $P(x)$. Of course, some of these roots may be complex numbers. A complex root $a + bi$, $b \neq 0$, may be paired with its conjugate $a - bi$ to provide the quadratic factor

$$[x - (a + bi)][x - (a - bi)] = x^2 - 2ax + a^2 + b^2 \tag{11}$$

which has real coefficients. Thus, a *quadratic* factor with real coefficients results from each pair of complex conjugate roots; a *linear* factor with real coefficients results from each real root. Further, the discriminant of the quadratic factor in Equation (11) is $-4b^2$ and is therefore always negative, which shows that the quadratic factor has no real roots.

POLYNOMIALS WITH COMPLEX COEFFICIENTS

Although the definition of a polynomial given in Section 3.1 permits the coefficients to be complex numbers, we have limited our examples to polynomials with real coefficients because beginning calculus courses adhere to this restriction. To round out our work, we point out that both the Linear Factor Theorem and the Fundamental Theorem of Algebra hold for polynomials with complex coefficients.

On the other hand, the Conjugate Roots Theorem may not hold if the polynomial $P(x)$ has complex coefficients. To see this, consider the polynomial

$$P(x) = x - (2 + i)$$

which has a complex coefficient and has the root $2 + i$. Note that the complex conjugate $2 - i$ is *not* a root of $P(x)$ and that, therefore, the Conjugate Roots Theorem fails to apply to $P(x)$.

EXAMPLE 5
Find a polynomial $P(x)$ of degree 2 that has the roots -1 and $1 - i$.

SOLUTION
Since -1 is a root of $P(x)$, $x + 1$ is a factor. Similarly, $[x - (1 - i)]$ is also a factor of $P(x)$. We can then write

$$P(x) = (x + 1)[x - (1 - i)]$$
$$= x^2 + ix - 1 + i$$

which is a polynomial of degree 2 (with complex coefficients) that has the desired roots.

EXERCISE SET 8.3
In Exercises 1–6 find a polynomial $P(x)$ of lowest degree that has the indicated roots.

1. $2, -4, 4$ 2. $5, -5, 1, -1$ 3. $-1, -2, -3$ 4. $-3, \sqrt{2}, -\sqrt{2}$
5. $4, 1 \pm \sqrt{3}$ 6. $1, 2, 2 \pm \sqrt{2}$

In Exercises 7–10 find the polynomial $P(x)$ of lowest degree that has the indicated roots and satisfies the given condition.

7. $\frac{1}{2}, \frac{1}{2}, -2; P(2) = 3$ 8. $3, 3, -2, 2; P(4) = 12$
9. $\sqrt{2}, -\sqrt{2}, 4; P(-1) = 5$ 10. $\frac{1}{2}, -2, 5; P(0) = 5$

In Exercises 11–18 find the roots of the given equation.

11. $(x - 3)(x + 1)(x - 2) = 0$ 12. $(x - 3)(x^2 - 3x - 4) = 0$
13. $(x + 2)(x^2 - 16) = 0$ 14. $(x^2 - x)(x^2 - 2x + 5) = 0$
15. $(x^2 + 3x + 2)(2x^2 + x) = 0$ 16. $(x^2 + x + 4)(x - 3)^2 = 0$
17. $(x - 5)^3(x + 5)^2 = 0$ 18. $(x + 1)^2(x + 3)^4(x - 2) = 0$

In Exercises 19–22 find a polynomial that has the indicated roots and no others.

19. -2 of multiplicity 3
20. 1 of multiplicity 2, -4 of multiplicity 1
21. $\frac{1}{2}$ of multiplicity 2, -1 of multiplicity 2
22. -1 of multiplicity 2, 0 and 2 each of multiplicity 1

In Exercises 23–28 find a polynomial that has the indicated roots and no others.

23. $1 + 3i, -2$ 24. $1, -1, 2 - i$ 25. $1 + i, 2 - i$ 26. $-2, 3, 1 + 2i$
27. -2 is a root of multiplicity 2, $3 - 2i$
28. 3 is a triple root, $-i$

In Exercises 29–34 use the given root(s) to help in writing the given equation as a product of linear and quadratic factors with real coefficients.

29. $x^3 - 7x^2 + 16x - 10 = 0$; $3 - i$

30. $x^3 + x^2 - 7x + 65 = 0$; $2 + 3i$

31. $x^4 + 4x^3 + 13x^2 + 18x + 20 = 0$; $-1 - 2i$

32. $x^4 + 3x^3 - 5x^2 - 29x - 30 = 0$; $-2 + i$

33. $x^5 + 3x^4 - 12x^3 - 42x^2 + 32x + 120 = 0$; $-3 - i$, -2

34. $x^5 - 8x^4 + 29x^3 - 54x^2 + 48x - 16 = 0$; $2 + 2i$, 2

35. Write a polynomial $P(x)$ with complex coefficients that has the root $a + bi$, $b \neq 0$, and does not have $a - bi$ as a root.

36. Prove that a polynomial equation of degree 4 with real coefficients has 4 real roots, 2 real roots, or no real roots.

37. Prove that a polynomial equation of odd degree with real coefficients has at least one real root.

8.4 DESCARTES'S RULE OF SIGNS

In this section we will restrict our investigation to polynomials with real coefficients. Our objective is to obtain some information concerning the number of positive real roots and the number of negative real roots of such polynomials.

If the terms of a polynomial with real coefficients are written in descending order, then a **variation in sign** occurs whenever two successive terms have opposite signs. In determining the number of variations in sign, we ignore terms with zero coefficients. The polynomial

$$4x^5 - 3x^4 - 2x^2 + 1$$

has two variations in sign. The French mathematician René Descartes (1596–1650), who provided us with the foundations of analytic geometry, also gave us a theorem that relates the nature of the real roots of polynomials to the variations in sign. The proof of Descartes's theorem is outlined in Exercises 19–24.

Descartes's Rule of Signs	If $P(x)$ is a polynomial with real coefficients, then (i) the number of positive roots either is equal to the number of variations in sign of $P(x)$ or is less than the number of variations in sign by an even number, and (ii) the number of negative roots either is equal to the number of variations in sign of $P(-x)$ or is less than the number of variations in sign by an even number.

If it is determined that a polynomial of degree n has r real roots, then the remaining $n - r$ roots must be complex numbers.

To apply Descartes's Rule of Signs to the polynomial

$$P(x) = 3x^5 + 2x^4 - x^3 + 2x - 3$$

we first note that there are 3 variations in sign as indicated. Thus, either there are 3 positive roots or there is 1 positive root. Next, we form $P(-x)$,

$$P(-x) = 3(-x)^5 + 2(-x)^4 - (-x)^3 + 2(-x) - 3$$
$$= -3x^5 + 2x^4 + x^3 - 2x - 3$$

which can be obtained by negating the coefficients of the odd-power terms. We see that $P(-x)$ has two variations in sign and conclude that $P(x)$ has either 2 negative roots or no negative roots.

EXAMPLE 1

Use Descartes's Rule of Signs to analyze the roots of the equation

$$2x^5 + 7x^4 + 3x^2 - 2 = 0$$

SOLUTION

Since

$$P(x) = 2x^5 + 7x^4 + 3x^2 - 2$$

has 1 variation in sign, there is precisely 1 positive root. The polynomial $P(-x)$ is formed—

$$P(-x) = -2x^5 + 7x^4 + 3x^2 - 2$$

and is seen to have 2 variations in sign, so that $P(-x)$ has either 2 negative roots or no negative roots. Since $P(x)$ has 5 roots, the possibilities are

1 positive root, 2 negative roots, 2 complex roots

1 positive root, 0 negative roots, 4 complex roots

PROGRESS CHECK

Use Descartes's Rule of Signs to analyze the nature of the roots of the equation

$$x^6 + 5x^4 - 4x^2 - 3 = 0$$

ANSWER

1 positive root, 1 negative root, 4 complex roots

EXAMPLE 2

Use Descartes's Rule of Signs, the Rational Root Theorem, and the depressed equation to write

$$P(x) = 3x^4 + 2x^3 + 2x^2 + 2x - 1$$

as a product of linear and quadratic factors with real coefficients such that the quadratic factors have no real roots.

FERMAT'S LAST THEOREM

If you were asked to find natural numbers a, b, and c that satisfy the equation

$$a^2 + b^2 = c^2$$

you would have no trouble coming up with "triplets" such as (3, 4, 5) and (5, 12, 13). In fact, there are an infinite number of solutions, since any multiple of (3, 4, 5) such as (6, 8, 10) is also a solution.

Generalizing the above problem, suppose we seek natural numbers a, b, and c that satisfy the equation

$$a^n + b^n = c^n$$

for integer values of $n > 2$. Pierre Fermat, a great French mathematician of the seventeenth century, stated that there are no natural numbers a, b, and c that satisfy this equation for any integer $n > 2$. This seductively simple conjecture is known as Fermat's Last Theorem. Fermat wrote in his notebook that he had a proof but that it was too long to include in the margin. We know that the theorem is true for $n < 30,000$, but a proof of the general theorem or a counterexample has eluded mathematicians for three hundred years!

a^n

$+$

b^n

$=$

c^n

SOLUTION

We first use the Rational Root Theorem to list the possible rational roots.

possible numerators: ± 1 (factors of 1)
possible denominators: ± 1, ± 3 (factors of 3)
possible rational roots: ± 1, $\pm \frac{1}{3}$

Next, we note that $P(x)$ has real coefficients and that Descartes's Rule of Signs may therefore be employed. Since $P(x)$ has 1 variation in sign, there is precisely 1 positive real root. If this real root is a rational number, it must be either $+1$ or $+\frac{1}{3}$. Trying $+1$, we quickly see that $P(1) = 8$ and that $+1$ is not a root. Using synthetic division,

$$
\begin{array}{r|rrrrr}
\frac{1}{3} & 3 & 2 & 2 & 2 & -1 \\
  &   & 1 & 1 & 1 & 1 \\
\hline
  & 3 & 3 & 3 & 3 & 0
\end{array}
$$

coefficients of depressed equation

we see that $\frac{1}{3}$ is a root and the depressed equation is

$$Q_1(x) = 3x^3 + 3x^2 + 3x + 3 = 0$$

which has the same roots as

$$Q_2(x) = x^3 + x^2 + x + 1 = 0$$

Since any root of $Q_2(x)$ is also a root of $P(x)$ and since we have removed the only positive root, we know that $Q_2(x)$ cannot have any positive roots. (Verify that $Q_2(x)$ has no variations in sign!) However, forming

$$Q_2(-x) = -x^3 + x^2 - x + 1$$

we see that $Q_2(x)$ has at least 1 negative root. By the Rational Root Theorem, the only possible rational roots of $Q_2(x)$ are ± 1. Using synthetic division,

$$
\begin{array}{r|rrrr}
-1 & 1 & 1 & 1 & 1 \\
   &   & -1 & 0 & -1 \\
\hline
   & 1 & 0 & 1 & 0
\end{array}
$$

coefficients of depressed equation

we verify that -1 is indeed a root. Finally, we note that the depressed equation $x^2 + 1 = 0$ has no real roots, since the discriminant is negative. Thus,

$$P(x) = 3x^4 + 2x^3 + 2x^2 + 2x - 1 = 3\left(x - \frac{1}{3}\right)(x + 1)(x^2 + 1)$$

EXERCISE SET 8.4

In Exercises 1–12 use Descartes's Rule of Signs to analyze the nature of the roots of the given equation. List all possibilities.

1. $3x^4 - 2x^3 + 6x^2 + 5x - 2 = 0$
2. $2x^6 + 5x^5 + x^3 - 6 = 0$
3. $x^6 + 2x^4 + 4x^2 + 1 = 0$
4. $3x^3 - 2x + 2 = 0$
5. $x^5 - 4x^3 + 7x - 4 = 0$
6. $2x^3 - 5x^2 + 8x - 2 = 0$
7. $5x^3 + 2x^2 + 7x - 1 = 0$
8. $x^5 + 6x^4 - x^3 - 2x - 3 = 0$
9. $x^4 - 2x^3 + 5x^2 + 2 = 0$
10. $3x^4 - 2x^3 - 1 = 0$
11. $x^8 + 7x^3 + 3x - 5 = 0$
12. $x^7 + 3x^5 - x^3 - x + 2 = 0$

In Exercises 13–18 use Descartes's Rule of Signs, the Rational Root Theorem, and the depressed equation to find all roots of the given equation.

13. $x^4 - 6x^3 + 10x^2 - 6x + 9 = 0$
14. $2x^4 - 3x^3 + 5x^2 - 6x + 2 = 0$
15. $x^4 - 6x^2 + 8 = 0$
16. $x^4 - 4x^3 + 7x^2 - 6x + 2 = 0$
17. $4x^4 + 4x^3 - 3x^2 - 4x - 1 = 0$
18. $x^5 + x^4 - 7x^3 - 11x^2 - 8x - 12 = 0$

19. Prove that if $P(x)$ is a polynomial with real coefficients and r is a positive root of $P(x)$, then the depressed equation

$$Q(x) = \frac{P(x)}{(x - r)}$$

has at least one fewer variation in sign than $P(x)$. (*Hint:* Assume the leading coefficient of $P(x)$ to be positive and use synthetic division to obtain $Q(x)$. Note that the coefficients of $Q(x)$ remain positive at least until there is a variation in sign in $P(x)$.)

20. Prove that if $P(x)$ is a polynomial with real coefficients, then the number of positive roots is not greater than the number of variations in sign in $P(x)$. (*Hint:* Let $r_1, r_2, \ldots, r_k$ be the positive roots of $P(x)$, and let

$$P(x) = (x - r_1)(x - r_2) \ldots (x - r_k)Q(x)$$

Use the result of Exercise 19 to show that $Q(x)$ has at least k fewer variations in sign than does $P(x)$.)

21. Prove that if $r_1, r_2, \ldots, r_k$ are positive numbers, then

$$P(x) = (x - r_1)(x - r_2) \ldots (x - r_k)$$

has alternating signs. (*Hint:* Use the result of Exercise 20.)

22. Prove that the number of variations in sign of a polynomial with real coefficients is even if the first and last coefficients have the same sign and is odd if they are of opposite sign.

23. Prove that if the number of positive roots of the polynomial $P(x)$ with real coefficients is less than the number of variations in sign, then it is less by an even number. (*Hint:* Write $P(x)$ as a product of linear factors corresponding to the positive and negative roots, and of quadratic factors corresponding to complex roots. Apply the results of Exercises 21 and 22.)

24. Prove that the positive roots of $P(-x)$ correspond to the negative roots of $P(x)$, that is, if $a > 0$ is a root of $P(-x)$, then $-a$ is a root of $P(x)$.

TERMS AND SYMBOLS

complex conjugate (p. 334)
$\overline{z}$ (p. 335)
complex plane (p. 337)
real axis (p. 337)
imaginary axis (p. 337)
absolute value of a complex number (p. 338)

trigonometric form (p. 339)
polar form (p. 339)
modulus (p. 339)
argument (p. 339)
principal argument (p. 339)

De Moivre's theorem (p. 342)
nth roots of a complex number (p. 343)

nth roots of unity (p. 344)
root of multiplicity k (p. 348)
variation in sign (p. 353)

KEY IDEAS FOR REVIEW

☐ The complex number $a + bi$ can be associated with the point $P(a, b)$. The trigonometric or polar form of the complex number $a + bi$ is given by

$$a + bi = r(\cos \theta + i \sin \theta)$$

where r is the length of the line segment OP and θ is the measure of the angle in standard position whose terminal side is OP.

☐ The trigonometric form of a complex number is useful since multiplication and division of complex numbers take on simple forms. In particular, exponentiation of complex numbers is handled by De Moivre's theorem, which states

$$[r(\cos \theta + i \sin \theta)]^n = r^n(\cos n\theta + i \sin n\theta)$$

☐ The complex number u is an nth root of the complex number z if $u^n = z$. De Moivre's theorem can be used to find a formula for determining u.

☐ The following theorems concern polynomials and their roots.

Linear Factor Theorem

A polynomial $P(x)$ of degree $n \geq 1$ can be written as the product of n linear factors.

Fundamental Theorem of Algebra

If $P(x)$ is a polynomial of degree $n \geq 1$, then $P(x)$ has precisely n roots among the complex numbers.

Conjugate Roots Theorem

If $a + bi$, $b \neq 0$, is a root of the polynomial $P(x)$ with real coefficients, then $a - bi$ is also a root of $P(x)$.

☐ Descartes's Rule of Signs tells us the maximum number of positive roots and the maximum number of negative roots of a polynomial $P(x)$ with real coefficients.

REVIEW EXERCISES

Solutions to exercises whose numbers are in color are in the Solutions section in the back of the book.

8.1 In Exercises 1–3 write the given quotient in the form $a + bi$.

1. $\dfrac{3 - 2i}{4 + 3i}$ 2. $\dfrac{2 + i}{-5i}$

3. $\dfrac{-5}{1 + i}$

In Exercises 4–6 write the reciprocal of the given complex number in the form $a + bi$.

4. $1 + 3i$ 5. $-4i$

6. $2 - 5i$

In Exercises 7–9 determine the absolute value of the given complex number.

7. $2 - i$ 8. $-3 + 2i$

9. $-4 - 5i$

8.2 In Exercises 10–13 convert from trigonometric to algebraic form and vice versa.

10. $-3 + 3i$

11. $2(\cos 90° + i \sin 90°)$

12. $\sqrt{2}(\cos 315° + i \sin 315°)$

13. -2

In Exercises 14–16 find the indicated product or quotient. Express the answer in trigonometric form.

14. $4(\cos 22° + i \sin 22°) \cdot 6(\cos 15° + i \sin 15°)$

15. $\dfrac{5(\cos 71° + i \sin 71°)}{3(\cos 50° + i \sin 50°)}$

16. $2(\cos 210° + i \sin 210°) \cdot (\cos 240° + i \sin 240°)$

In Exercises 17 and 18 use De Moivre's theorem to express the given number in the form $a + bi$.

17. $(3 - 3i)^5$

18. $[2(\cos 90° + i \sin 90°)]^3$

19. Express the two square roots of -9 in trigonometric form.

20. Determine all roots of the equation $x^3 - 1 = 0$.

8.3 In Exercises 21–23 find a polynomial of lowest degree that has the indicated roots.

21. $-3, -2, -1$ 22. $3, \pm\sqrt{-3}$

23. $-2, \pm\sqrt{3}, 1$

In Exercises 24–26 find a polynomial that has the indicated roots and no others.

24. $\frac{1}{2}$ of multiplicity 2, -1 of multiplicity 2

25. $i, -i$, each of multiplicity 2

26. -1 of multiplicity 3, 3 of multiplicity 1

8.4 In Exercises 27–30 use Descartes's Rule of Signs to determine the maximum number of positive and negative real roots of the given equation.

27. $x^4 - 2x - 1 = 0$

28. $x^5 - x^4 + 3x^3 - 4x^2 + x - 5 = 0$

29. $x^3 - 5 = 0$

30. $3x^4 - 2x^2 + 1 = 0$

In Exercises 31–32 find all roots of the given equation.

31. $6x^3 + 15x^2 - x - 10 = 0$

32. $2x^4 - 3x^3 - 10x^2 + 19x - 6 = 0$

PROGRESS TEST 8A

1. Write the quotient $\dfrac{1 - i}{3 + 2i}$ in the form $a + bi$.

2. Write the reciprocal of $-2 + i$ in the form $a + bi$.

3. Determine the absolute value of $3 - 4i$.

In Problems 4 and 5 find the indicated product or quotient. Express the answer in trigonometric form.

4. $\dfrac{1}{2}(\cos 14° + i \sin 14°) \cdot 10(\cos 72° + i \sin 72°)$

5. $\dfrac{3(\cos 85° + i \sin 85°)}{6(\cos 8° + i \sin 8°)}$

6. Use De Moivre's theorem to express

$$\left[\frac{1}{5}(\cos 120° + i \sin 120°)\right]^4$$

in the form $a + bi$.

7. Express the three cube roots of -27 in trigonometric form.

In Problems 8 and 9 find a polynomial of lowest degree that has the indicated roots.

8. $-2, 1, 3$ 9. $-1, 1, 3 \pm \sqrt{2}$

In Problems 10 and 11 find the roots of the given equation.

10. $(x^2 + 1)(x - 2) = 0$

11. $(x + 1)^2(x - 3x - 2) = 0$

In Problems 12–14 find a polynomial that has the indicated roots and no others.

12. -3 of multiplicity 2; 1 of multiplicity 3

13. $-\frac{1}{4}$ of multiplicity 2; i, $-i$, and 1

14. i, $1 + i$

15. If $2 + i$ is a root of $x^3 - 6x^2 + 13x - 10 = 0$, write the equation as a product of linear and quadratic factors with real coefficients.

In Problems 16 and 17 determine the maximum number of roots, of the type indicated, of the given equation.

16. $2x^5 - 3x^4 + 1 = 0$; positive real roots

17. $3x^4 + 2x^3 - 2x^2 - 1 = 0$; negative real roots

18. Find all roots of the equation $3x^4 + 7x^3 - 3x^2 + 7x - 6 = 0$

PROGRESS TEST 8B

1. Write the quotient $\dfrac{-1}{2 + 2i}$ in the form $a + bi$.

2. Write the reciprocal of $3 - 4i$ in the form $a + bi$.

3. Determine the absolute value of $-2 + 2i$.

In Problems 4 and 5 find the indicated product or quotient. Express the answer in trigonometric form.

4. $(\cos 125° + i \sin 125°) \cdot 5(\cos 125° + i \sin 125°)$

5. $\dfrac{\frac{1}{2}(\cos 67° + i \sin 67°)}{\frac{1}{4}(\cos 12° + i \sin 12°)}$

6. Use De Moivre's theorem to express $(-2i)^6$ in the form $a + bi$.

7. Determine all roots of the equation $x^3 + 1 = 0$.

In Problems 8 and 9 find a polynomial of lowest degree that has the indicated roots.

8. $-\frac{1}{2}, 1, 1, -1$ 9. $2, 1 \pm \sqrt{3}$

In Problems 10 and 11 find the roots of the given equation.

10. $(x^2 - 3x + 2)(x - 2)^2$

11. $(x^2 + 3x - 1)(x - 2)(x + 3)^2$

In Problems 12–14 find a polynomial that has the indicated roots and no others.

12. $\frac{1}{2}$ of multiplicity 3; -2 of multiplicity 1

13. -3 of multiplicity 2; $1 + i$, $1 - i$

14. $3 \pm \sqrt{-1}$; -1 of multiplicity 2

15. If $1 - i$ is a root of $2x^4 - x^3 - 4x^2 + 10x - 4 = 0$, write the equation as a product of linear and quadratic factors with real coefficients.

In Problems 16 and 17 determine the maximum number of roots, of the type indicated, of the given equation.

16. $3x^4 + 3x - 1 = 0$; positive real roots

17. $2x^4 + x^3 - 3x^2 + 2x + 1 = 0$; negative real roots

18. Find all roots of the equation $2x^4 - x^3 - 4x^2 + 2x = 0$.

CHAPTER 9

ANALYTIC GEOMETRY: THE CONIC SECTIONS

In 1637 the great French philosopher and scientist René Descartes developed an idea that the nineteenth-century British philosopher John Stuart Mill described as "the greatest single step ever made in the progress of the exact sciences." Descartes combined the techniques of algebra with those of geometry and created a new field of study called **analytic geometry.** Analytic geometry enables us to apply algebraic methods and equations to the solution of problems in geometry and, conversely, to obtain geometric representations of algebraic equations.

We will first develop a formula for the coordinates of the midpoint of a line segment. We will then use the distance and midpoint formulas as tools to illustrate the usefulness of analytic geometry by proving a number of general theorems from plane geometry.

The power of the methods of analytic geometry is also very well demonstrated, as we shall see in this chapter, in a study of the conic sections. We will find in the course of that study that (a) a geometric definition can be converted into an algebraic equation and (b) an algebraic equation can be classified by the type of graph it represents.

9.1
ANALYTIC GEOMETRY

We have previously seen that the length d of the line segment joining points $P_1(x_1, y_1)$ and $P_2(x_2, y_2)$ is given by

$$d = \sqrt{(x_2 - x_1)^2 + (y_2 - y_1)^2}$$

It is also possible to obtain a formula for the coordinates (x, y) of the midpoint P of the line segment whose endpoints are P_1 and P_2 (see Figure 1). Passing lines through P and P_2 parallel to the y-axis and a line through P_1 parallel to the x-axis results in the similar right triangles P_1AP and P_1BP_2. Using the fact that corresponding sides of similar triangles are in proportion, we can write

$$\frac{\overline{P_1P_2}}{\overline{P_2B}} = \frac{\overline{P_1P}}{\overline{PA}}$$

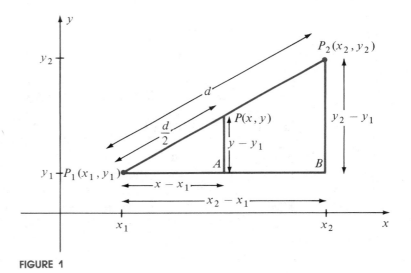

FIGURE 1

Since P is the midpoint of P_1P_2, the length of P_1P is $d/2$, so

$$\frac{d}{y_2 - y_1} = \frac{\dfrac{d}{2}}{y - y_1}$$

Solving for y we have

$$y = \frac{y_1 + y_2}{2}$$

Similarly,

$$\frac{\overline{P_1P_2}}{\overline{P_1B}} = \frac{\overline{P_1P}}{\overline{P_1A}} \quad \text{or} \quad \frac{d}{x_2 - x_1} = \frac{\dfrac{d}{2}}{x - x_1}$$

We solve for x to obtain

$$x = \frac{x_1 + x_2}{2}$$

We have established the following formula.

The Midpoint Formula	If $P(x, y)$ is the midpoint of the line segment whose endpoints are $P_1(x_1, y_1)$ and $P_2(x_2, y_2)$, then

$$x = \frac{x_1 + x_2}{2} \qquad y = \frac{y_1 + y_2}{2}$$

EXAMPLE 1

Find the midpoint of the line segment whose endpoints are $P_1(3, 4)$ and $P_2(-2, -6)$.

SOLUTION

If $P(x, y)$ is the midpoint, then

$$x = \frac{x_1 + x_2}{2} = \frac{3 + (-2)}{2} = \frac{1}{2}$$

$$y = \frac{y_1 + y_2}{2} = \frac{4 + (-6)}{2} = -1$$

Thus, the midpoint is $(\frac{1}{2}, -1)$.

PROGRESS CHECK

Find the midpoint of the line segment whose endpoints are the following.
(a) $(0, -4), (-2, -2)$ (b) $(-10, 4), (7, -5)$

ANSWERS

(a) $(-1, -3)$ (b) $(-\frac{3}{2}, -\frac{1}{2})$

The formulas for distance, midpoint of a line segment, and slope of a line are sufficient to allow us to demonstrate the beauty and power of analytic geometry. With these tools, we can prove theorems from plane geometry by placing the figures on a rectangular coordinate system.

EXAMPLE 2

Prove that the line joining the midpoints of two sides of a triangle is parallel to the third side and has length equal to one-half the third side.

SOLUTION

We place the triangle OAB in a convenient location, namely, with one vertex at the origin and one side on the positive x-axis (Figure 2). If Q and R are the

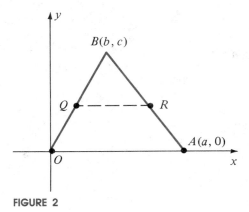

FIGURE 2

midpoints of OB and AB, then, by the midpoint formula, the coordinates of Q and R are

$$Q\left(\frac{b}{2}, \frac{c}{2}\right) \qquad R\left(\frac{a+b}{2}, \frac{c}{2}\right)$$

We see that the line joining Q and R has slope 0, since the difference of the y-coordinates is

$$\frac{c}{2} - \frac{c}{2} = 0$$

But side OA also has slope 0, which proves that $\overline{QR}$ is parallel to OA.
Applying the distance formula to $\overline{QR}$, we have

$$\overline{QR} = \sqrt{(x_2 - x_1)^2 + (y_2 - y_1)^2}$$

$$= \sqrt{\left(\frac{a+b}{2} - \frac{b}{2}\right)^2 + \left(\frac{c}{2} - \frac{c}{2}\right)^2}$$

$$= \sqrt{\left(\frac{a}{b}\right)^2} = \frac{a}{2}$$

Since $\overline{OA}$ has length a, we have shown that $\overline{QR}$ is one-half of $\overline{OA}$.

PROGRESS CHECK

Prove that the midpoint of the hypotenuse of a right triangle is equidistant from all three vertices. (*Hint:* Place the triangle so that two legs coincide with the positive x- and y-axes. Find the coordinates of the midpoint of the hypotenuse by the midpoint formula. Finally, compute the distance from the midpoint to each vertex by the distance formula.)

EXERCISE SET 9.1

In Exercises 1–12 find the midpoint of the line segment whose endpoints are given.

1. (2, 6), (3, 4)
2. (1, 1), (−2, 5)
3. (2, 0), (0, 5)
4. (−3, 0), (−5, 2)
5. (−2, 1), (−5, −3)
6. (2, 3), (−1, 3)
7. (0, −4), (0, 3)
8. (1, −3), (3, 2)
9. (−1, 3), (−1, 6)
10. (3, 2), (0, 0)
11. (1, −1), (−1, 1)
12. (2, 4), (2, −4)

13. Prove that the medians from the equal angles of an isosceles triangle are of equal length. (*Hint:* Place the triangle so that its vertices are at the points $A(-a, 0)$, $B(a, 0)$, and $C(0, b)$.)

14. Show that the midpoints of the sides of a rectangle are the vertices of a rhombus (a quadrilateral with four equal sides). (*Hint:* Place the rectangle so that its vertices are at the points $(0, 0)$, $(a, 0)$, $(0, b)$, and (a, b).)

15. Prove that a triangle with two equal medians is isosceles.

16. Show that the sum of the squares of the lengths of the medians of a triangle equals three-fourths the sum of the squares of the lengths of the sides. (*Hint:* Place the triangle so that its vertices are at the points $(-a, 0)$, $(b, 0)$, and $(0, c)$.)

17. Prove that the diagonals of a rectangle are equal in length. (*Hint:* Place the rectangle so that its vertices are at the points $(0, 0)$, $(a, 0)$, $(0, b)$, and (a, b).)

9.2
THE CIRCLE

The conic sections provide us with an outstanding opportunity to illustrate the double-edged power of analytic geometry. We will see that a geometric figure defined as a set of points can often be described analytically by an algebraic equation; conversely, we can start with an algebraic equation and use graphing procedures to study the properties of the curve.

First, let's see how the term conic section originates. If we pass a plane through a cone at various angles, as shown in Figure 3, the intersections are

Circle

Parabola

Ellipse

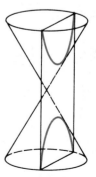
Hyperbola

FIGURE 3

called **conic sections.** In exceptional cases the intersection of a plane and a cone is a point, a line, or a pair of lines.

Let's begin with the geometric definition of a circle.

> A **circle** is the set of all points in a plane that are at a given distance from a fixed point. The fixed point is called the **center** of the circle and the given distance is called the **radius.**

Using the methods of analytic geometry, we place the center at a point (h, k) as in Figure 4. If $P(x, y)$ is a point on the circle, then the distance from P to the center (h, k) must be equal to the radius r. By the distance formula,

$$\sqrt{(x - h)^2 + (y - k)^2} = r$$

or

$$(x - h)^2 + (y - k)^2 = r^2$$

Since $P(x, y)$ is any point on the circle, we say that

$$(x - h)^2 + (y - k)^2 = r^2$$

is the **standard form of the equation of the circle** with center (h, k) and radius r.

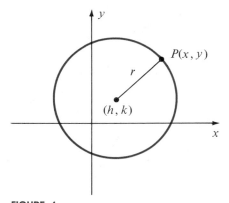

FIGURE 4

EXAMPLE 1

Write the equation of the circle with center at $(2, -5)$ and radius 3.

SOLUTION

Substituting $h = 2$, $k = -5$, and $r = 3$ in the equation

$$(x - h)^2 + (y - k)^2 = r^2$$

yields

$$(x - 2)^2 + (y + 5)^2 = 9$$

EXAMPLE 2

Find the center and radius of the circle whose equation is

$$(x + 1)^2 + (y - 3)^2 = 4$$

SOLUTION

Since the standard form is

$$(x - h)^2 + (y - k)^2 = r^2$$

we must have

$$x - h = x + 1 \qquad y - k = y - 3 \qquad r^2 = 4$$

Solving, we find that

$$h = -1 \qquad k = 3 \qquad r = 2$$

The center is at $(-1, 3)$ and the radius is 2.

PROGRESS CHECK

Find the center and radius of the circle whose equation is

$$(x - \tfrac{1}{2})^2 + (y + 5)^2 = 15$$

ANSWER

center $(\tfrac{1}{2}, -5)$, radius $\sqrt{15}$

When we are given the equation of a circle in the **general form**

$$Ax^2 + Ay^2 + Dx + Ey + F = 0, \quad A \neq 0$$

in which the coefficients of x^2 and y^2 are the same, we may rewrite the equation in standard form. The process involves completing the square in each variable. (If necessary, review Section 2.6.)

EXAMPLE 3

Write the equation of the circle $2x^2 + 2y^2 - 12x + 16y - 31 = 0$ in standard form.

SOLUTION

Grouping the terms in x and y and factoring produces

$$2(x^2 - 6x) + 2(y^2 + 8y) = 31$$

Completing the square in both x and y, we have

$$2(x^2 - 6x + 9) + 2(y^2 + 8y + 16) = 31 + 18 + 32$$
$$2(x - 3)^2 + 2(y + 4)^2 = 81$$

Note that the quantities 18 and 32 were added to the right-hand side because each factor is multiplied by 2. The last equation can be written as

$$(x - 3)^2 + (y + 4)^2 = \frac{81}{2}$$

This is the standard form of the equation of the circle with center at $(3, -4)$ and radius $9\sqrt{2}/2$.

PROGRESS CHECK

Write the equation of the circle $4x^2 + 4y^2 - 8x + 4y = 103$ in standard form, and determine the center and radius.

ANSWER

$(x - 1)^2 + \left(y + \dfrac{1}{2}\right)^2 = 27$, center $\left(1, -\dfrac{1}{2}\right)$, radius $\sqrt{27}$

EXAMPLE 4

Write the equation $3x^2 + 3y^2 - 6x + 15 = 0$ in standard form.

SOLUTION

Regrouping, we have

$$3(x^2 - 2x) + 3y^2 = -15$$

We then complete the square in x and y:

$$3(x^2 - 2x + 1) + 3y^2 = -15 + 3$$
$$3(x - 1)^2 + 3y^2 = -12$$
$$(x - 1)^2 + y^2 = -4$$

Since $r^2 = -4$ is an impossible situation, the graph of the equation is not a circle. Note that the left-hand side of the equation in standard form is a sum of squares and is therefore nonnegative, while the right-hand side is negative. Thus, there are no real values of x and y that satisfy the equation. This is an example of an equation that does not have a graph!

PROGRESS CHECK

Write the equation $x^2 + y^2 - 12y + 36 = 0$ in standard form, and analyze its graph.

ANSWER

The standard form is $x^2 + (y - 6)^2 = 0$. The equation is that of a "circle" with center at $(0, 6)$ and radius of 0. The "circle" is actually the point $(0, 6)$.

EXERCISE SET 9.2

In each of the following find an equation of the circle with center at (h, k) and radius r.

1. $(h, k) = (2, 3)$, $r = 2$
2. $(h, k) = (-3, 0)$, $r = 3$
3. $(h, k) = (-2, -3)$, $r = \sqrt{5}$
4. $(h, k) = (2, -4)$, $r = 4$
5. $(h, k) = (0, 0)$, $r = 3$
6. $(h, k) = (0, -3)$, $r = 2$
7. $(h, k) = (-1, 4)$, $r = 2\sqrt{2}$
8. $(h, k) = (2, 2)$, $r = 2$

In each of the following find the center and radius of the circle with the given equation.

9. $(x - 2)^2 + (y - 3)^2 = 16$
10. $(x + 2)^2 + y^2 = 9$
11. $(x - 2)^2 + (y + 2)^2 = 4$
12. $\left(x + \dfrac{1}{2}\right)^2 + (y - 2)^2 = 8$
13. $(x + 4)^2 + \left(y + \dfrac{3}{2}\right)^2 = 18$
14. $x^2 + (y - 2)^2 = 4$
15. $\left(x - \dfrac{1}{3}\right)^2 + y^2 = -\dfrac{1}{9}$
16. $(x - 1)^2 + \left(y - \dfrac{1}{2}\right)^2 = 3$

Write the equation of each given circle in standard form, and determine the center and radius if they exist.

17. $x^2 + y^2 + 4x - 8y + 4 = 0$
18. $x^2 + y^2 - 2x + 6y - 15 = 0$
19. $2x^2 + 2y^2 - 6x - 10y + 6 = 0$
20. $2x^2 + 2y^2 + 8x - 12y - 8 = 0$
21. $2x^2 + 2y^2 - 4x - 5 = 0$
22. $4x^2 + 4y^2 - 2y + 7 = 0$
23. $3x^2 + 3y^2 - 12x + 18y + 15 = 0$
24. $4x^2 + 4y^2 + 4x + 4y - 4 = 0$

Write the given equation in standard form, and determine if the graph of the equation is a circle, a point, or neither.

25. $x^2 + y^2 - 6x + 8y + 7 = 0$
26. $x^2 + y^2 + 4x + 6y + 5 = 0$
27. $x^2 + y^2 + 3x - 5y + 7 = 0$
28. $x^2 + y^2 - 4x - 6y - 13 = 0$
29. $2x^2 + 2y^2 - 12x - 4 = 0$
30. $2x^2 + 2y^2 + 4x - 4y + 25 = 0$
31. $2x^2 + 2y^2 - 6x - 4y - 2 = 0$
32. $2x^2 + 2y^2 - 10y + 6 = 0$
33. $3x^2 + 3y^2 + 12x - 4y - 20 = 0$
34. $x^2 + y^2 + x + y = 0$
35. $4x^2 + 4y^2 + 12x - 20y + 38 = 0$
36. $4x^2 + 4y^2 - 12x - 36 = 0$

9.3
THE PARABOLA

We begin our study of the parabola with the geometric definition.

A **parabola** is the set of all points that are equidistant from a given point and a given line.

The given point is called the **focus** and the given line is called the **directrix** of the parabola. In Figure 5 all points P on the parabola are equidistant from the focus F and the directrix L, that is, $\overline{PF} = \overline{PQ}$. The line through the focus that is perpendicular to the directrix is called the **axis of the parabola** (or simply the **axis**), and the parabola is seen to be symmetric with respect to the axis. The point V (Figure 5), where the parabola intersects its axis, is called the **vertex** of the parabola. The vertex, then, is the point from which the parabola opens. Note that the vertex is the point on the parabola that is closest to the directrix.

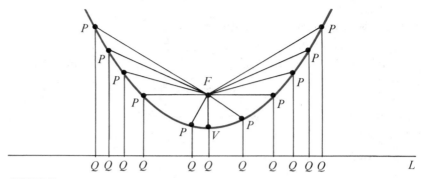

FIGURE 5

We can apply the methods of analytic geometry to find an equation of the parabola. We choose the y-axis as the axis of the parabola and the origin as the vertex (Figure 6). Since the vertex is on the parabola, it is equidistant from the focus and the directrix. Thus, if the coordinates of the focus F are $(0, p)$, then the equation of the directrix is $y = -p$. We then let $P(x, y)$ be any point on the parabola, and we equate the distance from P to the focus F and the distance from P to the directrix L. Using the distance formula,

$$\overline{PF} = \overline{PQ}$$
$$\sqrt{(x - 0)^2 + (y - p)^2} = \sqrt{(x - x)^2 + (y + p)^2}$$

Squaring both sides,

$$x^2 + y^2 - 2py + p^2 = y^2 + 2py + p^2$$
$$x^2 = 4py$$

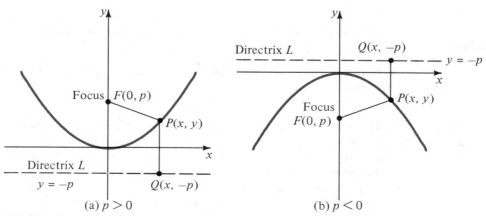

FIGURE 6

We have obtained an important form of the equation of a parabola.

Standard Form of the Equation of a Parabola	$$x^2 = 4py$$ is the standard form of the equation of a parabola whose vertex is at the origin, whose focus is at $(0, p)$, and whose axis is vertical.

Conversely, it can be shown that the graph of the equation $x^2 = 4py$ is a parabola. Note that substituting $-x$ for x leaves the equation unchanged, verifying symmetry with respect to the y-axis. If $p > 0$, the parabola opens upward as shown in Figure 6a, while if $p < 0$, the parabola opens downward, as shown in Figure 6b.

EXAMPLE 1
Determine the focus and directrix of the parabola $x^2 = 8y$, and sketch its graph.

SOLUTION
The equation of the parabola is of the form

$$x^2 = 4py = 8y$$

so $p = 2$. The equation of the directrix is $y = -p = -2$, and the focus is at $(0, p) = (0, 2)$. Since $p > 0$, the parabola opens upward. The graph of the parabola is shown in Figure 7.

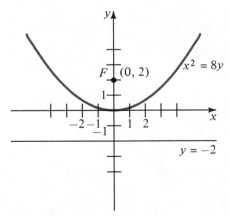

FIGURE 7

PROGRESS CHECK
Determine the focus and directrix of the parabola $x^2 = -3y$.

ANSWER
focus at $(0, -\frac{3}{4})$, directrix $y = \frac{3}{4}$

EXAMPLE 2
Find the equation of the parabola with vertex at $(0, 0)$ and focus at $(0, -\frac{3}{2})$.

SOLUTION
Since the focus is at $(0, p)$, we have $p = -\frac{3}{2}$. The equation of the parabola is
$$x^2 = 4py = 4(-\tfrac{3}{2}y) = -6y$$

PROGRESS CHECK
Find the equation of the parabola with vertex at $(0, 0)$ and focus at $(0, 3)$.

ANSWER
$x^2 = 12y$

If we place the parabola as shown in Figure 8, we can proceed as above to obtain the following result.

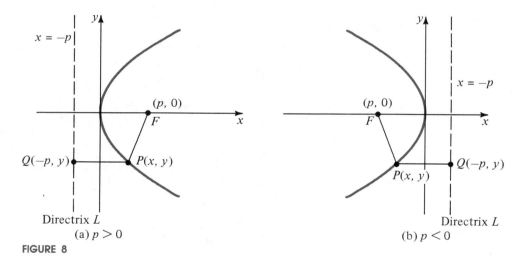

(a) $p > 0$ (b) $p < 0$

FIGURE 8

Standard Form of the Equation of a Parabola	$$y^2 = 4px$$ is the standard form of the equation of a parabola whose vertex is at the origin, whose focus is at $(p, 0)$, and whose axis is horizontal.

Note that substituting $-y$ for y leaves this equation unchanged, verifying symmetry with respect to the x-axis. If $p > 0$, the parabola opens to the right, as shown in Figure 8a, while if $p < 0$, the parabola opens to the left, as shown in Figure 8b.

**DEVICES WITH A
PARABOLIC SHAPE**

The properties of the parabola are used in the design of some important devices. For example, by rotating a parabola about its axis, we obtain a **parabolic reflector,** a shape used in the headlight of an automobile. In the accompanying figure, the light source (the bulb) is placed at the focus of the parabola. The headlight is coated with a reflecting material, and the rays of light bounce back in lines that are parallel to the axis of the parabola. This permits a headlight to disperse light in front of the auto where it is needed.

A reflecting telescope reverses the use of these same properties. Here, the rays of light from a distant star, which are nearly parallel to the axis of the parabola, are reflected by the mirror to the focus (see accompanying figure). The eyepiece is placed at the focus, where the rays of light are gathered.

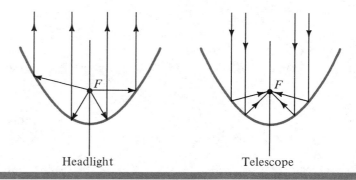

Headlight   Telescope

EXAMPLE 3

Find the equation of the parabola with vertex at $(0, 0)$ and directrix $x = \frac{1}{2}$.

SOLUTION

The directrix is $x = -p$, so $p = -\frac{1}{2}$. The equation of the parabola is then

$$y^2 = 4px = -2x$$

EXAMPLE 4

Find the equation of the parabola that has the x-axis as its axis, has vertex at $(0, 0)$, and passes through the point $(-2, 3)$.

SOLUTION

Since the axis of the parabola is the x-axis, the equation of the parabola is $y^2 = 4px$. The parabola passes through the point $(-2, 3)$, so the coordinates of this point must satisfy the equation of the parabola. Thus,

$$y^2 = 4px$$
$$(3)^2 = 4p(-2)$$
$$4p = -\frac{9}{2}$$

and the equation of the parabola is

$$y^2 = 4px = -\frac{9}{2}x$$

PROGRESS CHECK

Find the equation of the parabola that has the y-axis as its axis, has vertex at $(0, 0)$, and passes through the point $(1, -2)$.

ANSWER

$$x^2 = -\frac{1}{2}y$$

EXERCISE SET 9.3

In Exercises 1–8 determine the focus and directrix of the given parabola, and sketch the graph.

1. $x^2 = 4y$
2. $x^2 = -4y$
3. $y^2 = 2x$
4. $y^2 = -\frac{3}{2}x$

5. $x^2 + 5y = 0$
6. $2y^2 - 3x = 0$
7. $y^2 - 12x = 0$
8. $x^2 - 9y = 0$

In Exercises 9–20 determine the equation of the parabola that has its vertex at the origin and that satisfies the given conditions.

9. Focus at $(1, 0)$

10. Focus at $(0, -3)$

11. Directrix $x = -\frac{3}{2}$

12. Directrix $y = \frac{5}{2}$

13. Axis is the x-axis, and parabola passes through the point $(2, 1)$.

14. Axis is the y-axis, and parabola passes through the point $(4, -2)$.

15. Axis is the x-axis, and $p = -\frac{5}{4}$.

16. Axis is the y-axis, and $p = 2$.

17. Focus at $(-1, 0)$ and directrix $x = 1$

18. Focus at $\left(0, -\frac{5}{2}\right)$ and directrix $y = \frac{5}{2}$

19. Axis is the x-axis, and parabola passes through the point $(4, 2)$.

20. Axis is the y-axis, and parabola passes through the point $(2, 4)$.

In Exercises 21–24 determine whether the given parabola opens upward, downward, to the left, or to the right.

21. $4x^2 + y = 0$

22. $4x^2 - y = 0$

23. $2x + y^2 = 0$

24. $2x - 5y^2 = 0$

9.4
THE ELLIPSE

The geometric definition of an ellipse is as follows.

An **ellipse** is the set of all points the sum of whose distances from two fixed points is a constant.

The fixed points are called the **foci** of the ellipse. An ellipse may be constructed in the following way. Place a thumbtack at each of the foci F_1 and F_2, and attach one end of a string to each of the thumbtacks. Hold a pencil tight against the string, as shown in Figure 9, and move the pencil. The point P will describe an ellipse, since the sum of the distances from P to the foci is always a constant, namely, the length of the string.

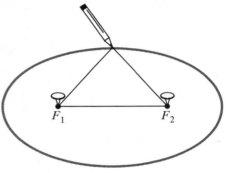

FIGURE 9

We will use Figure 10 to obtain an equation for an ellipse. The line segment joining the foci F_1 and F_2 is called the **major axis** of the ellipse, and the midpoint of the major axis is called the **center** of the ellipse. The line segment through the center and perpendicular to the major axis is called the **minor axis.** The points at which the ellipse intersects the major axis (V_1 and V_2 in Figure 10) are called the **vertices** of the ellipse.

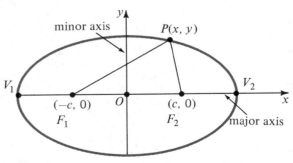

FIGURE 10

It is easiest to derive the equation of an ellipse when the center is at the origin and the major axis is one of the coordinate axes. We first consider the case when the major axis is the x-axis. If the focus F_2 is at $(c, 0)$, then the other focus F_1 is at $(-c, 0)$, as in Figure 10. Let $P(x, y)$ be a point on the ellipse, and let the constant sum of the distances from P to the foci be denoted by $2a$. Then we have

$$\overline{F_1P} + \overline{F_2P} = 2a$$

Using the distance formula, we may rewrite this as

$$\sqrt{(x - c)^2 + (y - 0)^2} + \sqrt{(x + c)^2 + (y - 0)^2} = 2a$$

or

$$\sqrt{(x - c)^2 + y^2} = 2a - \sqrt{(x + c)^2 + y^2}$$

Squaring both sides, we obtain

$$(x - c)^2 + y^2 = 4a^2 - 4a\sqrt{(x + c)^2 + y^2} + (x + c)^2 + y^2$$
$$x^2 - 2cx + c^2 + y^2 = 4a^2 - 4a\sqrt{(x + c)^2 + y^2} + x^2 + 2cx + c^2 + y^2$$

Simplifying, we have

$$a\sqrt{(x + c)^2 + y^2} = a^2 + cx$$

Squaring both sides, we now have

$$a^2[(x + c)^2 + y^2] = (a^2 + cx)^2$$
$$a^2[x^2 + 2cx + c^2 + y^2] = a^4 + 2a^2cx + c^2x^2$$
$$a^2x^2 + 2a^2cx + a^2c^2 + a^2y^2 = a^4 + 2a^2cx + c^2x^2$$
$$(a^2 - c^2)x^2 + a^2y^2 = a^2(a^2 - c^2) \tag{1}$$

Since the vertex V_1 lies on the ellipse and since the sum of the distances from any point on the ellipse to the foci is $2a$, we must have

$$\overline{F_1V_1} + \overline{F_2V_1} = 2a$$

From Figure 10,

$$\overline{F_1F_2} = 2c$$

and

$$\overline{F_1V_1} + \overline{F_2V_1} > \overline{F_1F_2}$$

Substituting the values from the above equations yields

$$2a > 2c$$

so $a > c$ and, therefore, $a^2 > c^2$. Dividing both sides of Equation (1) by $a^2(a^2 - c^2)$, we obtain

$$\frac{x^2}{a^2} + \frac{a^2y^2}{a^2 - c^2} = 1$$

WHISPERING GALLERIES	The domed roof in the accompanying figure has the shape of an ellipse that has been rotated about its major axis. It can be shown, using basic laws of physics, that a sound uttered at one focus will be reflected to the other focus, where it will be clearly heard. This property of such rooms is known as the "whispering gallery effect."

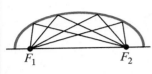

Famous whispering galleries include the dome of St. Paul's Cathedral, London; St. John Lateran, Rome; the Salle des Cariatides in the Louvre, Paris; and the original House of Representatives (now the National Statuary Hall in the United States Capitol), Washington, D.C.

Letting

$$b^2 = a^2 - c^2, \quad b > 0 \tag{2}$$

we obtain the **standard form** of the equation of an ellipse with center at $(0, 0)$ and foci at $(-c, 0)$ and $(c, 0)$.

Standard Form of the Equation of an Ellipse	$$\frac{x^2}{a^2} + \frac{y^2}{b^2} = 1, \quad b \le a \tag{3}$$

Conversely, we can show that the graph of an equation of the form given in Equation (3) is an ellipse with center at $(0, 0)$ and foci at $(-c, 0)$ and $(c, 0)$.

The vertices of the ellipse are the x-intercepts of the ellipse, and these values may be obtained by letting $y = 0$ in Equation (3). We have $x^2/a^2 = 1$, so $x = \pm a$. Thus, V_1 is the point $(-a, 0)$ and V_2 is the point $(a, 0)$; the length of the major axis is $2a$ (Figure 11). When $x = 0$, we obtain $y = \pm b$, so the ellipse intersects the y-axis at the points $(0, -b)$ and $(0, b)$. Thus, the length of the minor axis is $2b$. Observe also that since $c \ge 0$, $b^2 \le a^2$, so $b \le a$. Thus, the major axis is longer than the minor axis. Figure 11 shows the ellipse given by Equation (3), along with the relationships between a, b, and c.

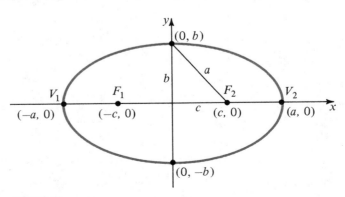

FIGURE 11

Since Equation (3) remains unchanged when we replace x with $-x$, or y with $-y$, we conclude that the ellipse is symmetric with respect to both the x-axis and the y-axis and, therefore, with respect to the origin.

EXAMPLE 1

Discuss and sketch the graph of the equation

$$9x^2 + 25y^2 = 225$$

SOLUTION

We obtain the standard form by dividing both sides by 225, obtaining

$$\frac{x^2}{25} + \frac{y^2}{9} = 1$$

which is in the form of Equation (3). This is the equation of an ellipse whose major axis is the x-axis. We have $a^2 = 25$ and $b^2 = 9$, so $a = 5$ and $b = 3$. Since $c^2 = a^2 - b^2$, we have

$$c^2 = 25 - 9 = 16$$

so $c = 4$. Thus, the foci are at $(-4, 0)$ and $(4, 0)$. The vertices are at $(-5, 0)$ and $(5, 0)$. The ellipse is sketched in Figure 12.

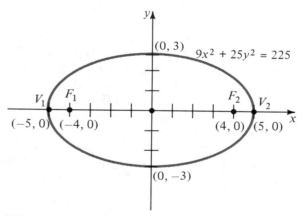

FIGURE 12

EXAMPLE 2

Find the equation of an ellipse with center at $(0, 0)$, a focus at $(-3, 0)$, and a vertex at $(6, 0)$.

SOLUTION

Since a focus is at $(-3, 0)$ and the center is at the origin, $c = 3$. Since a vertex is at $(6, 0)$, $a = 6$. We obtain b^2 by using Equation (2):

$$b^2 = a^2 - c^2 = 6^2 - 3^2 = 27$$

Then the equation of the ellipse is

$$\frac{x^2}{36} + \frac{y^2}{27} = 1$$

When the major axis of an ellipse is located on the y-axis and the center of the ellipse is at the origin, the foci F_1 and F_2 are at $(0, -c)$ and $(0, c)$ as shown in Figure 13. If the constant sum of the distances is again denoted by $2a$, a similar derivation shows that the standard form of the equation of such an ellipse is as follows.

Standard Form of the Equation of an Ellipse	$\dfrac{x^2}{b^2} + \dfrac{y^2}{a^2} = 1, \quad b \le a$	(4)

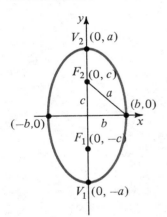

FIGURE 13

The length of the major axis is $2a$, and the length of the minor axis is $2b$. Thus, *the major axis is always the longer axis.*

EXAMPLE 3

Discuss and sketch the graph of the equation

$$9x^2 + 4y^2 = 36$$

SOLUTION

We obtain the standard form by dividing both sides by 36, obtaining

$$\frac{x^2}{4} + \frac{y^2}{9} = 1$$

which is of the form in Equation (4). Hence, the graph of this equation is an ellipse. We have $a^2 = 9$ and $b^2 = 4$, so $a = 3$ and $b = 2$. From Equation (2) we have

$$c^2 = a^2 - b^2 = 9 - 4 = 5$$
$$c = \sqrt{5}$$

The major axis of this ellipse lies along the y-axis; the foci are at $(0, -\sqrt{5})$ and $(0, \sqrt{5})$; the vertices are at $(0, -3)$ and $(0, 3)$. The ellipse is sketched in Figure 14.

PROGRESS CHECK

Discuss and sketch the graph of the equation $9x^2 + y^2 = 9$, indicating the foci and vertices.

ANSWER

See Figure 15.

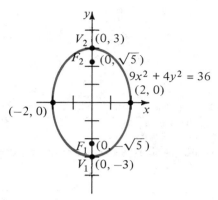

FIGURE 14

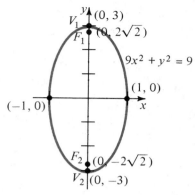

FIGURE 15

EXAMPLE 4

Is the major axis of the ellipse

$$25x^2 + 4y^2 = 100$$

along the x-axis or along the y-axis?

SOLUTION

We first obtain the standard form of the equation by dividing both sides by 100:

$$\frac{x^2}{4} + \frac{y^2}{25} = 1$$

Since we must have $a \geq b$, the equation is of the form

$$\frac{x^2}{b^2} + \frac{y^2}{a^2} = 1$$

and the major axis is along the y-axis.

If $a = b$, then the equation of the ellipse becomes

$$\frac{x^2}{a^2} + \frac{y^2}{a^2} = 1$$

or

$$x^2 + y^2 = a^2$$

which we identify as a circle with center at the origin and radius a. The circle is seen to be a special case of an ellipse in which the major and minor axes are equal.

EXERCISE SET 9.4

In Exercises 1–10 sketch and discuss the graph of the given equation, giving the coordinates of the foci and vertices.

1. $25x^2 + 16y^2 - 400 = 0$

2. $x^2 + 4y^2 - 16 = 0$

3. $9x^2 + y^2 - 36 = 0$

4. $3x^2 + y^2 - 81 = 0$

5. $25x^2 + 64y^2 - 1600 = 0$

6. $9x^2 + y^2 - 36 = 0$

7. $4x^2 + y^2 - 9 = 0$

8. $5x^2 + 12y^2 - 45 = 0$

9. $2x^2 + 10y^2 - 40 = 0$

10. $25x^2 + 169y^2 - 4225 = 0$

In Exercises 11–24 find an equation of the ellipse satisfying the given conditions.

11. Center at $(0, 0)$, focus at $(0, 3)$, and vertex at $(0, 7)$

12. Center at $(0, 0)$, focus at $(2, 0)$, and vertex at $(-4, 0)$

13. Foci at $(0, \pm\sqrt{5})$ and vertices at $(0, \pm 3)$

14. Foci at $(0, \pm\sqrt{7})$ and vertices at $(0, \pm 4)$

15. Foci at $(\pm\sqrt{15}, 0)$ and vertex at $(4, 0)$

16. Foci at $(\pm 2\sqrt{6}, 0)$ and vertex at $(-5, 0)$

17. Focus at $(0, 2\sqrt{2})$ and vertices at $(0, \pm 3)$

18. Focus at $(0, -2\sqrt{15})$ and vertices at $(0, \pm 8)$

19. Foci at $(\pm 4, 0)$ and vertices at $(\pm 5, 0)$

20. Foci at $(\pm 5, 0)$ and vertices at $(\pm 6, 0)$

21. Foci at $(\pm 4\sqrt{5}, 0)$, length of major axis is 18

22. Foci at $(\pm 6, 0)$, length of minor axis is $\sqrt{13}$

23. Vertices at $(\pm 7, 0)$ and passing through the point
$\left(1, \dfrac{6}{7}\sqrt{3}\right)$.

24. Vertices at $(0, \pm 1)$ and passing through the point
$\left(\dfrac{1}{4}, \dfrac{\sqrt{3}}{2}\right)$.

In Exercises 25–28 determine whether the major axis of the given ellipse lies along the x-axis or along the y-axis.

25. $7x^2 + 4y^2 - 81 = 0$

26. $16x^2 + y^2 - 2 = 0$

27. $x^2 + 4y^2 - 6 = 0$

28. $4x^2 + 5y^2 - 1 = 0$

29. The **eccentricity** of an ellipse is defined as the ratio c/a. Describe the general shape of the ellipse when its eccentricity is (a) almost 1, (b) almost 0.

9.5
THE HYPERBOLA

The hyperbola is the remaining conic section that we will consider in this chapter.

A **hyperbola** is the set of all points the difference of whose distances from two fixed points is a positive constant.

The two fixed points are called the **foci** of the hyperbola. The line through the foci is called the **transverse axis,** the midpoint of the line segment between the foci is called the **center,** and the line through the center and perpendicular to the transverse axis is called the **conjugate axis.** The two separate parts of the hyperbola are called its **branches** (Figure 16).

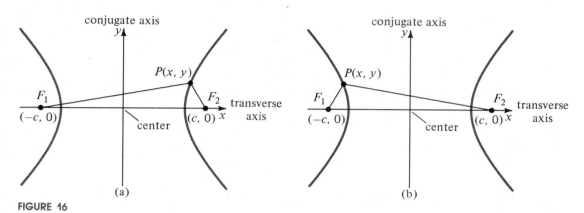

FIGURE 16

It is easiest to derive the equation of a hyperbola when the center is at the origin and the transverse axis is one of the coordinate axes. If the foci lie on the x-axis and one focus F_2 is at $(c, 0)$, $c > 0$, then the other focus F_1 is at $(-c, 0)$. (See Figures 16a and 16b.) Let $P(x, y)$ be a point on the hyperbola, and let the constant difference of the distances from P to the foci be denoted by $2a$. If P is on the right branch, we have

$$\overline{PF}_1 - \overline{PF}_2 = 2a \tag{1a}$$

whereas if P is on the left branch, we have

$$\overline{PF}_2 - \overline{PF}_1 = 2a \tag{1b}$$

Both of these equations can be expressed by the single equation

$$\left|\overline{PF}_1 - \overline{PF}_2\right| = 2a$$

or

$$\left|\sqrt{(x + c)^2 + y^2} - \sqrt{(x - c)^2 + y^2}\right| = 2a \tag{2}$$

Squaring both sides of Equation (2), we have

$$(x + c)^2 + y^2 - 2\sqrt{[(x + c)^2 + y^2][(x - c)^2 + y^2]} + (x - c)^2 + y^2 = 4a^2$$

We next expand this equation, cancel terms, isolate the radical on the left, and square both sides, obtaining

$$(c^2 - a^2)x^2 - a^2y^2 = a^2(c^2 - a^2)$$

We now focus on triangle F_1PF_2 in Figure 16a. Since the sum of the lengths of two sides of a triangle is always greater than the length of the third side, we have

$$\overline{PF_1} < \overline{PF_2} + \overline{F_1F_2}$$
$$\overline{PF_1} - \overline{PF_2} < \overline{F_1F_2}$$
$$2a < 2c \qquad \text{by Equation (1a)}$$

so $a < c$. Similarly, if P is on the left branch (Figure 16b), we again find that $a < c$. Thus, $a^2 < c^2$. Dividing both sides of Equation (3) by $a^2(c^2 - a^2)$, we obtain

$$\frac{x^2}{a^2} - \frac{y^2}{c^2 - a^2} = 1$$

Letting

$$b^2 = c^2 - a^2 > 0 \tag{4}$$

we obtain the standard form of the equation of a hyperbola with center at $(0, 0)$ and foci at $(-c, 0)$ and $(c, 0)$:

Standard Form of the Equation of a Hyperbola

$$\frac{x^2}{a^2} - \frac{y^2}{b^2} = 1 \tag{5}$$

Conversely, it can be shown that the graph of an equation of the form given in Equation (5) is a hyperbola with center at $(0, 0)$ and foci at $(-c, 0)$ and $(c, 0)$.

The **vertices** of the hyperbola are the x-intercepts of the hyperbola, and these values may be obtained by letting $y = 0$ in Equation (5). We have

$$\frac{x^2}{a^2} = 1 \text{ or } x = \pm a$$

Thus, V_1 is the point $(-a, 0)$ and V_2 is the point $(a, 0)$ (Figure 17a). When $x = 0$, we obtain the equation $-\frac{y^2}{b^2} = 1$, which has no solutions, so there are no y-intercepts. Since Equation (5) remains unchanged when we replace x with $-x$ and y with $-y$, we conclude that the hyperbola is symmetric with respect to both the x-axis and the y-axis and, therefore, with respect to the origin.

To see why there are no points on the hyperbola when $|x| < a$, we proceed as follows. Solve Equation (5) for y, obtaining

$$y = \pm\frac{b}{a}\sqrt{x^2 - a^2}$$

The radical is defined only if $x^2 - a^2 \geq 0$. That is, there is no value of y when $x^2 < a^2$ or when $|x| < a$. Thus, there are no points on the hyperbola for

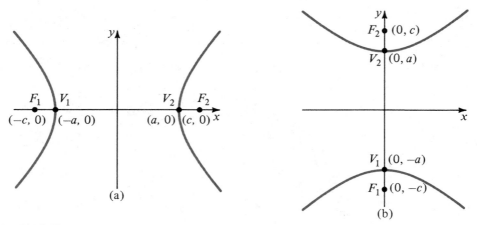

FIGURE 17

$-a < x < a$. Since $x^2 - a^2 \geq 0$ when $x \geq a$ or $x \leq -a$, we obtain the two branches shown in Figure 17a.

When the foci of a hyperbola are located on the y-axis and the center of the hyperbola is at the origin, the foci are at $(0, -c)$ and $(0, c)$ as shown in Figure 17b. If the constant difference of the distances is again denoted by $2a$, a similar derivation shows that the standard form of the equation of such a hyperbola is as follows:

Standard Form of the Equation of a Hyperbola	$$\dfrac{y^2}{a^2} - \dfrac{x^2}{b^2} = 1$$	(6)

EXAMPLE 1

Discuss and sketch the graph of the equation $4x^2 - 9y^2 = 36$.

SOLUTION

We obtain the standard form by dividing both sides by 36, which yields

$$\frac{x^2}{9} - \frac{y^2}{4} = 1$$

which is of the form in Equation (5). This is the equation of a hyperbola whose foci lie on the x-axis. We have $a^2 = 9$ and $b^2 = 4$, so $a = 3$, and the vertices are at $(-3, 0)$ and $(3, 0)$. From Equation (4) we obtain

$$c^2 = a^2 + b^2 = 9 + 4 = 13$$

Then $c = \pm\sqrt{13}$, and the foci are at $F_1(-\sqrt{13}, 0)$ and $F_2(\sqrt{13}, 0)$. The hyperbola is sketched in Figure 18a.

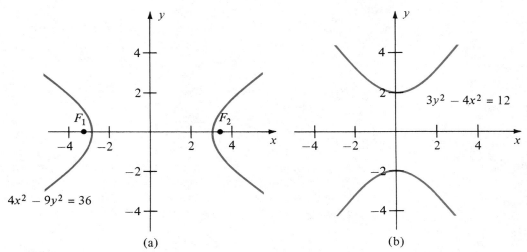

FIGURE 18

EXAMPLE 2

Discuss and sketch the graph of the equation $3y^2 - 4x^2 = 12$.

SOLUTION

We obtain the standard form by dividing both sides by 12, which yields

$$\frac{y^2}{4} - \frac{x^2}{3} = 1$$

which is of the form in Equation (6). This is the equation of a hyperbola whose foci lie on the y-axis. We have $a^2 = 4$ and $b^2 = 3$, so $a = 2$, and the vertices are at $(0, -2)$ and $(0, 2)$. From Equation (4) we obtain

$$c^2 = a^2 + b^2 = 4 + 3 = 7$$

so $c = \pm\sqrt{7}$. The foci are at $(0, -\sqrt{7})$ and $(0, \sqrt{7})$. The hyperbola is sketched in Figure 18b.

PROGRESS CHECK

Discuss the graph of the equation $16x^2 - y^2 = 16$.

ANSWER

Hyperbola; $a = 1$, $b = 4$, vertices at $(-1, 0)$ and $(1, 0)$, foci at $(-\sqrt{17}, 0)$ and $(\sqrt{17}, 0)$.

EXAMPLE 3

Find the equation of a hyperbola with center at $(0, 0)$, a focus at $(0, 4)$, and a vertex at $(0, 3)$.

SOLUTION

Since a focus is at (0, 4), the foci lie on the y-axis, and $c = 4$. Since a vertex is at (0, 3), $a = 3$. We obtain b^2 by using Equation (4):

$$b^2 = c^2 - a^2 = 16 - 9 = 7$$

Then the equation of the hyperbola is

$$\frac{y^2}{9} - \frac{x^2}{7} = 1$$

ASYMPTOTES

There is a feature of the hyperbola that distinguishes it from the parabola and the ellipse. By examining the corresponding values of y for values of x that are very far away from the origin, we obtain certain lines that enable us to sketch a given hyperbola readily. Solving Equation (5) for y, we obtain

$$y = \pm\frac{b}{a}\sqrt{x^2 - a^2}$$

$$y = \pm\frac{b}{a}\sqrt{x^2\left(1 - \frac{a^2}{x^2}\right)}$$

$$y = \pm\frac{b}{a}x\sqrt{1 - \frac{a^2}{x^2}} \tag{7}$$

For large values of $|x|$, the term a^2/x^2 in Equation (7) is very small, and the graph of Equation (5) approaches the asymptotes

$$y = \pm\frac{b}{a}x$$

A similar discussion shows that the asymptotes of the hyperbola given by Equation (6) are

$$y = \pm\frac{a}{b}x$$

The asymptotes are shown in Figure 19.

An easy way to obtain the asymptotes of a hyperbola whose equation is in standard form is to replace the 1 on the right side with 0 and then to solve for y in terms of x.

To summarize,

Asymptotes of the Hyperbola	$\dfrac{x^2}{a^2} - \dfrac{y^2}{b^2} = 1$ has asymptotes $y = \pm\dfrac{b}{a}x$
	$\dfrac{y^2}{a^2} - \dfrac{x^2}{b^2} = 1$ has asymptotes $y = \pm\dfrac{a}{b}x$

The asymptotes of a hyperbola can be readily constructed. We consider the hyperbola given by Equation (5). Plot the points (a, b), $(-a, b)$, $(-a, -b)$, and $(a, -b)$, which form a rectangle. The lines determined by the diagonals of this rectangle have slopes b/a and $-b/a$ (Figure 19) and are thus the asymptotes of the hyperbola. The hyperbola is then sketched by using the asymptotes as an aid.

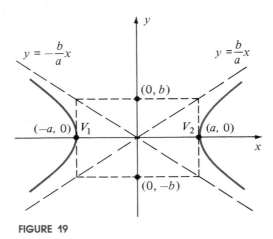

FIGURE 19

EXAMPLE 4
Find the asymptotes of the hyperbola

$$\frac{y^2}{9} - \frac{x^2}{4} = 1$$

SOLUTION
Setting the right-hand side equal to 0 and solving for y, we find that

$$y = \pm\frac{3}{2}x$$

so the asymptotes are

$$y = \frac{3}{2}x \quad \text{and} \quad y = -\frac{3}{2}x$$

EXAMPLE 5
Discuss and sketch the graph of the equation $25x^2 - 4y^2 = 100$, including the asymptotes of the hyperbola.

SOLUTION

We first write the standard form of the equation by dividing both sides by 100, obtaining

$$\frac{x^2}{4} - \frac{y^2}{25} = 1 \tag{8}$$

Then $a^2 = 4$ and $b^2 = 25$, so $a = 2$ and $b = 5$. The asymptotes are obtained by replacing the 1 on the right side of (8) with 0 and solving for y, obtaining

$$y = \pm\frac{5}{2}x$$

To construct the asymptotes, we first plot the four points $(2, 5)$, $(-2, 5)$, $(-2, -5)$, and $(2, -5)$. The diagonals of the rectangle determined by these four points determine the asymptotes of the hyperbola, which can then be sketched by using the asymptotes as aids. The graph is shown in Figure 20.

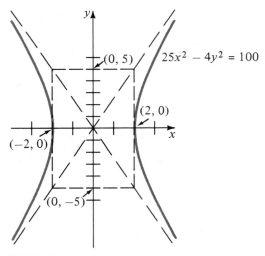

FIGURE 20

PROGRESS CHECK

Discuss and sketch the graph of the equation $4x^2 - 9y^2 = 144$, including the asymptotes of the hyperbola.

ANSWER

See Figure 21.

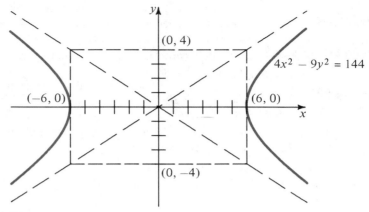

FIGURE 21

EXERCISE SET 9.5

In Exercises 1–8 sketch the graph of the given equation, giving the coordinates of the foci and vertices and including the asymptotes.

1. $x^2 - 9y^2 - 9 = 0$ 2. $9x^2 - y^2 - 36 = 0$ 3. $9y^2 - 4x^2 - 36 = 0$ 4. $9x^2 - 16y^2 - 144 = 0$
5. $x^2 - y^2 - 9 = 0$ 6. $y^2 - 4x^2 - 100 = 0$ 7. $-x^2 + 4y^2 - 100 = 0$ 8. $x^2 - 2y^2 + 18 = 0$

In Exercises 9–12 determine whether the foci of the given hyperbola lie on the x-axis or on the y-axis.

9. $2x^2 - 3y^2 - 5 = 0$ 10. $3x^2 - 3y^2 + 4 = 0$ 11. $y^2 - 4x^2 - 20 = 0$ 12. $4y^2 - 9x^2 + 36 = 0$

In Exercises 13–28 find the equation of the hyperbola satisfying the given conditions.

13. Center at (0, 0), focus at (0, 5), and vertex at (0, 4).

14. Center at (0, 0), focus at (0, 7), and vertex at (0, 5).

15. Foci at (±3, 0) and vertex at (2, 0).

16. Foci at (±4, 0) and vertex at (−1, 0).

17. Focus at (0, 9) and vertices at (0, ±1).

18. Focus at (0, 4) and vertices at (0, ±1).

19. Vertices at (0, ±3) and asymptote $y = x$.

20. Vertices at (±2, 0) and asymptote $y = -2x$.

21. Center at (0, 0), vertex at (0, −2), and asymptote $y = \frac{1}{2}x$.

22. Center at (0, 0), focus at (2, 0), and asymptote $y = -5x$.

23. Center at (0, 0), focus at (0, 3), and asymptote $y = 3x$.

24. Center at (0, 0), focus at (4, 0), and asymptote $y = x/5$.

25. Vertices at (0, ±4) and passing through the point (5, 5).

26. Vertices at (±2, 0) and passing through the point (3, 1).

27. Focus at (0, 3) and asymptotes $y = \pm x$.

28. Focus at (6, 0) and asymptotes $y = \pm 2x$.

29. The **eccentricity** of a hyperbola is defined as the quantity $e = c/a$. Since $c > a$, $e > 1$. Describe the general shape of a hyperbola for (a) e near 1 and (b) very large e.

9.6
TRANSLATION
OF AXES

In Section 2.4, horizontal and vertical shifts were used to aid in sketching the graph of a function. A technique known as translation of axes extends this idea to aid in analyzing and sketching graphs in general.

In Figure 22, the x and y coordinate axes are displayed; they intersect, as usual, at the origin O. Also displayed are an additional set of coordinate axes, x' and y', which are parallel to the x-axis and the y-axis, respectively, and which intersect at the point O'. We may think of the x'- and y'-axes as the result of shifting the x- and y-axes parallel to themselves until they intersect at the point O'. This process is called **translation of axes,** and we say that the x- and y-axes have been **translated.**

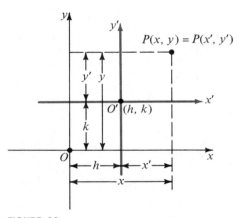

FIGURE 22

A point P in the plane has coordinates (x, y) with respect to the xy coordinate system and coordinates (x', y') with respect to the $x'y'$ coordinate system. There is a straightforward relationship between these pairs of coordinates. Let the origin O' of the $x'y'$ coordinate system have coordinates (h, k) with respect to the xy coordinate system. From Figure 22 we arrive at these useful formulas.

Translation of	$x = x' + h$ and $y = y' + k$	(1)
Axes Formulas	$x' = x - h$ and $y' = y - k$	(2)

EXAMPLE 1
The origin O' of the $x'y'$ coordinate system is at $(-2, 4)$.
(a) Express x' and y' in terms of x and y, respectively.
(b) Find the $x'y'$ coordinates of the point P whose xy coordinates are $(4, -6)$.

SOLUTION
(a) Substituting $h = -2$ and $k = 4$ in Equation (2), we obtain the translation formulas

$$x' = x + 2 \qquad y' = y - 4$$

(b) Substituting $x = 4$ and $y = -6$ yields

$$x' = 4 + 2 = 6 \qquad y' = -6 - 4 = -10$$

The $x'y'$ coordinates of P are $(6, -10)$.

PROGRESS CHECK
The origin O' of the $x'y'$ coordinate system is at $(-1, -2)$.
(a) Express x and y in terms of x' and y', respectively.
(b) Find the xy coordinates of the point P whose $x'y'$ coordinates are $(3, -3)$.

ANSWERS
(a) $x = x' - 1$, $y = y' - 2$ (b) $(2, -5)$

We can use the translation formulas to transform equations given in the xy coordinate system to equations in the $x'y'$ coordinate system, and vice versa. For example, in the $x'y'$ system, the equation of the circle with center at O' and radius r is

$$x'^2 + y'^2 = r^2$$

Substituting the translation formulas

$$x' = x - h \quad \text{and} \quad y' = y - k$$

we find that the equation of this circle in xy coordinates is

$$(x - h)^2 + (y - k)^2 = r^2$$

which is precisely the standard form of the equation of the circle with center at (h, k) and radius r, as discussed in Section 9.2.

EXAMPLE 2
Discuss and sketch the graph of the equation

$$x^2 - 4x + y^2 + 2y + 1 = 0$$

SOLUTION
We group the terms in x and in y

$$(x^2 - 4x \quad) + (y^2 + 2y \quad) = -1$$

in preparation for completing the square in each variable:

$$(x^2 - 4x + 4) + (y^2 + 2y + 1) = -1 + 4 + 1 = 4$$

(Note that the equation is balanced by adding $4 + 1$ to the right-hand side.) We then have

$$(x - 2)^2 + (y + 1)^2 = 4$$

which is the equation of a circle with center at $(2, -1)$ and radius 2. In terms of the $x'y'$ coordinate system with origin O' at $(2, -1)$, the equation becomes

$$x'^2 + y'^2 = 4$$

We see that the equation and analysis are simplified by translating the axes to the point $(h, k) = (2, -1)$ as shown in Figure 23.

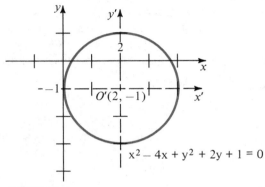

FIGURE 23

The technique of translation of axes can be applied to each of the conic sections. The results can be summarized as follows.

Standard Forms of the Conic Sections— Center or Vertex at (h, k)		
	Circle:	$(x - h)^2 + (y - k)^2 = r^2$ Center: (h, k) Radius: r
	Parabola:	$(x - h)^2 = 4p(y - k)$ Vertex: (h, k) Axis: $x = h$ $(y - k)^2 = 4p(x - h)$ Vertex: (h, k) Axis: $y = k$
	Ellipse:	$\dfrac{(x - h)^2}{a^2} + \dfrac{(y - k)^2}{b^2} = 1$ Center: (h, k)
	Hyperbola:	$\dfrac{(x - h)^2}{a^2} - \dfrac{(y - k)^2}{b^2} = 1$ $\dfrac{(y - k)^2}{a^2} - \dfrac{(x - h)^2}{b^2} = 1$ Center: (h, k)

If we write the equation of a conic in standard form, we can perform a translation of axes to the origin $O'(h, k)$ and then analyze and sketch the graph in the simplified form relative to the $x'y'$ coordinate system.

EXAMPLE 3

Sketch the graph of the equation

$$x^2 + 4x - 4y^2 + 24y - 48 = 0$$

SOLUTION

We rewrite the equation as

$$(x^2 + 4x \quad) - 4(y^2 - 6y \quad) = 48$$

and complete the square in both x and y.

$$(x^2 + 4x + 4) - 4(y^2 - 6y + 9) = 48 + 4 - 36$$
$$(x + 2)^2 - 4(y - 3)^2 = 16$$

Letting

$$x' = x + 2 \quad \text{and} \quad y' = y - 3$$

we rewrite the last equation as

$$x'^2 - 4y'^2 = 16$$

and after we divide both sides by 16,

$$\frac{x'^2}{16} - \frac{y'^2}{4} = 1$$

On the $x'y'$ coordinate system this is seen to be the equation of a hyperbola with center at $O'(-2, 3)$, $a = 4$, $b = 2$. Using the intercepts and asymptotes, the graph is sketched in Figure 24.

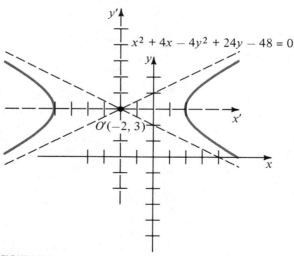

FIGURE 24

WARNING To complete the square in the equation

$$2(x^2 + 2x \qquad) - 3(y^2 - 4y \qquad) = 16$$

we must add 1 to the terms in x and 4 to the terms in y:

$$2(x^2 + 2x + 1) - 3(y^2 - 4y + 4) = 16 + 2 - 12$$

Note that adding 1 in the first parenthesis results in adding 2 to the left-hand side and is balanced by adding 2 to the right-hand side. Similarly, adding 4 in the second parenthesis results in adding -12 to the left-hand side, and this is also balanced on the right-hand side.

PROGRESS CHECK
(a)   Show that the graph of the equation

$$4x^2 + 16x + y^2 + 2y + 13 = 0$$

is an ellipse.
(b)   Show that the graph of the equation

$$4y^2 - 24y - 25x^2 - 50x - 89 = 0$$

is a hyperbola.

EXAMPLE 4
Sketch the graph of the equation

$$x^2 - 4x - 4y - 4 = 0$$

SOLUTION
We rewrite the equation as

$$(x^2 - 4x) = 4y + 4$$

and complete the square in x.

$$(x^2 - 4x + 4) = 4y + 4 + 4$$
$$(x - 2)^2 = 4y + 8 = 4(y + 2)$$

Letting

$$x' = x - 2 \quad \text{and} \quad y' = y + 2$$

we rewrite the last equation as

$$x'^2 = 4y'$$

On the $x'y'$ coordinate system this is seen to be the equation of a parabola with center at $O'(2, -2)$, $4p = 4$, so $p = 1$. The graph is sketched in Figure 25.

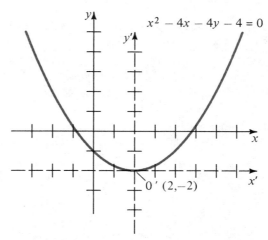

FIGURE 25

PROGRESS CHECK

Show that the graph of the equation

$$y^2 + 4y - 6x + 22 = 0$$

is a parabola.

It can be shown that the graph of the equation

$$Ax^2 + Cy^2 + Dx + Ey + F = 0$$

is a conic section—or a degenerate form of a conic section, such as a point, a line, a pair of lines, or no graph.

EXAMPLE 5

Identify the graph of the equation

$$x^2 - 4x + y^2 - 2y + 9 = 0$$

SOLUTION

We rewrite the equation as

$$(x^2 - 4x) + (y^2 - 2y) = -9$$

and complete the square in both x and y.

$$(x^2 - 4x + 4) + (y^2 - 2y + 1) = -9 + 4 + 1$$
$$(x - 2)^2 + (y + 1)^2 = -4$$

Since the sum of two squares is nonnegative, this equation has no graph.

EXERCISE SET 9.6

In Exercises 1–4 the origin O' of the $x'y'$ coordinate system is at $(-1, 4)$. Find the $x'y'$ coordinates of the point whose xy coordinates are given.

1. $(0, 0)$ 2. $(-2, 1)$ 3. $(4, 3)$ 4. $(-6, -2)$

In Exercises 5–8 the origin O' of the $x'y'$ coordinate system is at $(-3, 4)$. Find the xy coordinates of the point whose $x'y'$ coordinates are given.

5. $(0, 0)$ 6. $(-2, 1)$ 7. $(4, 3)$ 8. $(-6, -2)$

In Exercises 9–18 sketch the graph of the given equation.

9. $36x^2 - 100y^2 + 216x + 99 = 0$ 10. $x^2 - 4y^2 + 10x - 16y + 25 = 0$

11. $x^2 + 4x - y + 5 = 0$ 12. $2x^2 - 12x + y + 21 = 0$

13. $y^2 + 2x + 15 = 0$ 14. $16x^2 + 4y^2 + 12y - 7 = 0$

15. $x^2 + 4y^2 + 10x - 8y + 13 = 0$ 16. $9x^2 + 25y^2 - 36x + 50y - 164 = 0$

17. $x^2 + 9y^2 - 54y + 72 = 0$ 18. $x^2 - y^2 + 4x + 8y - 11 = 0$

In Exercises 19–26 identify the conic section whose equation is given.

19. $y^2 - 8x + 6y + 17 = 0$ 20. $4x^2 + 4y^2 - 12x + 16y - 11 = 0$

21. $4x^2 + y^2 + 24x - 4y + 24 = 0$ 22. $4x^2 - y^2 - 40x - 4y + 80 = 0$

23. $x^2 - y^2 + 6x + 4y - 4 = 0$ 24. $2x^2 + y^2 - 4x + 4y - 12 = 0$

25. $25x^2 - 16y^2 + 210x + 96y + 656 = 0$ 26. $x^2 - 10x + 8y + 1 = 0$

9.7
ROTATION OF AXES

In Section 9.6 we observed that the equation of a conic whose axis (or axes) is (or are) parallel to one of the coordinate axes can always be written in the form

$$Ax^2 + Cy^2 + Dx + Ey + F = 0 \qquad (1)$$

where A and C are not both zero. Conversely, the graph of such an equation is a conic section or a degenerate conic section (a point, a line, a pair of lines, or no graph). The particular conic section can be obtained by completing the square and translating axes.

If the axis (or axes) of a conic section is (or are) not parallel to the coordinate axes, it can be shown that the equation of the conic section will then be of the form

$$Ax^2 + Bxy + Cy^2 + Dx + Ey + F = 0, \quad B \neq 0 \qquad (2)$$

Notice that Equation (2) differs from Equation (1) in the presence of the "cross-product" term Bxy.

An equation of the form given in Equation (2), where A, B, and C are not all zero, is called the **general second-degree equation** in two variables. We shall find that the graph of the general second-degree equation is the same as that of Equation (1). To sketch the graph of Equation (2) when $B \neq 0$, we perform a transformation to eliminate the xy term, and we will show that this transformation is equivalent to a rotation of the coordinate axes.

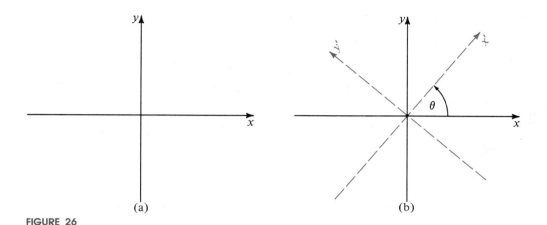

FIGURE 26

    In Figure 26a we show the original coordinate system, and in Figure 26b we obtain the new $x'y'$ coordinate system by rotating the x- and y-axes counterclockwise about the origin through an angle θ. Let P be a point in the plane. Then P has coordinates (x, y) in the original coordinate system and coordinates (x', y') in the rotated system. We can derive the relationship between the xy and $x'y'$ coordinate systems with the aid of Figure 27. Let r be the distance from the origin to P, and let α be the angle between the positive x'-axis and the line OP. Then

$$x' = r \cos \alpha \qquad y' = r \sin \alpha \tag{3}$$

and

$$x = r \cos (\theta + \alpha) \qquad y = r \sin (\theta + \alpha) \tag{4}$$

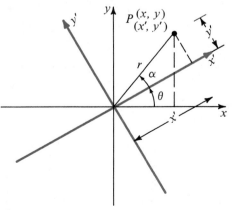

FIGURE 27

Using the formulas from Section 7.2, we can write the Equations in (4) as

$$x = r[\cos \theta \cos \alpha - \sin \theta \sin \alpha]$$
$$y = r[\sin \theta \cos \alpha + \cos \theta \sin \alpha]$$

and, by substituting the expressions in (3), we obtain these formulas.

Rotation of Axes Formulas	$x = x' \cos \theta - y' \sin \theta$ $y = x' \sin \theta + y' \cos \theta$	(5)

EXAMPLE 1

Suppose the coordinate axes are rotated counterclockwise through an angle $\theta = 45°$ to obtain the $x'y'$ coordinate system. Determine the equation of the curve

$$x^2 + xy + y^2 = 6$$

in the $x'y'$ system.

SOLUTION

Letting $\theta = 45°$ in Equation (5), we obtain the expressions

$$x = x' \cos 45° - y' \sin 45° = \frac{\sqrt{2}}{2}x' - \frac{\sqrt{2}}{2}y'$$

$$y = x' \sin 45° + y' \cos 45° = \frac{\sqrt{2}}{2}x' + \frac{\sqrt{2}}{2}y'$$

Substituting these expressions for x and y in the given equation we have

$$\left(\frac{\sqrt{2}}{2}x' - \frac{\sqrt{2}}{2}y'\right)^2 + \left(\frac{\sqrt{2}}{2}x' - \frac{\sqrt{2}}{2}y'\right)\left(\frac{\sqrt{2}}{2}x' + \frac{\sqrt{2}}{2}y'\right)$$
$$+ \left(\frac{\sqrt{2}}{2}x' + \frac{\sqrt{2}}{2}y'\right)^2 = 6$$

$$\frac{1}{2}x'^2 - x'y' + \frac{1}{2}y'^2 + \frac{1}{2}x'^2 - \frac{1}{2}y'^2 + \frac{1}{2}x'^2 + x'y' + \frac{1}{2}y'^2 = 6$$

$$\frac{3}{2}x'^2 + \frac{1}{2}y'^2 = 6$$

$$\frac{x'^2}{4} + \frac{y'^2}{12} = 1$$

which is the equation of an ellipse centered at the origin of the $x'y'$ system (Figure 28).

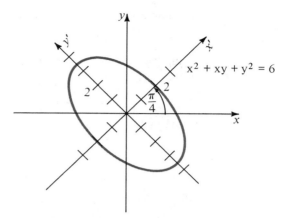

$x^2 + xy + y^2 = 6$

FIGURE 28

Observe that in Example 1 we obtained an equation of the curve in the $x'y'$ system that had no cross-product term. This was accomplished by rotating the x and y coordinate axes in the counterclockwise direction through an angle θ until they were parallel to the axis or axes of the given curve. Our objective, then, is to determine the appropriate angle θ such that the substitutions

$$x = x' \cos \theta - y' \sin \theta$$
$$y = x' \sin \theta + y' \cos \theta$$

in the general second-degree equation

$$Ax^2 + Bxy + Cy^2 + Dx + Ey + F = 0, \quad B \neq 0$$

will "drop out" the xy cross product. Substituting, we have

$$A(x' \cos \theta - y' \sin \theta)^2 + B(x' \cos \theta - y' \sin \theta)(x' \sin \theta + y' \cos \theta)$$
$$+ C(x' \sin \theta + y' \cos \theta)^2 + D(x' \cos \theta - y' \sin \theta)$$
$$+ E(x' \sin \theta + y' \cos \theta) + F = 0$$

or

$$A'x'^2 + B'x'y' + C'y'^2 + D'x' + E'y' + F' = 0$$

where (verify)

$$A' = A \cos^2 \theta + B \cos \theta \sin \theta + C \sin^2 \theta$$
$$B' = 2(C - A) \sin \theta \cos \theta + B(\cos^2 \theta - \sin^2 \theta)$$
$$C' = A \sin^2 \theta - B \sin \theta \cos \theta + C \cos^2 \theta$$
$$D' = D \cos \theta + E \sin \theta$$
$$E' = -D \sin \theta + E \cos \theta$$
$$F' = F$$

To eliminate the $x'y'$ term, we must have $B' = 0$ or

$$2(C - A) \sin \theta \cos \theta + B(\cos^2 \theta - \sin^2 \theta) = 0 \qquad (6)$$

Using the double-angle formulas (see Section 7.3), we can rewrite Equation (6) as

$$(C - A) \sin 2\theta + B \cos 2\theta = 0$$

which yields

$$\cot 2\theta = \frac{A - C}{B}, \quad B \neq 0 \qquad (7)$$

Thus, by rotating the coordinate axes counterclockwise through the angle θ satisfying Equation (7), we obtain an equation in x' and y' that has no cross-product term. It is easy to see that we can always choose θ so that $0 < \theta < \pi/2$, and we will always seek an angle θ that satisfies this condition.

EXAMPLE 2
Discuss and sketch the graph of the equation $xy = 1$.

SOLUTION
We describe and illustrate the steps of the procedure.

Rotation of Axes	
Step 1. Determine the coefficients A, B, and C.	*Step 1.* $A = 0$ $B = 1$ $C = 0$
Step 2. Determine θ from the equation $$\cot 2\theta = \frac{A - C}{B}$$	*Step 2.* $$\cot 2\theta = \frac{A - C}{B} = 0$$ Then $$2\theta = \frac{\pi}{2}$$ $$\theta = \frac{\pi}{4} \text{ or } 45°$$

Step 3. Form

$$x = x' \cos \theta - y' \sin \theta$$
$$y = x' \sin \theta + y' \cos \theta$$

Step 4. Substitute the expressions of Step 3 for x and y in the original equation.

Step 5. Analyze the equation in terms of the $x'y'$ coordinate system. Sketch the graph.

Step 3.

$$x = x' \cos 45° - y' \sin 45°$$
$$= \frac{\sqrt{2}}{2}x' - \frac{\sqrt{2}}{2}y'$$
$$y = x' \sin 45° + y' \cos 45°$$
$$= \frac{\sqrt{2}}{2}x' + \frac{\sqrt{2}}{2}y'$$

Step 4.

$$xy = 1$$
$$\left(\frac{\sqrt{2}}{2}x' - \frac{\sqrt{2}}{2}y' \right)\left(\frac{\sqrt{2}}{2}x' + \frac{\sqrt{2}}{2}y' \right) = 1$$
$$\frac{x'^2}{2} - \frac{y'^2}{2} = 1$$

Step 5. The equation is that of a hyperbola whose foci and intercepts are on the x'-axis. See Figure 29.

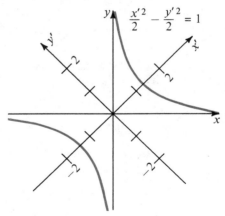

$$\frac{x'^2}{2} - \frac{y'^2}{2} = 1$$

FIGURE 29

PROGRESS CHECK
Discuss and sketch the graph of the equation

$$31x^2 + 10\sqrt{3}xy + 21y^2 = 144$$

ANSWER
$$\frac{x'^2}{4} + \frac{y'^2}{9} = 1 \quad \text{(Figure 30)}$$

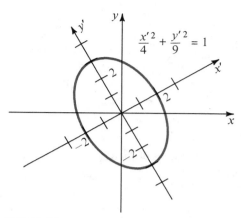

$$\frac{x'^2}{4} + \frac{y'^2}{9} = 1$$

FIGURE 30

In the examples of rotation of axes considered so far, we first determined cot 2θ, then the angle θ, and finally $\sin\theta$ and $\cos\theta$. This is easy to do when the angle 2θ is one of the familiar angles. When it is not readily clear how to obtain the angle 2θ from its cotangent, we find $\cos 2\theta$ and then use the half-angle formulas (see Section 7.3):

$$\sin\theta = \sqrt{\frac{1 - \cos 2\theta}{2}} \qquad \cos\theta = \sqrt{\frac{1 + \cos 2\theta}{2}}$$

Note that θ is always a first-quadrant angle and we can therefore take positive values for $\sin\theta$ and $\cos\theta$.

EXAMPLE 3
Discuss and sketch the graph of the equation

$$97x^2 + 192xy + 153y^2 - 80x + 60y - 125 = 0$$

SOLUTION
We have $A = 97$, $B = 192$, and $C = 153$. Then

$$\cot 2\theta = \frac{A - C}{B} = \frac{97 - 153}{192} = -\frac{56}{192} = -\frac{7}{24}$$

Using the triangle of Figure 31, $\cos 2\theta = -7/25$, so

$$\cos\theta = \sqrt{\frac{1 + \cos 2\theta}{2}} = \sqrt{\frac{1 - 7/25}{2}} = \frac{3}{5}$$

$$\sin\theta = \sqrt{\frac{1 - \cos 2\theta}{2}} = \sqrt{\frac{1 + 7/25}{2}} = \frac{4}{5}$$

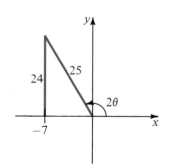

FIGURE 31

Then the equations of rotation are

$$x = x' \cos \theta - y' \sin \theta = \frac{3}{5}x' - \frac{4}{5}y'$$

$$y = x' \sin \theta + y' \cos \theta = \frac{4}{5}x' + \frac{3}{5}y'$$

Substituting these expressions in the original equation, we have

$$97\left(\frac{3}{5}x' - \frac{4}{5}y'\right)^2 + 192\left(\frac{3}{5}x' - \frac{4}{5}y'\right)\left(\frac{4}{5}x' + \frac{3}{5}y'\right)$$

$$+ 153\left(\frac{4}{5}x' + \frac{3}{5}y'\right)^2 - 80\left(\frac{3}{5}x' - \frac{4}{5}y'\right) + 60\left(\frac{4}{5}x' + \frac{3}{5}y'\right) - 125 = 0$$

Expanding and simplifying, we obtain

$$9x'^2 + y'^2 + 4y' = 5$$

(Verify that this is correct.) Completing the square in y',

$$9x'^2 + (y' + 2)^2 = 9$$

or

$$x'^2 + \frac{(y' + 2)^2}{9} = 1$$

The graph is an ellipse whose center in the $x'y'$ system is at $(0, -2)$; $a = 3$, $b = 1$, and the major axis is along the y'-axis (Figure 32).

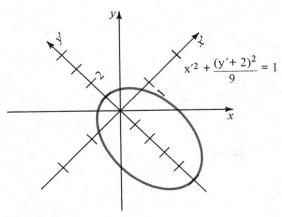

FIGURE 32

PROGRESS CHECK

Perform a rotation of axes so that the equation

$$x^2 + 2xy + y^2 + 4\sqrt{2}x - 4\sqrt{2}y = 0$$

does not have a cross-product term in the $x'y'$ coordinate system.

ANSWER

$x'^2 = 4y'$

EXERCISE SET 9.7

In Exercises 1–10 the coordinate axes are rotated counterclockwise through an angle θ to obtain the $x'y'$ coordinate system. Determine the equation of the given curve in the $x'y'$ system.

1. $\theta = 45°$; $5x^2 - 6xy + 5y^2 - 36 = 0$

2. $\theta = 30°$; $x^2 + y^2 = 25$

3. $\theta = 30°$; $x - \sqrt{3}y - 4 = 0$

4. $\theta = 45°$; $x^2 + 4xy + y^2 + \sqrt{2}x - 3\sqrt{2}y + 6 = 0$

5. $\theta = \cot^{-1}\left(\dfrac{4}{3}\right)$; $135x^2 + 65y^2 + 240xy - 270x -$
   $390y - 225 = 0$

6. $\theta = \cot^{-1}\left(\dfrac{7}{24}\right)$; $1201x^2 + 336xy + 674y^2 - 350x -$
   $1200y = 625$

7. $\theta = \tan^{-1}\left(\dfrac{24}{7}\right)$; $9x^2 - 24xy + 16y^2 - 120x -$
   $90y = 0$

8. $\theta = 30°$; $x^2 + 2\sqrt{3}xy - y^2 + 6x = 0$

9. $\theta = 30°$; $-4x^2 + 2\sqrt{3}xy - 6y^2 - 6y = 0$

10. $\theta = 90°$; $x^2 - y^2 + 6x - 5y + 1 = 0$

In Exercises 11–20 perform a rotation of axes to obtain an equation without an xy term. Sketch the graph of the given equation.

11. $9x^2 + 4xy + 6y^2 - 5 = 0$

12. $154x^2 + 240xy + 54y^2 + 117 = 0$

13. $9x^2 + y^2 + 6xy - 10\sqrt{10}x + 10\sqrt{10}y + 90 = 0$

14. $4x^2 + 8xy + 4y^2 + 25\sqrt{2}x + 23\sqrt{2}y + 38 = 0$

15. $52x^2 - 72xy + 73y^2 - 120x - 90y + 125 = 0$

16. $5x^2 - 6xy + 5y^2 - 30\sqrt{2}x + 18\sqrt{2}y + 82 = 0$

17. $3x^2 + 4xy - \dfrac{38}{\sqrt{5}}x - \dfrac{4}{\sqrt{5}}y + 11 = 0$

18. $5x^2 + 12xy - 12\sqrt{13}x = 36$

19. $73x^2 + 72xy + 52y^2 + 510x + 320y + 825 = 0$

20. $8x^2 - 16xy + 8y^2 + 33\sqrt{2}x - 31\sqrt{2}y + 70 = 0$

21. Show that $B^2 - 4AC = B'^2 - 4A'C'$ for any angle θ. (*Hint:* Substitute Equations (5) in Equation (2).)

22. Show that $A + C = A' + C'$ for any angle θ. (*Hint:* Substitute Equations (5) in Equation (2).)

9.8
POLAR COORDINATES

Thus far we have represented a point P in the plane by the ordered pair (x, y), where x and y represent the distances of P from the y-axis and the x-axis, respectively. In this section we discuss the polar coordinate system, another useful way of representing points in the plane.

We start with a point O, called the **origin** or **pole,** draw a fixed ray, called the **polar axis,** with an endpoint at O, and select a unit of length for measuring distance. For any point P in the plane other than O, we draw a ray from O to P (Figure 33). If $r = \overline{OP}$, the length of the segment OP, and θ is the angle between

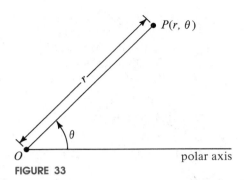

FIGURE 33

the polar axis and the ray OP, then r and θ are called the **polar coordinates** of P, and P is denoted by $P(r, \theta)$, or simply by (r, θ). As usual, the angle θ is positive if it is measured in a counterclockwise direction and negative if measured in a clockwise direction. The angle θ may be expressed in degrees or in radians.

EXAMPLE 1

Plot the points with the given polar coordinates.

(a) $P(3, 45°)$ (b) $P\left(3, \dfrac{\pi}{6}\right)$

(c) $P(2, -120°)$ (d) $P\left(2, -\dfrac{\pi}{6}\right)$

SOLUTION

See Figure 34.

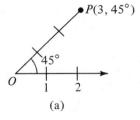

(a)

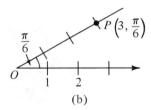

(b)

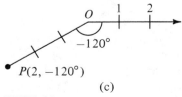

(c)

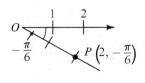

(d)

FIGURE 34

There is an important difference between the polar coordinates and the rectangular coordinates of a point P. The rectangular coordinates of P are unique. However, P does not have unique polar coordinates. In fact, P has infinitely many polar coordinates since $(r, \theta + 2\pi n)$ designates the same point for all integer values of n.

PROGRESS CHECK

Give three other polar coordinate representations for the point $(3, -30°)$.

ANSWER

Possibilities are $(3, 330°)$, $(3, -390°)$, $(3, 690°)$.

When P is the origin, we find that $r = 0$, so any angle θ can be assigned to the ray OP. We therefore say that the pole has polar coordinates $(0, \theta)$, where θ is any angle.

It will also be convenient to consider polar coordinates where r is negative. We shall consider the point $(-r, \theta)$ to be the point $(r, \theta + 180°)$ where r is a positive real number.

EXAMPLE 2

Find polar coordinates for the point $P\left(-2, \dfrac{\pi}{4}\right)$ so that r is nonnegative.

SOLUTION

We have

$$P\left(-2, \frac{\pi}{4}\right) = P\left(2, \frac{\pi}{4} + \pi\right) = P\left(2, \frac{5}{4}\pi\right)$$

Moreover, the given point can also be designated as $P\left(2, \dfrac{5}{4}\pi + 2\pi n\right)$ for all integer values of n.

PROGRESS CHECK

Which of the following polar coordinates represent the point $(-3, -45°)$?
(a) $(3, 45°)$ (b) $(3, 135°)$ (c) $(-3, 315°)$ (d) $(3, 225°)$

ANSWER

(b) and (c)

RECTANGULAR AND POLAR COORDINATES

It is often convenient to be able to convert beween polar coordinates and rectangular coordinates. To obtain the relationship between the two coordinate systems, we let the polar axis coincide with the positive x-axis and the pole with the origin of the rectangular system. From Figure 35 we obtain the following relations.

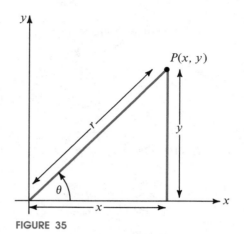

FIGURE 35

Conversion Equations	$x = r \cos \theta \qquad y = r \sin \theta$	(1)
	$\tan \theta = \dfrac{y}{x} \qquad r = \sqrt{x^2 + y^2}$	(2)

EXAMPLE 3
Find the rectangular coordinates of the point P whose polar coordinates are $(3, 225°)$.

SOLUTION
We let $r = 3$ and $\theta = 225°$ in Equation (1), obtaining

$$x = 3 \cos 225° = 3\left(-\frac{\sqrt{2}}{2}\right) = -\frac{3}{2}\sqrt{2}.$$

$$y = 3 \sin 225° = 3\left(-\frac{\sqrt{2}}{2}\right) = -\frac{3}{2}\sqrt{2}$$

Thus, the rectangular coordinates of P are $\left(-\frac{3}{2}\sqrt{2}, -\frac{3}{2}\sqrt{2}\right)$.

Converting from rectangular to polar coordinates is a bit more involved, since a point has many polar coordinate representations.

EXAMPLE 4
Find polar coordinates of the point P whose rectangular coordinates are $(\sqrt{3}, -1)$.

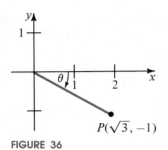

FIGURE 36

SOLUTION

The given point is shown in Figure 36. Using the equations in (2) with $x = \sqrt{3}$ and $y = -1$, we have

$$r = \sqrt{x^2 + y^2} = \sqrt{4} = 2$$

$$\tan \theta = \frac{y}{x} = \frac{-1}{\sqrt{3}} = -\frac{\sqrt{3}}{3}$$

Since P lies in the fourth quadrant, θ may be taken as $-30°$ or as $330°$. Thus, two possible polar coordinate representations for P are

$$(2, -30°) \quad \text{and} \quad (2, 330°)$$

PROGRESS CHECK

(a) Find the rectangular coordinates of the point P whose polar coordinates are $\left(2, \dfrac{\pi}{3}\right)$.

(b) Find polar coordinates of the point P whose rectangular coordinates are $(-1, -1)$.

ANSWERS

(a) $(1, \sqrt{3})$ (b) $(\sqrt{2}, 5\pi/4)$ (not unique)

POLAR EQUATIONS

An equation in the variables r and θ is called a **polar equation.** A **solution** of a polar equation is an ordered pair (a, b) that satisfies the polar equation when $r = a$ and $\theta = b$ are substituted in the equation. Just as the graph of an equation in x and y is obtained by plotting all the solutions of the equation on a rectangular coordinate system, the **graph of a polar equation** is obtained by plotting all the solutions of the equation on a **polar coordinate system.** When sketching the graph of a polar equation, it is convenient to superimpose the rectangular system on the polar coordinate system so that the origin coincides with the pole and the positive x-axis coincides with the polar axis.

EXAMPLE 5

Sketch the graph of the polar equation $r = 4 \cos \theta$.

SOLUTION

We form the following table of values.

θ	0	$\dfrac{\pi}{6}$	$\dfrac{\pi}{4}$	$\dfrac{\pi}{3}$	$\dfrac{\pi}{2}$	$\dfrac{2}{3}\pi$	$\dfrac{3}{4}\pi$	$\dfrac{5}{6}\pi$	π
r	4	$2\sqrt{3}$	$2\sqrt{2}$	2	0	-2	$-2\sqrt{2}$	$-2\sqrt{3}$	-4

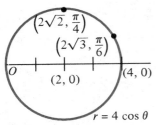

FIGURE 37

Plotting the points in this table, we seem to obtain a circle of radius 2, with center at (2, 0), shown in Figure 37. As θ takes values from π to 2π, we obtain the same circle again. In Example 9 we shall show that the graph of this polar equation is indeed a circle of radius 2, with center at (2, 0) in the xy coordinate system.

EXAMPLE 6

Sketch the graph of the polar equation $r = 1 + \sin \theta$.

SOLUTION

We form the following table of values.

θ	0	$\dfrac{\pi}{6}$	$\dfrac{\pi}{3}$	$\dfrac{\pi}{2}$	$\dfrac{2}{3}\pi$	$\dfrac{5}{6}\pi$	π	$\dfrac{7}{6}\pi$	$\dfrac{4}{3}\pi$	$\dfrac{3}{2}\pi$	$\dfrac{5}{3}\pi$	$\dfrac{11}{6}\pi$	2π
r	1	$\dfrac{3}{2}$	$1+\dfrac{\sqrt{3}}{2}$	2	$1+\dfrac{\sqrt{3}}{2}$	$\dfrac{3}{2}$	1	$\dfrac{1}{2}$	$1-\dfrac{\sqrt{3}}{2}$	0	$1-\dfrac{\sqrt{3}}{2}$	$\dfrac{1}{2}$	1

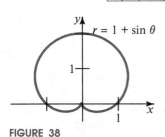

FIGURE 38

Plotting these points, we obtain the heart-shaped curve shown in Figure 38, which is called a **cardioid.**

EXAMPLE 7

Sketch the graph of the polar equation $r = 2 \sin 2\theta$.

SOLUTION

Instead of using a table of values, we proceed as follows. When $\theta = 0, r = 0$. As θ increases from 0 to $\pi/4$, r increases from 0 to 2, since $\sin 2\theta$ increases from 0 to 1. As θ increases from $\pi/4$ to $\pi/2$, $\sin 2\theta$ decreases from 1 to 0, so r decreases from 2 to 0. Plotting these points, we obtain the loop shown in Figure 39a. Similarly, as θ increases from $\frac{1}{2}\pi$ to $\frac{3}{4}\pi$, r decreases from 0 to -2, and as θ increases from $\frac{3}{4}\pi$ to π, we find that r increases from -2 to 0. These points yield the second loop, shown in Figure 39b. As θ increases from π to 2π, we obtain two additional loops. The final figure, shown in Figure 39c, is called a **four-leaved rose.**

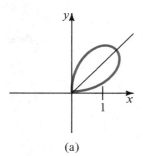

(a)

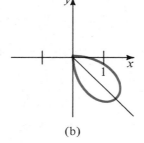

(b)

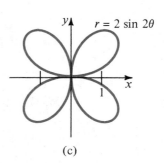

(c)

FIGURE 39

Many interesting curves are obtained as the graphs of polar equations, and some of these are given as exercises.

PROGRESS CHECK

Sketch the graph of the given polar equation.
(a) $r = 2 \sin \theta$ (b) $r = 1 + 2 \cos \theta$ (c) $r = \cos 3\theta$

ANSWER
See Figure 40.

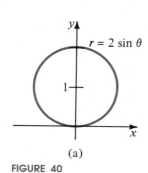

(a)

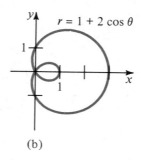

(b)

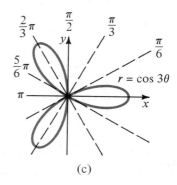

(c)

FIGURE 40

WARNING When an equation is given in the variables r and θ, we assume that the graph is to be sketched using polar coordinates. Don't use rectangular coordinates to sketch the equation

$$r = 2 \cos 2\theta$$

The conversion formulas enable us to transform polar equations to equations in x and y, and vice versa.

EXAMPLE 8
Transform the polar equation $r = 4 \cos \theta$ to an equation in x and y.

SOLUTION
This equation was plotted in Example 6. Multiplying both sides of the given equation by r, we have

$$r^2 = 4r \cos \theta$$

or

$$x^2 + y^2 = 4x$$

Completing the square in the last equation, we have

$$(x - 2)^2 + y^2 = 4$$

whose graph is a circle of radius 2 with center at (2, 0). This result verifies our solution of Example 6.

PROGRESS CHECK

Transform the rectangular equation $x = 5$ to a polar equation.

ANSWER

$r \cos \theta = 5$

EXERCISE SET 9.8

1. Plot the points with the given polar coordinates.
   (a) $(4, 30°)$ (b) $(-2, 60°)$
   (c) $\left(5, \dfrac{\pi}{4}\right)$ (d) $\left(-3, -\dfrac{\pi}{2}\right)$

2. Plot the points with the given polar coordinates.
   (a) $(2, 225°)$ (b) $(-4, -150°)$
   (c) $\left(5, \dfrac{2}{3}\pi\right)$ (d) $\left(-4, -\dfrac{3}{4}\pi\right)$

3. For each point given in polar coordinates, give two other polar coordinate representations.
   (a) $(6, 135°)$ (b) $(-2, 120°)$
   (c) $\left(4, \dfrac{5}{6}\pi\right)$ (d) $\left(-4, -\dfrac{7}{4}\pi\right)$

4. For each point given in polar coordinates, give two other polar coordinate representations.
   (a) $(4, 315°)$ (b) $(-3, -150°)$
   (c) $\left(1, \dfrac{11}{6}\pi\right)$ (d) $\left(-1, -\dfrac{3}{2}\pi\right)$

5. For each point given in polar coordinates, give a polar coordinate representation with $r \geq 0$.
   (a) $(-2, 30°)$ (b) $(-4, -60°)$
   (c) $\left(-3, \dfrac{2}{3}\pi\right)$ (d) $\left(-1, -\dfrac{7}{6}\pi\right)$

6. For each point given in polar coordinates, give a polar coordinate representation with $r \geq 0$.
   (a) $(-2, 60°)$ (b) $(-3, 45°)$
   (c) $\left(-5, \dfrac{7}{6}\pi\right)$ (d) $\left(-4, -\dfrac{2}{3}\pi\right)$

7. Which of the following polar coordinates represent the point $(2, -30°)$?
   (a) $(-2, 150°)$ (b) $(-2, 330°)$
   (c) $(2, 330°)$ (d) $(2, 510°)$

8. Which of the following polar coordinates represent the point $\left(4, \dfrac{2}{3}\pi\right)$?
   (a) $\left(4, \dfrac{5}{3}\pi\right)$ (b) $\left(-4, \dfrac{5}{3}\pi\right)$
   (c) $\left(4, -\dfrac{4}{3}\pi\right)$ (d) $\left(4, -\dfrac{2}{3}\pi\right)$

9. Find the rectangular coordinates of the points with the given polar coordinates.
   (a) $(5, 330°)$ (b) $(2, 270°)$
   (c) $\left(4, \dfrac{\pi}{6}\right)$ (d) $\left(-3, -\dfrac{2}{3}\pi\right)$

10. Find the rectangular coordinates of the points with the given polar coordinates.
    (a) $(1, 315°)$ (b) $(3, 150°)$
    (c) $\left(2, \dfrac{3}{4}\pi\right)$ (d) $\left(-4, -\dfrac{4}{3}\pi\right)$

11. Find polar coordinates of the points with the given rectangular coordinates.
    (a) $(-2, 2)$ (b) $(1, -\sqrt{3})$
    (c) $(\sqrt{3}, 1)$ (d) $(-4, -4)$

12. Find polar coordinates of the points with the given rectangular coordinates.
    (a) $(-1, -\sqrt{3})$ (b) $(-\sqrt{3}, 1)$
    (c) $(3, -3)$ (d) $(3, 3)$

In Exercises 13–26 sketch the graph of the given polar equation.

13. $\theta = 45°$ 14. $\theta = \dfrac{2}{3}\pi$ 15. $r = 3$ 16. $r = 1 - \sin\theta$

17. $r = 2\sin\theta$ 18. $r = 2$ 19. $r = -3\cos\theta$ 20. $r = 2 + 4\sin\theta$

21. $r = 3\sin 5\theta$ 22. $r = 3\cos 4\theta$ 23. $r^2 = 1 + \sin 2\theta$ 24. $r^2 = 4\sin 2\theta$

25. $r = \theta,\ \theta \geq 0$ 26. $r\theta = 2$

In Exercises 27–32, transform the given polar equation to an equation in x and y.

27. $r = 4$ 28. $\theta = \dfrac{\pi}{4}$ 29. $r + 2\sin\theta = 0$ 30. $r = 3\sec\theta$

31. $r\cos\theta = 2$ 32. $r = 2\tan\theta$

In Exercises 33–38 transform the given rectangular equation to a polar equation.

33. $x^2 + y^2 = 25$ 34. $y = 3$ 35. $x = -5$ 36. $y^2 = 2x$

37. $y = 3x$ 38. $x^2 - y^2 = 9$

9.9 PARAMETRIC EQUATIONS

There are situations in which it is convenient to describe movement along a curve as a function of time. We may then consider each of the coordinates of a point $P(x, y)$ to be a function of time t, that is,

$$x = f(t) \qquad y = g(t)$$

These are called **parametric equations,** and the variable t is called a **parameter.** The equations may hold for all real values of t or, quite commonly, for all real values of t in an interval $[a, b]$. The following example illustrates how the graph of a curve described in parametric form can be sketched.

EXAMPLE 1

Sketch the curve whose parametric equations are

$$x = 2t \qquad y = t^2$$

for $1 \leq t \leq 4$.

TABLE 1

t	1	$\dfrac{3}{2}$	2	$\dfrac{5}{2}$	3	$\dfrac{7}{2}$	4
$x = 2t$	2	3	4	5	6	7	8
$y = t^2$	1	$\dfrac{9}{4}$	4	$\dfrac{25}{4}$	9	$\dfrac{49}{4}$	16
(x, y)	$(2, 1)$	$\left(3, \dfrac{9}{4}\right)$	$(4, 4)$	$\left(5, \dfrac{25}{4}\right)$	$(6, 9)$	$\left(7, \dfrac{49}{4}\right)$	$(8, 16)$

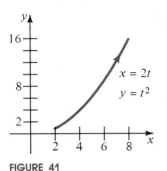

FIGURE 41

SOLUTION
For each value of t in the interval $[1, 4]$ we obtain corresponding values of x and y. Some of these values are shown in Table 1. Plotting the points (x, y) and connecting them by a smooth curve, we obtain the curve shown in Figure 41. The arrow on the curve indicates the direction of motion along the curve as the parameter t increases from 1 to 4.

To see if we can further identify the curve in Example 1, we eliminate the parameter t. Thus,

$$y = t^2 = \left(\frac{x}{2}\right)^2 = \frac{x^2}{4}$$

which is the familiar rectangular equation of a parabola.

Observe that a curve frequently has many different parametric equations. For example, the curve given in Example 1 can also be described by the parametric equations

$$x = 3t \quad y = \frac{9}{4}t^2, \quad -2 \leq t \leq 2$$

One of the main advantages of describing a curve by means of parametric equations with time t as the parameter is that as we move along the curve, we know *when* we arrive at each point on the curve.

EXAMPLE 2
Discuss and sketch the curve whose parametric equations are

$$x = r \cos t \quad y = r \sin t, \quad 0 \leq t \leq 2\pi$$

where r is a positive constant.

SOLUTION
We show that the curve is the circle of radius r centered at the origin. We have

$$x^2 + y^2 = r^2 \cos^2 t + r^2 \sin^2 t$$
$$= r^2(\cos^2 t + \sin^2 t)$$
$$= r^2$$

Thus, if $P(x, y)$ is a point on the curve, then $x^2 + y^2 = r^2$, so P lies on the circle of radius r centered at the origin. Observe that as t increases from 0 to 2π, the point P starts at $(r, 0)$ and traverses the circle in a counterclockwise direction, ending at $(r, 0)$, as shown in Figure 42a.

Consider now the curve whose parametric equations are

$$x = r \cos t \quad y = r \sin t, \quad 0 \leq t \leq 4\pi$$

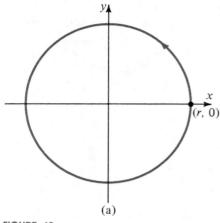

(a)

(b)

FIGURE 42

In this case, the parametric equations describe the circle of radius r centered at the origin, but the circle is now traversed twice in a counterclockwise direction. Similarly, the parametric equations

$$x = r \cos t \quad y = r \sin t, \quad 0 \le t \le \pi$$

describe the upper half of the circle of radius r centered at the origin (Figure 42b).

PROGRESS CHECK

Discuss the curve whose parametric equations are

$$x = 2 + 3t \quad y = 1 - 3t, \quad t \ge 0$$

Eliminate the parameter t.

ANSWER

The curve consists of all points on the line $y = 3 - x$ for which $x \ge 2$.

EXAMPLE 3

Discuss the curve whose parametric equations are

$$x = 4 \cos t \quad y = 3 \sin t, \quad 0 < t < 2\pi$$

SOLUTION

We eliminate the parameter t by first writing

$$\frac{x}{4} = \cos t \quad \frac{y}{3} = \sin t$$

and then using the familiar trigonometric identity

$$\sin^2 t + \cos^2 t = 1$$

or

$$\left(\frac{x}{4}\right)^2 + \left(\frac{y}{3}\right)^2 = 1$$

$$\frac{x^2}{16} + \frac{y^2}{9} = 1$$

whose graph is an ellipse. The point P that traverses the curve starts when $t = 0$, so P starts at (4, 0) and moves in a counterclockwise direction. The y-intercept (0, 3) is reached when $t = \pi/2$.

EXERCISE SET 9.9

In Exercises 1–12 sketch the curve with the given parametric equations, and find a rectangular equation for the curve by eliminating the parameter.

1. $x = 4t - 1$, $y = 2t$
2. $x = \sin t$, $y = \cos t$
3. $x = t^2$, $y = t^3$
4. $x = 3 \sin t + 1$, $y = 2 \cos t + 2$
5. $x = \dfrac{t}{2}$, $y = t^3$
6. $x = 2t$, $y = 3t$
7. $x = \dfrac{3t}{1 + t^3}$, $y = \dfrac{3t^2}{1 + t^3}$
8. $x = 3\sqrt{t - 3}$, $y = 2\sqrt{4 - t}$
9. $x = 4 \sec t$, $y = 3 \tan t$
10. $x = 2 \cos t$, $y = 2 \cos\left(\dfrac{1}{2}t\right)$
11. $x = 2 \sin t$, $y = 3 \cos t$
12. $x = \sin 3t$, $y = \cos 3t$

In Exercises 13–18 eliminate the parameter t to show that the given parametric equations describe the stated curve.

13. $x = \sec t$, $y = \tan t$; hyperbola
16. $x = 3t^2 - 1$, $y = t^2 + 1$; line
14. $x = 5 \sin t - 5$, $y = 4 \cos t + 4$; ellipse
17. $x = 2t + 1$, $y = -t^2$; parabola
15. $x = t$, $y = \dfrac{2}{t}$; hyperbola
18. $x = \sqrt{\dfrac{t - 1}{t + 1}}$, $y = \dfrac{1}{\sqrt{t + 1}}$; ellipse

TERMS AND SYMBOLS

standard form of the equation of an ellipse (p. 377 and p. 379)

hyperbola (p. 381)

foci of a hyperbola (p. 382)

transverse axis (p. 382)

conjugate axis (p. 382)

standard forms of the equations of a hyperbola (p. 383 and p. 384)

vertices (p. 383)

asymptotes of a hyperbola (p. 386)

translation of axes formulas (p. 390)

general second-degree equation (p. 396)

rotation of axes formulas (p. 398)

pole (p. 404)

polar axis (p. 404)

polar coordinates (p. 405)

conversion equations between polar and rectangular coordinates (p. 407)

polar equation (p. 408)

solution of a polar equation (p. 408)

graph of a polar equation (p. 408)

parametric equation (p. 412)

parameter (p. 412)

KEY IDEAS FOR REVIEW

☐ The midpoint of the line segment joining the points $P_1(x_1, y_1)$ and $P_2(x_2, y_2)$ has coordinates

$$\left(\frac{x_1 + x_2}{2}, \frac{y_1 + y_2}{2} \right)$$

☐ Theorems from plane geometry can be proved using the methods of analytic geometry. In general, place the given geometric figure in a convenient position relative to the origin and axes. The distance formula, the midpoint formula, and the computation of slope are the basic tools to apply in proving a theorem.

☐ The conic sections represent the possible intersections of a plane and a cone. The conic sections are the circle, parabola, ellipse, and hyperbola. (In special cases, these may reduce to a point, a line, two lines, or no graph.)

☐ Each conic section has a geometric definition that can be used to derive a second-degree equation in two variables whose graph corresponds to the conic.

☐ The equation of a conic with an axis parallel to one of the coordinate axes can always be written as

$$Ax^2 + Cy^2 + Dx + Ey + F = 0$$

where A and C are not both zero. Conversely, the graph of such an equation is a conic section or a degenerate conic section (a point, a line, a pair of lines, or no graph). The particular conic section can be found by completing the square and translating axes.

☐ If no axis of a conic section is parallel to a coordinate axis, its equation is of the form

$$Ax^2 + Bxy + Cy^2 + Dx + Ey + F = 0, \quad B \neq 0$$

Conversely, the graph of such an equation, where A, B, and C are not all zero, is a conic section, a line, a pair of lines, or no graph. The particular conic section can be obtained by first rotating axes to eliminate the xy term and then translating the resulting axes.

☐ Polar coordinates specify a point P in the plane in terms of an angle θ and the distance r from the origin.

☐ Parametric equations provide an alternative way to describe a curve in the plane.

REVIEW EXERCISES

Solutions to exercises whose numbers are in color are in the Solutions section in the back of the book.

9.1 In Exercises 1 and 2 find the midpoint of the line segment whose endpoints are given.

1. $(-5, 4)$, $(3, -6)$ 2. $(-2, 0)$, $(-3, 5)$

3. Find the coordinates of the point P_2 if $(2, 2)$ are the coordinates of the midpoint of the line segment joining $P_1(-6, -3)$ and P_2.

4. Use the distance formula to show that $P_1(-1, 2)$, $P_2(4, 3)$, $P_3(1, -1)$, and $P_4(-4, -2)$ are the coordinates of a parallelogram.

5. Show that the points $A(-8, 4)$, $B(5, 3)$, and $C(2, -2)$ are the vertices of a right triangle.

6. Find an equation of the perpendicular bisector of the line segment joining the points $A(-4, -3)$ and $B(1, 3)$. (The perpendicular bisector passes through the midpoint of AB and is perpendicular to AB.)

9.2 7. Write an equation of the circle whose center is at $(-5, 2)$ and whose radius is 4.

8. Write an equation of the circle whose center is at $(-3, 3)$ and whose radius is 2.

In Exercises 9 and 10 determine the center and radius of the circle with the given equation.

9. $(x - 2)^2 + (y + 3)^2 = 9$

10. $x^2 + y^2 + 4x - 6y = -10$

9.3 In Exercises 11 and 12 determine the focus and directrix of the given parabola, and sketch the graph.

11. $x^2 = -\dfrac{2}{3}y$ 12. $3y^2 + 2x = 0$

In Exercises 13 and 14 determine the equation of the parabola with its vertex at the origin that satisfies the given conditions.

13. directrix $y = \dfrac{7}{4}$

14. axis the y-axis; parabola passing through the point $\left(1, \dfrac{5}{2}\right)$

9.4 In Exercises 15 and 16 sketch and discuss the graph of the given ellipse, giving the coordinates of the foci and vertices.

15. $81x^2 + 4y^2 = 9$

16. $4x^2 + y^2 + 8x - 2y = 4$

In Exercises 17 and 18 find an equation of the ellipse satisfying the given conditions.

17. foci at $(\pm 3, 0)$, vertices at $(\pm 9, 0)$

18. foci at $(0, \pm 2)$, length of minor axis $= 2$

9.5 In Exercises 19 and 20 sketch and discuss the graph of the given hyperbola, giving the coordinates of the foci and vertices and including the asymptotes.

19. $4x^2 - y^2 - 2y = 4$

20. $7y^2 - x^2 = 4$

In Exercises 21 and 22 find an equation of the hyperbola with center at the origin that satisfies the given conditions.

21. foci at $(0, \pm 7)$, vertices at $(0, \pm 2)$

22. vertex at $(2, 0)$; hyperbola passing through the point $(3, 1)$

In Exercises 23–34 identify the conic section whose equation is given.

9.6 23. $2x^2 - 4x + y^2 = 0$

24. $4y + x^2 - 2x = 1$

25. $x^2 - 2x + y^2 - 4y = -6$

26. $-x^2 - 4x + y^2 + 4 = 0$

27. $y^2 + 2x - 4 = x^2 - 2x$

28. $x^2 - 4x + 2y = 6 + y^2$

9.7 29. $xy + 2x - 2y = 6$

30. $x^2 + y^2 + xy + x - y = 3$

31. $2x^2 - y^2 + 4xy - 2x + 3y = 6$

32. $x^2 + 4xy + 4y^2 - 3x = 6$

33. $13x^2 + 10xy + 13y^2 = 72$

34. $13x^2 - 32xy + 37y^2 - 4\sqrt{5}x - 2\sqrt{5}y = 40$

9.8 In Exercises 35 and 36 sketch the graph of the given polar equation.

35. $r = \sec \theta \tan \theta$

36. $r = |\theta|$

37. Transform the equation $r = 3 \csc \theta$ to an equation in x and y.

38. Transform the equation $x^2 + 9y^2 = 9$ to a polar equation.

9.9 39. Sketch the curve with parametric equations
$$x = t^2 - 2t + 1$$
$$y = (t - 1)^4 - 1$$

40. Show that the curve with parametric equations
$$x = \frac{\sqrt{t - 1}}{\sqrt{t + 2}} \qquad y = \frac{1}{\sqrt{t + 2}}$$
is an ellipse by eliminating the parameter t.

PROGRESS TEST 9A

1. Find the midpoint of the line segment whose end-points are $(2, 4)$ and $(-2, 4)$.

2. Find the coordinates of the point P if $(-3, 3)$ are the coordinates of the midpoint of the line segment joining P and Q $(-5, 4)$.

3. By using the slope of a line, show that the points $A(-3, -1)$, $B(-5, 4)$, $C(2, 6)$, and $D(4, 1)$ determine a parallelogram.

4. Write an equation of the circle of radius 6 whose center is at $(2, -3)$.

5. Determine the center and radius of the circle $x^2 + y^2 - 2x + 4y = 1$.

6. Determine the vertex and axis of the parabola $x^2 + 6x + 2y + 7 = 0$, and sketch the graph.

In Problems 7–9 write the given equation in standard form and determine the intercepts.

7. $x^2 + 4y^2 = 4$
8. $4y^2 - 9x^2 = 36$
9. $4x^2 - 4y^2 = 1$

10. Use the intercepts and asymptotes of the hyperbola $9x^2 - y^2 = 9$ to sketch its graph.

In Problems 11–15 identify the conic section.

11. $x^2 + 9y^2 - 4x + 6y + 4 = 0$
12. $x^2 - 3x + 4y = 2$
13. $3x^2 + 6xy + 3y^2 - 4x + 5y = 12$
14. $x^2 - 2xy + y^2 = 4$
15. $x^2 + xy + y^2 = 1$
16. Sketch the graph of the polar equation $r = 4 \csc \theta$.
17. Transform the equation $r = \sec^2 \theta/2$ to an equation in x and y.
18. Transform the equation $x^2 + xy + y^2 = 1$ to a polar equation.
19. Sketch the curve with the parametric equations
$$x = \sqrt{t}, \quad y = t - 1$$
20. Identify the curve with the parametric equations
$$x = 4 \sin t + 1, \quad y = 3 \cos t - 1$$
by eliminating the parameter t.

PROGRESS TEST 9B

1. Find the midpoint of the line segment whose end-points are $(-5, -3)$ and $(4, 1)$.

2. Find the coordinates of the point A if $(-2, -\frac{1}{2})$ are the coordinates of the midpoint of the line segment joining A and $B(3, -2)$.

3. Show that the diagonals of the quadrilateral whose vertices are $P(-3, 1)$, $Q(-1, 4)$, $R(5, 0)$, and $S(3, -3)$ are equal.

4. Write an equation of the circle of radius 5 whose center is at $(-2, -5)$.

5. Determine the center and radius of the circle $4x^2 + 4y^2 - 4x - 8y = 35$.

6. Determine the vertex and axis of the parabola $9x^2 + 18y - 6x + 7 = 0$, and sketch the graph.

In Problems 7–9 write the given equation in standard form and determine the intercepts.

7. $5x^2 + 9y^2 = 25$
8. $7x^2 + 6y^2 = 21$
9. $y^2 + 3x^2 = 9$

10. Use the intercepts and asymptotes of the hyperbola $4y^2 - x^2 = 1$ to sketch its graph.

In Problems 11–15 identify the conic section.

11. $5y^2 - 4x^2 - 6x + 2 = 0$
12. $x^2 + y^2 + 4x - 6y = -13$
13. $-x^2 - 2xy + y^2 = -1$
14. $2x^2 - 5xy + 2y^2 = 0$
15. $-x^2 + 4\sqrt{3}xy + 3y^2 = 7$
16. Sketch the graph of the polar equation $r = \csc^2 \theta/2$.
17. Transform the equation $r = -\sin \theta + 4 \cos \theta$ to an equation in x and y.
18. Transform the equation $x^2 - xy - y^2 = 1$ to a polar equation.
19. Sketch the curve with the parametric equations
$$x = \sin t, \quad y = \cos 2t$$
20. Identify the curve with the parametric equations
$$x = t^2 - 2t, \quad y = t + 1$$
by eliminating the parameter t.

CHAPTER 10

SYSTEMS OF EQUATIONS AND INEQUALITIES

Many problems in business and engineering require the solution of systems of equations and inequalities. In fact, systems of linear equations and inequalities occur with such frequency that mathematicians and computer scientists have devoted considerable energy to devising methods for their solution. With the aid of large-scale computers it is possible to solve systems involving thousands of equations or inequalities, a task that previous generations would not have dared tackle.

We begin with the study of the methods of substitution and elimination, methods that are applicable to all types of systems. We then introduce graphical methods for solving systems of linear inequalities and apply this technique to linear programming problems, a type of optimization problem.

The method of Gaussian elimination is used to introduce the material on matrices and determinants, which serves as an introduction to linear algebra. We will show that matrices and determinants provide neat schemes for automating the computational procedures for solving systems of linear equations.

10.1
SYSTEMS OF
EQUATIONS

A pile of 9 coins consists of nickels and quarters. If the total value of the coins is $1.25, how many of each type of coin are there?

The natural way to approach this problem is to let

$$x = \text{the number of nickels}$$

and

$$y = \text{the number of quarters}$$

that is, to use two variables. The requirements can then be expressed as

$$x + y = 9$$
$$5x + 25y = 125$$

This is an example of a **system of equations,** and we seek values of x and y that satisfy *both* equations. An ordered pair (a, b) such that $x = a$, $y = b$ satisfies both equations is called a **solution** of the system. Thus,

$$x = 5 \quad y = 4$$

is a solution because substituting in the equations of the system gives

$$5 + 4 = 9$$
$$5(5) + 25(4) = 125$$

SOLVING BY
SUBSTITUTION

If we can use one of the equations of a system to express one variable in terms of the other variable, then we can *substitute* this expression in the other equation.

EXAMPLE 1
Solve the system of equations.

$$x^2 + y^2 = 25$$
$$x + y = -1$$

SOLUTION
From the second equation we have

$$y = -1 - x$$

Substituting for y in the first equation,

$$x^2 + (-1 - x)^2 = 25$$
$$x^2 + 1 + 2x + x^2 = 25$$
$$2x^2 + 2x - 24 = 0$$
$$x^2 + x - 12 = 0$$
$$(x + 4)(x - 3) = 0$$

which yields $x = -4$ and $x = 3$. Substituting these values for x in the equation $x + y = -1$, we obtain the corresponding values of y.

$$x = -4: \quad -4 + y = -1 \qquad x = 3: \quad 3 + y = -1$$
$$y = 3 \qquad\qquad\qquad y = -4$$

Thus, $x = -4$, $y = 3$ and $x = 3$, $y = -4$ are solutions of the system of equations.

Note that the equations represent a circle and a line; the algebraic solution tells us that they intersect in two points. See Figure 1.

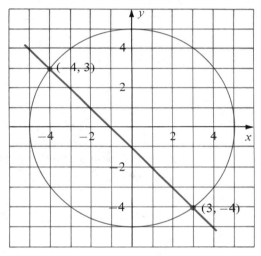

FIGURE 1

PROGRESS CHECK

Solve the system of equations.

(a) $x^2 + 3y^2 = 12$ (b) $x^2 + y^2 = 34$

  $x + 3y = 6$ $x - y = 2$

ANSWERS

(a) $x = 3$, $y = 1$; $x = 0$, $y = 2$

(b) $x = -3$, $y = -5$; $x = 5$, $y = 3$

It is possible for a system of equations to have no solutions. Surprisingly, a system of equations may even have an infinite number of solutions. The following terminology is used to distinguish these situations.

Consistent and Inconsistent Systems	• A **consistent** system of equations has one or more solutions. • An **inconsistent** system of equations has no solutions.

EXAMPLE 2

Solve the system of equations.

(a) $\begin{aligned} x^2 - 2x - y + 3 &= 0 \\ x + y - 1 &= 0 \end{aligned}$ (b) $\begin{aligned} x + 4y &= 10 \\ -2x - 8y &= -20 \end{aligned}$

SOLUTION

(a) Solving the second equation for y, we have

$$y = 1 - x$$

and substituting in the first equation yields

$$x^2 - 2x - (1 - x) + 3 = 0$$
$$x^2 - x + 2 = 0$$

Since the discriminant of this quadratic equation is negative, the equation has no real roots. But any solution of the system of equations must satisfy this quadratic equation. We can therefore conclude that the system is inconsistent. The graphs of the equations are a parabola and a line that do not intersect (see Figure 2).

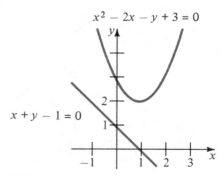

FIGURE 2

(b) Solving the first equation for x, we have

$$x = 10 - 4y$$

and substituting in the second equation gives

$$-2(10 - 4y) - 8y = -20$$
$$-20 + 8y - 8y = -20$$
$$-20 = -20$$

The substitution procedure has resulted in an identity, indicating that any solution of the first equation will also satisfy the second equation. Since there are an infinite number of ordered pairs $x = a$, $y = b$ satisfying the first equation, the system is consistent and has an infinite number of solutions.

PROGRESS CHECK

Solve by substitution.

(a) $\quad 3x - y = \quad 7$ $\qquad$ (b) $\quad -5x + 2y = -4$

$\qquad -9x + 3y = -22$ $\qquad\qquad \dfrac{5}{2}x - y = \quad 2$

ANSWERS

(a) no solution $\qquad$ (b) any point on the line $-5x + 2y = -4$

WARNING $\qquad$ The expression for x or y obtained from an equation *must not be substituted in the same equation*. From the first equation of the system

$$x + 2y = -1$$
$$3x^2 + y = \quad 2$$

we obtain

$$x = -1 - 2y$$

Substituting *(incorrectly)* in the same equation would result in

$$(-1 - 2y) + 2y = -1$$
$$-1 = -1$$

The substitution $x = -1 - 2y$ must be made in the *second* equation.

SOLVING BY ELIMINATION

The method of elimination seeks to combine the equations of a system in such a way as to *eliminate* one of the variables.

EXAMPLE 3

Solve the system.

$$4x^2 + \quad 9y^2 = 36$$
$$-9x^2 + 18y^2 = \quad 4$$

SOLUTION

We can employ the method of elimination to obtain an equation that has just one variable. If we multiply the first equation by -2 and add the result to the second equation, we have

$$
\begin{aligned}
-8x^2 - 18y^2 &= -72 \\
\underline{-9x^2 + 18y^2} &= \underline{4} \\
-17x^2 &= -68 \\
x^2 &= 4 \\
x &= \pm 2
\end{aligned}
$$

We can now substitute $+2$ and -2 for x in either of the original equations. Using the first equation,

$$
\begin{array}{ll}
4x^2 + 9y^2 = 36 & \qquad 4x^2 + 9y^2 = 36 \\
4(2)^2 + 9y^2 = 36 & \qquad 4(-2)^2 + 9y^2 = 36 \\
\quad y = \pm \dfrac{2}{3}\sqrt{5} & \qquad \quad y = \pm \dfrac{2}{3}\sqrt{5}
\end{array}
$$

We then have four solutions: $x = 2$, $y = 2\sqrt{5}/3$; $x = 2$, $y = -2\sqrt{5}/3$; $x = -2$, $y = 2\sqrt{5}/3$; $x = -2$, $y = -2\sqrt{5}/3$. The graphs of the equations are an ellipse and a hyperbola that intersect in four points as shown in Figure 3.

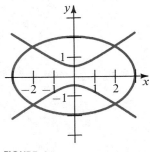

FIGURE 3

PROGRESS CHECK

Find the real solutions of the system.

$$
\begin{aligned}
x^2 - 4x + y^2 - 4y &= 1 \\
x^2 - 4x \phantom{{}+ y^2} + y &= -5
\end{aligned}
$$

ANSWER

$x = 2$, $y = -1$ (The parabola is tangent to the circle.)

SYSTEMS OF LINEAR EQUATIONS

If the equations of a system are of the first degree in x and y, we call this a **system of linear equations** or simply a **linear system.** The methods of substitution and elimination can be used to solve a linear system.

When we graph two linear equations on the same set of coordinate axes, there are three possibilities:

(a)   The two lines intersect in a point (Figure 4a). The system is consistent and has a unique solution, the point of intersection.

(b)   The two lines are parallel (Figure 4b). Since the lines do not intersect, the linear system has no solution and is inconsistent.

(c)   The equations are different forms of the same line (Figure 4c). The system is consistent and has an infinite number of solutions, namely, any point on the line.

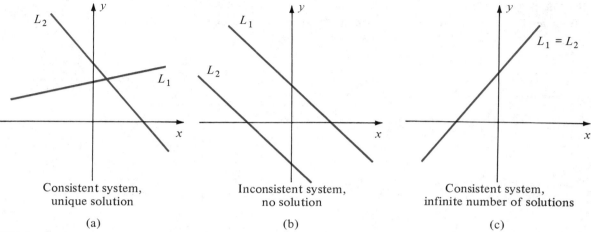

Consistent system,
unique solution

(a)

Inconsistent system,
no solution

(b)

Consistent system,
infinite number of solutions

(c)

FIGURE 4

EXAMPLE 4

If 3 sulfa pills and 4 penicillin pills cost 69 cents, while 5 sulfa pills and 2 penicillin pills cost 73 cents, what is the cost of each type of pill?

SOLUTION

Using two variables, we let

$$x = \text{the cost of each sulfa pill}$$
$$y = \text{the cost of each penicillin pill}$$

Then

$$3x + 4y = 69$$
$$5x + 2y = 73$$

We multiply the second equation by -2 and add to eliminate y:

$$
\begin{aligned}
3x + 4y &= 69 \\
-10x - 4y &= -146 \\
\hline
7x &= -77 \\
x &= 11
\end{aligned}
$$

Substituting in the first equation, we have

$$3(11) + 4y = 69$$
$$4y = 36$$
$$y = 9$$

Thus, each sulfa pill costs 11 cents and each penicillin pill costs 9 cents.

EXERCISE SET 10.1

In Exercises 1–10 solve the system of equations by the method of substitution.

1. $x + y = 1$
   $x - y = 3$

2. $x + 2y = 8$
   $3x - 4y = 4$

3. $x^2 + y^2 = 13$
   $2x - y = 4$

4. $x^2 + 4y^2 = 32$
   $x + 2y = 0$

5. $y^2 - x = 0$
   $y - 4x = -3$

6. $xy = -4$
   $4x - y = 8$

7. $x^2 - 2x + y^2 = 3$
   $2x + y = 4$

8. $4x^2 + y^2 = 4$
   $x - y = 3$

9. $xy = 1$
   $x - y + 1 = 0$

10. $\frac{1}{2}x - \frac{3}{2}y = 4$
    $\frac{3}{2}x + y = 1$

In Exercises 11–20 solve the system of equations by the method of elimination.

11. $x + 2y = 1$
    $5x + 2y = 13$

12. $x - 4y = -7$
    $2x + 3y = -8$

13. $25y^2 - 16x^2 = 400$
    $9y^2 - 4x^2 = 36$

14. $x^2 - y^2 = 3$
    $x^2 + y^2 = 5$

15. $4x^2 + 9y^2 = 72$
    $4x - 3y^2 = 0$

16. $x^2 + y^2 + 2y = 9$
    $y - 2x = 4$

17. $3x - y = 4$
    $6x - 2y = -8$

18. $2x + 3y = -2$
    $-3x - 5y = 4$

19. $2y^2 - x^2 = -1$
    $4y^2 + x^2 = 25$

20. $x^2 + 4y^2 = 25$
    $4x^2 + y^2 = 25$

In Exercises 21–32 determine whether the system is consistent (C) or inconsistent (I). If the system is consistent, find all solutions.

21. $2x + 2y = 6$
    $3x + 3y = 6$

22. $2x + y = 2$
    $3x - y = 8$

23. $y^2 - 8x^2 = 9$
    $y^2 + 3x^2 = -31$

24. $4y^2 + 3x^2 = 24$
    $3y^2 - 2x^2 = 35$

25. $3x + 3y = 9$
    $2x + 2y = -6$

26. $x - 4y = -7$
    $2x - 8y = -4$

27. $3x - y = 18$
    $\frac{3}{2}x - \frac{1}{2}y = 9$

28. $2x + y = 6$
    $x + \frac{1}{2}y = 3$

29. $x^2 - 3xy - 2y^2 - 2 = 0$
    $x - y - 2 = 0$

30. $3x^2 + 8y^2 = 20$
    $x^2 + 4y^2 = 10$

31. $2x - 2y = 4$
    $x - y = 8$

32. $2x - 3y = 8$
    $4x - 6y = 16$

In Exercises 33–42 use a pair of equations to solve the given problem.

33. A pile of 34 coins worth $4.10 consists of nickels and quarters. Find the number of each type of coin.

34. Car A can travel 20 kilometers per hour faster than car B. If car A travels 240 kilometers in the same time that car B travels 200 kilometers, what is the speed of each car?

35. How many pounds of nuts worth $2.10 per pound and how many pounds of raisins worth $0.90 per pound must be mixed to obtain a mixture of two pounds that is worth $1.62 per pound?

36. A part of $8000 was invested at an annual interest of 7% and the remainder at 8%. If the total interest received at the end of one year is $590, how much was invested at each rate?

37. The owner of a service station sold 1325 gallons of gasoline and collected 200 ration tickets. If Type-A ration tickets are used to purchase 10 gallons of gasoline and Type-B are used to purchase 1 gallon of gasoline, how many of each type of ration ticket did the station collect?

38. A bank is paying 12% annual interest on one-year certificates, and treasury notes are paying 10% annual interest. An investor received $620 interest at the end

of one year by investing a total of $6000. How much was invested at each rate?

39. The sum of the squares of the sides of a rectangle is 100 square meters. If the area of the rectangle is 48 square meters, find the length of each side of the rectangle.

40. Find the dimensions of a rectangle with an area of 30 square feet and a perimeter of 22 feet.

41. Find two numbers such that their product is 20 and their sum is 9.

42. Find two numbers such that the sum of their squares is 65 and their sum is 11.

10.2 SYSTEMS OF LINEAR INEQUALITIES

GRAPHING LINEAR INEQUALITIES

When we draw the graph of a linear equation, say

$$y = 2x - 1$$

we can readily see that the graph of the line divides the plane into two regions, which we call **half-planes** (see Figure 5). If, in the equation $y = 2x - 1$, we

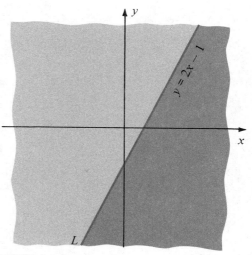

FIGURE 5

replace the equals sign with any of the symbols $<$, $>$, $\leq$, or $\geq$, we have a **linear inequality in two variables.** By the **graph of a linear inequality** such as

$$y < 2x - 1$$

we mean the set of all points whose coordinates satisfy the inequality. Thus, the point $(4, 2)$ lies on the graph of $y < 2x - 1$, since

$$2 < (2)(4) - 1 = 7$$

shows that $x = 4$, $y = 2$ satisfies the inequality. However, the point $(1, 5)$ does *not* lie on the graph of $y < 2x - 1$ since

$$5 < (2)(1) - 1 = 1$$

is not true. Since the coordinates of every point on the line L in Figure 5 satisfy the *equation* $y = 2x - 1$, we readily see that the coordinates of those points in the half-plane below the line must satisfy the *inequality* $y < 2x - 1$. Similarly, the coordinates of those points in the half-plane above the line must satisfy the *inequality* $y > 2x - 1$. This suggests that the graph of a linear inequality in two variables is a half-plane, and it leads to a straightforward method for graphing linear inequalities.

EXAMPLE 1
Sketch the graph of the inequality $x + y \geq 1$.

SOLUTION

Graphing Linear Inequalities

Step 1. Replace the inequality sign with an equals sign and plot the line.
(a) If the inequality is $\leq$ or $\geq$, plot a solid line (points on the line will satisfy the inequality).
(b) If the inequality is $<$ or $>$, plot a dashed line (points on the line will not satisfy the inequality).

Step 1. $x + y = 1$

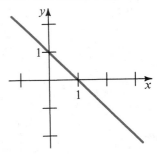

Step 2. Choose any point that is not on the line as a test point. If the origin is not on the line, it is the most convenient choice.

Step 2. Choose $(0, 0)$ as a test point.

Step 3. Substitute the coordinates of the test point into the inequality.
(a) If the test point satisfies the inequality, then the coordinates of every point in the half-plane that contains the test point will satisfy the inequality.

Step 3. Substituting $(0, 0)$ in

$$x + y \geq 1$$

gives

$$0 + 0 \geq 1 \quad (?)$$
$$0 \geq 1$$

which is false.

(b) If the test point does not satisfy the inequality, then the half-plane on the other side of the line contains all the points satisfying the inequality.

Since $(0, 0)$ is in the half-plane below the line and does not satisfy the inequality, all the points above the line will satisfy the inequality. See Figure 6.

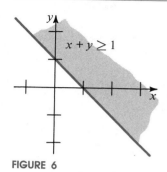

FIGURE 6

EXAMPLE 2

Sketch the graph of the inequality $2x - 3y > 6$.

SOLUTION

We first graph the line $2x - 3y = 6$. We draw a dashed or broken line to indicate that $2x - 3y = 6$ is not part of the graph (see Figure 7). Since $(0, 0)$ is not on the line, we can use it as a test point.

$$2x - 3y > 6$$
$$2(0) - 3(0) > 6 \quad (?)$$
$$0 - 0 > 6 \quad (?)$$
$$0 > 6$$

is false. Since $(0, 0)$ is in the half-plane above the line, the graph consists of the half-plane below the line.

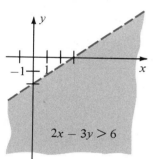

FIGURE 7

PROGRESS CHECK

Graph the inequalities.

(a) $y \leq 2x + 1$ (b) $y + 3x > -2$ (c) $y \geq -x + 1$

ANSWERS

(a)

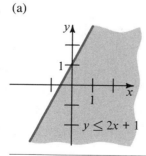

(b)

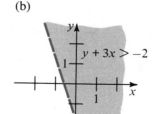

(c)

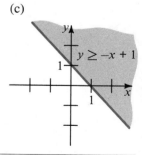

EXAMPLE 3

Graph the inequalities.

(a) $y > x$ (b) $2x \geq 5$

SOLUTION

(a) Since the origin lies on the line $y = x$, we choose another test point, say $(0, 1)$ above the line. Since $(0, 1)$ does satisfy the inequality, the graph of the inequality is the half-plane above the line. See Figure 8a.

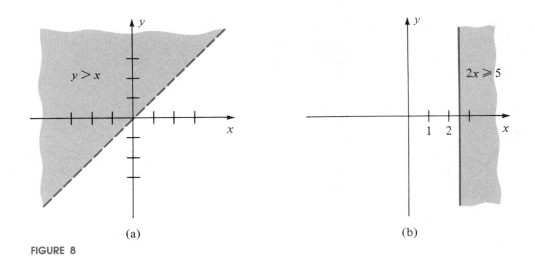

(a)

(b)

FIGURE 8

(b)   The graph of $2x = 5$ is a vertical line, and the graph of $2x \geq 5$ is the half-plane to the right of the line and also the line itself. See Figure 8b.

PROGRESS CHECK
Graph the inequalities.
(a) $2y \geq 7$ (b) $x < -2$ (c) $1 \leq y < 3$

ANSWERS
(a)

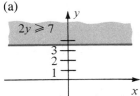

(b)

(c)

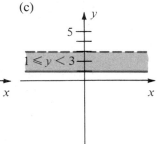

SYSTEMS OF LINEAR INEQUALITIES

We may also consider **systems of linear inequalities** in two variables, x and y. Examples of such systems are

$$2x - 3y > 6 \qquad 2x - 5y \leq 12$$
$$x + 2y < 2 \qquad 2x + y \leq 18$$
$$x \geq 0$$
$$y \geq 0$$

The **solution of a system of linear inequalities** consists of all ordered pairs (a, b) such that the substitution $x = a$, $y = b$ satisfies *all* the inequalities. Thus, the ordered pair $(2, 1)$ is a solution of the system

$$2x - 3y \le 2$$
$$x + y \le 6$$

since the substitution $x = 2$, $y = 1$ satisfies both inequalities.

$$(2)(2) - (3)(1) = 1 \le 2$$
$$2 + 1 = 3 \le 6$$

We can graph the solution set of a system of linear inequalities by graphing the solution set of each inequality and marking that portion of the graph that satisfies *all* the inequalities.

EXAMPLE 4

Graph the solution set of the system.

$$2x - 3y \le 2$$
$$x + y \le 6$$

SOLUTION

In Figure 9 we have graphed the solution set of each of the inequalities. The cross-hatched region indicates those points that satisfy both inequalities and is therefore the solution set of the system of inequalities.

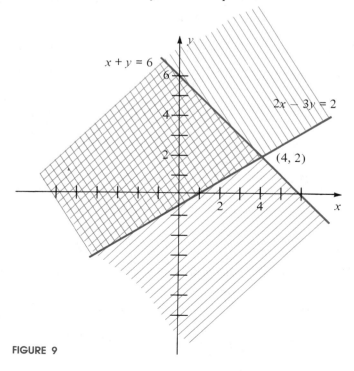

FIGURE 9

EXAMPLE 5
Graph the solution set of the system.

$$x + y < 2$$
$$2x + 3y \geq 9$$
$$x \geq 1$$

SOLUTION
See Figure 10. Since there are no points satisfying *all* the inequalities, we conclude that the system is inconsistent and has no solutions.

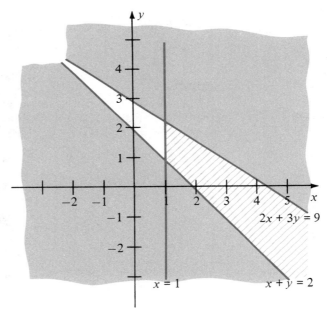

FIGURE 10

PROGRESS CHECK
Graph the solution set of the given system.

(a) $x + y \geq 3$ (b) $2x + y \leq 4$
      $x + 2y < 8$ $x + y \leq 3$
                            $x \geq 0$
                            $y \geq 0$

ANSWERS

(a)   (b)

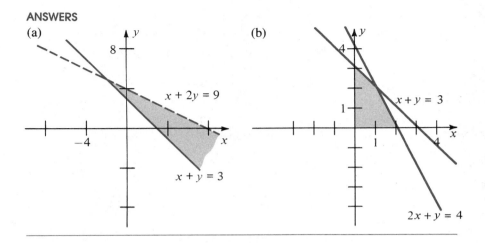

EXAMPLE 6

A dietitian at a university is planning a menu for a meal to consist of two primary foods, A and B, whose nutritional contents are shown in the table. The dietitian insists that the meal provide at most 12 units of fat, at least 2 units of carbohydrate, and at least 1 unit of protein. If x and y represent the number of grams of food types A and B, respectively, write a system of linear inequalities expressing the restrictions. Graph the solution set.

	Nutritional Content in Units per Gram		
	Fat	Carbohydrate	Protein
A	2	2	0
B	3	1	1

SOLUTION

The number of units of fat contained in the meal is $2x + 3y$, so x and y must satisfy the inequality

$$2x + 3y \leq 12 \quad \text{fat requirement}$$

Similarly, the requirements for carbohydrate and protein result in the inequalities

$$2x + y \geq 2 \quad \text{carbohydrate requirement}$$
$$y \geq 1 \quad \text{protein requirement}$$

Of course, we must also have $x \geq 0$, since negative quantities of food type A would make no sense. The system of linear inequalities is then

$$2x + 3y \leq 12$$
$$2x + y \geq 2$$
$$x \geq 0$$
$$y \geq 1$$

and the graph is shown in Figure 11.

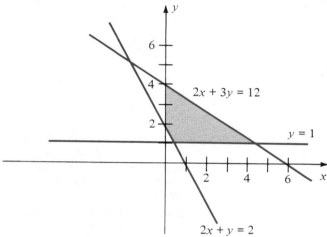

FIGURE 11

EXERCISE SET 10.2

Graph the solution set of the given inequality in the following exercises.

1. $y \leq x + 2$	2. $y \geq x + 3$	3. $y > x - 4$	4. $y < x - 5$
5. $y \leq 4 - x$	6. $y \geq 2 - x$	7. $y > x$	8. $y \leq 2x$
9. $3x - 5y > 15$	10. $2y - 3x < 12$	11. $x \leq 4$	12. $3x > -2$
13. $y > -3$	14. $5y \leq 25$	15. $x > 0$	16. $y < 0$
17. $-2 \leq x \leq 3$	18. $-6 < y < -2$		

19. A steel producer makes two types of steel, regular and special. A ton of regular steel requires 2 hours in the open-hearth furnace and a ton of special steel requires 5 hours. Let x and y denote the number of tons of regular and special steel, respectively, made per day. If the open-hearth furnace is available at most 15 hours per day, write an inequality that must be satisfied by x and y. Graph this inequality.

20. A patient is placed on a diet that restricts caloric intake to 1500 calories per day. The patient plans to eat x ounces of cheese, y slices of bread, and z apples on the first day of the diet. If cheese contains 100 calories per ounce, bread 110 calories per slice, and apples 80 calories each, write an inequality that must be satisfied by x, y, and z.

Graph the solution set of the system of linear inequalities.

21. $2x - y \leq 3$
    $2x + 3y \geq -3$

22. $x - y \leq 4$
    $2x + y \geq 6$

23. $3x - y \geq -7$
    $3x + y \leq -2$

24. $3x - 2y > 1$
    $2x + 3y \leq 18$

25. $3x - 2y \geq -4$
    $2x - y \leq 5$
    $y \geq 1$

26. $2x - y \geq -3$
    $x + y \leq 5$
    $y \geq 1$

27. $2x - y \leq 5$
    $x + 2y \geq 1$
    $x \geq 0$
    $y \geq 0$

28. $-x + 3y \leq 2$
    $4x + 3y \leq 18$
    $x \geq 0$
    $y \geq 0$

29. $3x + y \leq 6$
    $x - 2y \leq -1$
    $x \geq 2$

30. $x - y \geq -2$
    $x + y \geq -5$
    $y \geq 0$

31. $3x - 2y \leq -6$
    $8x + 3y \leq 24$
    $5x + 4y \geq 20$
    $x \geq 0$
    $y \geq 0$

32. $2x + 3y \geq 18$
    $x + 3y \geq 12$
    $4x + 3y \geq 24$
    $x \geq 0$
    $y \geq 0$

33. A farmer has 10 quarts of milk and 15 quarts of cream, which he will use to make ice cream and yogurt. Each quart of ice cream requires 0.4 quart of milk and 0.2 quart of cream, and each quart of yogurt requires 0.2 quart of milk and 0.4 quart of cream. Graph the set of points representing the possible production of ice cream and of yogurt.

34. A coffee packer uses Jamaican and Colombian coffee to prepare a mild blend and a strong blend. Each pound of mild blend contains $\frac{1}{2}$ pound of Jamaican coffee and $\frac{1}{2}$ pound of Colombian coffee, and each pound of the strong blend requires $\frac{1}{4}$ pound of Jamaican coffee and $\frac{3}{4}$ pound of Colombian coffee. The packer has available 100 pounds of Jamaican coffee and 125 pounds of Colombian coffee. Graph the set of points representing the possible production of the two blends.

35. A trust fund of $100,000 that has been established to provide university scholarships must adhere to certain restrictions.

    (a) No more than half of the fund may be invested in common stocks.

    (b) No more than $35,000 may be invested in preferred stocks.

    (c) No more than $60,000 may be invested in all types of stocks.

    (d) The amount invested in common stocks cannot be more than twice the amount invested in preferred stocks.

    Graph the solution set representing the possible investments in common and preferred stocks.

36. An institution serves a luncheon consisting of two dishes, A and B, whose nutritional content in grams per unit served is given in the accompanying table.

	Fat	Carbohydrate	Protein
A	1	1	2
B	2	1	6

The luncheon is to provide no more than 10 grams of fat, no more than 7 grams of carbohydrate, and at least 6 grams of protein. Graph the solution set of possible quantities of dishes A and B.

10.3 LINEAR PROGRAMMING (Optional)

Let's pose the following problem.

A lot is zoned for an apartment building to consist of no more than 40 apartments, totaling no more than 45,000 square feet. A builder is planning to construct 1-bedroom apartments, each of which will require 1000 square feet and will rent for $200 per month, and 2-bedroom apartments, each of which will utilize 1500 square feet and will rent for $280 per month. If all available apartments can be rented, how many apartments of each type should be built to maximize the builder's monthly rental revenue?

If we let x denote the number of 1-bedroom units and y denote the number of 2-bedroom units, the accompanying table displays the information given in the problem.

	Number of units	Square feet	Rental
1-bedroom	x	1,000	$200
2-bedroom	y	1,500	280
Total	40	45,000	z

Using the methods of the previous section, we can translate the **constraints** or requirements upon the variables x and y into a system of inequalities. The total number of apartments is $x + y$, so we have

$$x + y \le 40 \quad \text{number of units constraint}$$

Since each 1-bedroom apartment occupies 1000 square feet of space, x apartments will occupy $1000x$ square feet of space. Similarly, the 2-bedroom apartments will require $1500y$ square feet of space. The total amount of space needed is $1000x + 1500y$, so we must have

$$1000x + 1500y \le 45,000 \quad \text{square footage constraint}$$

Moreover, since x and y denote the number of apartments to be built, we must have $x \ge 0$, $y \ge 0$. Thus, we have obtained the following system of inequalities:

$$
\begin{aligned}
x + \quad y &\le \quad 40 && \text{number of units constraint} \\
1000x + 1500y &\le 45,000 && \text{square footage constraint} \\
x &\ge \quad 0 && \text{need for number of apartments} \\
y &\ge \quad 0 && \text{to be nonnegative}
\end{aligned}
$$

We can graph the solution set of this system of linear inequalities as in Figure 12.

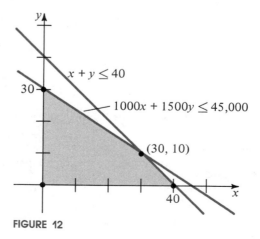

FIGURE 12

But the problem as stated asks that we *maximize* the monthly rental $z = 200x + 280y$, a requirement that we have never before seen in a mathematical problem of this sort! It is this requirement to **optimize,** that is, to seek a maximum or a minimum value of a linear expression, that characterizes a linear programming problem.

Linear Programming Problem	A **linear programming problem** seeks the optimal (either the largest or the smallest) value of a linear expression called the **objective function** while satisfying constraints that can be formulated as a system of linear inequalities.

Returning to our apartment builder, we can state the linear programming problem in this way:

$$\begin{aligned} \text{maximize} \qquad & z = 200x + 280y \\ \text{subject to} \qquad & x + \quad y \le \quad 40 \\ & 1000x + 1500y \le 45{,}000 \\ & x \ge \quad 0 \\ & y \ge \quad 0 \end{aligned}$$

Then the coordinates of each point of the solution set shown in Figure 12 are a **feasible solution,** that is, the coordinates give us ordered pairs (a, b) that satisfy the system of linear inequalities. But which points provide us with values of x and y that maximize the rental income z? For example, the points $(40, 0)$ and $(15, 20)$ are feasible solutions yielding these results for z:

x	y	$z = 200x + 280y$
40	0	8000
15	20	8600

Clearly, building 15 1-bedroom and 20 2-bedroom units yields a higher rental revenue than building 40 1-bedroom units, but is there a solution that will yield a still higher value for z?

Before providing the key to solving linear programming problems, we first must note that the solution set is bounded by straight lines, and we use the term **vertex** to denote an intersection point of any two boundary lines. We are then ready to state the following theorem.

Fundamental Theorem of Linear Programming	If a linear programming problem has an optimal solution, that solution occurs at a vertex of the set of feasible solutions.

With this result, the builder need only examine the vertices of the solution set of Figure 12, rather than considering each of the infinite number of feasible solutions—a bewildering task! We then evaluate the objective function $z = 200x + 280y$ for the coordinates of the vertices $(0, 0)$, $(0, 30)$, $(40, 0)$, and $(30, 10)$.

x	y	$z = 200x + 280y$
0	0	0
0	30	8400
40	0	8000
30	10	8800

Since the largest value of z is 8800 and this value corresponds to $x = 30$, $y = 10$, the builder finds that the optimal strategy is to build 30 1-bedroom and 10 2-bedroom units.

We can now illustrate the steps in solving a linear programming problem.

EXAMPLE 1

Solve the linear programming problem

$$\text{minimize} \qquad z = x - 4y$$
$$\text{subject to} \qquad x + 2y \le 10$$
$$-x + 4y \le 8$$
$$x \ge 0$$
$$y \ge 1$$

SOLUTION

Linear Programming	
Step 1. Sketch the solution set of the system of linear inequalities.	*Step 1*. 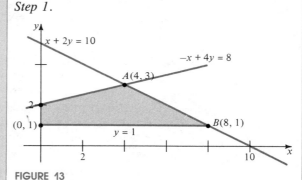 FIGURE 13
Step 2. Determine all vertices of the solution set.	*Step 2*. The vertices $(0, 1)$ and $(0, 2)$ are the y-intercepts of the lines whose equations are $y = 1$ and $-x + 4y = 8$, respectively. The vertex B in Figure 13 is the intersection of the lines $y = 1$ and $x + 2y = 10$ and is seen to be $(8, 1)$. The vertex A of

Figure 13 is the intersection of the lines whose equations are

$$-x + 4y = 8$$

and

$$x + 2y = 10$$

Solving the system of equations (try elimination) yields the vertex $A(4, 3)$.

Step 3. Evaluate the objective function for the coordinates of each vertex.

Step 3.

Vertex	x	y	$z = x - 4y$
(0, 1)	0	1	-4
(0, 2)	0	2	-8
(8, 1)	8	1	4
(4, 3)	4	3	-8

Step 4. The point or points providing the optimal value of the objective function are solutions of the linear programming problem.

Step 4. The minimal value of the objective function is -8, which occurs at the vertices (0, 2) and (4, 3). Thus, $x = 0$, $y = 2$ and $x = 4$, $y = 3$ are both solutions of the linear programming problem.

Linear programming problems occur in real-life situations with great frequency. In certain industries these problems can involve thousands of variables and hundreds of constraints. Obviously, the method of graphical solution we presented for two variables cannot be used. A solution method known as the simplex algorithm was first devised by George Dantzig in 1947. Despite the sophistication of this approach, the number of calculations required becomes unmanageably large for hand computation for even relatively small numbers of constraints. Fortunately, the discovery of the simplex algorithm occurred at the time electronic computers made their initial appearance. Since then industries such as oil refining and steel production have used linear programming to determine the optimum use of their facilities.

EXERCISE SET 10.3

In Exercises 1–8 find the minimum value and the maximum value of the linear expression, subject to the given constraints. Indicate coordinates of the vertices at which the minimum and maximum values occur.

1. $x - \dfrac{1}{2}y$ subject to

$$3x - y \leq 1$$
$$x \geq 0$$
$$x \leq 5$$
$$y \geq 0$$

2. $2x + y$ subject to

$$x + y \leq 4$$
$$x \geq 1$$
$$y \geq 2$$

3. $\frac{1}{2}x - 2y$ subject to

$$x + 2y \leq 6$$
$$3y - 2x \leq 2$$
$$x \geq 0$$
$$y \geq 0$$

4. $0.2x + 0.8y$ subject to

$$3y + x \leq 8$$
$$3x - 5y \geq 2$$
$$x \geq 0$$
$$y \geq 0$$

5. $2x - y$ subject to

$$y - x \leq 0$$
$$4y + 3x \geq 6$$
$$x \leq 4$$

6. $x + 3y$ subject to

$$2x + y \geq 2$$
$$4x + 5y \leq 40$$
$$x \geq 0$$
$$y \geq 1$$
$$y \leq 6$$

7. $2x - y$ subject to

$$2y - x \leq 8$$
$$x + 2y \geq 12$$
$$5x + 2y \leq 44$$
$$x \geq 3$$

8. $y - x$ subject to

$$2y - 5x \leq 10$$
$$5x + 6y \leq 50$$
$$5x + y \leq 20$$
$$x \geq 0$$
$$y \geq 1$$

9. A firm has budgeted $1500 for display space at a toy show. Two types of display booths are available: "preferred space" costs $18 per square foot, with a minimum rental of 60 square feet, and "regular space" costs $12 per square foot, with a minimum rental of 30 square feet. It is estimated that there will be 120 visitors for each square foot of "preferred space" and 60 visitors for each square foot of "regular space." How should the firm allot its budget to maximize the number of potential clients that will visit the booths?

10. A company manufactures an 8-bit computer and a 16-bit computer. To meet existing orders, it must schedule at least 50 8-bit computers for the next production cycle and can produce no more than 150 8-bit computers. The manufacturing facilities are adequate to produce no more than 300 16-bit computers, but the total number of computers that can be produced cannot exceed 400. The profit on each 8-bit computer is $310; on each 16-bit computer the profit is $275. Find the number of computers of each type that should be manufactured to maximize profit.

11. Swift Truckers is negotiating a contract with Better Spices, which uses two sizes of containers: large, 4-cubic-foot containers weighing 10 pounds and small,

2-cubic-foot containers weighing 8 pounds. Swift Truckers will use a vehicle that can handle a maximum load of 3280 pounds and a cargo size of up to 1000 cubic feet. The firms have agreed on a shipping rate of 50 cents for each large container and 30 cents for each small container. How many containers of each type should Swift place on a truck to maximize income?

12. A bakery makes both yellow cake and white cake. Each pound of yellow cake requires $\frac{1}{4}$ pound of flour and $\frac{1}{4}$ pound of sugar; each pound of white cake requires $\frac{1}{3}$ pound of flour and $\frac{1}{5}$ pound of sugar. The baker finds that 100 pounds of flour and 80 pounds of sugar are available. If yellow cake sells for $3 per pound and white cake sells for $2.50 per pound, how many pounds of each cake should the bakery produce to maximize income, assuming that all cakes baked can be sold?

13. A shop sells a mixture of Java and Colombian coffee beans for $4 per pound. The shopkeeper has allocated $1000 for buying fresh beans and finds that he must pay $1.50 per pound for Java beans and $2 per pound for Colombian beans. In a satisfactory mixture the weight of Colombian beans will be at least twice and no more than four times the weight of the Java beans.

How many pounds of each type of coffee bean should be ordered to maximize the profit if all the mixture can be sold?

14. A pension fund plans to invest up to $50,000 in U.S. Treasury bonds yielding 12% interest per year and corporate bonds yielding 15% interest per year. The fund manager is told to invest a minimum of $25,000 in the Treasury bonds and a minimum of $10,000 in the corporate bonds, with no more than $\frac{1}{4}$ of the total investment to be in corporate bonds. How much should the manager invest in each type of bond to achieve a maximum amount of annual interest? What is the maximum interest?

15. A farmer intends to plant crops A and B on all or part of a 100-acre field. Seed for crop A costs $6 per acre, and labor and equipment costs $20 per acre. For crop B, seed costs $9 per acre, and labor and equipment costs $15 per acre. The farmer cannot spend more than $810 for seed and $1800 for labor and equipment. If the income per acre is $150 for crop A and $175 for crop B, how many acres of each crop should be planted to maximize total income?

16. The farmer in Exercise 15 finds that a worldwide surplus in crop B reduces the income to $140 per acre while the income for crop A remains steady at $150 per acre. How many acres of each crop should be planted to maximize total income?

17. In preparing food for the college cafeteria, a dietitian will combine Volume Pack A and Volume Pack B. Each pound of Volume Pack A costs $2.50 and contains 4 units of carbohydrate, 3 units of protein, and 5 units of fat. Each pound of Volume Pack B costs $1.50 and contains 3 units of carbohydrate, 4 units of protein, and 1 unit of fat. If minimum monthly requirements are 60 units of carbohydrates, 52 units of protein, and 42 units of fat, how many pounds of each food pack will the dietitian use to minimize costs?

18. A lawn service uses a riding mower that cuts a 5000-square-foot area per hour and a smaller mower that cuts a 3000-square-foot area per hour. Surprisingly, each mower uses $\frac{1}{2}$ gallon of gasoline per hour. Near the end of a long summer day, the supervisor finds that both mowers are empty and that there remains 0.6 gallon of gasoline in the storage cans. To conclude the day at a sensible point, at least 4000 square feet of lawn must still be mowed. If the cost of operating the riding mower is $9 per hour and the cost of operating the smaller mower is $5 per hour, how much of the remaining gasoline should be allocated to each mower to do the job at the least possible cost?

10.4
SYSTEMS OF LINEAR EQUATIONS IN THREE UNKNOWNS

The method of substitution and the method of elimination can both be applied to systems of linear equations in three unknowns and, more generally, to systems of linear equations in any number of unknowns. There is yet another method, ideally suited for computers, which we will now apply to solving linear systems in three unknowns.

GAUSSIAN ELIMINATION AND TRIANGULAR FORM

In solving equations, we found it convenient to transform an equation into an equivalent equation having the same solution set. Similarly, we can attempt to transform a system of equations into another system, called an **equivalent system,** that has the same solution set. In particular, the objective of **Gaussian elimination** is to transform a linear system into an equivalent system in triangular form such as

$$3x - y + 3z = -11$$
$$2y + z = 2$$
$$2z = -4$$

A linear system is in **triangular form** when the only nonzero coefficient of x appears in the first equation, the only nonzero coefficients of y appear in the first and second equations, and so on.

Note that when a linear system is in triangular form, the last equation immediately yields the value of an unknown. In our example, we see that

$$2z = -4$$
$$z = -2$$

Substituting $z = -2$ in the second equation yields

$$2y + (-2) = 2$$
$$y = 2$$

Finally, substituting $z = -2$ and $y = 2$ in the first equation yields

$$3x - (2) + 3(-2) = -11$$
$$3x = -3$$
$$x = -1$$

This process of **back-substitution** thus allows us to solve a linear system quickly when it is in triangular form.

The challenge, then, is to find a means of transforming a linear system into triangular form. We now offer (without proof) a list of operations that transform a system of linear equations into an equivalent system.

(1) Interchange any two equations.
(2) Multiply an equation by a nonzero constant.
(3) Replace an equation by the sum of itself plus a constant times another equation.

Using these operations we can now demonstrate the method of Gaussian elimination.

EXAMPLE 1
Solve the linear system.

$$2y - z = -5$$
$$x - 2y + 2z = 9$$
$$2x - 3y + 3z = 14$$

SOLUTION

Gaussian Elimination	
Step 1. (a) If necessary, interchange equations to obtain a nonzero coefficient for x in the first equation.	*Step 1.* (a) Interchanging the first two equations yields $$x - 2y + 2z = 9$$ $$2y - z = -5$$ $$2x - 3y + 3z = 14$$

(b) Replace the second equation with the sum of itself plus an appropriate multiple of the first equation, which will result in a zero coefficient for x.
(c) Replace the third equation with the sum of itself plus an appropriate multiple of the first equation, which will result in a zero coefficient for x.

(b) The coefficient of x in the second equation is already 0.

(c) Replace the third equation with the sum of itself plus -2 times the first equation.

$$x - 2y + 2z = 9$$
$$2y - z = -5$$
$$y - z = -4$$

Step 2. Apply the procedures of Step 1 to the second and third equations.

Step 2. Replace the third equation with the sum of itself and $-\frac{1}{2}$ times the second equation.

$$x - 2y + 2z = 9$$
$$2y - z = -5$$
$$-\frac{1}{2}z = -\frac{3}{2}$$

Step 3. The system is now in triangular form. The solution is obtained by back-substitution.

Step 3. From the third equation,

$$-\frac{1}{2}z = -\frac{3}{2}$$
$$z = 3$$

Substituting this value of z in the second equation, we have

$$2y - (3) = -5$$
$$y = -1$$

Substituting for y and for z in the first equation, we have

$$x - 2(-1) + 2(3) = 9$$
$$x + 8 = 9$$
$$x = 1$$

The solution is $x = 1$, $y = -1$, $z = 3$.

PROGRESS CHECK
Solve by Gaussian elimination.

(a)
$$2x - 4y + 2z = 1$$
$$3x + y + 3z = 5$$
$$x - y - 2z = -8$$

(b)
$$-2x + 3y - 12z = -17$$
$$3x - y - 15z = 11$$
$$-x + 5y + 3z = -9$$

ANSWERS

(a) $x = -\dfrac{3}{2}, y = \dfrac{1}{2}, z = 3$ (b) $x = 5, y = -1, z = \dfrac{1}{3}$

CONSISTENT AND INCONSISTENT SYSTEMS

The graph of a linear equation in three unknowns is a plane in three-dimensional space. A system of three linear equations in three unknowns corresponds to three planes (Figure 14). If the planes intersect in a point P (Figure 14a), the coordinates of the point P are a solution of the system and can be found by Gaussian

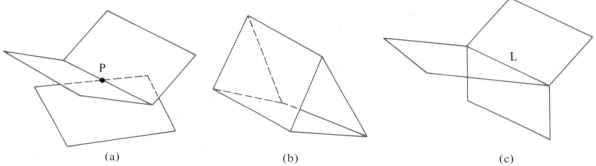

(a)   (b)   (c)

FIGURE 14

elimination. The cases of no solution and of an infinite number of solutions are signaled as follows.

Consistent and Inconsistent Systems
• If Gaussian elimination results in an equation of the form $$0x + 0y + 0z = c, \quad c \neq 0$$ then the system is inconsistent (Figure 14b). • If Gaussian elimination results in no equation of the type above but results in an equation of the form $$0x + 0y + 0z = 0$$ then the system is consistent and has an infinite number of solutions (Figure 14c). • Otherwise, the system is consistent and has a unique solution.

EXAMPLE 2
Solve the linear system.

$$\begin{aligned}
x - 2y + 2z &= -4 \\
x + y - 7z &= 8 \\
-x - 4y + 16z &= -20
\end{aligned}$$

SOLUTION

Replacing the second equation by itself minus the first equation, and replacing the third equation by itself plus the first equation, we have

$$x - 2y + 2z = -4$$
$$3y - 9z = 12$$
$$-6y + 18z = -24$$

Replacing the third equation of the last system by itself plus 2 times the second equation results in the system

$$x - 2y + 2z = -4$$
$$3y - 9z = 12$$
$$0x + 0y + 0z = 0$$

in which the last equation indicates that the system is consistent and has an infinite number of solutions. If we solve the second equation of the last system for y, we have

$$y = 3z + 4$$

Then, solving the first equation for x, we have

$$x = 2y - 2z - 4$$
$$= 2(3z + 4) - 2z - 4 \quad \text{Substituting for } y$$
$$= 4z + 4$$

The equations

$$x = 4z + 4$$
$$y = 3z + 4$$

yield a solution of the original system for every real value of z. For example, if $z = 0$, then $x = 4$, $y = 4$, $z = 0$ satisfies the original system; if $z = -2$, then $x = -4$, $y = -2$, $z = -2$ is another solution.

PROGRESS CHECK

(a) Verify that the linear system

$$x - 2y + z = 3$$
$$2x + y - 2z = -1$$
$$-x - 8y + 7z = 5$$

is consistent.

(b) Verify that the linear system

$$2x + y + 2z = 1$$
$$x - 4y + 7z = -4$$
$$x - y + 3z = -1$$

has an infinite number of solutions.

EXERCISE SET 10.4

Solve by Gaussian elimination. Indicate if the system is inconsistent or has an infinite number of solutions.

1. $x + 2y + 3z = -6$
   $2x - 3y - 4z = 15$
   $3x + 4y + 5z = -8$

2. $2x + 3y + 4z = -12$
   $x - 2y + z = -5$
   $3x + y + 2z = 1$

3. $x + y + z = 1$
   $x + y - 2z = 3$
   $2x + y + z = 2$

4. $2x - y + z = 3$
   $x - 3y + z = 4$
   $-5x - 2z = -5$

5. $x + y + z = 2$
   $x - y + 2z = 3$
   $3x + 5y + 2z = 6$

6. $x + y + z = 0$
   $x + y = 3$
   $y + z = 1$

7. $x + 2y + z = 7$
   $x + 2y + 3z = 11$
   $2x + y + 4z = 12$

8. $4x + 2y - z = 5$
   $3x + 3y + 6z = 1$
   $5x + y - 8z = 8$

9. $x + y + z = 2$
   $x + 2y + z = 3$
   $x + y - z = 2$

10. $x + y - z = 2$
    $x + 2y + z = 3$
    $x + y + 4z = 3$

11. $2x + y + 3z = 8$
    $-x + y + z = 10$
    $x + y + z = 12$

12. $2x - 3z = 4$
    $x + 4y - 5z = -6$
    $3x + 4y - z = -2$

13. $x + 3y + 7z = 1$
    $3x - y - 5z = 9$
    $2x + y + z = 4$

14. $2x - y + z = 2$
    $3x + y + 2z = 3$
    $x + y - z = -1$

15. $x - 2y + 3z = -2$
    $x - 5y + 9z = 4$
    $2x - y = 6$

16. $x + 2y - 2z = 8$
    $5y - z = 6$
    $-2x + y + 3z = -2$

17. $x - 2y + z = -5$
    $2x + z = -10$
    $y - z = 15$

18. $2y - 3z = 4$
    $x + 2z = -2$
    $x - 8y + 14z = -18$

19. A special low-calorie diet consists of dishes A, B, and C. Each unit of A has 2 grams of fat, 1 gram of carbohydrate, and 3 grams of protein. Each unit of B has 1 gram of fat, 2 grams of carbohydrate, and 1 gram of protein. Each unit of C has 1 gram of fat, 2 grams of carbohydrate, and 3 grams of protein. The diet must provide exactly 10 grams of fat, 14 grams of carbohydrate, and 18 grams of protein. How much of each dish should be used?

20. A furniture manufacturer makes chairs, coffee tables, and dining room tables. Each chair requires 2 minutes of sanding, 2 minutes of staining, and 4 minutes of varnishing. Each coffee table requires 5 minutes of sanding, 4 minutes of staining, and 3 minutes of varnishing. Each dining room table requires 5 minutes of sanding, 4 minutes of staining, and 6 minutes of varnishing. The sanding benches, staining benches, and varnishing benches are available 6, 5, and 6 hours per day, respectively. How many of each type of furniture can be made if all facilities are used to capacity?

21. A manufacturer produces 12", 16", and 19" television sets that require assembly, testing, and packing. The 12" sets each require 45 minutes to assemble, 30 minutes to test, and 10 minutes to package. The 16" sets each require 1 hour to assemble, 45 minutes to test, and 15 minutes to package. The 19" sets each require $1\frac{1}{2}$ hours to assemble, 1 hour to test, and 15 minutes to package. If the assembly line operates for $17\frac{3}{4}$ hours per day, the test facility is used for $12\frac{1}{2}$ hours per day, and the packing equipment is used for $3\frac{3}{4}$ hours per day, how many of each type of set can be produced?

10.5
MATRICES AND
LINEAR SYSTEMS

We have already studied several methods for solving a linear system such as

$$2x + 3y = -7$$
$$3x - y = 17$$

This system can be displayed by a **matrix,** which is simply a rectangular array of mn real numbers arranged in m horizontal rows and n vertical columns. The numbers are called the **entries** or **elements** of the matrix and are enclosed within brackets. Thus,

$$A = \begin{bmatrix} 2 & 3 & -7 \\ 3 & -1 & 17 \end{bmatrix} \begin{matrix} \leftarrow \\ \leftarrow \end{matrix} \text{rows}$$
$$\uparrow \quad \uparrow \quad \uparrow$$
$$\text{columns}$$

is a matrix consisting of two rows and three columns, whose entries are obtained from the two given equations. In general, a matrix of m rows and n columns is said to be of **dimension m by n,** written $m \times n$. The matrix A is seen to be of dimension 2×3. If the numbers of rows and columns of a matrix are both equal to n, the matrix is called a **square matrix** of **order** n.

EXAMPLE 1

(a)
$$A = \begin{bmatrix} -1 & 4 \\ 0.1 & -2 \end{bmatrix}$$

is a 2×2 matrix. Since matrix A has two rows and two columns, it is a square matrix of order 2.

(b)
$$B = \begin{bmatrix} 4 & -5 \\ -2 & 1 \\ 3 & 0 \end{bmatrix}$$

has three rows and two columns and is a 3×2 matrix.

(c)
$$C = [-8 \quad 6 \quad 1]$$

is a 1×3 matrix and is called a **row matrix** since it has precisely one row.

(d)
$$D = \begin{bmatrix} 2 \\ -4 \end{bmatrix}$$

is a 2×1 matrix and is called a **column matrix** since it has precisely one column.

There is a convenient way of denoting a general $m \times n$ matrix, using "double subscripts."

$$A = \begin{bmatrix} a_{11} & a_{12} & \cdots & a_{1j} & \cdots & a_{1n} \\ a_{21} & a_{22} & \cdots & a_{2j} & \cdots & a_{2n} \\ \cdot & \cdot & & \cdot & & \cdot \\ \cdot & \cdot & & \cdot & & \cdot \\ a_{i1} & a_{i2} & \cdots & a_{ij} & \cdots & a_{in} \\ \cdot & \cdot & & \cdot & & \cdot \\ \cdot & \cdot & & \cdot & & \cdot \\ a_{m1} & a_{m2} & \cdots & a_{mj} & \cdots & a_{mn} \end{bmatrix} \begin{array}{l} \leftarrow \text{first row} \\ \leftarrow \text{second row} \\ \\ \\ \leftarrow i\text{th row} \\ \\ \\ \leftarrow m\text{th row} \end{array}$$

$$\begin{array}{cccc} \uparrow & \uparrow & \uparrow & \uparrow \\ \text{first} & \text{second} & j\text{th} & n\text{th} \\ \text{column} & \text{column} & \text{column} & \text{column} \end{array}$$

Thus, a_{ij} is the entry in the ith row and jth column of the matrix A. It is customary to write $A = [a_{ij}]$ to indicate that a_{ij} is the entry in row i and column j of matrix A.

EXAMPLE 2

Let

$$A = \begin{bmatrix} 3 & -2 & 4 & 5 \\ 9 & 1 & 2 & 0 \\ -3 & 2 & -4 & 8 \end{bmatrix}$$

Matrix A is of dimension 3×4. The element a_{12} is found in the first row and second column and is seen to be -2. Similarly, we see that $a_{31} = -3$, $a_{33} = -4$, and $a_{34} = 8$.

PROGRESS CHECK

Let

$$B = \begin{bmatrix} 4 & 8 & 1 \\ 2 & -5 & 3 \\ -8 & 6 & -4 \\ 0 & 1 & -1 \end{bmatrix}$$

Find (a) b_{11} (b) b_{23} (c) b_{31} (d) b_{42}

ANSWERS

(a) 4   (b) 3   (c) -8 (d) 1

If we begin with the system of linear equations

$$2x + 3y = -7$$
$$3x - y = 17$$

the matrix

$$\begin{bmatrix} 2 & 3 \\ 3 & -1 \end{bmatrix}$$

in which the first column is formed from the coefficients of x and the second column is formed from the coefficients of y, is called the **coefficient matrix.** The matrix

$$\begin{bmatrix} 2 & 3 & | & -7 \\ 3 & -1 & | & 17 \end{bmatrix}$$

which includes the column consisting of the right-hand sides of the equations separated by a dashed line, is called the **augmented matrix.**

EXAMPLE 3

Write a system of linear equations that corresponds to the augmented matrix.

$$\begin{bmatrix} -5 & 2 & -1 & | & 15 \\ 0 & -2 & 1 & | & -7 \\ \frac{1}{2} & 1 & -1 & | & 3 \end{bmatrix}$$

SOLUTION

We attach the unknown x to the first column, the unknown y to the second column, and the unknown z to the third column. The resulting system is

$$-5x + 2y - z = 15$$
$$- 2y + z = -7$$
$$\tfrac{1}{2}x + y - z = 3$$

Now that we have seen how a matrix can be used to represent a system of linear equations, we next proceed to show how routine operations on that matrix can yield the solution of the system. These ''matrix methods'' are simply a clever streamlining of the methods already studied in this chapter.

In Section 4 of this chapter we used three elementary operations to transform a system of linear equations into triangular form. When applying the same procedures to a matrix, we speak of rows, columns, and elements instead of equations, variables, and coefficients. The three elementary operations that yield an equivalent system now become the **elementary row operations.**

Elementary Row Operations	The following elementary row operations transform an augmented matrix into an equivalent system.
	(1) Interchange any two rows.
	(2) Multiply each element of any row by a constant $k \neq 0$.
	(3) Replace each element of a given row by the sum of itself plus k times the corresponding element of any other row.

The method of Gaussian elimination introduced in Section 4 of this chapter can now be restated in terms of matrices. By use of elementary row operations we seek to transform an augmented matrix into a matrix for which $a_{ij} = 0$ when $i > j$. The resulting matrix will have the following appearance for a system of three linear equations in three unknowns.

$$\begin{bmatrix} * & * & * & | & * \\ 0 & * & * & | & * \\ 0 & 0 & * & | & * \end{bmatrix}$$

Since this matrix represents a linear system in triangular form, back-substitution will provide a solution of the original system. We will illustrate the process with an example.

EXAMPLE 4

Solve the system.

$$x - y + 4z = 4$$
$$2x + 2y - z = 2$$
$$3x - 2y + 3z = -3$$

SOLUTION

We describe and illustrate the steps of the procedure.

Gaussian Elimination	
Step 1. Form the augmented matrix.	*Step 1.* The augmented matrix is $$\begin{bmatrix} 1 & -1 & 4 & \mid & 4 \\ 2 & 2 & -1 & \mid & 2 \\ 3 & -2 & 3 & \mid & -3 \end{bmatrix}$$
Step 2. If necessary, interchange rows to make sure that a_{11}, the first element of the first row, is nonzero. We call a_{11} the **pivot element** and row 1 the **pivot row.**	*Step 2.* We see that $a_{11} = 1 \neq 0$. The pivot element is a_{11} and is shown in color.
Step 3. Arrange to have 0 as the first element of every row below row 1. This is done by replacing row 2, row 3, and so on by the sum of itself and an appropriate multiple of row 1.	*Step 3.* To make $a_{21} = 0$, replace row 2 by the sum of itself and (-2) times row 1; to make $a_{31} = 0$, replace row 3 by the sum of itself and (-3) times row 1. $$\begin{bmatrix} 1 & -1 & 4 & \mid & 4 \\ 0 & 4 & -9 & \mid & -6 \\ 0 & 1 & -9 & \mid & -15 \end{bmatrix}$$
Step 4. Repeat the process defined by Steps 2 and 3, allowing row 2, row 3, and so on to play the role of the first row. Thus row 2, row 3, and so on serve as the pivot rows.	*Step 4.* Since $a_{22} = 4 \neq 0$, it will serve as the next pivot element and is shown in color. To make $a_{32} = 0$, replace row 3 by the sum of itself and $(-\frac{1}{4})$ times row 2. $$\begin{bmatrix} 1 & -1 & 4 & \mid & 4 \\ 0 & 4 & -9 & \mid & -6 \\ 0 & 0 & -\frac{27}{4} & \mid & -\frac{27}{2} \end{bmatrix}$$
Step 5. The corresponding linear system is in triangular form. Solve by back-substitution.	*Step 5.* The third row of the final matrix yields $$-\frac{27}{4}z = -\frac{27}{2}$$ $$z = 2$$

Substituting $z = 2$, we obtain from the second row of the final matrix

$$4y - 9z = -6$$
$$4y - 9(2) = -6$$
$$y = 3$$

Substituting $y = 3$, $z = 2$, we obtain from the first row of the final matrix

$$x - y + 4z = 4$$
$$x - 3 + 4(2) = 4$$
$$x = -1$$

The solution is $x = -1$, $y = 3$, $z = 2$.

PROGRESS CHECK

Solve the linear system by matrix methods.

$$2x + 4y - z = 0$$
$$x - 2y - 2z = 2$$
$$-5x - 8y + 3z = -2$$

ANSWER

$x = 6$, $y = -2$, $z = 4$

Note that we have described the process of Gaussian elimination in a manner that will apply to any augmented matrix that is $n \times (n + 1)$; that is, Gaussian elimination may be used on any system of n linear equations in n unknowns that has a unique solution.

It is also permissible to perform elementary row operations in clever ways to simplify the arithmetic. For instance, you may wish to interchange rows, or to multiply a row by a constant to obtain a pivot element equal to 1. We will illustrate these ideas with an example.

EXAMPLE 5

Solve by matrix methods.

$$2y + 3z = 4$$
$$4x + y + 8z + 15w = -14$$
$$x - y + 2z = 9$$
$$-x - 2y - 3z - 6w = 10$$

SOLUTION

We begin with the augmented matrix and perform a sequence of elementary row operations. The pivot element is shown in color.

$$\begin{bmatrix} 0 & 2 & 3 & 0 & | & 4 \\ 4 & 1 & 8 & 15 & | & -14 \\ 1 & -1 & 2 & 0 & | & 9 \\ -1 & -2 & -3 & -6 & | & 10 \end{bmatrix}$$ Augmented matrix.
Note that $a_{11} = 0$.

$$\begin{bmatrix} 1 & -1 & 2 & 0 & | & 9 \\ 4 & 1 & 8 & 15 & | & -14 \\ 0 & 2 & 3 & 0 & | & 4 \\ -1 & -2 & -3 & -6 & | & 10 \end{bmatrix}$$ Interchanged rows 1 and 3 so that $a_{11} = 1$.

$$\begin{bmatrix} 1 & -1 & 2 & 0 & | & 9 \\ 0 & 5 & 0 & 15 & | & -50 \\ 0 & 2 & 3 & 0 & | & 4 \\ 0 & -3 & -1 & -6 & | & 19 \end{bmatrix}$$ To make $a_{21} = 0$, replaced row 2 by the sum of itself and (-4) times row 1.
To make $a_{41} = 0$, replaced row 4 by the sum of itself and row 1.

$$\begin{bmatrix} 1 & -1 & 2 & 0 & | & 9 \\ 0 & 1 & 0 & 3 & | & -10 \\ 0 & 2 & 3 & 0 & | & 4 \\ 0 & -3 & -1 & -6 & | & 19 \end{bmatrix}$$ Multiplied row 2 by $\frac{1}{5}$ so that $a_{22} = 1$.

$$\begin{bmatrix} 1 & -1 & 2 & 0 & | & 9 \\ 0 & 1 & 0 & 3 & | & -10 \\ 0 & 0 & 3 & -6 & | & 24 \\ 0 & 0 & -1 & 3 & | & -11 \end{bmatrix}$$ To make $a_{32} = 0$, replaced row 3 by the sum of itself and (-2) times row 2.
To make $a_{42} = 0$, replaced row 4 by the sum of itself and 3 times row 2.

$$\begin{bmatrix} 1 & -1 & 2 & 0 & | & 9 \\ 0 & 1 & 0 & 3 & | & -10 \\ 0 & 0 & -1 & 3 & | & -11 \\ 0 & 0 & 3 & -6 & | & 24 \end{bmatrix}$$ Interchanged rows 3 and 4 so that the next pivot will be $a_{33} = -1$.

$$\begin{bmatrix} 1 & -1 & 2 & 0 & | & 9 \\ 0 & 1 & 0 & 3 & | & -10 \\ 0 & 0 & -1 & 3 & | & -11 \\ 0 & 0 & 0 & 3 & | & 9 \end{bmatrix}$$ To make $a_{43} = 0$, replaced row 4 by the sum of itself and 3 times row 3.

The last row of the matrix indicates that

$$3w = -9$$
$$w = -3$$

The remaining variables are found by back-substitution.

Third row of final matrix	Second row of final matrix	First row of final matrix
$-z + 3w = -11$	$y + 3w = -10$	$x - y + 2z = 9$
$-z + 3(-3) = -11$	$y + 3(-3) = -10$	$x - (-1) + 2(2) = 9$
$z = 2$	$y = -1$	$x = 4$

The solution is $x = 4$, $y = -1$, $z = 2$, $w = -3$.

There is an important variant of Gaussian elimination known as **Gauss–Jordan elimination.** The objective is to transform a linear system into a form that yields a solution without back-substitution. For a 3×3 system that has a unique solution, the final matrix and equivalent linear system will look like this.

$$\begin{bmatrix} 1 & 0 & 0 & | & c_1 \\ 0 & 1 & 0 & | & c_2 \\ 0 & 0 & 1 & | & c_3 \end{bmatrix} \qquad \begin{matrix} x + 0y + 0z = c_1 \\ 0x + \ y + 0z = c_2 \\ 0x + 0y + \ z = c_3 \end{matrix}$$

The solution is then seen to be $x = c_1$, $y = c_2$, and $z = c_3$.

The execution of the Gauss–Jordan method is essentially the same as that of Gaussian elimination except that

(a) the pivot elements are always required to be equal to 1, and

(b) all elements in a column, other than the pivot element, are forced to be 0.

These objectives are accomplished by the use of elementary row operations, as illustrated in the following example.

EXAMPLE 6
Solve the linear system by the Gauss–Jordan method.

$$\begin{aligned} x - 3y + 2z &= 12 \\ 2x + \ y - 4z &= -1 \\ x + 3y - 2z &= -8 \end{aligned}$$

SOLUTION
We begin with the augmented matrix. At each stage, the pivot element is shown in color and is used to force all elements in that column (other than the pivot element itself) to be zero.

$$\begin{bmatrix} 1 & -3 & 2 & | & 12 \\ 2 & 1 & -4 & | & -1 \\ 1 & 3 & -2 & | & -8 \end{bmatrix} \qquad \text{Pivot element is } a_{11}.$$

$$\begin{bmatrix} 1 & -3 & 2 & | & 12 \\ 0 & 7 & -8 & | & -25 \\ 0 & 6 & -4 & | & -20 \end{bmatrix}$$

To make $a_{21} = 0$, replaced row 2 by the sum of itself and -2 times row 1.
To make $a_{31} = 0$, replaced row 3 by the sum of itself and -1 times row 1.

$$\begin{bmatrix} 1 & -3 & 2 & | & 12 \\ 0 & 1 & -4 & | & -5 \\ 0 & 6 & -4 & | & -20 \end{bmatrix}$$

Replaced row 2 by the sum of itself and -1 times row 3 to yield the next pivot, $a_{22} = 1$.

$$\begin{bmatrix} 1 & 0 & -10 & | & -3 \\ 0 & 1 & -4 & | & -5 \\ 0 & 0 & 20 & | & 10 \end{bmatrix}$$

To make $a_{12} = 0$, replaced row 1 by the sum of itself and 3 times row 2.
To make $a_{32} = 0$, replaced row 3 by the sum of itself and -6 times row 2.

$$\begin{bmatrix} 1 & 0 & -10 & | & -3 \\ 0 & 1 & -4 & | & -5 \\ 0 & 0 & 1 & | & \frac{1}{2} \end{bmatrix}$$

Multiplied row 3 by $\frac{1}{20}$ so that $a_{33} = 1$.

$$\begin{bmatrix} 1 & 0 & 0 & | & 2 \\ 0 & 1 & 0 & | & -3 \\ 0 & 0 & 1 & | & \frac{1}{2} \end{bmatrix}$$

To make $a_{13} = 0$, replaced row 1 by the sum of itself and 10 times row 3.
To make $a_{23} = 0$, replaced row 2 by the sum of itself and 4 times row 3.

We can see the solution directly from the final matrix: $x = 2$, $y = -3$, and $z = \frac{1}{2}$.

EXERCISE SET 10.5

In Exercises 1–6 state the dimension of each matrix.

1. $\begin{bmatrix} 3 & -1 \\ 2 & 4 \end{bmatrix}$

2. $[1 \quad 2 \quad 3 \quad -1]$

3. $\begin{bmatrix} 4 & 2 & 3 \\ 5 & -1 & 4 \\ 2 & 3 & 6 \\ -8 & -1 & 2 \end{bmatrix}$

4. $\begin{bmatrix} -1 \\ 3 \\ 2 \end{bmatrix}$

5. $\begin{bmatrix} 4 & 2 & 1 \\ 3 & 1 & 5 \\ -4 & -2 & 3 \end{bmatrix}$

6. $\begin{bmatrix} 3 & -1 & 2 & 6 \\ 2 & 8 & 4 & 1 \end{bmatrix}$

7. Given

$$A = \begin{bmatrix} 3 & -4 & -2 & 5 \\ 8 & 7 & 6 & 2 \\ 1 & 0 & 9 & -3 \end{bmatrix}$$

find (a) a_{12} (b) a_{22} (c) a_{23} (d) a_{34}

8. Given

$$B = \begin{bmatrix} -5 & 6 & 8 \\ 4 & 1 & 3 \\ 0 & 2 & -6 \\ -3 & 9 & 7 \end{bmatrix}$$

find (a) b_{13} (b) b_{21} (c) b_{33} (d) b_{42}

In Exercises 9–12 write the coefficient matrix and the augmented matrix for each given linear system.

9. $3x - 2y = 12$
     $5x + y = -8$

10. $3x - 4y = 15$
     $4x - 3y = 12$

11. $\frac{1}{2}x + y + z = 4$
     $2x - y - 4z = 6$
     $4x + 2y - 3z = 8$

12. $2x + 3y - 4z = 10$
     $-3x + y = 12$
     $5x - 2y + z = -8$

In Exercises 13–16 write the linear system whose augmented matrix is given.

13. $\begin{bmatrix} \frac{3}{2} & 6 & | & -1 \\ 4 & 5 & | & 3 \end{bmatrix}$

14. $\begin{bmatrix} 4 & 0 & | & 2 \\ -7 & 8 & | & 3 \end{bmatrix}$

15. $\begin{bmatrix} 1 & 1 & 3 & | & -4 \\ -3 & 4 & 0 & | & 8 \\ 2 & 0 & 7 & | & 6 \end{bmatrix}$

16. $\begin{bmatrix} 4 & 8 & 3 & | & 12 \\ 1 & -5 & 3 & | & -14 \\ 0 & 2 & 7 & | & 18 \end{bmatrix}$

In Exercises 17–20 the augmented matrix corresponding to a linear system has been transformed to the given matrix by elementary row operations. Find a solution of the original linear system.

17. $\begin{bmatrix} 1 & 2 & 0 & | & 3 \\ 0 & 1 & -2 & | & 4 \\ 0 & 0 & 1 & | & 2 \end{bmatrix}$

18. $\begin{bmatrix} 1 & 0 & 2 & | & -1 \\ 0 & 1 & 3 & | & 2 \\ 0 & 0 & 1 & | & 5 \end{bmatrix}$

19. $\begin{bmatrix} 1 & -2 & 1 & | & 3 \\ 0 & 1 & 3 & | & 2 \\ 0 & 0 & 1 & | & -4 \end{bmatrix}$

20. $\begin{bmatrix} 1 & -4 & 2 & | & -4 \\ 0 & 1 & 3 & | & -2 \\ 0 & 0 & 1 & | & 5 \end{bmatrix}$

In Exercises 21–30 solve the given linear system by applying Gaussian elimination to the augmented matrix.

21. $x - 2y = -4$
      $2x + 3y = 13$

22. $2x + y = -1$
     $3x - y = -7$

23. $x + y + z = 4$
      $2x - y + 2z = 11$
      $x + 2y + z = 3$

24. $x - y + z = -5$
      $3x + y + 2z = -5$
      $2x - y - z = -2$

25. $2x + y - z = 9$
      $x - 2y + 2z = -3$
     $3x + 3y + 4z = 11$

26. $2x + y - z = -2$
     $-2x - 2y + 3z = 2$
      $3x + y - z = -4$

27. $-x - y + 2z = 9$
      $x + 2y - 2z = -7$
     $2x - y + z = -9$

28. $4x + y - z = -1$
      $x - y + 2z = 3$
     $-x + 2y - z = 0$

29.
$$\begin{aligned}
x + y - z + 2w &= 0 \\
2x + y \quad\ - w &= -2 \\
3x \quad\ + 2z \quad &= -3 \\
-x + 2y \quad\ + 3w &= 1
\end{aligned}$$

30.
$$\begin{aligned}
2x + y \quad\ - 3w &= -7 \\
3x \quad\ + 2z + w &= 0 \\
-x + 2y \quad\ + 3w &= 10 \\
-2x - 3y + 2z - w &= 7
\end{aligned}$$

31–40. Solve the linear systems of Exercises 21–30 by Gauss–Jordan elimination applied to the augmented matrix.

10.6
MATRIX OPERATIONS AND APPLICATIONS (Optional)

After defining a new type of mathematical entity, it is useful to define operations using this entity. It is common practice to begin with a definition of equality.

Equality of Matrices

Two matrices are equal if they are of the same dimension and their corresponding entries are equal.

EXAMPLE 1

Solve for all unknowns.

$$\begin{bmatrix} -2 & 2x & 9 \\ y-1 & 3 & -4s \end{bmatrix} = \begin{bmatrix} z & 6 & 9 \\ -4 & r & 7 \end{bmatrix}$$

SOLUTION

Equating corresponding elements, we must have

$$\begin{aligned}
-2 = z \quad &\text{or} \quad z = -2 \\
2x = 6 \quad &\text{or} \quad x = 3 \\
y - 1 = -4 \quad &\text{or} \quad y = -3 \\
3 = r \quad &\text{or} \quad r = 3 \\
-4s = 7 \quad &\text{or} \quad s = -\tfrac{7}{4}
\end{aligned}$$

Matrix addition can be performed only when the matrices are of the same dimension.

Matrix Addition

The sum of two $m \times n$ matrices A and B is the $m \times n$ matrix obtained by adding the corresponding elements of A and B.

EXAMPLE 2

Given the following matrices,

$$A = [2 \quad -3 \quad 4] \qquad B = [5 \quad 3 \quad 2]$$

$$C = \begin{bmatrix} 1 & 6 & -1 \\ -2 & 4 & 5 \end{bmatrix} \qquad D = \begin{bmatrix} 16 & 2 & 9 \\ 4 & -7 & -1 \end{bmatrix}$$

find (if possible): (a) $A + B$ (b) $A + D$ (c) $C + D$

SOLUTION

(a) Since A and B are both 1×3 matrices, they can be added, giving

$$A + B = [2 + 5 \quad -3 + 3 \quad 4 + 2] = [7 \quad 0 \quad 6]$$

(b) Matrices A and D are not of the same dimension and cannot be added.

(c) C and D are both 2×3 matrices. Thus,

$$C + D = \begin{bmatrix} 1 + 16 & 6 + 2 & -1 + 9 \\ -2 + 4 & 4 + (-7) & 5 + (-1) \end{bmatrix} = \begin{bmatrix} 17 & 8 & 8 \\ 2 & -3 & 4 \end{bmatrix}$$

Matrices are a natural way of writing the information displayed in a table.

TABLE 1

TV Sets	Boston	Miami	Chicago
17″	140	84	25
19″	62	17	48

For example, Table 1 displays the current inventory of the Quality TV Company at its various outlets. The same data is displayed by the matrix

$$S = \begin{bmatrix} 140 & 84 & 25 \\ 62 & 17 & 48 \end{bmatrix}$$

where we understand the columns to represent the cities and the rows to represent the sizes of the television sets. If the matrix

$$M = \begin{bmatrix} 30 & 46 & 15 \\ 50 & 25 & 60 \end{bmatrix}$$

specifies the number of sets of each size received at each outlet the following month, then the matrix

$$T = S + M = \begin{bmatrix} 170 & 130 & 40 \\ 112 & 42 & 108 \end{bmatrix}$$

gives the revised inventory.

Suppose the salespeople at each outlet are told that half of the revised inventory is to be placed on sale. To determine the number of sets of each size to be placed on sale, we need to multiply each element of the matrix T by 0.5. When working with matrices, we call a real number such as 0.5 a **scalar** and define **scalar multiplication** as follows.

Scalar Multiplication	To multiply a matrix A by a scalar c, multiply each entry of A by c.

EXAMPLE 3
The matrix Q

$$Q = \begin{matrix} & \text{Regular} & \text{Unleaded} & \text{Premium} \\ & \begin{bmatrix} 130 & 250 & 60 \\ 110 & 180 & 40 \end{bmatrix} & & \begin{matrix} \text{City A} \\ \text{City B} \end{matrix} \end{matrix}$$

shows the quantity (in thousands of gallons) of the principal types of gasolines stored by a refiner at two different locations. It is decided to increase the quantity of each type of gasoline stored at each site by 10%. Use scalar multiplication to determine the desired inventory levels.

SOLUTION
To increase each entry of matrix Q by 10%, we compute the scalar product $1.1Q$.

$$1.1Q = 1.1 \begin{bmatrix} 130 & 250 & 60 \\ 110 & 180 & 40 \end{bmatrix}$$

$$= \begin{bmatrix} 1.1(130) & 1.1(250) & 1.1(60) \\ 1.1(110) & 1.1(180) & 1.1(40) \end{bmatrix} = \begin{bmatrix} 143 & 275 & 66 \\ 121 & 198 & 44 \end{bmatrix}$$

We denote $A + (-1)B$ by $A - B$ and refer to this as the **difference** of A and B.

Matrix Subtraction	The difference of two $m \times n$ matrices A and B is the $m \times n$ matrix obtained by subtracting each entry of B from the corresponding entry of A.

EXAMPLE 4

Using the matrices C and D of Example 2, find $C - D$.

SOLUTION

By definition,

$$C - D = \begin{bmatrix} 1 - 16 & 6 - 2 & -1 - 9 \\ -2 - 4 & 4 - (-7) & 5 - (-1) \end{bmatrix} = \begin{bmatrix} -15 & 4 & -10 \\ -6 & 11 & 6 \end{bmatrix}$$

MATRIX MULTIPLICATION

We will use the Quality TV Company again, this time to help us arrive at a definition of matrix multiplication. Suppose

$$S = \begin{array}{c} \\ \\ \end{array} \begin{matrix} \text{Boston} & \text{Miami} & \text{Chicago} \\ \begin{bmatrix} 60 & 85 & 70 \\ 40 & 100 & 20 \end{bmatrix} & \begin{matrix} 17'' \\ 19'' \end{matrix} \end{matrix}$$

is a matrix representing the supply of television sets at the end of the year. Further, suppose the cost of each 17" set is \$80 and the cost of each 19" set is \$125. To find the total cost of the inventory at each outlet, we need to multiply the number of 17" sets by \$80, the number of 19" sets by \$125, and sum the two products. If we let

$$C = [80 \quad 125]$$

be the cost matrix, we seek to define the product

$$[80 \quad 125] \begin{bmatrix} 60 & 85 & 70 \\ 40 & 100 & 20 \end{bmatrix}$$

so that the result will be a matrix displaying the total cost at each outlet. To find the total cost at the Boston outlet, we need to calculate

$$(80)(60) + (125)(40) = 9800$$

$$[80 \quad 125] \begin{bmatrix} 60 & 85 & 70 \\ 40 & 100 & 20 \end{bmatrix}$$

At the Miami outlet, the total cost is

$$(80)(85) + (125)(100) = 19{,}300$$

$$[80 \quad 125] \begin{bmatrix} 60 & 85 & 70 \\ 40 & 100 & 20 \end{bmatrix}$$

At the Chicago outlet, the total cost is

$$(80)(70) + (125)(20) = 8100$$

$$[80 \quad 125] \begin{bmatrix} 60 & 85 & 70 \\ 40 & 100 & 20 \end{bmatrix}$$

The total cost at each outlet can then be displayed by the 1×3 matrix

$$[9800 \quad 19{,}300 \quad 8100]$$

which is the product of C and S. Thus,

$$CS = [80 \quad 125] \begin{bmatrix} 60 & 85 & 70 \\ 40 & 100 & 20 \end{bmatrix}$$

$$= [(80)(60) + (125)(40) \quad (80)(85) + (125)(100) \quad (80)(70) + (125)(20)]$$

$$= [9800 \quad 19{,}300 \quad 8100]$$

Our example illustrates the process for multiplying a matrix by a row matrix. If the matrix C had more than one row, we would repeat the process using each row of C. Here is an example.

EXAMPLE 5

Find the product AB if

$$A = \begin{bmatrix} 2 & 1 \\ 3 & 5 \end{bmatrix} \qquad B = \begin{bmatrix} 4 & -6 & -2 & 4 \\ 2 & 0 & 1 & -5 \end{bmatrix}$$

SOLUTION

$$AB = \begin{bmatrix} (2)(4) + (1)(2) & (2)(-6) + (1)(0) & (2)(-2) + (1)(1) & (2)(4) + (1)(-5) \\ (3)(4) + (5)(2) & (3)(-6) + (5)(0) & (3)(-2) + (5)(1) & (3)(4) + (5)(-5) \end{bmatrix}$$

$$= \begin{bmatrix} 10 & -12 & -3 & 3 \\ 22 & -18 & -1 & -13 \end{bmatrix}$$

PROGRESS CHECK

Find the product AB if

$$A = \begin{bmatrix} -2 & -1 & 2 \\ 4 & 3 & 1 \end{bmatrix} \qquad B = \begin{bmatrix} 5 & -4 \\ 3 & 1 \\ -1 & 0 \end{bmatrix}$$

ANSWER

$$AB = \begin{bmatrix} -15 & 7 \\ 28 & -13 \end{bmatrix}$$

It is important to note that the product AB of an $m \times n$ matrix A and an $n \times r$ matrix B exists only when the number of columns of A equals the number of rows of B (see Figure 15). The product AB will then be of dimension $m \times r$.

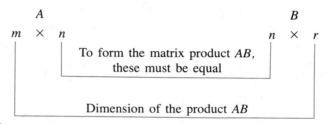

FIGURE 15

EXAMPLE 6
Given the matrices

$$A = \begin{bmatrix} 1 & -1 \\ 2 & 3 \end{bmatrix} \quad B = \begin{bmatrix} 5 & -3 \\ -2 & 2 \end{bmatrix} \quad C = \begin{bmatrix} 3 & -1 & -2 \\ 1 & 0 & 4 \end{bmatrix} \quad D = \begin{bmatrix} 1 \\ 2 \\ 3 \end{bmatrix}$$

(a)   Show that $AB \neq BA$.
(b)   Determine the dimension of AC.

SOLUTION

(a) $$AB = \begin{bmatrix} (1)(5) + (-1)(-2) & (1)(-3) + (-1)(2) \\ (2)(5) + (3)(-2) & (2)(-3) + (3)(2) \end{bmatrix} = \begin{bmatrix} 7 & -5 \\ 4 & 0 \end{bmatrix}$$

$$BA = \begin{bmatrix} (5)(1) + (-3)(2) & (5)(-1) + (-3)(3) \\ (-2)(1) + (2)(2) & (-2)(-1) + (2)(3) \end{bmatrix} = \begin{bmatrix} -1 & -14 \\ 2 & 8 \end{bmatrix}$$

Since the corresponding elements of AB and BA are not equal, $AB \neq BA$.
(b)   The product of a 2×2 matrix and a 2×3 matrix is a 2×3 matrix.

PROGRESS CHECK
If possible, find the dimension of CD and of CB, using the matrices of Example 6.

ANSWER
2×1; not defined

We saw in Example 6 that $AB \neq BA$, that is, the commutative law does not hold for matrix multiplication. However, the associative law $A(BC) = (AB)C$ does hold when the dimensions of A, B, and C permit us to find the necessary products.

PROGRESS CHECK
Verify that $A(BC) = (AB)C$ for the matrices A, B, and C of Example 6.

MATRICES AND LINEAR SYSTEMS

Matrix multiplication provides a convenient shorthand means of writing a linear system. For example, the linear system

$$2x - y - 2z = 3$$
$$3x + 2y + z = -1$$
$$x + y - 3z = 14$$

can be expressed as

$$AX = B$$

where

$$A = \begin{bmatrix} 2 & -1 & -2 \\ 3 & 2 & 1 \\ 1 & 1 & -3 \end{bmatrix} \quad X = \begin{bmatrix} x \\ y \\ z \end{bmatrix} \quad B = \begin{bmatrix} 3 \\ -1 \\ 14 \end{bmatrix}$$

To verify this, simply form the matrix product AX and then apply the definition of matrix equality to the matrix equation $AX = B$.

EXAMPLE 7
Write out the linear system $AX = B$ if

$$A = \begin{bmatrix} -2 & 3 \\ 1 & 4 \end{bmatrix} \quad X = \begin{bmatrix} x \\ y \end{bmatrix} \quad B = \begin{bmatrix} 16 \\ -3 \end{bmatrix}$$

SOLUTION
Equating corresponding elements of the matrix equation $AX = B$ yields

$$-2x + 3y = 16$$
$$x + 4y = -3$$

EXERCISE SET 10.6

1. For what values of a, b, c, and d are the matrices A and B equal?

$$A = \begin{bmatrix} a & b \\ 6 & -2 \end{bmatrix} \qquad B = \begin{bmatrix} 3 & -4 \\ c & d \end{bmatrix}$$

2. For what values of a, b, c, and d are the matrices A and B equal?

$$A = \begin{bmatrix} a+b & 2c \\ a & c-d \end{bmatrix} \qquad B = \begin{bmatrix} -1 & 6 \\ 5 & 10 \end{bmatrix}$$

In Exercises 3–18 the following matrices are given.

$$A = \begin{bmatrix} 2 & 3 & 1 \\ -3 & 4 & 1 \end{bmatrix} \qquad B = \begin{bmatrix} 2 & -1 \\ 3 & 2 \\ 4 & 1 \end{bmatrix} \qquad C = \begin{bmatrix} 1 & 2 & 3 \\ 4 & -1 & 2 \\ 3 & 2 & 5 \end{bmatrix} \qquad D = \begin{bmatrix} -3 & 2 \\ 4 & 1 \end{bmatrix}$$

$$E = \begin{bmatrix} 1 & -3 & 2 \\ 3 & 2 & 4 \\ 1 & 1 & 2 \end{bmatrix} \qquad F = \begin{bmatrix} 1 & 3 \\ -2 & 4 \end{bmatrix} \qquad G = \begin{bmatrix} -2 & 4 & 2 \\ 1 & 0 & 3 \end{bmatrix}$$

If possible, compute the indicated matrix.

3. $C + E$ 4. $C - E$ 5. $2A + 3G$ 6. $3G - 4A$

7. $A + F$ 8. $2B - D$ 9. AB 10. BA

11. $CB + D$ 12. $EB - FA$ 13. $DF + AB$ 14. $AC + 2DG$

15. $DA + EB$ 16. $FG + B$ 17. $2GE - 3A$ 18. $AB + FG$

19. If $A = \begin{bmatrix} -2 & 3 \\ 2 & -3 \end{bmatrix}$, $B = \begin{bmatrix} -1 & 3 \\ 2 & 0 \end{bmatrix}$, $C = \begin{bmatrix} -4 & -3 \\ 0 & -4 \end{bmatrix}$, show that $AB = AC$.

20. If $A = \begin{bmatrix} 1 & 2 \\ 3 & 2 \end{bmatrix}$ and $B = \begin{bmatrix} 2 & -1 \\ -3 & 4 \end{bmatrix}$, show that $AB \neq BA$.

21. If $A = \begin{bmatrix} -2 & 3 \\ 2 & -3 \end{bmatrix}$ and $B = \begin{bmatrix} 3 & 6 \\ 2 & 4 \end{bmatrix}$, show that $AB = \begin{bmatrix} 0 & 0 \\ 0 & 0 \end{bmatrix}$.

22. If $A = \begin{bmatrix} 0 & 1 \\ 1 & 0 \end{bmatrix}$, show that $A \cdot A = \begin{bmatrix} 1 & 0 \\ 0 & 1 \end{bmatrix}$.

23. If $I = \begin{bmatrix} 1 & 0 & 0 \\ 0 & 1 & 0 \\ 0 & 0 & 1 \end{bmatrix}$ and $A = \begin{bmatrix} a_{11} & a_{12} & a_{13} \\ a_{21} & a_{22} & a_{23} \\ a_{31} & a_{32} & a_{33} \end{bmatrix}$, show that $AI = A$ and $IA = A$.

24. Pesticides are sprayed on plants to eliminate harmful insects. However, some of the pesticide is absorbed by the plant, and the pesticide is then absorbed by herbivores (plant-eating animals such as cows) when they eat the plants that have

been sprayed. Suppose that we have three pesticides and four plants and that the amounts of pesticide absorbed by the different plants are given by the matrix

$$
\begin{array}{cccc}
\text{Plant 1} & \text{Plant 2} & \text{Plant 3} & \text{Plant 4}
\end{array}
$$

$$
A = \begin{bmatrix} 3 & 2 & 4 & 3 \\ 6 & 5 & 2 & 4 \\ 4 & 3 & 1 & 5 \end{bmatrix}
\begin{array}{l} \text{Pesticide 1} \\ \text{Pesticide 2} \\ \text{Pesticide 3} \end{array}
$$

where a_{ij} denotes the amount of pesticide i in milligrams that has been absorbed by plant j. Thus, plant 4 has absorbed 5 mg of pesticide 3. Now suppose that we have three herbivores and that the numbers of plants eaten by these animals are given by the matrix

$$
\begin{array}{ccc}
\text{Herbivore 1} & \text{Herbivore 2} & \text{Herbivore 3}
\end{array}
$$

$$
B = \begin{bmatrix} 18 & 30 & 20 \\ 12 & 15 & 10 \\ 16 & 12 & 8 \\ 6 & 4 & 12 \end{bmatrix}
\begin{array}{l} \text{Plant 1} \\ \text{Plant 2} \\ \text{Plant 3} \\ \text{Plant 4} \end{array}
$$

How much of pesticide 2 has been absorbed by herbivore 3?

25. What does entry $(2, 3)$ in the matrix product AB of Exercise 24 represent?

In Exercises 26–29 indicate the matrices A, X, and B so that the matrix equation $AX = B$ is equivalent to the given linear system.

26.
$$7x - 2y = 6$$
$$-2x + 3y = -2$$

27.
$$3x + 4y = -3$$
$$3x - y = 5$$

28.
$$5x + 2y - 3z = 4$$
$$2x - \frac{1}{2}y + z = 10$$
$$x + y - 5z = -3$$

29.
$$3x - y + 4z = 5$$
$$2x + 2y + \frac{3}{4}z = -1$$
$$x - \frac{1}{4}y + z = \frac{1}{2}$$

In Exercises 30–33 write out the linear system that is represented by the matrix equation $AX = B$.

30.
$$A = \begin{bmatrix} 2 & -1 \\ -3 & 4 \end{bmatrix} \quad X = \begin{bmatrix} x \\ y \end{bmatrix} \quad B = \begin{bmatrix} -2 \\ 10 \end{bmatrix}$$

31.
$$A = \begin{bmatrix} 1 & -5 \\ 4 & 3 \end{bmatrix} \quad X = \begin{bmatrix} x_1 \\ x_2 \end{bmatrix} \quad B = \begin{bmatrix} 0 \\ 2 \end{bmatrix}$$

32.
$$A = \begin{bmatrix} 1 & 7 & -2 \\ 3 & 6 & 1 \\ -4 & 2 & 0 \end{bmatrix} \quad X = \begin{bmatrix} x \\ y \\ z \end{bmatrix} \quad B = \begin{bmatrix} 3 \\ -3 \\ 2 \end{bmatrix}$$

33.
$$A = \begin{bmatrix} 4 & 5 & -2 \\ 0 & 3 & -1 \\ 0 & 0 & 2 \end{bmatrix} \quad X = \begin{bmatrix} x_1 \\ x_2 \\ x_3 \end{bmatrix} \quad B = \begin{bmatrix} 2 \\ -5 \\ 4 \end{bmatrix}$$

34. The $m \times n$ matrix all of whose elements are zero is called the **zero matrix** and is denoted by O. Show that $A + O = A$ for every $m \times n$ matrix A.

35. The square matrix of order n such that $a_{ii} = 1$ and $a_{ij} = 0$ when $i \neq j$ is called the **identity matrix** of order n and is denoted by I_n. (Note: The definition indicates that the diagonal elements are all equal to 1 and all elements off the diagonal are 0.) Show that $AI_n = I_nA$ for every square matrix A of order n.

36. The matrix B, each of whose entries is the negative of the corresponding entry of matrix A, is called the **additive inverse** of the matrix A. Show that $A + B = O$ where O is the zero matrix (see Exercise 34).

10.7
INVERSES OF
MATRICES (Optional)

If $a \neq 0$, then the linear equation $ax = b$ can be solved easily by multiplying both sides by the reciprocal of a. Thus, we obtain $x = \dfrac{1}{a} \cdot b$. It would be nice if we could multiply both sides of the matrix equation $AX = B$ by the "reciprocal of A." Unfortunately, a matrix has *no* reciprocal. However, we shall discuss a notion that, for a square matrix, provides an analogue of the reciprocal of a real number and will enable us to solve the linear system in a manner distinct from the Gauss–Jordan method discussed in Section 5 of this chapter.

In this section we confine our attention to square matrices. The $n \times n$ matrix

$$
I_n = \begin{bmatrix}
1 & 0 & 0 \ldots 0 \\
0 & 1 & 0 \ldots 0 \\
\cdot & \cdot & \cdot & \cdot \\
\cdot & \cdot & \cdot & \cdot \\
\cdot & \cdot & \cdot & \cdot \\
0 & 0 & 0 \ldots 1
\end{bmatrix}
$$

which has 1s on the main diagonal and 0s elsewhere, is called the **identity matrix.** Examples of identity matrices are

$$
I_2 = \begin{bmatrix} 1 & 0 \\ 0 & 1 \end{bmatrix} \quad
I_3 = \begin{bmatrix} 1 & 0 & 0 \\ 0 & 1 & 0 \\ 0 & 0 & 1 \end{bmatrix} \quad
I_4 = \begin{bmatrix} 1 & 0 & 0 & 0 \\ 0 & 1 & 0 & 0 \\ 0 & 0 & 1 & 0 \\ 0 & 0 & 0 & 1 \end{bmatrix}
$$

If A is any $n \times n$ matrix, we can show that

$$
AI_n = I_nA = A
$$

(see Exercise 35, Section 10.6). Thus, I_n is the matrix analogue of the real number 1.

An $n \times n$ matrix A is called **invertible** or **nonsingular** if we can find an $n \times n$ matrix B such that

$$
AB = BA = I_n
$$

The matrix B is called an **inverse** of A.

EXAMPLE 1
Let

$$
A = \begin{bmatrix} 2 & 1 \\ 3 & 2 \end{bmatrix} \quad \text{and} \quad B = \begin{bmatrix} 2 & -1 \\ -3 & 2 \end{bmatrix}
$$

Since

$$AB = BA = \begin{bmatrix} 1 & 0 \\ 0 & 1 \end{bmatrix} \quad \text{(verify this)}$$

we conclude that A is an invertible matrix and that B is an inverse of A. Of course, if B is an inverse of A, then A is an inverse of B.

If an $n \times n$ matrix A has an inverse, then it can be shown that it can only have one inverse. We denote the inverse of A by A^{-1}. Thus, we have

$$AA^{-1} = I_n \quad \text{and} \quad A^{-1}A = I_n$$

Note that the products AA^{-1} and $A^{-1}A$ yield the *identity matrix,* and that the products $a \cdot \dfrac{1}{a}$ and $\dfrac{1}{a} \cdot a$ yield the *identity element.* For this reason, A^{-1} may be thought of as the matrix analogue of the reciprocal $\dfrac{1}{a}$ of the real number a.

PROGRESS CHECK

Verify that the matrices

$$A = \begin{bmatrix} 4 & 5 \\ 2 & 2 \end{bmatrix} \quad \text{and} \quad B = \begin{bmatrix} -1 & \dfrac{5}{2} \\ 1 & -2 \end{bmatrix}$$

are inverses of each other.

WARNING If $a \neq 0$ is a real number, then a^{-1} has the property that $aa^{-1} = a^{-1}a = 1$. Since $a^{-1} = \dfrac{1}{a}$, we may refer to a^{-1} as the inverse *or* reciprocal of a. However, the matrix A^{-1} is the inverse of the $n \times n$ matrix A, since $AA^{-1} = A^{-1}A = I_n$, but cannot be referred to as the reciprocal of A, since *matrix division is not defined.*

We now develop a practical method for finding the inverse of an invertible matrix. Suppose we want to find the inverse of the matrix

$$A = \begin{bmatrix} 1 & 3 \\ 2 & 5 \end{bmatrix}$$

Let the inverse be denoted by

$$B = \begin{bmatrix} b_1 & b_2 \\ b_3 & b_4 \end{bmatrix}$$

Then we must have

$$AB = I_2 \tag{1}$$

and

$$BA = I_2 \tag{2}$$

Equation (1) now becomes

$$\begin{bmatrix} 1 & 3 \\ 2 & 5 \end{bmatrix} \begin{bmatrix} b_1 & b_2 \\ b_3 & b_4 \end{bmatrix} = \begin{bmatrix} 1 & 0 \\ 0 & 1 \end{bmatrix}$$

or

$$\begin{bmatrix} b_1 + 3b_3 & b_2 + 3b_4 \\ 2b_1 + 5b_3 & 2b_2 + 5b_4 \end{bmatrix} = \begin{bmatrix} 1 & 0 \\ 0 & 1 \end{bmatrix}$$

Since two matrices are equal if and only if their corresponding entries are equal, we have

$$\begin{aligned} b_1 + 3b_3 &= 1 \\ 2b_1 + 5b_3 &= 0 \end{aligned} \tag{3}$$

and

$$\begin{aligned} b_2 + 3b_4 &= 0 \\ 2b_2 + 5b_4 &= 1 \end{aligned} \tag{4}$$

We solve the linear systems (3) and (4) by Gauss–Jordan elimination. We begin with the augmented matrices of the linear systems and perform a sequence of elementary row operations as follows.

(3)	(4)	
$\begin{bmatrix} 1 & 3 & \mid & 1 \\ 2 & 5 & \mid & 0 \end{bmatrix}$	$\begin{bmatrix} 1 & 3 & \mid & 0 \\ 2 & 5 & \mid & 1 \end{bmatrix}$	Augmented matrices of (3) and (4).
$\begin{bmatrix} 1 & 3 & \mid & 1 \\ 0 & -1 & \mid & -2 \end{bmatrix}$	$\begin{bmatrix} 1 & 3 & \mid & 0 \\ 0 & -1 & \mid & 1 \end{bmatrix}$	To make $a_{21} = 0$, replaced row 2 by the sum of itself and -2 times row 1.
$\begin{bmatrix} 1 & 3 & \mid & 1 \\ 0 & 1 & \mid & 2 \end{bmatrix}$	$\begin{bmatrix} 1 & 3 & \mid & 0 \\ 0 & 1 & \mid & -1 \end{bmatrix}$	Multiplied row 2 by -1 to obtain $a_{22} = 1$.
$\begin{bmatrix} 1 & 0 & \mid & -5 \\ 0 & 1 & \mid & 2 \end{bmatrix}$	$\begin{bmatrix} 1 & 0 & \mid & 3 \\ 0 & 1 & \mid & -1 \end{bmatrix}$	To make $a_{12} = 0$, replaced row 1 by the sum of itself and -3 times row 2.

Thus, $b_1 = -5$ and $b_3 = 2$ is the solution of (3), and $b_2 = 3$ and $b_4 = -1$ is the solution of (4). We can check that

$$B = \begin{bmatrix} -5 & 3 \\ 2 & -1 \end{bmatrix}$$

also satisfies the requirement $BA = I_2$ of Equation (2).

Observe that the linear systems (3) and (4) have the same coefficient matrix and that an identical sequence of elementary row operations was performed in the Gauss–Jordan elimination. This suggests that we can solve the systems *at the same time*. We simply write the coefficient matrix A and next to it list the right-hand sides of (3) and (4) to obtain the matrix

$$\begin{bmatrix} 1 & 3 & \vdots & 1 & 0 \\ 2 & 5 & \vdots & 0 & 1 \end{bmatrix} \tag{5}$$

Note that the columns to the right of the dashed line in (5) form the identity matrix I_2. Performing the same sequence of elementary row operations on matrix (5) as we did on matrices (3) and (4) yields

$$\begin{bmatrix} 1 & 0 & \vdots & -5 & 3 \\ 0 & 1 & \vdots & 2 & -1 \end{bmatrix} \tag{6}$$

Then A^{-1} is the matrix to the right of the dashed line in (6).

The procedure outlined for the 2×2 matrix A applies in general. Thus, we have the following method for finding the inverse of an invertible $n \times n$ matrix A.

Computing A^{-1}

Step 1. Form the $n \times 2n$ matrix $[A \mid I_n]$ by adjoining the identity matrix I_n to the given matrix A.

Step 2. Apply elementary row operations to the matrix $[A \mid I_n]$ to transform the matrix A to I_n.

Step 3. The final matrix is of the form $[I_n \mid B]$ where B is A^{-1}.

EXAMPLE 2
Find the inverse of

$$A = \begin{bmatrix} 1 & 2 & 3 \\ 2 & 5 & 7 \\ 1 & 1 & 1 \end{bmatrix}$$

SOLUTION

We form the 3×6 matrix $[A \mid I_3]$ and transform it by elementary row operations to the form $[I_3 \mid A^{-1}]$. The pivot element at each stage is shown in color.

$$\begin{bmatrix} 1 & 2 & 3 & \vdots & 1 & 0 & 0 \\ 2 & 5 & 7 & \vdots & 0 & 1 & 0 \\ 1 & 1 & 1 & \vdots & 0 & 0 & 1 \end{bmatrix}$$ Matrix A augmented by I_3.

$$\begin{bmatrix} 1 & 2 & 3 & \vdots & 1 & 0 & 0 \\ 0 & 1 & 1 & \vdots & -2 & 1 & 0 \\ 0 & -1 & -2 & \vdots & -1 & 0 & 1 \end{bmatrix}$$ To make $a_{21} = 0$, replaced row 2 by the sum of itself and -2 times row 1. To make $a_{31} = 0$, replaced row 3 by the sum of itself and -1 times row 1.

$$\begin{bmatrix} 1 & 0 & 1 & \vdots & 5 & -2 & 0 \\ 0 & 1 & 1 & \vdots & -2 & 1 & 0 \\ 0 & 0 & -1 & \vdots & -3 & 1 & 1 \end{bmatrix}$$ To make $a_{12} = 0$, replaced row 1 by the sum of itself and -2 times row 2. To make $a_{32} = 0$, replaced row 3 by the sum of itself and row 2.

$$\begin{bmatrix} 1 & 0 & 1 & \vdots & 5 & -2 & 0 \\ 0 & 1 & 1 & \vdots & -2 & 1 & 0 \\ 0 & 0 & 1 & \vdots & 3 & -1 & -1 \end{bmatrix}$$ Multiplied row 3 by -1.

$$\begin{bmatrix} 1 & 0 & 0 & \vdots & 2 & -1 & 1 \\ 0 & 1 & 0 & \vdots & -5 & 2 & 1 \\ 0 & 0 & 1 & \vdots & 3 & -1 & -1 \end{bmatrix}$$ To make $a_{13} = 0$, replaced row 1 by the sum of itself and -1 times row 3. To make $a_{23} = 0$, replaced row 2 by the sum of itself and -1 times row 3.

The final matrix is of the form $[I_3 \mid A^{-1}]$; that is,

$$A^{-1} = \begin{bmatrix} 2 & -1 & 1 \\ -5 & 2 & 1 \\ 3 & -1 & -1 \end{bmatrix}$$

We now have a practical method for finding the inverse of an invertible matrix, but we don't know whether a given square matrix *has* an inverse. It can be shown that if the preceding procedure is carried out with the matrix $[A \mid I_n]$ and we arrive at a point at which all possible candidates for the next pivot element are zero, then the matrix is not invertible and we may stop our calculations.

EXAMPLE 3

Find the inverse of

$$A = \begin{bmatrix} 1 & 2 & 6 \\ 0 & 0 & 2 \\ -3 & -6 & -9 \end{bmatrix}$$

SOLUTION

We begin with $[A \mid I_3]$.

$$\begin{bmatrix} 1 & 2 & 6 & | & 1 & 0 & 0 \\ 0 & 0 & 2 & | & 0 & 1 & 0 \\ -3 & -6 & -9 & | & 0 & 0 & 1 \end{bmatrix}$$

$$\begin{bmatrix} 1 & 2 & 6 & | & 1 & 0 & 0 \\ 0 & 0 & 2 & | & 0 & 1 & 0 \\ 0 & 0 & 9 & | & 3 & 0 & 1 \end{bmatrix}$$
To make $a_{31} = 0$, replaced row 3 by the sum of itself and 3 times row 1.

Note that $a_{22} = a_{32} = 0$ in the last matrix. We cannot perform any elementary row operations upon rows 2 and 3 that will produce a nonzero pivot element for a_{22}. We conclude that the matrix A does not have an inverse.

PROGRESS CHECK

Show that the matrix A is not invertible.

$$A = \begin{bmatrix} 1 & 2 & -3 \\ 3 & 2 & 1 \\ 5 & 6 & -5 \end{bmatrix}$$

SOLVING LINEAR SYSTEMS

Consider a linear system of n equations in n unknowns.

$$
\begin{aligned}
a_{11}x_1 + a_{12}x_2 + \cdots + a_{1n}x_n &= b_1 \\
a_{12}x_2 + a_{22}x_2 + \cdots + a_{2n}x_n &= b_2 \\
& \vdots \\
a_{n1}x_1 + a_{n2}x_2 + \cdots + a_{nn}x_n &= b_n
\end{aligned}
\tag{7}
$$

As has already been pointed out in Section 6 of this chapter, we can write the linear system (7) in matrix form as

$$AX = B \tag{8}$$

where

$$A = \begin{bmatrix} a_{11} & a_{12}...a_{1n} \\ a_{21} & a_{22}...a_{2n} \\ \cdot & \cdot \\ \cdot & \cdot \\ \cdot & \cdot \\ a_{n1} & a_{n2}...a_{nn} \end{bmatrix} \quad X = \begin{bmatrix} x_1 \\ x_2 \\ \cdot \\ \cdot \\ \cdot \\ x_n \end{bmatrix} \quad B = \begin{bmatrix} b_1 \\ b_2 \\ \cdot \\ \cdot \\ \cdot \\ b_n \end{bmatrix}$$

CODED MESSAGES

Cryptography is the study of methods for encoding and decoding messages. One of the very simplest techniques for doing this involves the use of the inverse of a matrix.

First, attach a different number to every letter of the alphabet. For example, we can let A be 1, B be 2, and so on, as shown in the accompanying table. Suppose that we then want to send the message

ALGEBRA WORKS

A	B	C	D	E	F	G
↕	↕	↕	↕	↕	↕	↕
1	2	3	4	5	6	7
H	I	J	K	L	M	N
↕	↕	↕	↕	↕	↕	↕
8	9	10	11	12	13	14
O	P	Q	R	S	T	
↕	↕	↕	↕	↕	↕	
15	16	17	18	19	20	
U	V	W	X	Y	Z	
↕	↕	↕	↕	↕	↕	
21	22	23	24	25	26	

Substituting for each letter, we send the message

$$1, 12, 7, 5, 2, 18, 1, 23, 15, 18, 11, 19 \tag{1}$$

Unfortunately, this simple code can be easily cracked. A better method involves the use of matrices.

Break the message (1) into four 3×1 matrices:

$$X_1 = \begin{bmatrix} 1 \\ 12 \\ 7 \end{bmatrix} \quad X_2 = \begin{bmatrix} 5 \\ 2 \\ 18 \end{bmatrix} \quad X_3 = \begin{bmatrix} 1 \\ 23 \\ 15 \end{bmatrix} \quad X_4 = \begin{bmatrix} 18 \\ 11 \\ 19 \end{bmatrix}$$

The sender and receiver jointly select an invertible 3×3 matrix such as

$$A = \begin{bmatrix} 1 & 1 & 2 \\ 1 & 1 & 1 \\ 1 & 0 & 1 \end{bmatrix}$$

The sender forms the 3×1 matrices

$$AX_1 = \begin{bmatrix} 27 \\ 20 \\ 8 \end{bmatrix} \quad AX_2 = \begin{bmatrix} 43 \\ 25 \\ 23 \end{bmatrix} \quad AX_3 = \begin{bmatrix} 54 \\ 39 \\ 16 \end{bmatrix} \quad AX_4 = \begin{bmatrix} 67 \\ 48 \\ 37 \end{bmatrix}$$

and sends the message

$$27, 20, 8, 43, 25, 23, 54, 39, 16, 67, 48, 37 \tag{2}$$

To decode the message, the receiver uses the inverse of matrix A,

$$A^{-1} = \begin{bmatrix} -1 & 1 & 1 \\ 0 & 1 & -1 \\ 1 & -1 & 0 \end{bmatrix}$$

and forms

$$A^{-1}\begin{bmatrix} 27 \\ 20 \\ 8 \end{bmatrix} = X_1 \quad A^{-1}\begin{bmatrix} 43 \\ 25 \\ 23 \end{bmatrix} = X_2 \quad A^{-1}\begin{bmatrix} 54 \\ 39 \\ 16 \end{bmatrix} = X_3 \quad A^{-1}\begin{bmatrix} 67 \\ 48 \\ 37 \end{bmatrix} = X_4$$

which, of course, is the original message (2) and which can be understood by using the accompanying table.

If the receiver sends back the message

$$46, 37, 29, 50, 39, 30, 75, 52, 37$$

what is the response?

Suppose now that the coefficient matrix A is invertible so that we can compute A^{-1}. Multiplying both sides of (8) by A^{-1} we have

$$A^{-1}(AX) = A^{-1}B$$
$$(A^{-1}A)X = A^{-1}B \qquad \text{Associative law}$$
$$I_n X = A^{-1}B \qquad A^{-1}A = I_n$$
$$X = A^{-1}B \qquad I_n X = X$$

Thus, we have the following result.

If $AX = B$ is a linear system of n equations in n unknowns and if the coefficient matrix A is invertible, then the system has exactly one solution, given by

$$X = A^{-1}B$$

WARNING Since matrix multiplication is not commutative, you must be careful to write the solution to the system $AX = B$ as $X = A^{-1}B$ and *not* $X = BA^{-1}$.

EXAMPLE 4

Solve the linear system by finding the inverse of the coefficient matrix.

$$x + 2y + 3z = -3$$
$$2x + 5y + 7z = 4$$
$$x + y + z = 5$$

SOLUTION

The coefficient matrix

$$A = \begin{bmatrix} 1 & 2 & 3 \\ 2 & 5 & 7 \\ 1 & 1 & 1 \end{bmatrix}$$

is the matrix whose inverse was obtained in Example 2 as

$$A^{-1} = \begin{bmatrix} 2 & -1 & 1 \\ -5 & 2 & 1 \\ 3 & -1 & -1 \end{bmatrix}$$

Since

$$B = \begin{bmatrix} -3 \\ 4 \\ 5 \end{bmatrix}$$

we obtain the solution of the given system as

$$X = A^{-1}B = \begin{bmatrix} 2 & -1 & 1 \\ -5 & 2 & 1 \\ 3 & -1 & -1 \end{bmatrix} \begin{bmatrix} -3 \\ 4 \\ 5 \end{bmatrix} = \begin{bmatrix} -5 \\ 28 \\ -18 \end{bmatrix}$$

Thus $x = -5$, $y = 28$, $z = -18$.

PROGRESS CHECK

Solve the linear system by finding the inverse of the coefficient matrix.

$$\begin{aligned} x - 2y + z &= 1 \\ x + 3y + 2z &= 2 \\ -x \quad\quad + z &= -11 \end{aligned}$$

ANSWER

$x = 7$, $y = 1$, $z = -4$

The inverse of the coefficient matrix is especially useful when we need to solve a number of linear systems

$$AX = B_1, \; AX = B_2, \; \ldots, \; AX = B_k$$

where the coefficient matrix is the same and the right-hand side changes.

EXAMPLE 5

A steel producer makes two types of steel, regular and special. A ton of regular steel requires 2 hours in the open-hearth furnace and 5 hours in the soaking pit; a ton of special steel requires 2 hours in the open-hearth furnace and 3 hours in the soaking pit. How many tons of each type of steel can be manufactured daily if
(a) the open-hearth furnace is available 8 hours per day and the soaking pit is available 15 hours per day?
(b) the open-hearth furnace is available 9 hours per day and the soaking pit is available 15 hours per day?

SOLUTION

Let

$$x = \text{the number of tons of regular steel to be made}$$
$$y = \text{the number of tons of special steel to be made}$$

Then the total amount of time required in the open-hearth furnace is

$$2x + 2y$$

Similarly, the total amount of time required in the soaking pit is

$$5x + 3y$$

If we let b_1 and b_2 denote the number of hours that the open-hearth furnace and the soaking pit, respectively, are available per day, then we have

$$2x + 2y = b_1$$
$$5x + 3y = b_2$$

or

$$\begin{bmatrix} 2 & 2 \\ 5 & 3 \end{bmatrix} \begin{bmatrix} x \\ y \end{bmatrix} = \begin{bmatrix} b_1 \\ b_2 \end{bmatrix}$$

Then

$$\begin{bmatrix} x \\ y \end{bmatrix} = \begin{bmatrix} 2 & 2 \\ 5 & 3 \end{bmatrix}^{-1} \begin{bmatrix} b_1 \\ b_2 \end{bmatrix}$$

We find (verify) the inverse of the coefficient matrix to be

$$\begin{bmatrix} 2 & 2 \\ 5 & 3 \end{bmatrix}^{-1} = \begin{bmatrix} -\frac{3}{4} & \frac{1}{2} \\ \frac{5}{4} & -\frac{1}{2} \end{bmatrix}$$

(a) We are given $b_1 = 8$ and $b_2 = 15$. Then

$$\begin{bmatrix} x \\ y \end{bmatrix} = \begin{bmatrix} -\frac{3}{4} & \frac{1}{2} \\ \frac{5}{4} & -\frac{1}{2} \end{bmatrix} \begin{bmatrix} 8 \\ 15 \end{bmatrix} = \begin{bmatrix} \frac{3}{2} \\ \frac{5}{2} \end{bmatrix}$$

That is, $\frac{3}{2}$ tons of regular steel and $\frac{5}{2}$ tons of special steel can be manufactured daily.

(b) We are given $b_1 = 9$ and $b_2 = 15$. Then

$$\begin{bmatrix} x \\ y \end{bmatrix} = \begin{bmatrix} -\frac{3}{4} & \frac{1}{2} \\ \frac{5}{4} & -\frac{1}{2} \end{bmatrix} \begin{bmatrix} 9 \\ 15 \end{bmatrix} = \begin{bmatrix} \frac{3}{4} \\ \frac{15}{4} \end{bmatrix}$$

That is, $\frac{3}{4}$ tons of regular steel and $\frac{15}{4}$ tons of special steel can be manufactured daily.

EXERCISE SET 10.7

In Exercises 1–4 determine whether the matrix B is the inverse of the matrix A.

1. $A = \begin{bmatrix} 2 & \frac{1}{2} \\ -1 & 3 \end{bmatrix}$ $B = \begin{bmatrix} 1 & -1 \\ -2 & 4 \end{bmatrix}$

2. $A = \begin{bmatrix} 3 & -1 \\ -2 & 2 \end{bmatrix}$ $B = \begin{bmatrix} \frac{1}{2} & \frac{1}{4} \\ \frac{1}{2} & \frac{3}{4} \end{bmatrix}$

3.
$$A = \begin{bmatrix} 1 & 2 & 2 \\ -1 & 3 & 0 \\ 0 & 2 & 1 \end{bmatrix} \quad B = \begin{bmatrix} 3 & 2 & -6 \\ 1 & 1 & -2 \\ -2 & -2 & 5 \end{bmatrix}$$

4.
$$A = \begin{bmatrix} 1 & 0 & -2 \\ 2 & 1 & 3 \\ -4 & 1 & 2 \end{bmatrix} \quad B = \begin{bmatrix} 1 & 2 & -2 \\ -2 & -4 & 1 \\ 0 & 1 & -1 \end{bmatrix}$$

In Exercises 5–10 find the inverse of the given matrix.

5. $\begin{bmatrix} -1 & 5 \\ 2 & -4 \end{bmatrix}$

6. $\begin{bmatrix} 2 & 0 \\ -1 & -2 \end{bmatrix}$

7. $\begin{bmatrix} -1 & 1 \\ -2 & 1 \end{bmatrix}$

8. $\begin{bmatrix} 2 & 1 & 0 \\ 1 & 1 & 0 \\ 1 & 1 & 1 \end{bmatrix}$

9. $\begin{bmatrix} 1 & -2 & 3 \\ -1 & 3 & -4 \\ 0 & 5 & -4 \end{bmatrix}$

10. $\begin{bmatrix} 1 & 1 & 0 \\ 1 & 0 & 0 \\ 1 & 2 & 2 \end{bmatrix}$

In Exercises 11–18 find the inverse, if possible.

11. $\begin{bmatrix} 1 & 3 \\ -1 & 4 \end{bmatrix}$

12. $\begin{bmatrix} 6 & -4 \\ 9 & -6 \end{bmatrix}$

13. $\begin{bmatrix} 1 & 1 & 3 \\ 2 & -8 & -4 \\ -1 & 2 & 0 \end{bmatrix}$

14. $\begin{bmatrix} 8 & 7 & -1 \\ -5 & -5 & 1 \\ -4 & -4 & 1 \end{bmatrix}$

15. $\begin{bmatrix} 2 & 0 \\ 0 & -3 \end{bmatrix}$

16. $\begin{bmatrix} -1 & 0 & 0 \\ 0 & 4 & 0 \\ 0 & 0 & 2 \end{bmatrix}$

17. $\begin{bmatrix} 1 & 0 & -1 \\ 2 & 1 & 0 \\ 0 & 1 & 1 \end{bmatrix}$

18. $\begin{bmatrix} 1 & 0 & -3 & 0 \\ 0 & 1 & 0 & 0 \\ -1 & 0 & 4 & 0 \\ 2 & 0 & -6 & 1 \end{bmatrix}$

In Exercises 19–24 solve the given linear system by finding the inverse of the coefficient matrix.

19. $2x + y = 5$
$x - 3y = 6$

20. $2x - 3y = -5$
$3x + y = -13$

21. $3x + y - z = 2$
$x - 2y = 8$
$3y + z = -8$

22. $3x + 2y - z = 10$
$2x - y + z = -1$
$-x + y - 2z = 5$

23. $2x - y + 3z = -11$
$3x - y + z = -5$
$x + y + z = -1$

24. $2x + 3y - 2z = 13$
$4x + 2y + z = 3$
$y - z = 5$

25–42. Solve the linear systems of Section 4 of this chapter, Exercises 1–18, by finding the inverse of the coefficient matrix.

43. Solve the linear systems $AX = B_1$ and $AX = B_2$ given
$$A^{-1} = \begin{bmatrix} 3 & -2 & 4 \\ 2 & -1 & 0 \\ 0 & 4 & 1 \end{bmatrix} \quad B_1 = \begin{bmatrix} 1 \\ -1 \\ 5 \end{bmatrix} \quad B_2 = \begin{bmatrix} 4 \\ 3 \\ -2 \end{bmatrix}$$

44. Solve the linear systems $AX = B_1$ and $AX = B_2$ given
$$A^{-1} = \begin{bmatrix} 1 & 0 & -1 \\ 1 & 2 & 0 \\ -1 & -1 & 3 \end{bmatrix} \quad B_1 = \begin{bmatrix} 2 \\ -3 \\ 2 \end{bmatrix} \quad B_2 = \begin{bmatrix} 4 \\ -3 \\ -5 \end{bmatrix}$$

45. Show that the matrix

$$\begin{bmatrix} a & b & c \\ 0 & 0 & 0 \\ d & e & f \end{bmatrix}$$

   is not invertible.

46. A trustee decides to invest $30,000 in two mortgages, which yield 10% and 15% per year, respectively. How should the $30,000 be invested in the two mortgages if the total annual interest is to be

   (a) $3600?   (b) $4000?   (c) $5000?

   (*Hint:* Some of these investment objectives cannot be attained.)

**10.8
DETERMINANTS AND
CRAMER'S RULE**

In this section we will define a determinant and will develop manipulative skills for evaluating determinants. We will then show that determinants have important applications and can be used to solve linear systems.

Associated with every square matrix A is a number called the **determinant** of A, denoted by $|A|$. If A is the 2×2 matrix

$$A = \begin{bmatrix} a_{11} & a_{12} \\ a_{21} & a_{22} \end{bmatrix}$$

then $|A|$ is said to be a **determinant of second order** and is defined by the rule

$$|A| = \begin{vmatrix} a_{11} & a_{12} \\ a_{21} & a_{22} \end{vmatrix} = a_{11}a_{22} - a_{21}a_{12}$$

EXAMPLE 1
Compute the real number represented by

$$\begin{vmatrix} 4 & -5 \\ 3 & -1 \end{vmatrix}$$

SOLUTION
We apply the rule for a determinant of second order.

$$\begin{vmatrix} 4 & -5 \\ 3 & -1 \end{vmatrix} = (4)(-1) - (3)(-5) = 11$$

PROGRESS CHECK
Compute the real number represented by

(a) $\begin{vmatrix} -6 & 2 \\ -1 & -2 \end{vmatrix}$ (b) $\begin{vmatrix} \frac{1}{2} & \frac{1}{4} \\ -4 & -2 \end{vmatrix}$

ANSWERS
(a)   14   (b)   0

To simplify matters, when we want to compute the determinant of a matrix we will say "evaluate the determinant." This is not technically correct, however, since a determinant *is* a real number.

MINORS AND COFACTORS

The rule for evaluating a determinant of order 3 is

$$\begin{vmatrix} a_{11} & a_{12} & a_{13} \\ a_{21} & a_{22} & a_{23} \\ a_{31} & a_{32} & a_{33} \end{vmatrix} = a_{11}a_{22}a_{33} - a_{11}a_{32}a_{23} - a_{12}a_{21}a_{33} \\ + a_{12}a_{31}a_{23} + a_{13}a_{21}a_{32} - a_{13}a_{31}a_{22}$$

The situation becomes even more cumbersome for determinants of higher order! Fortunately, we don't have to memorize this rule; instead, we shall see that it is possible to evaluate a determinant of order 3 by reducing the problem to that of evaluating three determinants of order 2.

The **minor of an element** a_{ij} is the determinant of the matrix remaining after deleting the row and column in which the element a_{ij} appears. Given the matrix

$$\begin{bmatrix} 4 & 0 & -2 \\ 1 & -6 & 7 \\ -3 & 2 & 5 \end{bmatrix}$$

the minor of the element in row 2, column 3, is

$$\begin{vmatrix} 4 & 0 & -2 \\ 1 & -6 & 7 \\ -3 & 2 & 5 \end{vmatrix} = \begin{vmatrix} 4 & 0 \\ -3 & 2 \end{vmatrix} = 8 - 0 = 8$$

The **cofactor** of the element a_{ij} is the minor of the element a_{ij} multiplied by $(-1)^{i+j}$. Since $(-1)^{i+j}$ is $+1$ if $i + j$ is even and -1 if $i + j$ is odd, we see that the cofactor is the minor with a sign attached. The cofactor attaches the sign to the minor according to this pattern.

$$\begin{array}{ccccc} + & - & + & - & \ldots \\ - & + & - & + & \ldots \\ + & - & + & - & \ldots \\ - & + & - & + & \ldots \end{array}$$

EXAMPLE 2

Find the cofactor of each element in the first row of the matrix.

$$\begin{bmatrix} -2 & 0 & 12 \\ -4 & 5 & 3 \\ 7 & 8 & -6 \end{bmatrix}$$

SOLUTION

The cofactors are

$$(-1)^{1+1} \begin{vmatrix} -2 & 0 & 12 \\ -4 & 5 & 3 \\ 7 & 8 & -6 \end{vmatrix} = \begin{vmatrix} 5 & 3 \\ 8 & -6 \end{vmatrix} = -30 - 24 = -54$$

$$(-1)^{1+2} \begin{vmatrix} -2 & 0 & 12 \\ -4 & 5 & 3 \\ 7 & 8 & -6 \end{vmatrix} = - \begin{vmatrix} -4 & 3 \\ 7 & -6 \end{vmatrix} = -(24 - 21) = -3$$

$$(-1)^{1+3} \begin{vmatrix} -2 & 0 & 12 \\ -4 & 5 & 3 \\ 7 & 8 & -6 \end{vmatrix} = \begin{vmatrix} -4 & 5 \\ 7 & 8 \end{vmatrix} = -32 - 35 = -67$$

PROGRESS CHECK

Find the cofactor of each entry in the second column of the matrix.

$$\begin{bmatrix} 16 & -9 & 3 \\ -5 & 2 & 0 \\ -3 & 4 & -1 \end{bmatrix}$$

ANSWER

Cofactor of -9 is -5; cofactor of 2 is -7; cofactor of 4 is -15.

The cofactor is the key to the process of evaluating determinants of order 3 or higher.

Expansion by Cofactors	To evaluate a determinant, form the sum of the products obtained by multiplying each entry of any row or any column by its cofactor. This process is called **expansion by cofactors.**

Let's illustrate the process with an example.

EXAMPLE 3
Evaluate the determinant by cofactors.

$$\begin{vmatrix} -2 & 7 & 2 \\ 6 & -6 & 0 \\ 4 & 10 & -3 \end{vmatrix}$$

SOLUTION

Expansion by Cofactors	
Step 1. Choose a row or column about which to expand. In general, a row or column containing zeros will simplify the work.	*Step 1.* We will expand about column 3.
Step 2. Expand about the cofactors of the chosen row or column by multiplying each entry of the row or column by its cofactor.	*Step 2.* The expansion about column 3 is $$(2)(-1)^{1+3}\begin{vmatrix} 6 & -6 \\ 4 & 10 \end{vmatrix}$$ $$+(0)(-1)^{2+3}\begin{vmatrix} -2 & 7 \\ 4 & 10 \end{vmatrix}$$ $$+(-3)(-1)^{3+3}\begin{vmatrix} -2 & 7 \\ 6 & -6 \end{vmatrix}$$
Step 3. Evaluate the cofactors and form their sum.	*Step 3.* Using the rule for evaluating a determinant of order 2, we have $$(2)(1)[(6)(10)-(4)(-6)]+0$$ $$+(-3)(1)[(-2)(-6)-(6)(7)]$$ $$=2(60+24)-3(12-42)$$ $$=258$$

Note that expansion by cofactors of *any row or any column* will produce the same result. This important property of determinants can be used to simplify the arithmetic. The best choice of a row or column about which to expand is the one that has the most zero entries. The reason for this is that if an entry is zero, the entry times its cofactor will be zero, so we don't have to evaluate that cofactor.

PROGRESS CHECK
Evaluate the determinant of Example 3 by expanding about the second row.

ANSWER
258

EXAMPLE 4

Verify the rule for evaluating a determinant of order 3.

$$\begin{vmatrix} a_{11} & a_{12} & a_{13} \\ a_{21} & a_{22} & a_{23} \\ a_{31} & a_{32} & a_{33} \end{vmatrix} = \begin{aligned} & a_{11}a_{22}a_{33} - a_{11}a_{32}a_{23} - a_{12}a_{21}a_{33} \\ & + a_{12}a_{31}a_{23} + a_{13}a_{21}a_{32} - a_{13}a_{31}a_{22} \end{aligned}$$

SOLUTION

Expanding about the first row, we have

$$\begin{vmatrix} a_{11} & a_{12} & a_{13} \\ a_{21} & a_{22} & a_{23} \\ a_{31} & a_{32} & a_{33} \end{vmatrix} = a_{11} \begin{vmatrix} a_{22} & a_{23} \\ a_{32} & a_{33} \end{vmatrix} - a_{12} \begin{vmatrix} a_{21} & a_{23} \\ a_{31} & a_{33} \end{vmatrix} + a_{13} \begin{vmatrix} a_{21} & a_{22} \\ a_{31} & a_{32} \end{vmatrix}$$

$$= a_{11}(a_{22}a_{33} - a_{32}a_{23}) - a_{12}(a_{21}a_{33} - a_{31}a_{23}) + a_{13}(a_{21}a_{32} - a_{31}a_{22})$$

$$= a_{11}a_{22}a_{33} - a_{11}a_{32}a_{23} - a_{12}a_{21}a_{33} + a_{12}a_{31}a_{23} + a_{13}a_{21}a_{32} - a_{13}a_{31}a_{22}$$

PROGRESS CHECK

Show that the determinant is equal to zero.

$$\begin{vmatrix} a & b & c \\ a & b & c \\ d & e & f \end{vmatrix}$$

The process of expanding by cofactors works for determinants of any order. If we apply the method to a determinant of order 4, we will produce determinants of order 3; applying the method again will result in determinants of order 2.

EXAMPLE 5

Evaluate the determinant.

$$\begin{vmatrix} -3 & 5 & 0 & -1 \\ 1 & 2 & 3 & -3 \\ 0 & 4 & -6 & 0 \\ 0 & -2 & 1 & 2 \end{vmatrix}$$

SOLUTION

Expanding about the cofactors of the first column, we have

$$\begin{vmatrix} -3 & 5 & 0 & -1 \\ 1 & 2 & 3 & -3 \\ 0 & 4 & -6 & 0 \\ 0 & -2 & 1 & 2 \end{vmatrix} = -3 \begin{vmatrix} 2 & 3 & -3 \\ 4 & -6 & 0 \\ -2 & 1 & 2 \end{vmatrix} - 1 \begin{vmatrix} 5 & 0 & -1 \\ 4 & -6 & 0 \\ -2 & 1 & 2 \end{vmatrix}$$

Each determinant of order 3 can then be evaluated.

$$-3\begin{vmatrix} 2 & 3 & -3 \\ 4 & -6 & 0 \\ -2 & 1 & 2 \end{vmatrix} \begin{aligned} &= (-3)(-24) \\ &= 72 \end{aligned} \qquad -1\begin{vmatrix} 5 & 0 & -1 \\ 4 & -6 & 0 \\ -2 & 1 & 2 \end{vmatrix} \begin{aligned} &= (-1)(-52) \\ &= 52 \end{aligned}$$

The original determinant has the value $72 + 52 = 124$.

PROGRESS CHECK

Evaluate.

$$\begin{vmatrix} 0 & -1 & 0 & 2 \\ 3 & 0 & 4 & 0 \\ 0 & 5 & 0 & -3 \\ 1 & 0 & 1 & 0 \end{vmatrix}$$

ANSWER

7

CRAMER'S RULE

Determinants provide a convenient way of expressing formulas in many areas of mathematics, particularly in geometry. One of the best-known uses of determinants is in solving systems of linear equations, a procedure known as **Cramer's rule.**

In an earlier section we solved systems of linear equations by the method of elimination. Let's apply this method to the general system of two equations in two unknowns.

$$a_{11}x + a_{12}y = c_1 \qquad (1)$$
$$a_{21}x + a_{22}y = c_2 \qquad (2)$$

If we multiply Equation (1) by a_{22} and Equation (2) by $-a_{12}$ and add, we will eliminate y.

$$\begin{aligned} a_{11}a_{22}x + a_{12}a_{22}y &= \quad c_1a_{22} \\ \underline{-a_{21}a_{12}x - a_{12}a_{22}y} &= \underline{-c_2a_{12}} \\ a_{11}a_{22}x - a_{21}a_{12}x &= c_1a_{22} - c_2a_{12} \end{aligned}$$

Thus,

$$x(a_{11}a_{22} - a_{21}a_{12}) = c_1a_{22} - c_2a_{12}$$

or

$$x = \frac{c_1a_{22} - c_2a_{12}}{a_{11}a_{22} - a_{21}a_{12}}$$

Similarly, multiplying Equation (1) by a_{21} and Equation (2) by $-a_{11}$ and adding, we can eliminate x and solve for y.

$$y = \frac{c_2 a_{11} - c_1 a_{21}}{a_{11} a_{22} - a_{21} a_{12}}$$

The denominators in the expressions for x and y are identical and can be written as the determinant of the matrix

$$A = \begin{bmatrix} a_{11} & a_{12} \\ a_{21} & a_{22} \end{bmatrix}$$

If we apply this same idea to the numerators, we have

$$x = \frac{\begin{vmatrix} c_1 & a_{12} \\ c_2 & a_{22} \end{vmatrix}}{|A|} \qquad y = \frac{\begin{vmatrix} a_{11} & c_1 \\ a_{21} & c_2 \end{vmatrix}}{|A|} \qquad |A| \neq 0$$

What we have arrived at is Cramer's rule, which is a means of expressing the solution of a system of linear equations in determinant form.

The following example outlines the steps for using Cramer's rule.

EXAMPLE 6
Solve by Cramer's rule.

$$3x - y = 9$$
$$x + 2y = -4$$

SOLUTION

Cramer's Rule	
Step 1. Write the determinant of the coefficient matrix A.	*Step 1.* $$|A| = \begin{vmatrix} 3 & -1 \\ 1 & 2 \end{vmatrix}$$
Step 2. The numerator for x is the determinant of the matrix obtained from A by replacing the column of coefficients of x with the column of right-hand sides of the equations.	*Step 2.* $$x = \frac{\begin{vmatrix} 9 & -1 \\ -4 & 2 \end{vmatrix}}{|A|}$$
Step 3. The numerator for y is the determinant of the matrix obtained from A by replacing the column of coefficients of y with the column of right-hand sides of the equations.	*Step 3.* $$y = \frac{\begin{vmatrix} 3 & 9 \\ 1 & -4 \end{vmatrix}}{|A|}$$

Step 4. Evaluate the determinants to obtain the solution. If $|A| = 0$, Cramer's rule cannot be used.

Step 4.

$$|A| = 6 + 1 = 7$$

$$x = \frac{18 - 4}{7} = \frac{14}{7} = 2$$

$$y = \frac{12 - 9}{7} = -\frac{21}{7} = -3$$

PROGRESS CHECK
Solve by Cramer's rule.

$$2x + 3y = -4$$
$$3x + 4y = -7$$

ANSWER
$x = -5$, $y = 2$

The steps outlined in Example 6 can be applied to solve any system of linear equations in which the number of equations is the same as the number of variables and in which $|A| \neq 0$. Here is an example with three equations and three unknowns.

EXAMPLE 7
Solve by Cramer's rule.

$$3x \quad\quad + 2z = -2$$
$$2x - y \quad\quad = 0$$
$$\quad\quad 2y + 6z = -1$$

SOLUTION
We form the determinant of coefficients.

$$|A| = \begin{vmatrix} 3 & 0 & 2 \\ 2 & -1 & 0 \\ 0 & 2 & 6 \end{vmatrix}$$

Then

$$x = \frac{|A_1|}{|A|} \quad\quad y = \frac{|A_2|}{|A|} \quad\quad z = \frac{|A_3|}{|A|}$$

where A_1 is obtained from A by replacing its first column by the column of right-hand sides, A_2 is obtained from A by replacing the second column of A with the column of right-hand sides, and A_3 is obtained from A by replacing its third column with the column of right-hand sides. Thus

$$x = \frac{\begin{vmatrix} -2 & 0 & 2 \\ 0 & -1 & 0 \\ -1 & 2 & 6 \end{vmatrix}}{|A|} \quad\quad y = \frac{\begin{vmatrix} 3 & -2 & 2 \\ 2 & 0 & 0 \\ 0 & -1 & 6 \end{vmatrix}}{|A|} \quad\quad z = \frac{\begin{vmatrix} 3 & 0 & -2 \\ 2 & -1 & 0 \\ 0 & 2 & -1 \end{vmatrix}}{|A|}$$

Expanding by cofactors we calculate $|A| = -10$, $|A_1| = 10$, $|A_2| = 20$, and $|A_3| = -5$, obtaining

$$x = \frac{10}{-10} = -1 \quad y = \frac{20}{-10} = -2 \quad z = \frac{-5}{-10} = \frac{1}{2}$$

PROGRESS CHECK

Solve by Cramer's rule.

$$\begin{aligned} 3x \quad\quad - z &= 1 \\ -6x + 2y \quad\quad &= -5 \\ -4y + 3z &= 5 \end{aligned}$$

ANSWER

$$x = \frac{2}{3}, \; y = -\frac{1}{2}, \; z = 1$$

WARNING

(a)   Each equation of the linear system must be written in the form

$$Ax + By + Cz = k$$

before using Cramer's rule.

(b)   If $|A| = 0$, Cramer's rule cannot be used.

Determinants have significant theoretical importance but are not of much use for computational purposes. The matrix methods discussed in this chapter provide the basis for techniques better suited for computer implementation.

EXERCISE SET 10.8

In Exercises 1–6 evaluate the given determinant.

1. $\begin{vmatrix} 2 & -3 \\ 4 & 5 \end{vmatrix}$ 2. $\begin{vmatrix} 3 & 4 \\ -1 & 2 \end{vmatrix}$ 3. $\begin{vmatrix} -4 & 1 \\ 0 & 2 \end{vmatrix}$ 4. $\begin{vmatrix} 2 & 2 \\ 3 & 3 \end{vmatrix}$

5. $\begin{vmatrix} 0 & 0 \\ 1 & 3 \end{vmatrix}$ 6. $\begin{vmatrix} -4 & -1 \\ -2 & 3 \end{vmatrix}$

In Exercises 7–10 let

$$A = \begin{bmatrix} 3 & -1 & 2 \\ 4 & 1 & -3 \\ 5 & -2 & 0 \end{bmatrix}$$

7.   Compute the minor of each of the following elements.   8.   Compute the minor of each of the following elements.
     (a) a_{11} (b) a_{23} (c) a_{31} (d) a_{33} (a) a_{12} (b) a_{22} (c) a_{23} (d) a_{32}

9. Compute the cofactor of each of the following elements.

   (a) a_{11} (b) a_{23} (c) a_{31} (d) a_{33}

10. Compute the cofactor of each of the following elements.

   (a) a_{12} (b) a_{22} (c) a_{23} (d) a_{32}

In Exercises 11–20 evaluate the given determinant.

11. $\begin{vmatrix} 4 & -2 & 5 \\ 5 & 2 & 0 \\ 2 & 0 & 4 \end{vmatrix}$

12. $\begin{vmatrix} 4 & 1 & 2 \\ 0 & 2 & 3 \\ 0 & 0 & -4 \end{vmatrix}$

13. $\begin{vmatrix} -1 & 2 & 0 \\ 3 & 4 & 1 \\ 6 & 5 & 2 \end{vmatrix}$

14. $\begin{vmatrix} -1 & 3 & 2 \\ 0 & 7 & 7 \\ 2 & 1 & 3 \end{vmatrix}$

15. $\begin{vmatrix} 2 & 2 & 4 \\ 3 & 8 & 1 \\ 1 & 1 & 2 \end{vmatrix}$

16. $\begin{vmatrix} 0 & 1 & 3 \\ 2 & 5 & -1 \\ 4 & 2 & -2 \end{vmatrix}$

17. $\begin{vmatrix} 3 & 2 & 1 & 0 \\ -1 & -3 & -1 & 0 \\ 0 & 0 & 2 & 2 \\ 4 & 1 & 3 & 3 \end{vmatrix}$

18. $\begin{vmatrix} -1 & 2 & 4 & 0 \\ 3 & -2 & -3 & 0 \\ 0 & 4 & 2 & 5 \\ 0 & -3 & 1 & 4 \end{vmatrix}$

19. $\begin{vmatrix} 2 & -3 & 2 & -4 \\ 0 & 4 & -1 & 9 \\ 0 & 1 & 2 & 0 \\ 0 & 1 & 3 & -1 \end{vmatrix}$

20. $\begin{vmatrix} 1 & 1 & 0 & 1 \\ 0 & -1 & 4 & -1 \\ -2 & 3 & 1 & -4 \\ 0 & 2 & 0 & 2 \end{vmatrix}$

In Exercises 21–28 solve the given linear system by use of Cramer's rule.

21. $\begin{aligned} 2x + y + z &= -1 \\ 2x - y + 2z &= 2 \\ x + 2y + z &= -4 \end{aligned}$

22. $\begin{aligned} x - y + z &= -5 \\ 3x + y + 2z &= -5 \\ 2x - y - z &= -2 \end{aligned}$

23. $\begin{aligned} 2x + y - z &= 9 \\ x - 2y + 2z &= -3 \\ 3x + 3y + 4z &= 11 \end{aligned}$

24. $\begin{aligned} 2x + y - z &= -2 \\ -2x - 2y + 3z &= 2 \\ 3x + y - z &= -4 \end{aligned}$

25. $\begin{aligned} -x - y + 2z &= 7 \\ x + 2y - 2z &= -7 \\ 2x - y + z &= -4 \end{aligned}$

26. $\begin{aligned} 4x + y - z &= -1 \\ x - y + 2z &= 3 \\ -x + 2y - z &= 0 \end{aligned}$

27. $\begin{aligned} x + y - z + 2w &= 0 \\ 2x + y \quad\;\; - w &= -2 \\ 3x \quad\;\; + 2z \quad\quad &= -3 \\ -x + 2y \quad\;\; + 3w &= 1 \end{aligned}$

28. $\begin{aligned} 2x + y \quad\quad - 3w &= -7 \\ 3x \quad\;\; + 2z + w &= -1 \\ -x + 2y \quad\;\; + 3w &= 0 \\ -2x - 3y + 2z - w &= 8 \end{aligned}$

29. Show that

$$\begin{vmatrix} a_1 + b_1 & a_2 + b_2 \\ c & d \end{vmatrix} = \begin{vmatrix} a_1 & a_2 \\ c & d \end{vmatrix} + \begin{vmatrix} b_1 & b_2 \\ c & d \end{vmatrix}$$

30. Show that

$$\begin{vmatrix} ka_{11} & ka_{12} \\ a_{21} & a_{22} \end{vmatrix} = \begin{vmatrix} a_{11} & a_{12} \\ ka_{21} & ka_{22} \end{vmatrix} = k \begin{vmatrix} a_{11} & a_{12} \\ a_{21} & a_{22} \end{vmatrix}$$

31. Prove that if a row or column of a square matrix consists entirely of zeros, the determinant of the matrix is zero. (*Hint:* Expand by cofactors.)

32. Prove that if matrix B is obtained by multiplying each element of a row of a square matrix A by a constant k, then $|B| = k|A|$.

33. Prove that if A is an $n \times n$ matrix and $B = kA$, where k is a constant, then $|B| = k^n|A|$.

34. Prove that if matrix B is obtained from a square matrix A by interchanging the rows and columns of A, then $|B| = |A|$.

TERMS AND SYMBOLS

solution of a system of equations (p. 420)
method of substitution (p. 420)
consistent system (p. 422)
inconsistent system (p. 422)
method of elimination (p. 423)
linear system (p. 424)
half-plane (p. 427)
linear inequality in two variables (p. 427)
graph of a linear inequality (p. 427)
systems of linear inequalities (p. 430)

solution of a system of linear inequalities (p. 431)
constraints (p. 436)
linear programming problem (p. 437)
objective function (p. 437)
feasible solution (p. 437)
vertex (p. 437)
Gaussian elimination (p. 441)
triangular form (p. 441)
back-substitution (p. 442)
matrix (p. 446)
entries, elements (p. 446)
dimension (p. 447)
square matrix (p. 447)

order (p. 447)
row matrix (p. 447)
column matrix (p. 447)
$[a_{ij}]$ (p. 447)
coefficient matrix (p. 448)
augmented matrix (p. 448)
elementary row operations (p. 449)
pivot element (p. 450)
pivot row (p. 450)
Gauss–Jordan elimination (p. 453)
equality of matrices (p. 456)
scalar (p. 458)
scalar multiplication (p. 458)

zero matrix (p. 463)
identity matrix, I_n (p. 463)
additive inverse (p. 463)
invertible or nonsingular matrix (p. 465)
matrix inverse (p. 465)
A^{-1} (p. 466)
determinant (p. 476)
minor (p. 477)
cofactor (p. 477)
expansion by cofactors (p. 478)
Cramer's rule (p. 481)

KEY IDEAS FOR REVIEW

☐ The method of substitution involves solving an equation for one variable and substituting the result into another equation.

☐ The method of elimination involves multiplying an equation by a nonzero constant so that when it is added to a second equation, a variable drops out.

☐ A consistent system of equations has one or more real solutions; an inconsistent system has no real solutions.

☐ The graph of a pair of linear equations in two variables is two straight lines, which may either (a) intersect in a point, (b) be parallel, or (c) be the same line. If the two straight lines intersect, the coordinates of the point of intersection are a solution of the system of linear equations. If the lines are distinct and do not intersect, the system is inconsistent.

☐ When using any method of solution, it is possible to detect the special cases when lines are parallel or reduce to the same line.

☐ It is often easier and more natural to set up word problems using two or more variables.

☐ The solution of a system of linear inequalities can be found graphically as the region satisfying all the inequalities.

☐ To solve a linear programming problem, it is only necessary to consider the vertices of the region of feasible solutions.

☐ Gaussian elimination is a systematic way of transforming a linear system to triangular form. A linear system in triangular form is easily solved by back-substitution.

☐ A matrix is a rectangular array of numbers.

☐ Systems of linear equations can be conveniently handled in matrix notation. By dropping the names of the variables, matrix notation focuses on the coefficients and the right-hand side of the system. The elementary row operations are then seen to be an abstraction of those operations that produce equivalent systems of equations.

☐ Gaussian elimination and Gauss–Jordan elimination both involve the use of elementary row operations on the augmented matrix corresponding to a linear system. In the case of a system of three equations with three unknowns, the final matrices will be of this form:

$$\begin{bmatrix} * & * & * & | & * \\ 0 & * & * & | & * \\ 0 & 0 & * & | & * \end{bmatrix} \qquad \begin{bmatrix} 1 & 0 & 0 & | & c_1 \\ 0 & 1 & 0 & | & c_2 \\ 0 & 0 & 1 & | & c_3 \end{bmatrix}$$

Gaussian elimination   Gauss–Jordan elimination

If Gauss–Jordan elimination is used, the solution can be read from the final matrix; if Gaussian elimination is used, back-substitution is then performed with the final matrix.

☐ A linear system can be written in the form $AX = B$ where A is the coefficient matrix, X is a column matrix of the unknowns, and B is the column matrix of the right-hand sides.

☐ The sum and difference of two matrices A and B can be formed only if A and B are of the same dimension. The product AB can be formed only if the number of columns of A is the same as the number of rows of B.

☐ The $n \times n$ matrix B is said to be the inverse of the $n \times n$ matrix A if $AB = I_n$ and $BA = I_n$. We denote the inverse of A by A^{-1}. The inverse can be computed by using elementary row operations to transform the matrix $[A \mid I_n]$ to the form $[I_n \mid B]$. Then $B = A^{-1}$.

☐ If the linear system $AX = B$ has a unique solution, then $X = A^{-1}B$.

☐ Associated with every square matrix is a number called a determinant. The rule for evaluating a determinant of order 2 is

$$\begin{vmatrix} a & b \\ c & d \end{vmatrix} = ad - bc$$

☐ For determinants of order greater than 2, the method of expansion by cofactors may be used to reduce the problem to that of evaluating determinants of order 2.

☐ When expanding by cofactors, choose the row or column that contains the most zeros. This will ease the arithmetic burden.

☐ Cramer's rule provides a means for solving a linear system by expressing the value of each unknown as a quotient of determinants.

REVIEW EXERCISES

Solutions to exercises whose numbers are in color are in the Solutions section in the back of the book.

10.1 In Exercises 1–5 solve the given system by the method of substitution.

1. $3x - 2y = 4$
   $2x + y = -2$

2. $2x + 3y = -7$
   $-x + 2y = 7$

3. $x^2 + y^2 = 25$
   $x + 3y = 5$

4. $y^2 = 2x - 1$
   $x - y = 2$

5. $2x - y = 0$
   $x - 3y = \dfrac{7}{4}$

In Exercises 6–10 solve the given system by the method of elimination.

6. $5x - 2y = 14$
   $-x - 3y = 4$

7. $2x + 3y = -1$
   $-3x + 4y = -\dfrac{11}{4}$

8. $x^2 + y^2 = 9$
$y = x^2 + 3$

9. $y^2 = 4x$
$y^2 + x - 2y = 12$

10. $\dfrac{1}{3}x + \dfrac{1}{2}y = -1$
$\dfrac{1}{2}x - \dfrac{1}{4}y = \dfrac{5}{2}$

11. A manufacturer of faucets finds that the supply S and demand D are related to price p as follows.

$$S = 3p + 2$$
$$D = -2p + 17$$

Find the equilibrium price, that is, the price at which supply equals demand, and the number of faucets sold at that price.

10.2 In Exercises 12–17 graph the solution set of the linear inequality or system of linear inequalities.

12. $x - 2y \le 5$

13. $2x + y > 4$

14. $2x + 3y \le 2$
$x - y \ge 1$

15. $x - 2y \ge 4$
$2x - y \le 2$

16. $2x + 3y \le 6$
$x \ge 0$
$y \ge 1$

17. $2x + y \le 4$
$2x - y \le 3$
$x \ge 0$
$y \ge 0$

10.3 In Exercises 18 and 19 solve the given linear programming problem.

18. maximize $z = 5y - x$
subject to $\quad 8y - 3x \le 36$
$\qquad\qquad 6x + y \le 30$
$\qquad\qquad\qquad y \ge 1$
$\qquad\qquad\qquad x \ge 0$

19. minimize $z = x + 4y$
subject to $\quad 4x - y \ge 8$
$\qquad\qquad 4x + y \le 24$
$\qquad\qquad 5y + 4x \ge 32$

10.4 In Exercises 20–22 use Gaussian elimination to solve the given linear systems.

20. $-3x - y + z = 12$
$\quad 2x + 5y - 2z = -9$
$\quad -x + 4y + 2z = 15$

21. $3x + 2y - z = -8$
$\quad 2x \quad\;\; + 3z = 5$
$\quad x - 4y \quad\;\; = -4$

22. $\quad 5x - y + 2z = 10$
$-2x + 3y - z = -7$
$\quad 3x \quad\;\; + 2z = 7$

10.5 Exercises 23–26 refer to the matrix

$$A = \begin{bmatrix} -1 & 4 & 2 & 0 & 8 \\ 2 & 0 & -3 & -1 & 5 \\ 4 & -6 & 9 & 1 & -2 \end{bmatrix}$$

23. Determine the dimension of the matrix A.

24. Find a_{24}.　　25. Find a_{31}.

26. Find a_{15}.

Exercises 27 and 28 refer to the linear system

$$3x - 7y = 14$$
$$x + 4y = 6$$

27. Write the coefficient matrix of the linear system.

28. Write the augmented matrix of the linear system.

In Exercises 29 and 30 write a linear system corresponding to the augmented matrix.

29. $\begin{bmatrix} 4 & -1 & | & 3 \\ 2 & 5 & | & 0 \end{bmatrix}$　　30. $\begin{bmatrix} -2 & 4 & 5 & | & 0 \\ 6 & -9 & 4 & | & 0 \\ 3 & 2 & -1 & | & 0 \end{bmatrix}$

In Exercises 31 and 32 use back-substitution to solve the linear systems corresponding to the given augmented matrix.

31. $\begin{bmatrix} 1 & -2 & | & 7 \\ 0 & 1 & | & -4 \end{bmatrix}$　　32. $\begin{bmatrix} 1 & -2 & 2 & | & -9 \\ 0 & 1 & 3 & | & -8 \\ 0 & 0 & 1 & | & -3 \end{bmatrix}$

In Exercises 33 and 34 use matrix methods to solve the given linear system.

33. $x + y = 2$
$2x - 4y = -5$

34. $2x - y - 2z = 3$
$-2x + 3y + z = 3$
$2y - z = 6$

10.6 In Exercises 35 and 36 solve for x.

35. $\begin{bmatrix} 5 & -1 \\ 3 & 2x \end{bmatrix} = \begin{bmatrix} 5 & -1 \\ 3 & -6 \end{bmatrix}$

36. $\begin{bmatrix} 6 & x^2 \\ 4x & -2 \end{bmatrix} = \begin{bmatrix} 6 & 9 \\ -12 & -2 \end{bmatrix}$

Exercises 37–43 refer to the following matrices.

$$A = \begin{bmatrix} 2 & -1 \\ 3 & 2 \end{bmatrix} \qquad B = \begin{bmatrix} -1 & 5 \\ 4 & -3 \end{bmatrix}$$

$$C = \begin{bmatrix} -1 & 0 \\ 0 & 4 \\ 2 & -2 \end{bmatrix} \qquad D = \begin{bmatrix} 1 & 3 & 4 \\ -1 & 0 & -6 \end{bmatrix}$$

If possible, find the following.

37. $A + B$

38. $B - A$

39. $A + C$

40. $5D$

41. CD

42. DC

43. $-AB$

10.7 In Exercises 44 and 45 find the inverse of the given matrix.

44. $\begin{bmatrix} -2 & 3 \\ 1 & 4 \end{bmatrix}$

45. $\begin{bmatrix} 1 & 1 & -4 \\ -5 & -2 & 0 \\ 4 & 2 & -1 \end{bmatrix}$

In Exercises 46 and 47 solve the given system by finding the inverse of the coefficient matrix.

46. $\begin{aligned} 2x - y &= 1 \\ x + y &= 5 \end{aligned}$

47. $\begin{aligned} x + 2y - 2z &= -4 \\ 3x - y &= -2 \\ y + 4z &= 1 \end{aligned}$

10.8 In Exercises 48–51 evaluate the given determinant.

48. $\begin{vmatrix} 3 & 1 \\ -4 & 2 \end{vmatrix}$

49. $\begin{vmatrix} -1 & 2 \\ 0 & 6 \end{vmatrix}$

50. $\begin{vmatrix} 1 & -1 & 2 \\ 0 & 5 & 4 \\ 2 & 3 & 8 \end{vmatrix}$

51. $\begin{vmatrix} 1 & 2 & -1 \\ 0 & 3 & 4 \\ 0 & 0 & -1 \end{vmatrix}$

In Exercises 52–55 use Cramer's rule to solve the given linear system.

52. $\begin{aligned} 2x - y &= -3 \\ -2x + 3y &= 11 \end{aligned}$

53. $\begin{aligned} 3x - y &= 7 \\ 2x + 5y &= -18 \end{aligned}$

54. $\begin{aligned} 3x \quad\ + z &= 0 \\ x + y + z &= 0 \\ -3y + 2z &= -4 \end{aligned}$

55. $\begin{aligned} 2x + 3y + z &= -5 \\ 2y + 2z &= -3 \\ 4x + y - 2z &= -2 \end{aligned}$

PROGRESS TEST 10A

In Problems 1–3 find all real solutions of the given system.

1. $\begin{aligned} y^2 &= 5x \\ y^2 - x^2 &= 6 \end{aligned}$

2. $\begin{aligned} \frac{1}{4}x + \frac{1}{2}y &= 0 \\ 3x - 2y &= 8 \end{aligned}$

3. $\begin{aligned} x^2 + y^2 &= 25 \\ 4x^2 - y^2 &= 20 \end{aligned}$

4. An elegant men's shop is having a post-Christmas sale. All shirts are reduced to one low price, and all ties are reduced to an even lower price. A customer purchases 3 ties and 7 shirts, paying $135. Another customer selects 5 ties and 3 shirts, paying $95. What is the sale price of each tie and of each shirt?

In Problems 5 and 6 graph the solution set of the system of linear inequalities.

5. $\begin{aligned} x - 2y &\leq 1 \\ 3x + 2y &\geq 4 \end{aligned}$

6. $\begin{aligned} 2x + y &\leq 10 \\ -x + 3y &\leq 12 \\ x &\geq 0 \\ y &\geq 0 \end{aligned}$

7. Solve by Gaussian elimination.

$$\begin{aligned} 3x + 2y - z &= -4 \\ x - y + 3z &= 12 \\ 2x - y - 2z &= -20 \end{aligned}$$

Problems 8 and 9 refer to the matrix

$$A = \begin{bmatrix} -1 & 2 \\ -2 & 4 \\ 0 & 7 \end{bmatrix}$$

8. Find the dimension of the matrix A.

9. Find a_{31}.

10. Write the augmented matrix of the linear system

$$\begin{aligned} -7x \qquad + 6z &= 3 \\ 2y - z &= 10 \\ x - y + z &= 5 \end{aligned}$$

11. Write a linear system corresponding to the augmented matrix.

$$\begin{bmatrix} -5 & 2 & | & 4 \\ 3 & -4 & | & 4 \end{bmatrix}$$

12. Use back-substitution to solve the linear system corresponding to the augmented matrix

$$\begin{bmatrix} 1 & 1 & | & 0 \\ 0 & 1 & | & \frac{1}{2} \end{bmatrix}$$

13. Solve the linear system

$$\begin{aligned} -x + 2y &= -10 \\ \frac{1}{2}x + 2y &= -7 \end{aligned}$$

by applying Gaussian elimination to the augmented matrix.

14. Solve the linear system

$$\begin{aligned} 2x - y + 3z &= 2 \\ x + 2y - z &= 1 \\ -x + y + 4z &= 2 \end{aligned}$$

by applying Gauss–Jordan elimination to the augmented matrix.

15. Solve for x.

$$\begin{bmatrix} 2x - 1 & 0 \\ 1 & -3 \end{bmatrix} = \begin{bmatrix} 5 & 0 \\ 1 & -3 \end{bmatrix}$$

Problems 16–19 refer to the matrices

$$A = \begin{bmatrix} -4 & 0 & 3 \\ 6 & 2 & -3 \end{bmatrix} \quad B = \begin{bmatrix} -1 \\ -3 \end{bmatrix}$$

$$C = \begin{bmatrix} 4 & 2 \\ -2 & 0 \\ 3 & -1 \end{bmatrix} \quad D = \begin{bmatrix} 1 & -6 \\ 0 & 2 \\ 4 & -1 \end{bmatrix}$$

If possible, find the following.

16. $C - 2D$

17. AC

18. CB

19. BA

20. Find the inverse of the matrix.

$$\begin{bmatrix} -1 & 0 & 4 \\ 2 & 1 & -1 \\ 1 & -3 & 2 \end{bmatrix}$$

21. Solve the given linear system by finding the inverse of the coefficient matrix.

$$\begin{aligned} 3x - 2y &= -8 \\ 2x + 3y &= -1 \end{aligned}$$

In Problems 22 and 23 evaluate the given determinant.

22. $\begin{vmatrix} -6 & -2 \\ 2 & 1 \end{vmatrix}$

23. $\begin{vmatrix} 0 & -1 & 2 \\ 2 & -2 & 3 \\ 1 & 4 & 5 \end{vmatrix}$

24. Use Cramer's rule to solve the linear system.

$$\begin{aligned} x + 2y &= -2 \\ -2x - 3y &= 1 \end{aligned}$$

PROGRESS TEST 10B

In Problems 1–3 find all real solutions of the given system.

1. $\begin{aligned} x^2 - y^2 &= 9 \\ x^2 + y^2 &= 41 \end{aligned}$

2. $\begin{aligned} 2x - 3y &= 11 \\ 3x + 5y &= -12 \end{aligned}$

3. $\begin{aligned} x^2 + 3y^2 &= 12 \\ x + 3y &= 6 \end{aligned}$

4. A movie theater charges $3 admission for an adult and $1.50 for a child. If 600 tickets were sold and the total revenue received was $1350, how many tickets of each type were sold?

In Problems 5 and 6 graph the solution set of the system of linear inequalities.

5. $2x - 3y \geq 6$
   $3x + y \leq 3$

6. $2x + y \geq 4$
   $2x - 5y \leq 5$
   $y \geq 1$

7. Solve by Gaussian elimination.

$$\begin{aligned} x \quad + 2z &= \quad 7 \\ 3y + 4z &= -10 \\ -2x + y - 2z &= -14 \end{aligned}$$

Problems 8 and 9 refer to the matrix

$$B = \begin{bmatrix} -1 & 5 & 0 & 6 \\ 4 & -2 & 1 & -2 \end{bmatrix}$$

8. Determine the dimension of the matrix B.
9. Find b_{23}.
10. Write the coefficient matrix of the linear system.

$$\begin{aligned} 2x - 6y &= \quad 5 \\ x + 3y &= -2 \end{aligned}$$

11. Write a linear system corresponding to the augmented matrix

$$\begin{bmatrix} 16 & 0 & 6 & | & 10 \\ -4 & -2 & 5 & | & 8 \\ 2 & 3 & -1 & | & -6 \end{bmatrix}$$

12. Use back-substitution to solve the linear system corresponding to the augmented matrix.

$$\begin{bmatrix} 1 & 3 & -1 & | & 0 \\ 0 & 1 & -2 & | & 5 \\ 0 & 0 & 1 & | & -3 \end{bmatrix}$$

13. Solve the linear system

$$\begin{aligned} 2x + 3y &= -11 \\ 3x - 2y &= \quad 3 \end{aligned}$$

by applying Gaussian elimination to the augmented matrix.

14. Solve the linear system

$$\begin{aligned} 2x + y - 2z &= \quad 7 \\ -3x - 5y + 4z &= -3 \\ 5x + 4y \quad &= \quad 17 \end{aligned}$$

by applying Gauss–Jordan elimination to the augmented matrix.

15. Solve for y.

$$\begin{bmatrix} 2 & -5 & 1 \\ -3 & 1-y & 2 \end{bmatrix} = \begin{bmatrix} 2 & -5 & 1 \\ -3 & 6 & 2 \end{bmatrix}$$

Problems 16–19 refer to the matrices

$$A = \begin{bmatrix} 2 & -3 \\ -1 & 0 \\ 2 & 1 \end{bmatrix} \quad B = \begin{bmatrix} 1 & -2 & 3 \end{bmatrix} \quad C = \begin{bmatrix} 0 & -4 \\ 3 & 1 \\ -2 & 5 \end{bmatrix}$$

If possible, find the following.

16. BA
17. $2C + 3A$
18. CB
19. $BC - A$
20. Find the inverse of the matrix.

$$\begin{bmatrix} 2 & -5 & -2 \\ -1 & 3 & 0 \\ -2 & 0 & 4 \end{bmatrix}$$

21. Solve the given linear system by finding the inverse of the coefficient matrix.

$$\begin{aligned} 5x - 2y &= -6 \\ -2x + 2y &= \quad 3 \end{aligned}$$

In Problems 22 and 23 evaluate the given determinant.

22. $\begin{vmatrix} -2 & 4 \\ 3 & 5 \end{vmatrix}$

23. $\begin{vmatrix} -2 & 0 & 1 \\ 1 & 2 & 0 \\ 2 & -1 & -1 \end{vmatrix}$

24. Use Cramer's rule to solve the linear system.

$$\begin{aligned} x + y &= -1 \\ 2x - 4y &= -5 \end{aligned}$$

CHAPTER 11

SEQUENCES AND SERIES

The topics in this chapter are related in that they all involve the set of natural numbers. As you might expect, despite our return to a simpler number system, the approach and results will be more advanced than in earlier chapters. For example, in discussing sequences we will be dealing with functions whose domain is the set of natural numbers. Yet, sequences lead to considerations of series, and the underlying concepts of infinite series can be used as an introduction to calculus.

Another of the topics, mathematical induction, provides a means of proving certain theorems involving the natural numbers that appear to resist other means of proof. As an example, we will use mathematical induction to prove that the sum of the first n consecutive positive integers is $n(n + 1)/2$.

Yet another topic is the binomial theorem, which gives us a way to expand the expression $(a + b)^n$ where n is a natural number. One of the earliest results obtained in a calculus course requires the binomial theorem in its derivation.

11.1

SEQUENCES AND SIGMA NOTATION

INFINITE SEQUENCES

Can you see a pattern or relationship that describes this string of numbers?

$$1, 4, 9, 16, 25, \ldots$$

If we rewrite this string as

$$1^2, 2^2, 3^2, 4^2, 5^2, \ldots$$

it is clear that these are the squares of successive natural numbers. Each number in the string is called a **term.** We could write the nth term of the list as a function a defined by

$$a(n) = n^2$$

where n is a natural number. Such a string of numbers is called an infinite sequence, since the list is infinitely long.

An **infinite sequence** (often called simply a **sequence**) is a function whose domain is the set of all natural numbers.

The range of the function a is

$$a(1), a(2), a(3), \ldots, a(n), \ldots$$

which we write as

$$a_1, a_2, a_3, \ldots a_n, \ldots$$

That is, we indicate a sequence by using subscript notation rather than function notation. We say that a_1 is the **first term** of the sequence, a_2 is the **second term,** and so on, and we write the **nth term** as a_n where $a_n = a(n)$.

EXAMPLE 1
Write the first three terms and the tenth term of each of the sequences whose nth term is given.

(a) $a_n = n^2 + 1$ (b) $a_n = \dfrac{n}{n + 1}$ (c) $a_n = 2^n - 1$

SOLUTION
The first three terms are found by substituting $n = 1, 2,$ and 3 in the formula for a_n. The tenth term is found by substituting $n = 10$.

(a) $a_1 = 1^2 + 1 = 2$ $a_2 = 2^2 + 1 = 5$ $a_3 = 3^2 + 1 = 10$
$a_{10} = 10^2 + 1 = 101$

(b) $a_1 = \dfrac{1}{1 + 1} = \dfrac{1}{2}$ $a_2 = \dfrac{2}{2 + 1} = \dfrac{2}{3}$ $a_3 = \dfrac{3}{3 + 1} = \dfrac{3}{4}$

$a_{10} = \dfrac{10}{10 + 1} = \dfrac{10}{11}$

(c) $a_1 = 2^1 - 1 = 1$ $a_2 = 2^2 - 1 = 3$ $a_3 = 2^3 - 1 = 7$
$a_{10} = 2^{10} - 1 = 1023$

PROGRESS CHECK

Write the first three terms and the twelfth term of each of the sequences whose nth term is given.

(a) $a_n = 3(1 - n)$ (b) $a_n = n^2 + n + 1$ (c) $a_n = 5$

ANSWERS

(a) $a_1 = 0$, $a_2 = -3$, $a_3 = -6$, $a_{12} = -33$
(b) $a_1 = 3$, $a_2 = 7$, $a_3 = 13$, $a_{12} = 157$
(c) $a_1 = a_2 = a_3 = a_{12} = 5$

An infinite sequence is often defined by a formula expressing the nth term by reference to preceding terms. Such a sequence is said to be defined by a **recursive** formula.

EXAMPLE 2

Find the first four terms of the sequence defined by

$$a_n = a_{n-1} + 3 \quad \text{with} \quad a_1 = 2 \quad \text{and} \quad n \geq 2$$

SOLUTION

Any term of the sequence can be obtained if the preceding term is known. Of course, this recursive formulation requires a starting point, and we are indeed given a_1. Then

$$a_1 = 2$$
$$a_2 = a_1 + 3 = 2 + 3 = 5$$
$$a_3 = a_2 + 3 = 5 + 3 = 8$$
$$a_4 = a_3 + 3 = 8 + 3 = 11$$

PROGRESS CHECK

Find the first four terms of the infinite sequence

$$a_n = 2a_{n-1} - 1 \quad \text{with} \quad a_1 = -1 \quad \text{and} \quad n \geq 2$$

ANSWER

$-1, -3, -7, -15$

SUMMATION NOTATION

In the following sections of this chapter, we will seek the sum of terms of a sequence such as

$$a_1 + a_2 + a_3 + \cdots + a_m$$

Since sums occur frequently in mathematics, a special notation has been developed that is defined in the following way.

Summation Notation

$$\sum_{k=1}^{m} a_k = a_1 + a_2 + a_3 + \cdots + a_m$$

This is often referred to as **sigma notation,** since the Greek letter Σ indicates a sum of terms of the form a_k. The letter k is called the **index of summation** and always assumes successive integer values, starting with the value written under the Σ sign and ending with the value written above the Σ sign.

EXAMPLE 3

Evaluate (a) $\displaystyle\sum_{k=1}^{3} 2^k(k + 1)$ (b) $\displaystyle\sum_{i=2}^{4} (i^2 + 2)$.

SOLUTION
(a)   The terms are of the form

$$a_k = 2^k(k + 1)$$

and the sigma notation indicates that we want the sum of the terms a_1 through a_3. Forming the terms and adding,

$$\sum_{k=1}^{3} 2^k(k + 1) = 2^1(1 + 1) + 2^2(2 + 1) + 2^3(3 + 1)$$

$$= 4 + 12 + 32 = 48$$

(b)   Any letter may be used for the index of summation. Here, the letter i is used, and

$$\sum_{i=2}^{4} (i^2 + 2) = (2^2 + 2) + (3^2 + 2) + (4^2 + 2)$$

$$= 6 + 11 + 18 = 35$$

Note that the index of summation can begin with an integer value other than 1.

EXAMPLE 4
Write each sum using summation notation.

(a) $\dfrac{1}{2} + \dfrac{1}{2 \cdot 2} + \dfrac{1}{2 \cdot 3} + \dfrac{1}{2 \cdot 4} + \dfrac{1}{2 \cdot 5}$ (b) $\dfrac{2}{3} + \dfrac{3}{4} + \dfrac{4}{5} + \dfrac{5}{6}$

SOLUTION

(a)   The denominator of each term is of the form $2 \cdot k$, where k assumes integer values from 1 to 5. Then

$$\sum_{k=1}^{5} \frac{1}{2 \cdot k}$$

expresses the desired sum.

(b)   If the value of the numerator of a term is k, then the denominator is $k + 1$. Letting k range from 2 to 5,

$$\sum_{k=2}^{5} \frac{k}{k + 1}$$

expresses the desired sum.

PROGRESS CHECK

Write each sum using summation notation.

(a) $x_1^2 + x_2^2 + x_3^2 + \cdots + x_{20}^2$ (b) $2^3 + 3^4 + 4^5 + 5^6$

ANSWERS

(a) $\displaystyle\sum_{k=1}^{20} x_k^2$ (b) $\displaystyle\sum_{k=2}^{5} k^{k+1}$

If a sequence is defined by $a_n = c$, where c is a real constant, then

$$\sum_{k=1}^{r} a_k = a_1 + a_2 + \cdots + a_r$$

$$= c + c + \cdots + c$$

$$= rc$$

This leads to the rule:

For any real constant c,

$$\sum_{k=1}^{n} c = nc$$

EXAMPLE 5

Evaluate (a) $\displaystyle\sum_{j=1}^{20} 5$ (b) $\displaystyle\sum_{k=1}^{4} (k^2 - 2)$.

SOLUTION

(a) $\displaystyle\sum_{j=1}^{20} 5 = 20 \cdot 5 = 100$

(b) $\displaystyle\sum_{k=1}^{4} (k^2 - 2) = (1^2 - 2) + (2^2 - 2) + (3^2 - 2) + (4^2 - 2)$

$$= -1 + 2 + 7 + 14 = 22$$

The following are properties of sums expressed in sigma notation.

Properties of Sums	For the sequences $a_1, a_2, \ldots$, and $b_1, b_2, \ldots$, (i) $\displaystyle\sum_{k=1}^{n} (a_k + b_k) = \sum_{k=1}^{n} a_k + \sum_{k=1}^{n} b_k$ (ii) $\displaystyle\sum_{k=1}^{n} (a_k - b_k) = \sum_{k=1}^{n} a_k - \sum_{k=1}^{n} b_k$ (iii) $\displaystyle\sum_{k=1}^{n} ca_k = c\sum_{k=1}^{n} a_k, \quad c$ a constant

EXAMPLE 6

Use the properties of sums to evaluate $\displaystyle\sum_{k=1}^{4} (k^2 - 2)$.

SOLUTION

Rather than write out the terms as was done in Example 4b, we may write

$$\sum_{k=1}^{4} (k^2 - 2) = \sum_{k=1}^{4} k^2 - \sum_{k=1}^{4} 2 = \sum_{k=1}^{4} k^2 - 8$$
$$= 1^2 + 2^2 + 3^2 + 4^2 - 8 = 30 - 8 = 22$$

SERIES

A sum of terms of a sequence is called a **series.** We denote by S_n the sum of the first n terms of an infinite sequence where n is a natural number. Summation notation is very useful in handling a series. For example, given the sequence

$$a_1, a_2, a_3, \ldots, a_n, \ldots$$

AREAS BY RECTANGLES

Many textbooks introduce the integral calculus by the use of rectangles to approximate area. In the accompanying figure, we are interested in calculating the area under the curve of the function $f(x) = x^2$ that is bounded by the x-axis and the lines $x = a$ and $x = b$. The interval $[a, b]$ is divided into n subintervals of equal width, and a rectangle is erected in each interval as shown. We then seek to use the sum of the areas of the rectangles as an approximation to the area under the curve.

To calculate the area of a rectangle, we need to know the height and the width. Since the interval $[a, b]$ has been divided into n parts of equal width, we see that

$$\text{width of rectangle} = \frac{b - a}{n}$$

Next, note that the height of the rectangle whose left endpoint is at x_k is determined by the value of the function at that point; that is,

$$\text{height of rectangle} = f(x_k) = x_k^2$$

The area of a "typical" rectangle is then

$$\left(\frac{b - a}{n}\right)f(x_k) = \left(\frac{b - a}{n}\right)x_k^2$$

and the sum of the areas of the rectangles is neatly expressed in summation notation by

$$\sum_{k=1}^{n}\left(\frac{b - a}{n}\right)x_k^2 = \left(\frac{b - a}{n}\right)\sum_{k=1}^{n}x_k^2$$

Intuitively, we see that the greater the number of rectangles, the better our approximation will be, and this concept is pursued in calculus. The student is urged to let $a = 1$ and $b = 3$ in the accompanying figure and to use the method of approximating rectangles with $n = 2$, $n = 4$, and $n = 8$. The *exact* answer is 26/3 square units, and the approximations improve as n grows larger.

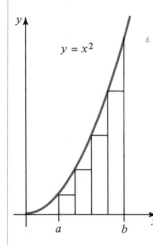

$y = x^2$

we have

$$S_1 = a_1$$
$$S_2 = a_1 + a_2$$
$$S_3 = a_1 + a_2 + a_3$$

and, in general,

$$S_n = \sum_{k=1}^{n} a_k = a_1 + a_2 + \cdots + a_n$$

The number S_n is called the **nth partial sum,** and the numbers

$$S_1, S_2, S_3, \ldots , S_n, \ldots$$

form a sequence called the **sequence of partial sums.** This type of sequence is studied in calculus courses, where methods are developed for analyzing infinite series.

EXAMPLE 7

Given the infinite sequence

$$a_n = n^2 - 1$$

find S_4.

SOLUTION

The first four terms of the sequence are

$$a_1 = 0 \quad a_2 = 3 \quad a_3 = 8 \quad a_4 = 15$$

Then the sum S_4 is given by

$$S_4 = \sum_{k=1}^{4} a_k = 0 + 3 + 8 + 15 = 26$$

If a series alternates in sign, then a multiplicative factor of $(-1)^k$ or $(-1)^{k+1}$ can be used to obtain the proper sign. For example, the series

$$-1^2 + 2^2 - 3^2 + 4^2$$

can be written in sigma notation as

$$\sum_{k=1}^{4} (-1)^k k^2$$

while the series

$$1^2 - 2^2 + 3^2 - 4^2$$

can be written as

$$\sum_{k=1}^{4} (-1)^{k+1} k^2$$

EXAMPLE 8

The terms of a sequence are of the form $a_k = \sqrt{k}$, and the terms are negated when k is even. Write an expression for the general term a_n and for the sum S_n in summation notation.

SOLUTION

If we multiply each term by $(-1)^{k+1}$, then the odd terms will be positive and the even terms will be negative. The general term a_n is then

$$a_n = (-1)^{n+1}\sqrt{n}$$

and the sequence is

$$\sqrt{1}, \ -\sqrt{2}, \ \sqrt{3}, \ -\sqrt{4}, \ \ldots, \ (-1)^{n+1}\sqrt{n}, \ \ldots$$

Finally, the sum S_n is given by

$$S_n = \sum_{k=1}^{n} (-1)^{k+1}\sqrt{k}$$

EXERCISE SET 11.1

In Exercises 1–12 find the first four terms and the twentieth term of the sequence whose nth term is given.

1. $a_n = 2n$
2. $a_n = 2n + 1$
3. $a_n = 4n - 3$
4. $a_n = 3n - 1$

5. $a_n = 5$
6. $a_n = 1 - \dfrac{1}{n}$
7. $a_n = \dfrac{n}{n+1}$
8. $a_n = \sqrt{n}$

9. $a_n = 2 + 0.1^n$
10. $a_n = \dfrac{n^2 - 1}{n^2 + 1}$
11. $a_n = \dfrac{n^2}{2n + 1}$
12. $a_n = \dfrac{2n + 1}{n^2}$

In Exercises 13–18 a sequence is defined recursively. Find the indicated term of the sequence.

13. $a_n = 2a_{n-1} - 1$, $a_1 = 2$; find a_4
14. $a_n = 3 - 3a_{n-1}$, $a_1 = -1$; find a_3

15. $a_n = \dfrac{1}{a_{n-1} + 1}$, $a_3 = 2$; find a_6
16. $a_n = \dfrac{n}{a_{n-1}}$, $a_2 = 1$; find a_5

17. $a_n = (a_{n-1})^2$, $a_1 = 2$; find a_4
18. $a_n = (a_{n-1})^{n-1}$, $a_1 = 2$; find a_4

In Exercises 19–26 find the indicated sum.

19. $\displaystyle\sum_{k=1}^{5} (3k - 1)$
20. $\displaystyle\sum_{k=1}^{5} (3 - 2k)$
21. $\displaystyle\sum_{k=1}^{6} (k^2 + 1)$
22. $\displaystyle\sum_{k=0}^{4} \dfrac{k}{k^2 + 1}$

23. $\displaystyle\sum_{k=3}^{5} \dfrac{k}{k-1}$
24. $\displaystyle\sum_{k=2}^{4} 4(2^k)$
25. $\displaystyle\sum_{j=1}^{4} 20$
26. $\displaystyle\sum_{i=1}^{10} 50$

In Exercises 27–36 use summation notation to express the sum. (The answer is not unique.)

27. $1 + 3 + 5 + 7 + 9$
28. $2 + 5 + 8 + 11 + 14$

29. $1 + 4 + 9 + 16 + 25$
30. $1 - 4 + 9 - 16 + 25$

31. $-1 + \dfrac{1}{\sqrt{2}} - \dfrac{1}{\sqrt{3}} + \dfrac{1}{\sqrt{4}}$
32. $\dfrac{1}{2 \cdot 4} + \dfrac{1}{2 \cdot 5} + \dfrac{1}{2 \cdot 6} + \dfrac{1}{2 \cdot 7}$

33. $\dfrac{1}{1^2 + 1} - \dfrac{2}{2^2 + 1} + \dfrac{3}{3^2 + 1} - \dfrac{4}{4^2 + 1}$
34. $2 - 4 + 8 - 16$

35. $1 + \dfrac{1}{x} + \dfrac{1}{x^2} + \dfrac{1}{x^3} + \cdots + \dfrac{1}{x^n}$
36. $\dfrac{1}{1 \cdot 2} + \dfrac{1}{2 \cdot 3} + \dfrac{1}{3 \cdot 4} + \dfrac{1}{4 \cdot 5} + \cdots + \dfrac{1}{49 \cdot 50}$

11.2
ARITHMETIC SEQUENCES

The sequence

$$2, 5, 8, 11, 14, 17, \ldots$$

is an example of a special type of sequence in which each successive term is obtained by adding a fixed number to the previous term.

Arithmetic Sequence

In an **arithmetic sequence** there is a real number d such that

$$a_n = a_{n-1} + d$$

for all $n > 1$. The number d is called the **common difference.**

An arithmetic sequence is also called an **arithmetic progression.** Returning to the sequence

$$2, 5, 8, 11, 14, 17, \ldots$$

the nth term can be defined recursively by

$$a_n = a_{n-1} + 3, \quad a_1 = 2$$

This is an arithmetic progression with the first term equal to 2 and a common difference of 3.

EXAMPLE 1
Write the first four terms of an arithmetic sequence whose first term is -4 and whose common difference is -3.

SOLUTION
Beginning with -4, we add the common difference -3 to obtain

$$-4 + (-3) = -7 \qquad -7 + (-3) = -10 \qquad -10 + (-3) = -13$$

Alternatively, we note that the sequence is defined by

$$a_n = a_{n-1} - 3, \quad a_1 = -4$$

which leads to the terms

$$a_1 = -4, \quad a_2 = -7, \quad a_3 = -10, \quad a_4 = -13$$

PROGRESS CHECK
Write the first four terms of an arithmetic sequence whose first term is 4 and whose common difference is $-\frac{1}{3}$.

ANSWER
$4, \dfrac{11}{3}, \dfrac{10}{3}, 3, \ldots$

EXAMPLE 2
Show that the sequence

$$a_n = 2n - 1$$

is an arithmetic sequence, and find the common difference.

SOLUTION
We must show that the sequence satisfies

$$a_n - a_{n-1} = d$$

for some real number d. We have

$$a_n = 2n - 1$$
$$a_{n-1} = 2(n - 1) - 1 = 2n - 3$$

so

$$a_n - a_{n-1} = 2n - 1 - (2n - 3) = 2$$

This demonstrates that we are dealing with an arithmetic sequence whose common difference is 2.

For a given arithmetic sequence, it's easy to find a formula for the nth term a_n in terms of n and the first term a_1. Since

$$a_2 = a_1 + d$$

and

$$a_3 = a_2 + d$$

we see that

$$a_3 = (a_1 + d) + d = a_1 + 2d$$

Similarly, we can show that

$$a_4 = a_3 + d = (a_1 + 2d) + d = a_1 + 3d$$
$$a_5 = a_4 + d = (a_1 + 3d) + d = a_1 + 4d$$

In general,

The nth term a_n of an arithmetic sequence is given by

$$a_n = a_1 + (n - 1)d$$

EXAMPLE 3

Find the seventh term of the arithmetic progression whose first term is 2 and whose common difference is 4.

SOLUTION

We substitute $n = 7$, $a_1 = 2$, $d = 4$ in the formula

$$a_n = a_1 + (n - 1)d$$

obtaining

$$a_7 = 2 + (7 - 1)4 = 2 + 24 = 26$$

PROGRESS CHECK

Find the 16th term of the arithmetic progression whose first term is -5 and whose common difference is $\frac{1}{2}$.

ANSWER

$$\frac{5}{2}$$

EXAMPLE 4

Find the 25th term of the arithmetic sequence whose first and 20th terms are -7 and 31, respectively.

SOLUTION

We can apply the given information to find d.

$$a_n = a_1 + (n - 1)d$$
$$a_{20} = a_1 + (20 - 1)d$$
$$31 = -7 + 19d$$
$$d = 2$$

Now we use the formula for a_n to find a_{25}.

$$a_n = a_1 + (n - 1)d$$
$$a_{25} = -7 + (25 - 1)2$$
$$a_{25} = 41$$

PROGRESS CHECK

Find the 60th term of the arithmetic sequence whose first and 10th terms are 3 and $-\frac{3}{2}$, respectively.

ANSWER

$$-\frac{53}{2}$$

ARITHMETIC SERIES

The series associated with an arithmetic sequence is called an **arithmetic series.** Since an arithmetic sequence has a common difference d, we can write the nth partial sum S_n as

$$S_n = a_1 + (a_1 + d) + (a_1 + 2d) + \cdots + (a_n - 2d) + (a_n - d) + a_n \quad (1)$$

where we write $a_2, a_3, \ldots$ in terms of a_1 and we write $a_{n-1}, a_{n-2}, \ldots$ in terms of a_n. Rewriting the right-hand side of Equation (1) in reverse order, we have

$$S_n = a_n + (a_n - d) + (a_n - 2d) + \cdots + (a_1 + 2d) + (a_1 + d) + a_1 \quad (2)$$

Summing the corresponding sides of Equations (1) and (2),

$$2S_n = (a_1 + a_n) + (a_1 + a_n) + (a_1 + a_n) + \cdots \qquad \text{Repeated } n \text{ times}$$
$$= n(a_1 + a_n)$$

Thus,

$$S_n = \frac{n}{2}(a_1 + a_n)$$

Since $a_n = a_1 + (n - 1)d$, we see that

$$S_n = \frac{n}{2}[a_1 + a_1 + (n - 1)d] \qquad \text{Substituting for } a_n$$

$$= \frac{n}{2}[2a_1 + (n - 1)d]$$

We now have two useful formulas.

Arithmetic Series

For an arithmetic series,

$$S_n = \frac{n}{2}(a_1 + a_n)$$

$$S_n = \frac{n}{2}[2a_1 + (n - 1)d]$$

The choice of which formula to use depends on the available information. The following examples illustrate the use of the formulas.

EXAMPLE 5

Find the sum of the first 30 terms of an arithmetic sequence whose first term is -20 and whose common difference is 3.

SOLUTION

We know that $n = 30$, $a_1 = -20$, and $d = 3$. Substituting in

$$S_n = \frac{n}{2}[2a_1 + (n - 1)d]$$

we obtain

$$S_{30} = \frac{30}{2}[2(-20) + (30 - 1)3]$$

$$= 15(-40 + 87)$$

$$= 705$$

PROGRESS CHECK

Find the sum of the first 10 terms of the arithmetic sequence whose first term is 2 and whose common difference is $-\frac{1}{2}$.

ANSWER

$$-\frac{5}{2}$$

EXAMPLE 6

The first term of an arithmetic series is 2, the last term is 58, and the sum is 450. Find the number of terms and the common difference.

SOLUTION

We have $a_1 = 2$, $a_n = 58$, and $S_n = 450$. Substituting in

$$S_n = \frac{n}{2}(a_1 + a_n)$$

we have

$$450 = \frac{n}{2}(2 + 58)$$

$$900 = 60n$$

$$n = 15$$

Now we substitute in

$$a_n = a_1 + (n - 1)d$$

$$58 = 2 + (14)d$$

$$56 = 14d$$

$$d = 4$$

PROGRESS CHECK

The first term of an arithmetic series is 6, the last term is 1, and the sum is 77/2. Find the number of terms and the common difference.

ANSWER

$$n = 11, \quad d = -\frac{1}{2}$$

EXERCISE SET 11.2

Write the next two terms of each of the following arithmetic sequences.

1. 3, 6, 9, 12, . . .

2. 2, −2, −6, −10, . . .

3. 0, $\frac{1}{4}$, $\frac{1}{2}$, $\frac{3}{4}$, . . .

4. $y - 4$, y, $y + 4$, $y + 8$, . . .

5. 0, log 10, log 100, log 1000, . . .

6. 4, $\frac{11}{2}$, 7, $\frac{17}{2}$, . . .

7. $\sqrt{5} - 2$, $\sqrt{5}$, $\sqrt{5} + 2$, $\sqrt{5} + 4$, . . .

8. 12, 8, 4, 0, . . .

Write the first four terms of the arithmetic sequence whose first term is a_1 and whose common difference is d.

9. $a_1 = 2$, $d = 4$

10. $a_1 = -2$, $d = -5$

11. $a_1 = 3$, $d = -\frac{1}{2}$

12. $a_1 = \frac{1}{2}$, $d = 2$

13. $a_1 = \frac{1}{3}$, $d = -\frac{1}{3}$

14. $a_1 = 6$, $d = \frac{5}{2}$

Find the specified term of the arithmetic sequence whose first term is a_1 and whose common difference is d.

15. $a_1 = 4$, $d = 3$; 8th term

16. $a_1 = -3$, $d = \frac{1}{4}$; 14th term

17. $a_1 = 14$, $d = -2$; 12th term

18. $a_1 = 6$, $d = -\frac{1}{3}$; 9th term

Given two terms of an arithmetic sequence, find the specified term.

19. $a_1 = -2$, $a_{20} = -2$; 24th term

20. $a_1 = \frac{1}{2}$, $a_{12} = 6$; 30th term

21. $a_1 = 0$, $a_{61} = 20$; 20th term

22. $a_1 = 23$, $a_{15} = -19$; 6th term

23. $a_1 = -\frac{1}{4}$, $a_{41} = 10$; 22nd term

24. $a_1 = -3$, $a_{18} = 65$; 30th term

Find the sum of the specified number of terms of the arithmetic sequence whose first term is a_1 and whose common difference is d.

25. $a_1 = 3$, $d = 2$; 20 terms

26. $a_1 = -4$, $d = \frac{1}{2}$; 24 terms

27. $a_1 = \frac{1}{2}$, $d = -2$; 12 terms

28. $a_1 = -3$, $d = -\frac{1}{3}$; 18 terms

29. $a_1 = 82$, $d = -2$; 40 terms

30. $a_1 = 6$, $d = 4$; 16 terms

31. How many terms of the arithmetic progression 2, 4, 6, 8, . . . add up to 930?

32. How many terms of the arithmetic progression 44, 41, 38, 35, . . . add up to 340?

33. The first term of an arithmetic series is 3, the last term is 90, and the sum is 1395. Find the number of terms and the common difference.

34. The first term of an arithmetic series is -3, the last term is $\frac{5}{2}$, and the sum is -3. Find the number of terms and the common difference.

35. The first term of an arithmetic series is $\frac{1}{2}$, the last term is $\frac{7}{4}$, and the sum is $\frac{27}{4}$. Find the number of terms and the common difference.

36. The first term of an arithmetic series is 20, the last term is -14, and the sum is 54. Find the number of terms and the common difference.

37. Find the sum of the first 16 terms of an arithmetic progression whose 4th and 10th terms are $-\frac{3}{4}$ and $\frac{1}{4}$, respectively.

38. Find the sum of the first 12 terms of an arithmetic progression whose 3rd and 6th terms are 9 and 18, respectively.

39. Show that the sum of the first n natural numbers is $n(n + 1)/2$.

40. Show that
$$1 + 3 + 5 + \cdots + (2n - 1) = n^2$$

11.3 GEOMETRIC SEQUENCES

The sequence

$$3, 6, 12, 24, 48, \ldots$$

in which each term after the first is obtained by multiplying the preceding one by 2, is an example of a geometric sequence.

Geometric Sequence

In a **geometric sequence** there is a real number r such that

$$a_n = ra_{n-1}$$

for all $n > 1$. The number r is called the **common ratio.**

A geometric sequence is also called a **geometric progression.** The common ratio r can be found by dividing any term a_k by the preceding term, a_{k-1}.

In a geometric sequence, the common ratio r is given by

$$r = \frac{a_k}{a_{k-1}}$$

Let's look at successive terms of a geometric sequence whose first term is a_1 and whose common ratio is r. We have

$$a_2 = ra_1$$
$$a_3 = ra_2 = r(ra_1) = r^2 a_1$$
$$a_4 = ra_3 = r(r^2 a_1) = r^3 a_1$$

FIBONACCI COUNTS THE RABBITS

Here is a problem that was first published in the year 1202.

A pair of newborn rabbits begins breeding at age one month and thereafter produces one pair of offspring per month. If we start with a newly born pair of rabbits, how many rabbits will there be at the beginning of each month? The problem was posed by Leonardo Fibonacci of Pisa, and the resulting sequence is known as a **Fibonacci sequence.**

The accompanying figure helps in analyzing the problem. At the beginning of month zero, we have the pair of newborn rabbits P_1. At the beginning of month 1, we still have the pair P_1, since the rabbits do not breed until age 1 month. At the beginning of month 2, the pair P_1 has the pair of offspring P_2. At the beginning of month 3, P_1 again has offspring, P_3, but P_2 does not breed during its first month. At the beginning of month 4, P_1 has offspring P_4, P_2 has offspring P_5, and P_3 does not breed during its first month.

If we let a_n denote the number of pairs of rabbits at the beginning of month n, we see that

Month	Pairs of Rabbits		
0	P_1		
1	P_1		
2	$P_1 \rightarrow P_2$		
3	$P_1 \rightarrow P_3$	P_2	
4	$P_1 \rightarrow P_4$	$P_2 \rightarrow P_5$	
	P_3		
5	$P_1 \rightarrow P_6$	$P_2 \rightarrow P_7$	
	$P_3 \rightarrow P_8$	P_4	P_5

$$a_0 = 1, \ a_1 = 1, \ a_2 = 2, \ a_3 = 3, \ a_4 = 5, \ a_5 = 8, \ldots$$

The sequence has the interesting property that each term is the sum of the two preceding terms; that is,

$$a_n = a_{n-1} + a_{n-2}$$

Strange as it seems, nature appears to be aware of the Fibonacci sequence. For example, arrangements of seeds on sunflowers and leaves on some trees are related to Fibonacci numbers. Stranger still, some researchers believe that cycle analysis, such as analysis of stock market prices, is also related in some way to Fibonacci numbers.

The pattern suggests that the exponent of r is one less than the subscript of a in the left-hand side.

The nth term of a geometric sequence is given by

$$a_n = a_1 r^{n-1}$$

Once again, mathematical induction is required to prove that the formula holds for all natural numbers.

EXAMPLE 1

Find the seventh term of the geometric sequence $-4, -2, -1, \ldots$.

SOLUTION

Since

$$r = \frac{a_k}{a_{k-1}}$$

we see that

$$r = \frac{a_3}{a_2} = \frac{-1}{-2} = \frac{1}{2}$$

Substituting $a_1 = -4$, $r = \frac{1}{2}$, and $n = 7$, we have

$$a_n = a_1 r^{n-1}$$

$$a_7 = (-4)\left(\frac{1}{2}\right)^{7-1} = (-4)\left(\frac{1}{2}\right)^6$$

$$= (-4)\left(\frac{1}{64}\right) = -\frac{1}{16}$$

PROGRESS CHECK

Find the sixth term of the geometric sequence 2, −6, 18, ...

ANSWER

−486

GEOMETRIC MEAN

In a geometric sequence, the terms between the first and last terms are called **geometric means.** We will illustrate the method of calculating such means.

EXAMPLE 2

Insert three geometric means between 3 and 48.

SOLUTION

The geometric sequence must look like this.

$$3, a_2, a_3, a_4, 48, \ldots$$

Thus, $a_1 = 3$, $a_5 = 48$, and $n = 5$. Substituting in

$$a_n = a_1 r^{n-1}$$
$$48 = 3r^4$$
$$r^4 = 16$$
$$r = \pm 2$$

Thus there are two geometric sequences with three geometric means between 3 and 48.

$$3, 6, 12, 24, 48, \ldots \quad r = 2$$
$$3, -16, 12, -24, 48, \ldots \quad r = -2$$

PROGRESS CHECK
Insert two geometric means between 5 and $\frac{8}{25}$.

ANSWER

$2, \dfrac{4}{5}$

GEOMETRIC SERIES

If $a_1, a_2, \ldots$ is a geometric sequence, then the nth partial sum

$$S_n = a_1 + a_2 + \cdots + a_n \tag{1}$$

is called a **geometric series.** Since each term of the series can be rewritten as $a_k = a_1 r^{k-1}$, we can rewrite Equation (1) as

$$S_n = a_1 + a_1 r + a_1 r^2 + \cdots + a_1 r^{n-2} + a_1 r^{n-1} \tag{2}$$

Multiplying each term in Equation (2) by r, we have

$$rS_n = a_1 r + a_1 r^2 + a_1 r^3 + \cdots + a_1 r^{n-1} + a_1 r^n \tag{3}$$

Subtracting Equation (2) from Equation (3) produces

$$rS_n - S_n = a_1 r^n - a_1$$
$$(r - 1)S_n = a_1(r^n - 1) \qquad \text{Factoring}$$
$$S_n = \frac{a_1(r^n - 1)}{r - 1} \qquad \begin{array}{l}\text{Dividing by } r - 1\\ \text{(if } r \neq 1)\end{array}$$

Changing the signs in both the numerator and the denominator gives us the following formula for the nth partial sum.

Geometric Series

In a geometric series with first term a_1 and common ratio $r \neq 1$,

$$S_n = \frac{a_1(1 - r^n)}{1 - r}$$

EXAMPLE 3
Find the sum of the first six terms of the geometric sequence whose first three terms are 12, 6, 3.

SOLUTION
The common ratio can be found by dividing any term by the preceding term.

$$r = \frac{a_k}{a_{k-1}} = \frac{a_2}{a_1} = \frac{6}{12} = \frac{1}{2}$$

Substituting $a_1 = 12$, $r = \frac{1}{2}$, $n = 6$ in the formula for S_n, we have

$$S_n = \frac{a_1(1 - r^n)}{1 - r} = \frac{12\left[1 - \left(\frac{1}{2}\right)^6\right]}{1 - \frac{1}{2}} = \frac{189}{8}$$

PROGRESS CHECK

Find the sum of the first five terms of the geometric sequence whose first three terms are 2, $-\frac{4}{3}$, $\frac{8}{9}$.

ANSWER

$$\frac{110}{81}$$

EXAMPLE 4

A father promises to give each child 2 cents on the first day and 4 cents on the second day and to continue doubling the amount each day for a total of 8 days. How much will each child receive on the last day? How much will each child have received in total after 8 days?

SOLUTION

The daily payout to each child forms a geometric sequence 2, 4, 8, . . . with $a_1 = 2$ and $r = 2$. The term a_8 is given by substituting in

$$a_n = a_1 r^{n-1}$$
$$a_8 = a_1 r^{8-1} = 2 \cdot 2^7 = 256$$

Thus, each child will receive \$2.56 on the last day. The total received by each child is given by

$$S_n = \frac{a_1(1 - r^n)}{1 - r}$$

$$S_8 = \frac{a_1(1 - r^8)}{1 - r} = \frac{2(1 - 2^8)}{1 - 2}$$

$$= \frac{2(1 - 256)}{-1} = 510$$

Each child will have received a total of \$5.10 after 8 days.

PROGRESS CHECK

A ball is dropped from a height of 64 feet. On each bounce, it rebounds half the height it fell (Figure 1). How high is the ball at the top of the fifth bounce? What is the total distance the ball has traveled at the top of the fifth bounce?

ANSWER
2 feet; 186 feet

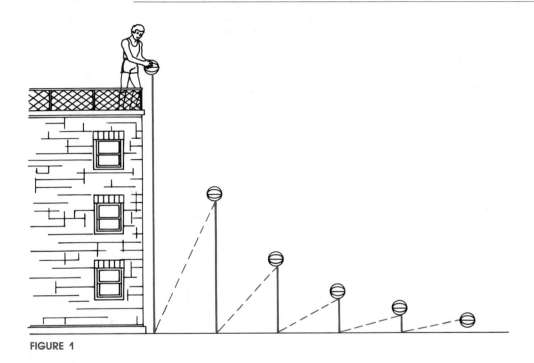

FIGURE 1

INFINITE GEOMETRIC SERIES

We now want to focus on a geometric series for which $|r| < 1$, say

$$\frac{1}{2} + \frac{1}{4} + \frac{1}{8} + \cdots + \frac{1}{2^n} + \cdots$$

To see how the sum increases as n increases, let's form a table of values of S_n.

n	1	2	3	4	5	6	7	8	9
S_n	0.500	0.750	0.875	0.938	0.969	0.984	0.992	0.996	0.998

We begin to suspect that S_n gets closer and closer to 1 as n increases. To see that this is really so, let's look at the formula

$$S_n = \frac{a_1(1 - r^n)}{1 - r}$$

when $|r| < 1$. When a number r that is less than 1 in absolute value is raised to higher and higher positive integer powers, the value of r^n gets smaller and smaller. Thus, the term r^n can be made as small as we like by choosing n sufficiently large. Since we are dealing with an infinite series, we say that "r^n approaches zero as n approaches infinity." We then replace r^n with 0 in the formula and denote the sum by S.

Sum of an Infinite Geometric Series

The sum S of the **infinite geometric series**

$$\sum_{k=0}^{\infty} a_1 r^k = a_1 + a_1 r + a_1 r^2 + \cdots + a_1 r^n + \cdots$$

is given by

$$S = \frac{a_1}{1 - r} \quad \text{when } |r| < 1$$

Applying this formula to the preceding series, we see that

$$S = \frac{\dfrac{1}{2}}{1 - \dfrac{1}{2}} = 1$$

which justifies the conjecture resulting from the examination of the above table. It is appropriate to remark that the ideas used in deriving the formula for an infinite geometric series have led us to the very border of the beginning concepts of calculus.

EXAMPLE 5
Find the sum of the infinite geometric series

$$\frac{3}{2} + 1 + \frac{2}{3} + \frac{4}{9} + \cdots$$

SOLUTION
The common ratio $r = \frac{2}{3}$. The sum of the infinite geometric series, with $|r| < 1$, is given by

$$S = \frac{a_1}{1 - r} = \frac{\dfrac{3}{2}}{1 - \dfrac{2}{3}} = \frac{9}{2}$$

PROGRESS CHECK
Find the sum of the infinite geometric series

$$4 - 1 + \frac{1}{4} - \frac{1}{16} + \cdots$$

ANSWER

$\dfrac{16}{5}$

The notation

$$0.6525\overline{52}$$

indicates a repeating decimal with a pattern in which 52 is repeated indefinitely. Every repeating decimal can be written as a rational number. We will apply the formula for the sum of an infinite geometric series to find the rational number equal to a repeating decimal.

EXAMPLE 6
Find the rational number that is equal to $0.6525\overline{52}$.

SOLUTION
Note that

$$0.6525\overline{52} = 0.6 + 0.052 + 0.00052 + 0.0000052 + \cdots$$

We treat the sum

$$0.052 + 0.00052 + 0.0000052 + \cdots$$

as an infinite geometric series with $a = 0.052$ and $r = 0.01$. Then

$$S = \frac{a}{1 - r} = \frac{0.052}{1 - 0.01} = \frac{0.052}{0.99} = \frac{52}{990}$$

and the repeating decimal is equal to

$$0.6 + \frac{52}{990} = \frac{6}{10} + \frac{52}{990} = \frac{646}{990} = \frac{323}{495}$$

PROGRESS CHECK
Write the repeating decimal $2.5454\overline{54}$ as a rational number.

ANSWER

$\dfrac{252}{99}$

EXERCISE SET 11.3

In Exercises 1–6 find the next term of the given geometric sequence.

1. 3, 6, 12, 24, . . .

2. −4, 12, −36, 108, . . .

3. −4, 3, $-\dfrac{9}{4}$, $\dfrac{27}{16}$, . . .

4. 2, −1, $\dfrac{1}{2}$, $-\dfrac{1}{4}$, . . .

5. 1.2, 0.24, 0.048, . . .

6. $\dfrac{1}{8}$, $\dfrac{1}{2}$, 2, 8, . . .

In Exercises 7–12 write the first four terms of the geometric sequence whose first term is a_1 and whose common ratio is r.

7. $a_1 = 3$, $r = 3$

8. $a_1 = -4$, $r = 2$

9. $a_1 = 4$, $r = \dfrac{1}{2}$

10. $a_1 = 16$, $r = -\dfrac{3}{2}$

11. $a_1 = -3$, $r = 2$

12. $a_1 = 3$, $r = -\dfrac{2}{3}$

In Exercises 13–24 use the information given about a geometric sequence to find the requested item.

13. $a_1 = 3$, $r = -2$; find a_8

14. $a_1 = 18$, $r = -\dfrac{1}{2}$; find a_6

15. $a_1 = 16$, $a_2 = 8$; find a_7

16. $a_1 = 15$, $a_2 = -10$; find a_6

17. $a_1 = 3$, $a_5 = \dfrac{1}{27}$; find a_7

18. $a_1 = 2$, $a_6 = \dfrac{1}{16}$; find a_3

19. $a_1 = \dfrac{16}{81}$, $a_6 = \dfrac{3}{2}$; find a_8

20. $a_4 = \dfrac{1}{4}$, $a_7 = 1$; find r

21. $a_2 = 4$, $a_8 = 256$; find r

22. $a_3 = 3$, $a_6 = -81$; find a_8

23. $a_1 = \dfrac{1}{2}$, $r = 2$, $a_n = 32$; find n

24. $a_1 = -2$, $r = 3$, $a_n = 162$; find n

25. Insert two geometric means between $\dfrac{1}{3}$ and 9.

26. Insert two geometric means between −3 and 192.

27. Insert two geometric means between 1 and $\dfrac{1}{64}$.

28. Insert three geometric means between $\dfrac{2}{3}$ and $\dfrac{32}{243}$.

In Exercises 29–32 find the requested partial sum for the geometric sequence whose first three terms are given.

29. 3, 1, $\dfrac{1}{3}$; find S_7

30. $\dfrac{1}{3}$, 1, 3; find S_6

31. −3, $\dfrac{6}{5}$, $-\dfrac{12}{25}$; find S_5

32. 2, $\dfrac{4}{3}$, $\dfrac{8}{9}$; find S_6

In Exercises 33–36 use the information given about a geometric sequence to find the requested partial sum.

33. $a_1 = 4$, $r = 2$; find S_8

34. $a_1 = -\dfrac{1}{2}$, $r = -3$; find S_{10}

35. $a_1 = 2$, $a_4 = -\dfrac{54}{8}$; find S_5

36. $a_1 = 64$, $a_7 = 1$; find S_6

37. A Christmas Club calls for savings of $5 in January, and twice as much in each successive month as in the previous month. How much money will have been saved by the end of November?

38. A city had 20,000 people in 1980. If the population increases 5% per year, how many people will the city have in 1990?

39. A city had 30,000 people in 1980. If the population increases 25% every 10 years, how many people will the city have in the year 2010?

40. For good behavior a child is offered a reward consisting of 1 cent on the first day, 2 cents on the second day, 4 cents on the third day, and so on. If the child behaves properly for two weeks, what is the total amount that the child will receive?

Evaluate the sum of each infinite geometric series.

41. $1 + \dfrac{1}{2} + \dfrac{1}{4} + \dfrac{1}{8} + \cdots$

42. $\dfrac{4}{5} + \dfrac{1}{5} + \dfrac{1}{20} + \dfrac{1}{80} + \cdots$

43. $1 - \dfrac{1}{3} + \dfrac{1}{9} - \dfrac{1}{27} + \cdots$

44. $\dfrac{1}{2} - \dfrac{1}{4} + \dfrac{1}{8} - \dfrac{1}{16} + \cdots$

45. $2 + \dfrac{1}{2} + \dfrac{1}{8} + \dfrac{1}{32} + \cdots$

46. $1 + 0.1 + 0.01 + 0.001 + \cdots$

47. $0.5 + (0.5)^2 + (0.5)^3 + (0.5)^4 + \cdots$

48. $\dfrac{2}{5} + \dfrac{4}{25} + \dfrac{8}{125} + \dfrac{16}{625} + \cdots$

49. $\dfrac{1}{3} - \dfrac{2}{9} + \dfrac{4}{27} - \dfrac{8}{81} + \cdots$

50. Find the rational number equal to $3.666\overline{6}$.

51. Find the rational number equal to $0.367\overline{67}$.

52. Find the rational number equal to $4.1414\overline{14}$.

53. Find the rational number equal to $0.325\overline{325}$.

11.4 MATHEMATICAL INDUCTION

Mathematical induction is a method of proof that serves as one of the most powerful tools available to the mathematician. Viewed another way, mathematical induction is a property of the natural numbers that enables us to prove theorems that would otherwise appear unmanageable.

We begin by considering the sums of consecutive odd integers

$$1 = 1$$
$$1 + 3 = 4$$
$$1 + 3 + 5 = 9$$
$$1 + 3 + 5 + 7 = 16$$
$$1 + 3 + 5 + 7 + 9 = 25$$

We instantly recognize that the sequence

$$1, 4, 9, 16, 25$$

consists of the squares of the integers 1, 2, 3, 4, and 5. Is this coincidental or do we have a general rule? Is the sum of the first n consecutive odd integers always equal to n^2? Curiosity leads us to try yet one more case.

$$1 + 3 + 5 + 7 + 9 + 11 = 36 = 6^2$$

Indeed, the sum of the first six odd integers is 6^2. This strengthens our *suspicion* that the result may hold in general, but we cannot possibly verify a theorem for *all* positive integers by testing one integer at a time. At this point we need to turn to the principle of mathematical induction.

Principle of Mathematical Induction	If a statement involving a natural number n (I) is true when $n = 1$ and (II) whenever it is true for $n = k$, is also true for $n = k + 1$, then the statement is true for all positive integer values of n.

Let's examine the logic of the principle of mathematical induction. Part (I) says that we must verify the statement for $n = 1$. Then, by Part (II), the statement is also true for $n = 1 + 1 = 2$. But Part (II) then implies that the statement must also be true for $n = 2 + 1 = 3$, and so on. The effect is similar to an endless string of dominoes whereby each domino causes the next to fall. Thus, it is plausible that the principle has established the validity of the statement for *all* positive integer values of n.

We outline the steps involved in applying the principle of mathematical induction in the following example.

EXAMPLE 1

Prove that the sum of the first n consecutive integers is given by $n(n + 1)/2$.

SOLUTION

Mathematical Induction	
Step 1. Verify that the statement is true for $n = 1$.	*Step 1.* The "sum" of the first integer is 1. Evaluating the formula for $n = 1$ yields $$\frac{1(1 + 1)}{2} = \frac{2}{2} = 1$$ which verifies the formula for $n = 1$.

Step 2. Assume the statement is true for $n = k$. Show it is true for $n = k + 1$.

Step 2. For $n = k$ we have

$$1 + 2 + 3 + \cdots + k = \frac{k(k + 1)}{2}$$

Adding the next consecutive integer, $k + 1$, to both sides, we obtain

$$1 + 2 + \cdots + k + (k + 1) = \frac{k(k + 1)}{2} + (k + 1)$$

$$= (k + 1)\left(\frac{k}{2} + 1\right)$$

$$= (k + 1)\left(\frac{k + 2}{2}\right)$$

$$= \frac{1}{2}(k + 1)(k + 2)$$

Thus, the formula holds for $n = k + 1$. By the principle of mathematical induction, it is then true for all positive integer values of n.

EXAMPLE 2

Prove that the sum of the first n consecutive odd integers is given by n^2.

SOLUTION

To verify the formula for $n = 1$, we need only observe that $1 = 1^2$.

The following table shows the correspondence between the natural numbers and the odd integers. We see that when $n = k$, the value of the nth consecutive

n	1	2	3	4	$\cdots$	k
nth odd integer	1	3	5	7	$\cdots$	$2k - 1$

odd integer is $2k - 1$. Since the formula is assumed to be true for $n = k$, we have

$$1 + 3 + 5 + \cdots + (2k - 1) = k^2$$

Adding the next consecutive odd integer, $2k + 1$, to both sides, we obtain

$$1 + 3 + \cdots + (2k - 1) + (2k + 1) = k^2 + (2k + 1)$$

or

$$1 + 3 + \cdots + (2k + 1) = (k + 1)^2$$

Thus, the sum of the first $k + 1$ consecutive odd integers is $(k + 1)^2$. By the principle of mathematical induction, the formula is true for all positive integer values of n.

The student should be aware that many of the theorems that were used in this book can be proved formally by using mathematical induction. Here is an example of a basic property of positive integer exponents that yields to this type of proof.

EXAMPLE 3

Prove that $(xy)^n = x^n y^n$ for all positive integer values of n.

SOLUTION

For $n = 1$, we have

$$(xy)^1 = xy = x^1 y^1$$

which verifies the validity of the statement for $n = 1$. Assuming the statement holds for $n = k$, we have

$$(xy)^k = x^k y^k$$

To show that the statement holds for $n = k + 1$, we write

$$
\begin{aligned}
(xy)^{k+1} &= (xy)^k (xy) && \text{Definition of exponents} \\
&= (x^k y^k)(xy) && \text{Statement holds for } n = k \\
&= (x^k x)(y^k y) && \text{Associative and commutative laws} \\
&= x^{k+1} y^{k+1} && \text{Definition of exponents}
\end{aligned}
$$

Thus, the statement holds for $n = k + 1$, and by the principle of mathematical induction the statement holds for all integer values of n.

EXERCISE SET 11.4

In Exercises 1–10 prove that the statement is true for all positive integer values of n by using the principle of mathematical induction.

1. $2 + 4 + 6 + \cdots + 2n = n(n + 1)$

2. $1^2 + 3^2 + 5^2 + \cdots + (2n - 1)^2 = \dfrac{n(2n + 1)(2n - 1)}{3}$

3. $2 + 5 + 8 + \cdots + (3n - 1) = \dfrac{n(3n + 1)}{2}$

4. $4 + 8 + 12 + \cdots + 4n = 2n(n + 1)$

5. $5 + 10 + 15 + \cdots + 5n = \dfrac{5n(n + 1)}{2}$

6. $1^2 + 2^2 + 3^2 + \cdots + n^2 = \dfrac{n(n + 1)(2n + 1)}{6}$

7. $1 \cdot 2 + 2 \cdot 3 + 3 \cdot 4 + \cdots + n(n + 1) = \dfrac{n(n + 1)(n + 2)}{3}$

8. $1^3 + 2^3 + 3^3 + \cdots + n^3 = \dfrac{n^2(n + 1)^2}{4}$

9. $1 + 5 + 9 + \cdots + (4n - 3) = n(2n - 1)$

10. $\left(\dfrac{x}{y}\right)^n = \dfrac{x^n}{y^n}$

11. Prove that the nth term a_n of an arithmetic progression whose first term is a_1 and whose common difference is d is given by $a_n = a_1 + (n - 1)d$.

12. Prove that the nth term a_n of a geometric progression whose first term is a_1 and whose common ratio is r is given by $a_n = a_1 r^{n-1}$.

13. Prove that $2 + 2^2 + 2^3 + \cdots + 2^n = 2^{n+1} - 2$.

14. Prove that $a + ar + ar^2 + \cdots + ar^{n-1} = \dfrac{a(1 - r^n)}{1 - r}$.

15. Prove that $x^n - 1$ is divisible by $x - 1$, $x \neq 1$. [*Hint:* Recall that divisibility requires the existence of a polynomial $Q(x)$ such that $x^n - 1 = (x - 1)Q(x)$.]

16. Prove that $x^n - y^n$ is divisible by $x - y$, $x \neq y$. [*Hint:* Note that $x^{n+1} - y^{n+1} = (x^{n+1} - xy^n) + (xy^n - y^{n+1})$.]

11.5
THE BINOMIAL THEOREM

By sequential multiplication by $(a + b)$ you may verify that

$$(a + b)^1 = a + b$$
$$(a + b)^2 = a^2 + 2ab + b^2$$
$$(a + b)^3 = a^3 + 3a^2b + 3ab^2 + b^3$$
$$(a + b)^4 = a^4 + 4a^3b + 6a^2b^2 + 4ab^3 + b^4$$
$$(a + b)^5 = a^5 + 5a^4b + 10a^3b^2 + 10a^2b^3 + 5ab^4 + b^5$$

The expression on the right-hand side of the equation is called the **expansion** of the left-hand side. If we were to predict the form of the expansion of $(a + b)^n$, where n is a natural number, the preceding example would lead us to conclude that it has the following properties.

(a) The expansion has $n + 1$ terms.

(b) The first term is a^n and the last term is b^n.

(c) The sum of the exponents of a and b in each term is n.

(d) In each successive term after the first, the exponent of a decreases by 1, and the exponent of b increases by 1.

(e)   The coefficients may be obtained from the following array, which is known as **Pascal's triangle.** Each number, with the exception of those at the ends of the rows, is the sum of the two nearest numbers in the row above. The numbers at the ends of the rows are always 1.

$$
\begin{array}{ccccccccccc}
 & & & & 1 & & 1 & & & & \\
 & & & 1 & & 2 & & 1 & & & \\
 & & 1 & & 3 & & 3 & & 1 & & \\
 & 1 & & 4 & & 6 & & 4 & & 1 & \\
1 & & 5 & & 10 & & 10 & & 5 & & 1
\end{array}
$$

Pascal's triangle is not a convenient means for determining the coefficients of the expansion when n is large. Here is an alternative method.

(e′)   The coefficient of any term (after the first) can be found by the following rule: In the preceding term, multiply the coefficient by the exponent of a and then divide by one more than the exponent of b.

EXAMPLE 1
Write the expansion of $(a + b)^6$.

SOLUTION
From Property (b) we know that the first term is a^6. Thus,

$$(a + b)^6 = a^6 + \cdots$$

From Property (e′) the next coefficient is

$$\frac{1 \cdot 6}{1} = 6$$

(since the exponent of b is 0). By Property (d) the exponents of a and b in this term are 5 and 1, respectively, so we have

$$(a + b)^6 = a^6 + 6a^5b + \cdots$$

Applying Property (e′) again, we find that the next coefficient is

$$\frac{6 \cdot 5}{2} = 15$$

and by Property (d) the exponents of a and b in this term are 4 and 2, respectively. Thus,

$$(a + b)^6 = a^6 + 6a^5b + 15a^4b^2 + \cdots$$

Continuing in this manner, we see that

$$(a + b)^6 = a^6 + 6a^5b + 15a^4b^2 + 20a^3b^3 + 15a^2b^4 + 6ab^5 + b^6$$

PROGRESS CHECK

Write the first five terms in the expansion of $(a + b)^{10}$.

ANSWER

$a^{10} + 10a^9b + 45a^8b^2 + 120a^7b^3 + 210a^6b^4$

The expansion of $(a + b)^n$ that we have described is called the **binomial theorem** or **binomial formula** and can be written

The Binomial Formula

$$(a + b)^n = a^n + \frac{n}{1}a^{n-1}b + \frac{n(n - 1)}{1 \cdot 2}a^{n-2}b^2 + \frac{n(n - 1)(n - 2)}{1 \cdot 2 \cdot 3}a^{n-3}b^3$$
$$+ \cdots + \frac{n(n - 1)(n - 2) \cdots (n - r + 1)}{1 \cdot 2 \cdot 3 \cdots r}a^{n-r}b^r + \cdots + b^n$$

The binomial formula can be proved by the method of mathematical induction discussed in the preceding section.

EXAMPLE 2

Find the expansion of $(2x - 1)^4$.

SOLUTION

Let $a = 2x$, $b = -1$, and apply the binomial formula.

$$(2x - 1)^4 = (2x)^4 + \frac{4}{1}(2x)^3(-1) + \frac{4 \cdot 3}{1 \cdot 2}(2x)^2(-1)^2$$
$$+ \frac{4 \cdot 3 \cdot 2}{1 \cdot 2 \cdot 3}(2x)(-1)^3 + (-1)^4$$
$$= 16x^4 - 32x^3 + 24x^2 - 8x + 1$$

PROGRESS CHECK

Find the expansion of $(x^2 - 2)^4$.

ANSWER

$x^8 - 8x^6 + 24x^4 - 32x^2 + 16$

FACTORIAL NOTATION

Note that the denominator of the coefficient in the binomial formula is always the product of the first n natural numbers. We use the symbol $n!$, which is read as **n factorial,** to indicate this type of product. For example,

$$4! = 4 \cdot 3 \cdot 2 \cdot 1 = 24$$
$$6! = 6 \cdot 5 \cdot 4 \cdot 3 \cdot 2 \cdot 1 = 720$$

and, in general,

---

n **Factorial**

$$n! = n(n-1)(n-2) \cdots 4 \cdot 3 \cdot 2 \cdot 1$$

Since

$$(n-1)! = (n-1)(n-2)(n-3) \cdots 4 \cdot 3 \cdot 2 \cdot 1$$

we see that for $n > 1$

$$n! = n(n-1)!$$

For convenience, we define 0! by

$$0! = 1$$

EXAMPLE 3

Evaluate each of the following.

(a) $\dfrac{5!}{3!}$

Since $5! = 5 \cdot 4 \cdot 3!$ we may write

$$\frac{5!}{3!} = \frac{5 \cdot 4 \cdot 3!}{3!} = 5 \cdot 4 = 20$$

(b) $\dfrac{9!}{8!} = \dfrac{9 \cdot 8!}{8!} = 9$

(c) $\dfrac{10!4!}{12!} = \dfrac{10!4!}{12 \cdot 11 \cdot 10!} = \dfrac{4!}{12 \cdot 11} = \dfrac{4 \cdot 3 \cdot 2 \cdot 1}{12 \cdot 11} = \dfrac{2}{11}$

(d) $\dfrac{n!}{(n-2)!} = \dfrac{n(n-1)(n-2)!}{(n-2)!} = n(n-1) = n^2 - n$

(e) $\dfrac{(2-2)!}{3!} = \dfrac{0!}{3 \cdot 2} = \dfrac{1}{6}$

PROGRESS CHECK

Evaluate each of the following.

(a) $\dfrac{12!}{10!}$ (b) $\dfrac{6!}{4!2!}$ (c) $\dfrac{10!8!}{9!7!}$

(d) $\dfrac{n!(n-1)!}{(n+1)!(n-2)!}$ (e) $\dfrac{8!}{6!(3-3)!}$

ANSWERS

(a) 132   (b) 15   (c) 80   (d) $\dfrac{n-1}{n+1}$ (e) 56

Here is what the binomial formula looks like in factorial notation.

$$(a+b)^n = a^n + \frac{n!}{1!(n-1)!}\, a^{n-1}b + \frac{n!}{2!(n-2)!}\, a^{n-2}b^2$$

$$+ \frac{n!}{3!(n-3)!}\, a^{n-3}b^3 + \cdots + \frac{n!}{r!(n-r)!}\, a^{n-r}b^r$$

$$+ \cdots + b^n$$

The symbol $\binom{n}{r}$, called the **binomial coefficient,** is defined in this way:

Binomial Coefficient	$\dbinom{n}{r} = \dfrac{n!}{r!(n-r)!}$

This symbol is useful in denoting the coefficients of the binomial expansion. Using this notation, the binomial formula can be written as

$$(a+b)^n = a^n + \binom{n}{1}a^{n-1}b + \binom{n}{2}a^{n-2}b^2 + \binom{n}{3}a^{n-3}b^3$$

$$+ \cdots + \binom{n}{r}a^{n-r}b^r + \cdots + b^n$$

Sometimes we merely want to find a certain term in the expansion of $(a+b)^n$. We shall use the following observation to answer this question. In the binomial formula for the expansion of $(a+b)^n$, b occurs in the second term, b^2 occurs in the third term, b^3 occurs in the fourth term, and, in general, b^k occurs in the $(k+1)$th term. The exponents of a and b must add up to n in each term. Since the exponent of b in the $(k+1)$th term is k, we conclude that the exponent of a must be $n-k$.

EXAMPLE 4
Find the fourth term in the expansion of $(x - 1)^5$.

SOLUTION
The exponent of b in the fourth term is 3, and the exponent of a is then $5 - 3 = 2$. From the binomial formula we see that the coefficient of the term a^2b^3 is

$$\binom{n}{3} = \binom{5}{3} = \frac{5!}{3!2!}$$

Since $a = x$ and $b = -1$, the fourth term is

$$\frac{5!}{3!2!}x^2(-1)^3 = -10x^2$$

PROGRESS CHECK
Find the third term in the expansion of

$$\left(\frac{x}{2} - 1\right)^8$$

ANSWER

$$\frac{7}{16}x^6$$

EXAMPLE 5
Find the term in the expansion of $(x^2 - y^2)^6$ that involves y^8.

SOLUTION
Since $y^8 = (-y^2)^4$, we seek that term which involves b^4 in the expansion of $(a + b)^6$. Thus, $b^4 = (-y^2)^4 = y^8$ occurs in the fifth term. In this term the exponent of a is $6 - 4 = 2$. By the binomial formula the corresponding coefficient is

$$\binom{6}{4} = \frac{6!}{4!2!} = 15$$

Since $a = x^2$ and $b = -y^2$, the desired term is

$$15(x^2)^2(-y^2)^4 = 15x^4y^8$$

PROGRESS CHECK
Find the term in the expansion of $(x^3 - \sqrt{2})^5$ that involves x^6.

ANSWER
$-20\sqrt{2}x^6$

EXERCISE SET 11.5

Expand and simplify.

1. $(3x + 2y)^5$
2. $(2a - 3b)^6$
3. $(4x - y)^4$
4. $\left(3 + \frac{1}{2}x\right)^4$

5. $(2 - xy)^5$
6. $(3a^2 + b)^4$
7. $(a^2b + 3)^4$
8. $(x - y)^7$

9. $(a - 2b)^8$
10. $\left(\frac{x}{y} + y\right)^6$
11. $\left(\frac{1}{3}x + 2\right)^3$
12. $\left(\frac{x}{y} + \frac{y}{x}\right)^5$

Find the first four terms in the given expansion and simplify.

13. $(2 + x)^{10}$
14. $(x - 3)^{12}$
15. $(3 - 2a)^9$
16. $(a^2 + b^2)^{11}$

17. $(2x - 3y)^{14}$
18. $\left(a - \frac{1}{a^2}\right)^8$
19. $(2x - yz)^{13}$
20. $\left(x - \frac{1}{y}\right)^{15}$

Evaluate.

21. $5!$
22. $7!$
23. $\frac{12!}{11!}$
24. $\frac{13!}{12!}$

25. $\frac{11!}{8!}$
26. $\frac{7!}{9!}$
27. $\frac{10!}{6!}$
28. $\frac{9!}{6!}$

29. $\frac{6!}{3!}$
30. $\binom{8}{5}$
31. $\binom{10}{6}$
32. $\frac{(n+1)!}{(n-1)!}$

In each expansion find only the term specified.

33. The fourth term in $(2x - 4)^7$.
34. The third term in $(4a + 3b)^{11}$.
35. The fifth term in $\left(\frac{1}{2}x - y\right)^{12}$.
36. The sixth term in $(3x - 2y)^{10}$.
37. The fifth term in $\left(\frac{1}{x} - 2\right)^9$.
38. The next to last term in $(a + 4b)^5$.
39. The middle term in $(x - 3y)^6$.
40. The middle term in $\left(2a + \frac{1}{2}b\right)^6$.
41. The term involving x^4 in $(3x + 4y)^7$.
42. The term involving x^6 in $(2x^2 - 1)^9$.
43. The term involving x^6 in $(2x^3 - 1)^9$.
44. The term involving x^8 in $\left(x^2 + \frac{1}{y}\right)^8$.
45. The term involving x^{12} in $\left(x^3 + \frac{1}{2}\right)^7$.
46. The term involving x^{-4} in $\left(y + \frac{1}{x^2}\right)^8$.
47. Evaluate $(1.3)^6$ to four decimal places by writing it as $(1 + 0.3)^6$ and using the binomial formula.
48. Using the method of Example 47, evaluate (a) $(3.4)^4$ (b) $(48)^5$ (*Hint*: 48 = 50 − 2.)

TERMS AND SYMBOLS

common difference (p. 502) common ratio (p. 508) mathematical induction $n!$ (p. 523)
arithmetic series (p. 505) geometric mean (p. 510) (p. 518) factorial (p. 523)
geometric sequence geometric series (p. 511) expansion of $(a + b)^n$ $\binom{n}{r}$ (p. 525)
(p. 508) infinite geometric series (p. 521)
geometric progression (p. 514) Pascal's triangle (p. 522) binomial coefficient
(p. 508) $0.537537\overline{537}$ (p. 515) binomial formula (p. 523) (p. 525)

KEY IDEAS FOR REVIEW

☐ An infinite sequence is a function whose domain is restricted to the set of natural numbers. We generally write a sequence by using subscript notation; that is, a_n replaces $a(n)$.

☐ An infinite sequence is defined recursively if each term is defined by reference to preceding terms.

☐ Sigma (Σ) or summation notation is a handy means of indicating a sum of terms. The values written below and above the Σ indicate the starting and ending values, respectively, of the index of summation.

☐ An arithmetic sequence has a common difference d between terms. We can define an arithmetic sequence recursively by writing $a_n = a_{n-1} + d$ and specifying a_1.

☐ A geometric sequence has a common ratio r between terms. We can define a geometric sequence recursively by writing $a_n = r a_{n-1}$ and specifying a_1.

☐ The formulas for the nth term of arithmetic and geometric sequences are

$$a_n = a_1 + (n - 1)d \quad \text{Arithmetic}$$
$$a_n = a_1 r^{n-1} \quad \text{Geometric}$$

☐ A series is the sum of the terms of a sequence.

☐ The formulas for the sum S_n of the first n terms of arithmetic and geometric sequences are

$$S_n = \frac{n}{2}(a_1 + a_n) \quad \text{Arithmetic}$$

$$S_n = \frac{n}{2}[2a_1 + (n - 1)d] \quad \text{Arithmetic}$$

$$S_n = \frac{a_1(1 - r^n)}{1 - r} \quad \text{Geometric}$$

☐ If the common ratio r satisfies $-1 < r < 1$, then the infinite geometric series has the sum S given by

$$S = \frac{a_1}{1 - r}$$

where a_1 is the first term of the series.

☐ Mathematical induction is useful in proving certain types of theorems involving natural numbers.

☐ The notation $n!$ indicates the product of the natural numbers 1 through n.

$$n! = n(n - 1)(n - 2) \cdots 2 \cdot 1 \quad \text{for } n \geq 1$$

$$0! = 1$$

☐ The binomial formula provides the terms of the expansion of $(a + b)^n$.

$$(a + b)^n = a^n + \frac{n!}{1!(n - 1)!}a^{n-1}b + \frac{n!}{2!(n - 2)!}a^{n-2}b^2$$

$$+ \frac{n!}{3!(n - 3)!}a^{n-3}b^3$$

$$+ \cdots + \frac{n!}{r!(n - r)!}a^{n-r}b^r + \cdots + b^n$$

☐ The notation $\binom{n}{r}$ is defined by

$$\binom{n}{r} = \frac{n!}{r!(n - r)!}$$

and is useful in writing out the binomial formula.

REVIEW EXERCISES

Solutions to exercises whose numbers are in color are in the Solutions section in the back of the book.

11.1 In Exercises 1 and 2 write the first three terms and the tenth term of the sequence whose nth term is given.

1. $a_n = n^2 + n + 1$ 2. $a_n = \dfrac{n^3 - 1}{n + 1}$

In Exercises 3 and 4 find the fifth term of the recursive sequence.

3. $a_n = n - a_{n-1}, \quad a_1 = 0$

4. $a_n = na_{n-1}, \quad a_1 = 1$

In Exercises 5–7 find the indicated sum.

5. $\displaystyle\sum_{k=1}^{4} (1 - 2k)$ 6. $\displaystyle\sum_{k=3}^{5} k(k + 1)$

7. $\displaystyle\sum_{i=1}^{5} 10$

In Exercises 8–10 express the sum in sigma notation.

8. $\dfrac{1}{3} + \dfrac{2}{4} + \dfrac{3}{5} + \dfrac{4}{6}$

9. $1 - x + x^2 - x^3 + x^4$

10. $\sin x + \sin 2x + \sin 3x + \cdots + \sin nx$

11.2 In Exercises 11 and 12 find the specified term of the arithmetic sequence whose first term is a_1 and whose common difference is d.

11. $a_1 = -2, \; d = 2$; 21st term

12. $a_1 = 6, \; d = -1$; 16th term

In Exercises 13 and 14, given two terms of an arithmetic sequence, find the specified term.

13. $a_1 = 4, \; a_{16} = 9$; 13th term

14. $a_1 = -4, \; a_{23} = -15$; 26th term

In Exercises 15 and 16 find the sum of the first 25 terms of the arithmetic sequence whose first term is a_1 and whose common difference is d.

15. $a_1 = -\dfrac{1}{3}, \; d = \dfrac{1}{3}$ 16. $a_1 = 6, \; d = -2$

11.3 In Exercises 17 and 18 determine the common ratio of the given geometric sequence.

17. $2, -6, 18, -54, \ldots$

18. $-\dfrac{1}{2}, \dfrac{3}{4}, -\dfrac{9}{8}, \dfrac{27}{16}, \ldots$

In Exercises 19 and 20, write the first four terms of the geometric sequence whose first term is a_1 and whose common ratio is r.

19. $a_1 = 5, \quad r = \dfrac{1}{5}$ 20. $a_1 = -2, \quad r = -1$

21. Find the sixth term of the geometric sequence $-4, 6, -9, \ldots$.

22. Find the eighth term of a geometric sequence for which $a_1 = -2$ and $a_5 = -32$.

23. Insert two geometric means between 3 and $1/72$.

24. Find the sum of the first six terms of the geometric progression whose first three terms are $\tfrac{1}{3}$, $\tfrac{1}{6}$, $\tfrac{1}{12}$.

25. Find the sum of the first six terms of the geometric progression for which $a_1 = -2$ and $r = 3$.

In Exercises 26 and 27 find the sum of the infinite geometric series.

26. $5 + \dfrac{5}{2} + \dfrac{5}{4} + \cdots$ 27. $3 - 2 + \dfrac{4}{3} - \cdots$

11.4 28. Use the principle of mathematical induction to show that

$$3 + 6 + 9 + \cdots + 3n = \frac{3n(n + 1)}{2}$$

is true for all positive integer values of n.

11.5 In Exercises 29–31 expand and simplify.

29. $(2x - y)^4$ 30. $\left(\dfrac{x}{2} - 2\right)^4$

31. $(x^2 + 1)^3$

In Exercises 32–37 evaluate the expression.

32. $6!$ 33. $\dfrac{13!}{11!2!}$

34. $\dfrac{(n - 1)!(n + 1)!}{n!n!}$ 35. $\dbinom{6}{4}$

36. $\dbinom{3}{0}$ 37. $\dbinom{10}{8}$

PROGRESS TEST 11A

1. Write the first four terms of a sequence whose nth term is $a_n = n/(n + 1)^2$.

2. Evaluate $\displaystyle\sum_{j=2}^{4} \frac{j}{j - 1}$.

3. Write the first four terms of the arithmetic sequence whose first term is -1 and whose common difference is $\frac{3}{2}$.

4. Find the 25th term of the arithmetic sequence whose first term is -4 and whose common difference is $\frac{1}{2}$.

5. Find the 15th term of an arithmetic sequence whose first and tenth terms are -1 and 26, respectively.

6. Find the sum of the first 10 terms of an arithmetic sequence whose first term is -4 and whose ninth term is 8.

7. Find the common ratio of the geometric sequence 12, 4, $\dfrac{4}{3}$, $\dfrac{4}{9}$, ...

8. Write the first four terms of the geometric sequence whose first term is $-\frac{2}{3}$ and whose common ratio is 2.

9. Find the tenth term of the sequence 2, -2, 2, ...

10. Insert two geometric means between -4 and 32.

11. Find the sum of the first seven terms of the geometric sequence whose first three terms are -8, 4, -2.

12. Find the sum of the infinite geometric series $-4 - \dfrac{4}{3} - \dfrac{4}{9} - \cdots$.

13. Use the principle of mathematical induction to show that $2 + 6 + 10 + \cdots + (4n - 2) = 2n^2$ is true for all positive integer values of n.

14. Find the first four terms in the expansion of $\left(a + \dfrac{1}{b}\right)^{10}$.

15. Evaluate $\dfrac{12!}{10!3!}$.

PROGRESS TEST 11B

1. Write the first four terms of a sequence whose nth term is
$$a_n = n^2 + \frac{2n}{n + 2}$$

2. Write the sum $2! + 3! + 4! + \cdots + n!$ in summation notation.

3. Write the first four terms of the arithmetic sequence whose first term is 6 and whose common difference is $-\frac{2}{3}$.

4. Find the sixth term of the arithmetic sequence whose first term is -5 and whose common difference is 3.

5. Find the 30th term of an arithmetic sequence whose first and 20th terms are 3 and -35, respectively.

6. The first term of an arithmetic series is -5, the last term is -2, and the sum is $-91/2$. Find the number of terms and the common difference.

7. Find the common ratio of the geometric sequence 20, 4, 0.8, 0.16.

8. Write the first four terms of the geometric sequence whose first term is -1 and whose common ratio is $-\frac{1}{4}$.

9. Find the sixth term of a geometric sequence for which $a_1 = 3$ and $a_4 = -\frac{1}{9}$.

10. Insert two geometric means between -6 and $-16/9$.

11. Find the sum of the first five terms of a geometric sequence if $a_1 = -8$ and $a_4 = -1$.

12. Find the sum of the infinite geometric progression $5 - 2 + \dfrac{4}{5} - \cdots$.

13. Use the principle of mathematical induction to show that
$$\frac{1}{1 \cdot 2} + \frac{1}{2 \cdot 3} + \frac{1}{3 \cdot 4} + \cdots + \frac{1}{n(n + 1)} = \frac{n}{n + 1}$$
is true for all positive integer values of n.

14. Find the third term in the expansion of $(2x - 1)^{10}$.

15. Evaluate $\dfrac{n \cdot n!}{(n + 1)!}$.

APPENDIX A
APPLICATIONS

Why the stress on learning algebra? Why are you required to learn the algebraic techniques presented in this book? One reason is that algebra provides the basic tools of mathematical manipulation that you will need in higher-level courses. In particular, a course in calculus cannot go smoothly for you if you are struggling with algebraic steps while the instructor is introducing new concepts.

But there is yet another reason for studying algebra. Many practical problems lead to equations or inequalities that must be solved. Once you have mastered the techniques of solving various algebraic forms, you have the potential to solve the applied problem.

Of course, you must first be able to translate a problem from words to algebra; that is, you must be able to go from words to an equation or inequality or a system of equations or inequalities. In reality, that's the tough part, since finding the solution is more or less a matter of rote application of the methods you have learned. But going from words to algebraic expressions is never a rote procedure. It requires careful interpretation of the words, a skill that comes only after practice, practice, and more practice.

We can't provide a set of rules that will enable you to translate *any* applied problem into algebraic expressions. We can, however, offer you a set of steps that will serve as a guide in "setting up" word problems.

Step 1. Read the problem carefully to understand what is required.
Step 2. Separate what is known from what is to be found.
Step 3. In many problems, an unknown quantity is the answer to a question such as "how much" or "how many." Let a symbol, say x, represent the unknown.
Step 4. If possible, represent other quantities in the problem in terms of x.
Step 5. Find the relationship in the problem that you can express as an equation (or an inequality).
Step 6. Solve and check.

The words and phrases in Table 1 should prove helpful in translating a word problem into an algebraic expression that can be solved.

Each problem in this appendix will lead to a linear or quadratic equation or inequality.

SIMPLE INTEREST

If a principal P is borrowed at a simple interest rate r, then the interest due at the end of each year is Pr, and the total interest I due at the end of t years is

$$I = Prt$$

Consequently, if S is the total amount owed at the end of t years, then

$$S = P + Prt$$

TABLE 1

Word or phrase	Algebraic symbol	Example	Algebraic expression
Sum	+	Sum of two numbers	$a + b$
Difference	−	Difference of two numbers	$a - b$
		Difference of a number and 3	$x - 3$
Product	× or ·	Product of two numbers	$a \cdot b$
Quotient	÷ or /	Quotient of two numbers	$\dfrac{a}{b}$ or a/b
Exceeds		a exceeds b by 3.	$a = b + 3$
More than		a is 3 more than b.	or
More of		There are 3 more of a than of b.	$a - 3 = b$
Twice		Twice a number	$2x$
		Twice the difference of x and 3	$2(x - 3)$
		3 more than twice a number	$2x + 3$
		3 less than twice a number	$2x - 3$
Is or equals	=	The sum of a number and 3 is 15.	$x + 3 = 15$

since both the principal and interest are to be repaid.

The basic formulas that we have derived for simple interest calculations are

$$I = Prt$$
$$S = P + Prt$$

EXAMPLE 1

A part of $7000 was borrowed at 6% simple annual interest and the remainder at 8%. If the total amount of interest due after 3 years is $1380, how much was borrowed at each rate?

SOLUTION

Let

$$n = \text{the amount borrowed at 6\%}$$

Then

$$7000 - n = \text{the amount borrowed at 8\%}$$

since the total amount is $7000. We can display the information in table form using the equation $I = Prt$.

	P	×	r	×	t	=	Interest
6% portion	n		0.06		3		$0.18n$
8% portion	$7000 - n$		0.08		3		$0.24(7000 - n)$

Note that we write the rate r in its decimal form, so $6\% = 0.06$ and $8\% = 0.08$.

Since the total interest of \$1380 is the sum of the interest from the two portions, we have

$$1380 = 0.18n + 0.24(7000 - n)$$
$$1380 = 0.18n + 1680 - 0.24n$$
$$0.06n = 300$$
$$n = 5000$$

We conclude that \$5000 was borrowed at 6% and \$2000 was borrowed at 8%.

DISTANCE (UNIFORM MOTION) PROBLEMS

Here is the key to the solution of distance problems.

Distance = rate × time

or

$$d = r \cdot t$$

The relationships that permit you to write an equation are sometimes obscured by the words. Here are some questions to ask as you set up a distance problem.

(a) Are there two distances that are equal? (Will two objects have traveled the same distance? Is the distance on a return trip the same as the distance going?)

(b) Is the sum (or difference) of two distances equal to a constant? (When two objects are traveling toward each other, they meet when the sum of the distances traveled by them equals the original distance between them.)

EXAMPLE 2

Two trains leave New York for Chicago. The first train travels at an average speed of 60 miles per hour. The second train, which departs an hour later, travels at an average speed of 80 miles per hour. How long will it take the second train to overtake the first train?

SOLUTION

Since we are interested in the time the second train travels, we choose to let

$$t = \text{the number of hours the second train travels}$$

Then

$$t + 1 = \text{the number of hours the first train travels}$$

since the first train departs 1 hour earlier.

	Rate	×	Time	=	Distance
First train	60		$t + 1$		$60(t + 1)$
Second train	80		t		$80t$

At the moment the second train overtakes the first, they must both have traveled the *same* distance. Thus,

$$60(t + 1) = 80t$$
$$60t + 60 = 80t$$
$$60 = 20t$$
$$3 = t$$

It takes the second train 3 hours to catch up with the first train.

LITERAL EQUATIONS

It is sometimes convenient to be able to turn a formula around, that is, to be able to solve for a different variable. For example, if we want to express the radius of a circle in terms of the circumference, we have

$$C = 2\pi r$$

Dividing by 2π, we get

$$\frac{C}{2\pi} = \frac{2\pi r}{2\pi}$$

$$\frac{C}{2\pi} = r$$

Now, given a value of C, we can determine a value of r.

EXAMPLE 3

If an amount P is borrowed at the simple annual interest rate r, then the amount S due at the end of t years is

$$S = P + Prt$$

Solve for P.

SOLUTION

$$S = P + Prt$$
$$S = P(1 + rt) \qquad \text{Common factor } P$$
$$\frac{S}{1 + rt} = P \qquad \text{Dividing both sides by } (1 + rt)$$

ECONOMIC ANALYSIS

A common application in business problems involves determining the best of two or more strategies. Deciding whether to purchase or to lease a piece of equipment is a good illustration.

EXAMPLE 4

A business can purchase a copier at a cost of $4000, or it can rent the same copier at a cost of $75 per month plus 2 cents per copy. In preparing a business plan, the comptroller estimates that the copier has a life of 5 years and would cost $20 per month to maintain. Above what volume of monthly use would it cost less to purchase the copier?

SOLUTION

Let x denote the number of copies made each month. Then the monthly cost of rental (in dollars) is

$$75 + 0.02x$$

and the total rental cost for 5 years (or 60 months) is

$$60(75 + 0.02x) \quad \text{or} \quad 4500 + 1.2x$$

The cost of purchase and maintenance for 5 years or 60 months is

$$4000 + 60(20) = 5200$$

so that purchase will cost less when

$$5200 < 4500 + 1.2x$$

Solving, we have

$$700 < 1.2x$$
$$583 < x$$

The firm should purchase the copier if it expects to make more than 583 copies per month.

WORK PROBLEMS

Work problems typically involve two or more people or machines working on the same task. The key to these problems is to express the *rate of work per unit of time,* whether it is an hour, a day, a week, or some other unit. For example, if a machine can do a job in 5 days, then

$$\text{rate of machine} = \frac{1}{5} \text{ job per day}$$

If this machine were used for 2 days, it would perform 2/5 of the job. In summary:

If a machine (or person) can complete a job in n days, then

$$\text{Rate of machine (or person)} = \frac{1}{n} \text{ job per day}$$

$$\text{Work done} = \text{Rate} \times \text{Time}$$

EXAMPLE 5

Working together, two cranes can unload a ship in 4 hours. The slower crane, working alone, requires 6 more hours than the faster crane to do the job. How long does it take each crane to do the job by itself?

SOLUTION

Let x = number of hours for the faster crane to do the job. Then $x + 6$ = number of hours for the slower crane to do the job. The rate of the faster crane is $1/x$, since this is the portion of the whole job completed in 1 hour; similarly, the rate of the slower crane is $1/(x + 6)$. We display this information in a table.

	Rate	×	Time	=	Work
Faster crane	$\dfrac{1}{x}$		4		$\dfrac{4}{x}$
Slower crane	$\dfrac{1}{x+6}$		4		$\dfrac{4}{x+6}$

When the two cranes work together, we must have

$$\begin{array}{c} \text{work done by} \\ \text{fast crane} \end{array} + \begin{array}{c} \text{work done by} \\ \text{slow crane} \end{array} = 1 \text{ whole job}$$

or

$$\frac{4}{x} + \frac{4}{x+6} = 1$$

To solve, we multiply by $x(x+6)$, obtaining

$$4(x+6) + 4x = x^2 + 6x$$
$$0 = x^2 - 2x - 24$$
$$0 = (x+4)(x-6)$$
$$x = -4 \quad \text{or} \quad x = 6$$

The solution $x = -4$ is rejected, since it makes no sense to speak of negative hours of work. Summarizing, we see that

$$x = 6 \quad \text{is the number of hours in which the fast crane can do the job alone.}$$

$$x + 6 = 12 \quad \text{is the number of hours in which the slow crane can do the job alone.}$$

EXERCISE SET FOR APPENDIX A

1. A bicycle store is closing out its entire stock of a certain brand of 3-speed and 10-speed models. The profit on a 3-speed bicycle is 11% of the sale price, and the profit on a 10-speed model is 22% of the sale price. If the entire stock will be sold for $16,000 and the profit on the entire stock will be 19%, how much will be obtained from the sale of each type of bicycle?

2. A film shop carrying black-and-white film and color film has $4000 in inventory. The profit on black-and-white film is 12%. The profit on color film is 21%. If all the film is sold, and if the profit on color film is $150 less than the profit on black-and-white film, how much was invested in each type of film?

3. A firm borrowed $12,000 at a simple annual interest rate of 8% for a period of 3 years. At the end of the first year, the firm found that its needs were reduced. The firm returned a portion of the original loan and retained the remainder until the end of the 3-year period. If the total interest paid was $1760, how much was returned at the end of the first year?

4. A finance company lent a certain amount of money to Firm A at 7% annual interest. An amount $100 less than the amount lent to Firm A was lent to Firm B at 8%, and an amount $200 more than the amount lent to Firm A was lent to Firm C at 8.5% for one year. If the total annual income is $126.50, how much was lent to each firm?

5. Two trucks leave Philadelphia for Miami. The first truck to leave travels at an average speed of 50 kilometers per hour. The second truck, which leaves 2 hours later, travels at an average speed of 55 kilometers per hour. How long will it take the second truck to overtake the first truck?

6. Jackie either drives or bicycles from home to school. Her average speed when driving is 36 miles per hour, and her average speed when bicycling is 12 miles per hour. If it takes her 1/2 hour less to drive to school than to bicycle, how long does it take her to go to school, and how far is the school from her home?

7. Professors Roberts and Jones, who live 676 miles apart, are exchanging houses and jobs for the summer. They start out for their new locations at exactly the same time, and they meet after 6.5 hours of driving. If their average speeds differ by 4 miles per hour, what are their average speeds?

8. Steve leaves school by moped for spring vacation. Forty minutes later his roommate, Frank, notices that Steve forgot to take his camera, so Frank decides to try to catch up with Steve by car. If Steve's average speed is 25 miles per hour and Frank averages 45 miles per hour, how long does it take Frank to overtake Steve?

9. A certain number is three times another. If the difference of their reciprocals is 8, find the numbers.

10. If 1/3 is subtracted from 3 times the reciprocal of a certain number, the result is 25/6. Find the number.

11. A 12-meter-long steel beam is to be cut into two pieces so that one piece will be 4 meters longer than the other. How long will each piece be?

12. A rectangular field whose length is 10 meters longer than its width is to be enclosed with exactly 100 meters of fencing material. What are the dimensions of the field?

13. An express train and a local train start out from the same point at the same time and travel in opposite directions. The express train travels twice as fast as the local train. If after 4 hours they are 480 kilometers apart, what is the average speed of each train?

14. A boat travels 20 kilometers upstream in the time that it would take the same boat to travel 30 kilometers downstream. If the rate of the stream is 5 kilometers per hour, find the speed of the boat in still water.

15. An airplane flying against the wind travels 300 miles in the time that it would take the same plane to travel 400 miles with the wind. If the wind speed is 20 miles per hour, find the speed of the airplane in still air.

In Exercises 16–23 solve for the indicated variable in terms of the remaining variables.

16. $V = \dfrac{1}{3}\pi r^2 h$ for h

17. $F = \dfrac{9}{5}C + 32$ for C

18. $S = \dfrac{1}{2}gt^2 + vt$ for v

19. $A = \dfrac{1}{2}h(b + b')$ for b

20. $A = P(1 + rt)$ for r

21. $\dfrac{1}{f} = \dfrac{1}{f_1} + \dfrac{1}{f_2}$ for f_2

22. $a = \dfrac{v_1 - v_0}{t}$ for v_0

23. $S = \dfrac{a - rL}{L - r}$ for L

24. An auto rental agency offers two options: $35 per day with no charge for mileage or $28 per day and 12 cents per mile. If m is the number of miles traveled per day, for what values of m is the fixed $35 daily rate the better deal?

25. The Green Lawn Company offers a seasonal lawn feeding program at a cost of 3 cents per square foot. To attract owners of larger properties, Green Lawn also offers an alternative rate of $600 plus 2 cents per square foot. If s is the number of square feet of lawn, for what values of s is the 2-cent rate preferable?

26. A graphic designer and her assistant working together can complete an advertising layout in 6 days. The assistant working alone could complete the job in 16 more days than the designer working alone. How long would it take each person to do the job alone?

27. A roofer and his assistant working together can finish a roofing job in 4 hours. The roofer working alone could finish the job in 6 fewer hours than the assistant working alone. How long would it take each person to do the job alone?

28. Working together, computers A and B can complete a data-processing job in 2 hours. Computer A working alone can do the job in 3 fewer hours than computer B working alone. How long does it take each computer to do the job by itself?

29. A number of students rented a car for $160 for a one-week camping trip. If another student had joined the original group, each person's share of expenses would have been reduced by $8. How many students were there in the original group?

APPENDIX B
LIMITS: A PREVIEW
OF CALCULUS

Calculus was developed in the seventeenth century by the British scientist Sir Isaac Newton and by the German mathematician Gottfried Wilhelm von Leibniz. One of the outstanding developments in the history of science and mathematics, calculus is the mathematics of change. Since everything around us is changing, calculus can be applied to the solution of a great many pressing problems in almost every field of human endeavor.

In this appendix we introduce the concept of a limit. This idea, which lies at the very heart of calculus, has important consequences.

In many applied problems, it is necessary to describe the behavior of a function f when x is near, but different from, a value a. For example, if x denotes the distance from an airplane to the ground, and $f(x)$ denotes the speed of the airplane at distance x, we are interested in knowing $f(a)$, the speed of the airplane at the time of landing. Moreover, we are interested in knowing how the speed $f(x)$ varies when x is near but different from a, since this information will enable the pilot to take the necessary steps for a smooth landing.

Let's begin by focusing on the function f defined by

$$f(x) = 2x + 1$$

How do the values $f(x)$ behave as x gets closer and closer to (but remains different from) $x = 2$? As shown in Figure 1, we consider two different ways in which x can get closer and closer to, or approach, the value 2: from the right of 2 (that is, from values greater than 2) and from the left of 2 (values less than 2).

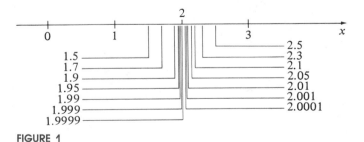

FIGURE 1

Table 1 shows the values of $f(x)$ as x approaches 2 from the right, and Table 2 shows these values as x approaches 2 from the left. We have plotted the data in Tables 1 and 2 on the graph shown in Figure 2.

TABLE 1 x approaches 2 from the right

x	3	2.5	2.3	2.1	2.05	2.01	2.001	2.0001
$f(x) = 2x + 1$	7	6	5.6	5.2	5.1	5.02	5.002	5.0002

TABLE 2 x approaches 2 from the left

x	1	1.5	1.7	1.9	1.95	1.99	1.999	1.9999
$f(x) = 2x + 1$	3	4	4.4	4.8	4.9	4.98	4.998	4.9998

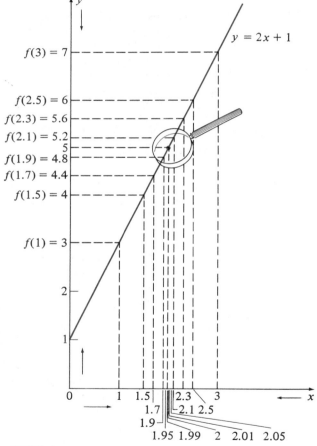

FIGURE 2

It is intuitively clear from Figure 2 that the values of $f(x) = 2x + 1$ get closer and closer to 5 as x approaches 2 from either side. The number 5 is called the **limit** of f as x approaches 2, and we write

$$\lim_{x \to 2} f(x) = 5 \quad \text{or} \quad \lim_{x \to 2} (2x + 1) = 5 \tag{1}$$

Here $x \to 2$ means "x approaches 2 (both from the right and from the left) but $x \neq 2$." Equation (1) is read, "the limit of $f(x)$ as x approaches, but remains different from, 2 is 5."

It is important to note that the value of the limit of f as $x \to 2$ in Equation (1) was obtained *intuitively*. It is possible to give a mathematically more precise definition of limit and then to carefully show that the above limit is indeed 5, but the intuitive approach is adequate for this introduction.

Let's examine the notion of a limit a little more closely.

A function f is said to **approach the limit L** as x approaches a if the values of $f(x)$ get closer and closer to the unique real number L as x gets closer and closer to (but remains different from) a. We write this statement as

$$\lim_{x \to a} f(x) = L$$

The definition of

$$\lim_{x \to a} f(x) = L$$

enables us to make the value of $f(x)$ as close to L as we want by taking x sufficiently close to (but different from) a.

PROGRESS CHECK

(a) Find $\lim_{x \to 3} (3x - 2)$ by completing the following tables of values.

x	3.4	3.3	3.2	3.1	3.05	3.01	3.001
$3x - 2$							

x	2.6	2.7	2.8	2.9	2.95	2.99	2.999
$3x - 2$							

 (b) Find $\lim_{x \to -2} (x^2 + x)$ by forming appropriate tables.

ANSWERS
(a) 7   (b) 2

You may have noticed that in Equation (1) we had $f(2) = 5$, so in *that* case, $\lim_{x \to 2} f(x)$ was simply $f(2)$. As we shall see in the following example, it is not always true that $\lim_{x \to a} f(x)$ is $f(a)$.

EXAMPLE 1
Given the function f defined by

$$f(x) = \frac{x^2 - 9}{x - 3}$$

find $\lim_{x \to 3} f(x)$.

SOLUTION
First, observe that this function is not defined for $x = 3$, because we cannot divide by zero. Thus, we cannot possibly find the limit as x approaches 3 merely by evaluating $f(3)$. The behavior of $f(x)$ as $x \to 3$ from the right and left is shown in Tables 3 and 4, respectively. Table 3 shows that the values of $f(x)$ approach 6 as x approaches 3 from the right. Table 4 shows that the values of $f(x)$ approach 6 as x approaches 3 from the left. Consequently,

$$\lim_{x \to 3} f(x) = 6 \quad \text{or} \quad \lim_{x \to 3} \frac{x^2 - 9}{x - 3} = 6$$

TABLE 3 $x \to 3$ from the right

x	4	3.5	3.1	3.01	3.001	3.0001
$f(x) = \dfrac{x^2 - 9}{x - 3}$	7	6.5	6.1	6.01	6.001	6.0001

TABLE 4 $x \to 3$ from the left

x	2	2.5	2.9	2.99	2.999	2.9999
$f(x) = \dfrac{x^2 - 9}{x - 3}$	5	5.5	5.9	5.99	5.999	5.9999

To see what is happening geometrically, we graph the function f. Observe that

$$f(x) = \frac{x^2 - 9}{x - 3} = \frac{(x + 3)(x - 3)}{(x - 3)} = x + 3, \qquad x \neq 3 \qquad (2)$$

Thus, if $x \neq 3$, the graph of f is the graph of $y = x + 3$, a straight line. If $x = 3$, $f(x)$ is not defined. Hence, the graph of f, shown in Figure 3, is a straight line with a hole at $x = 3$. As usual, we have marked the point where $x = 3$ with an open circle to indicate that it is not on the line. As the graph shows, as x approaches 3, the values of $f(x)$ approach the number 6.

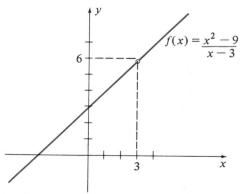

FIGURE 3

In the last two examples the value approached by $f(x)$ was the same finite number, whether x approached a from the right or from the left. This need not be the case, as the following examples show.

EXAMPLE 2

Let f be defined by

$$f(x) = \frac{|x|}{x}$$

Discuss $\lim_{x \to 0} f(x)$.

SOLUTION

First, observe that $f(0)$ is undefined. Moreover, we can write the given function as

$$f(x) = \begin{cases} \dfrac{x}{x} = 1 & \text{if } x > 0 \\[2em] \dfrac{-x}{x} = -1 & \text{if } x < 0 \end{cases}$$

As $x \to 0$ from the right, then $f(x)$ is always 1, so $f(x)$ approaches 1. If $x \to 0$ from the left, then $f(x)$ is always -1, so $f(x)$ approaches -1. Hence, the values of $f(x)$ do not approach a single finite number as $x \to 0$. This behavior is evident in the graph of f, which is shown in Figure 4.

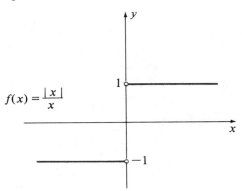

FIGURE 4

If the values of $f(x)$ do not approach a unique real number as x approaches the value a, we say that

$$\lim_{x \to a} f(x)$$

does not exist. Thus, for the function in Example 2,

$$\lim_{x \to 0} f(x) = \lim_{x \to 0} \frac{|x|}{x}$$

does not exist.

EXAMPLE 3

Consider the function f defined by

$$f(x) = \begin{cases} x + 2 & \text{if } x \leq 3 \\ \frac{1}{3}x + 5 & \text{if } x > 3 \end{cases} \tag{3}$$

Discuss $\lim_{x \to 3} f(x)$.

SOLUTION

The graph of this function is shown in Figure 5. Although $f(3)$ is defined, this limit does not exist. For, as $x \to 3$ from the right, then $f(x)$ approaches 6, and as $x \to 3$ from the left, then $f(x)$ approaches 5.

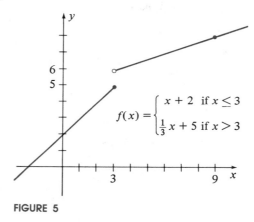

FIGURE 5

PROGRESS CHECK

Consider the function f defined by

$$f(x) = \begin{cases} 2x + 1 & \text{if } x \leq 1 \\ x + 3 & \text{if } x > 1 \end{cases}$$

Verify that $\lim_{x \to 1} f(x)$ does not exist.

One might be inclined to conjecture that when the rule for a function is given by several equations, the limit of the function does not exist at the values in the domain where the rule changes. This was the case in Example 3, where the rule for the function in Equation (3) changed at $x = 3$ and the limit as $x \to 3$ did not exist. We shall see in Example 4, however, that a function defined by two or more equations may have a limit at a value in the domain where the rule changes.

EXAMPLE 4

Consider the function f defined by

$$f(x) = \begin{cases} -\frac{2}{3}x + 6 & \text{if } x \geq 3 \\ x + 1 & \text{if } x < 3 \end{cases}$$

Discuss $\lim_{x \to 3} f(x)$.

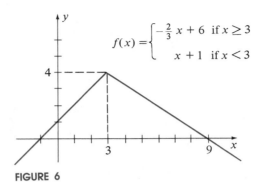

$$f(x) = \begin{cases} -\frac{2}{3}x + 6 & \text{if } x \geq 3 \\ x + 1 & \text{if } x < 3 \end{cases}$$

FIGURE 6

SOLUTION

The graph of this function is shown in Figure 6. By constructing appropriate tables, you can verify that

$$\lim_{x \to 3} f(x) = 4$$

It is possible also that, as $x \to a$, the values of $f(x)$ will not approach any real number. This phenomenon occurs in Example 5.

EXAMPLE 5

Determine whether $\lim_{x \to 0} f(x)$ exists for the function f defined by

$$f(x) = \frac{1}{x^2}$$

SOLUTION

The graph is shown in Figure 7.

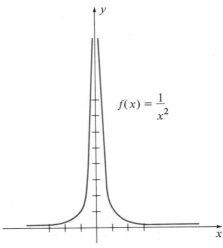

$$f(x) = \frac{1}{x^2}$$

FIGURE 7

To find $\lim_{x \to 0} f(x)$, we observe that as $x \to 0$ from the right, $f(x)$ gets larger and larger, and as $x \to 0$ from the left, $f(x)$ also gets larger and larger. This can also be seen from Tables 5 and 6. Since $f(x)$ approaches no real number as x approaches 0, we conclude that $\lim_{x \to 0} f(x)$ does not exist.

TABLE 5 $x \to 0$ from the right

x	2	1	0.7	0.5	0.2	0.1	0.05	0.01	0.001
$f(x) = \dfrac{1}{x^2}$	0.25	1	2.04	4	25	100	400	1×10^4	1×10^6

TABLE 6 $x \to 0$ from the left

x	-2	-1	-0.7	-0.5	-0.2	-0.1	-0.05	-0.01	-0.001
$f(x) = \dfrac{1}{x^2}$	0.25	1	2.04	4	25	100	400	1×10^4	1×10^6

EXERCISE SET FOR APPENDIX B

1. (a) Find $\lim_{x \to 4} (2x - 3)$ by completing the following tables of values.

x	4.4	4.2	4.1	4.01	4.001
$2x - 3$					

x	3.6	3.8	3.9	3.99	3.999
$2x - 3$					

   (b) Sketch the graph of $f(x) = 2x - 3$ and verify your conclusion.

2. (a) Find $\lim_{x \to 2} 1/(x - 2)$ by completing the following tables of values.

x	2.5	2.2	2.1	2.05	2.01	2.001	2.0001
$\dfrac{1}{x - 2}$							

x	1.5	1.7	1.9	1.95	1.99	1.999	1.9999
$\dfrac{1}{x - 2}$							

   (b) Sketch the graph of $f(x) = 1/(x - 2)$ and verify your conclusion.

In Exercises 3–14 find the limit, if it exists.

3. $\displaystyle\lim_{x\to3} 2x$

4. $\displaystyle\lim_{x\to-1} (2x + 3)$

5. $\displaystyle\lim_{x\to2} (3x^2 + 2x - 5)$

6. $\displaystyle\lim_{x\to0} \frac{2}{x}$

7. $\displaystyle\lim_{x\to-1} \frac{x^2 - 2x - 3}{x + 1}$

8. $\displaystyle\lim_{x\to2} \frac{x^2 - 16}{x - 4}$

9. $\displaystyle\lim_{x\to-2} (16 - x^2)$

10. $\displaystyle\lim_{x\to4} \frac{x + 1}{x^2 - 3x - 4}$

11. $\displaystyle\lim_{x\to0} \frac{x^2 - 2x}{x}$

12. $\displaystyle\lim_{x\to0} |x|$

13. $\displaystyle\lim_{x\to-6} \frac{x^2 - 36}{x + 6}$

14. $\displaystyle\lim_{h\to0} \frac{(x + h)^2 - x^2}{h}$

15. Let f be defined by

$$f(x) = \begin{cases} 2x + 1 & \text{if } x \le 2 \\ -\frac{5}{2}x + 10 & \text{if } x > 2 \end{cases}$$

   (a) Find $\displaystyle\lim_{x\to2} f(x)$.

   (b) Sketch the graph of $f(x)$.

16. Let f be defined by

$$f(x) = \begin{cases} x & \text{if } x \le 1 \\ 2x + 1 & \text{if } x > 1 \end{cases}$$

   (a) Find $\displaystyle\lim_{x\to1} f(x)$.

   (b) Sketch the graph of $f(x)$.

In Exercises 17–23 find $\displaystyle\lim_{x\to a} f(x)$, if it exists, and sketch the graph of $f(x)$.

17. $f(x) = \dfrac{x^2 - 16}{x - 4}$, $a = 4$

18. $f(x) = \dfrac{x^2 - 25}{x + 5}$, $a = -5$

19. $f(x) = \dfrac{x^2 + x - 2}{x + 2}$, $a = 1$

20. $f(x) = \dfrac{x - 4}{x^2 - 3x - 4}$, $a = -1$

21. $f(x) = \dfrac{1}{2x - 1}$, $a = \dfrac{1}{2}$

22. $f(x) = |x - 1|$, $a = 2$

23. $f(x) = \dfrac{|x - 2|}{x - 2}$, $a = 2$

In Exercises 24–29 use the graph to find the limit, if it exists.

24. $\lim\limits_{x \to 5} f(x)$

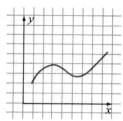

25. $\lim\limits_{x \to 3} f(x)$

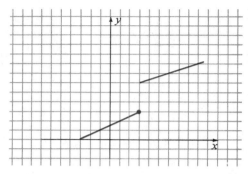

26. $\lim\limits_{x \to 3} f(x)$

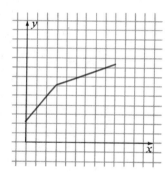

27. $\lim\limits_{x \to 2} f(x)$

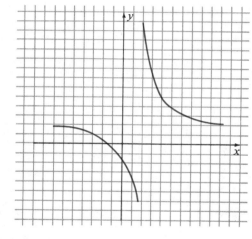

TABLES APPENDIX

TABLE I Exponentials and Their Reciprocals

x	e^x	e^{-x}	x	e^x	e^{-x}
0.00	1.0000	1.0000	1.4	4.0552	0.2466
0.01	1.0101	0.9900	1.5	4.4817	0.2231
0.02	1.0202	0.9802	1.6	4.9530	0.2019
0.03	1.0305	0.9704	1.7	5.4739	0.1827
0.04	1.0408	0.9608	1.8	6.0496	0.1653
0.05	1.0513	0.9512	1.9	6.6859	0.1496
0.06	1.0618	0.9418	2.0	7.3891	0.1353
0.07	1.0725	0.9324	2.1	8.1662	0.1225
0.08	1.0833	0.9231	2.2	9.0250	0.1108
0.09	1.0942	0.9139	2.3	9.9742	0.1003
0.10	1.1052	0.9048	2.4	11.023	0.0907
0.11	1.1163	0.8958	2.5	12.182	0.0821
0.12	1.1275	0.8869	2.6	13.464	0.0743
0.13	1.1388	0.8781	2.7	14.880	0.0672
0.14	1.1503	0.8694	2.8	16.445	0.0608
0.15	1.1618	0.8607	2.9	18.174	0.0550
0.16	1.1735	0.8521	3.0	20.086	0.0498
0.17	1.1853	0.8437	3.1	22.198	0.0450
0.18	1.1972	0.8353	3.2	24.533	0.0408
0.19	1.2092	0.8270	3.3	27.113	0.0369
0.20	1.2214	0.8187	3.4	29.964	0.0334
0.21	1.2337	0.8106	3.5	33.115	0.0302
0.22	1.2461	0.8025	3.6	36.598	0.0273
0.23	1.2586	0.7945	3.7	40.447	0.0247
0.24	1.2712	0.7866	3.8	44.701	0.0224
0.25	1.2840	0.7788	3.9	49.402	0.0202
0.26	1.2969	0.7711	4.0	54.598	0.0183
0.27	1.3100	0.7634	4.1	60.340	0.0166
0.28	1.3231	0.7558	4.2	66.686	0.0150
0.29	1.3364	0.7483	4.3	73.700	0.0136
0.30	1.3499	0.7408	4.4	81.451	0.0123
0.35	1.4191	0.7047	4.5	90.017	0.0111
0.40	1.4918	0.6703	4.6	99.484	0.0101
0.45	1.5683	0.6376	4.7	109.95	0.0091
0.50	1.6487	0.6065	4.8	121.51	0.0082
0.55	1.7333	0.5769	4.9	134.29	0.0074
0.60	1.8221	0.5488	5	148.41	0.0067
0.65	1.9155	0.5220	6	403.43	0.0025
0.70	2.0138	0.4966	7	1,096.6	0.0009
0.75	2.1170	0.4724	8	2,981.0	0.0003
0.80	2.2255	0.4493	9	8,103.1	0.0001
0.85	2.3396	0.4274	10	22,026	0.00005
0.90	2.4596	0.4066	11	59,874	0.00002
0.95	2.5857	0.3867	12	162,754	0.000006
1.0	2.7183	0.3679	13	442,413	0.000002
1.1	3.0042	0.3329	14	1,202,604	0.0000008
1.2	3.3201	0.3012	15	3,269,017	0.0000003
1.3	3.6693	0.2725			

TABLE II Common Logarithms

N	0	1	2	3	4	5	6	7	8	9
1.0	.0000	.0043	.0086	.0128	.0170	.0212	.0253	.0294	.0334	.0374
1.1	.0414	.0453	.0492	.0531	.0569	.0607	.0645	.0682	.0719	.0755
1.2	.0792	.0828	.0864	.0899	.0934	.0969	.1004	.1038	.1072	.1106
1.3	.1139	.1173	.1206	.1239	.1271	.1303	.1335	.1367	.1399	.1430
1.4	.1461	.1492	.1523	.1553	.1584	.1614	.1644	.1673	.1703	.1732
1.5	.1761	.1790	.1818	.1847	.1875	.1903	.1931	.1959	.1987	.2014
1.6	.2041	.2068	.2095	.2122	.2148	.2175	.2201	.2227	.2253	.2279
1.7	.2304	.2330	.2355	.2380	.2405	.2430	.2455	.2480	.2504	.2529
1.8	.2553	.2577	.2601	.2625	.2648	.2672	.2695	.2718	.2742	.2765
1.9	.2788	.2810	.2833	.2856	.2878	.2900	.2923	.2945	.2967	.2989
2.0	.3010	.3032	.3054	.3075	.3096	.3118	.3139	.3160	.3181	.3201
2.1	.3222	.3243	.3263	.3284	.3304	.3324	.3345	.3365	.3385	.3404
2.2	.3424	.3444	.3464	.3483	.3502	.3522	.3541	.3560	.3579	.3598
2.3	.3617	.3636	.3655	.3674	.3692	.3711	.3729	.3747	.3766	.3784
2.4	.3802	.3820	.3838	.3856	.3874	.3892	.3909	.3927	.3945	.3692
2.5	.3979	.3997	.4014	.4031	.4048	.4065	.4082	.4099	.4116	.4133
2.6	.4150	.4166	.4183	.4200	.4216	.4232	.4249	.4265	.4281	.4298
2.7	.4314	.4330	.4346	.4362	.4378	.4393	.4409	.4425	.4440	.4456
2.8	.4472	.4487	.4502	.4518	.4533	.4548	.4564	.4579	.4594	.4609
2.9	.4624	.4639	.4654	.4669	.4683	.4698	.4713	.4728	.4742	.4757
3.0	.4771	.4786	.4800	.4814	.4829	.4843	.4857	.4871	.4886	.4900
3.1	.4914	.4928	.4942	.4955	.4969	.4983	.4997	.5011	.5024	.5038
3.2	.5051	.5065	.5079	.5092	.5105	.5119	.5132	.5145	.5159	.5172
3.3	.5185	.5198	.5211	.5224	.5237	.5250	.5263	.5276	.5289	.5302
3.4	.5315	.5328	.5340	.5353	.5366	.5378	.5391	.5403	.5416	.5428
3.5	.5441	.5453	.5465	.5478	.5490	.5502	.5514	.5527	.5539	.5551
3.6	.5563	.5575	.5587	.5599	.5611	.5623	.5635	.5647	.5658	.5670
3.7	.5682	.5694	.5705	.5717	.5729	.5740	.5752	.5763	.5775	.5786
3.8	.5798	.5809	.5821	.5832	.5843	.5855	.5866	.5877	.5888	.5899
3.9	.5911	.5922	.5933	.5944	.5955	.5966	.5977	.5988	.5999	.6010
4.0	.6021	.6031	.6042	.6053	.6064	.6075	.6085	.6096	.6107	.6117
4.1	.6128	.6138	.6149	.6160	.6170	.6180	.6191	.6201	.6212	.6222
4.2	.6232	.6243	.6253	.6263	.6274	.6284	.6294	.6304	.6314	.6325
4.3	.6335	.6345	.6355	.6365	.6375	.6385	.6395	.6405	.6415	.6425
4.4	.6435	.6444	.6454	.6464	.6474	.6484	.6493	.6503	.6513	.6522
4.5	.6532	.6542	.6551	.6561	.6571	.6580	.6590	.6599	.6609	.6618
4.6	.6628	.6637	.6646	.6656	.6665	.6675	.6684	.6693	.6702	.6712
4.7	.6721	.6730	.6739	.6749	.6758	.6767	.6776	.6785	.6794	.6803
4.8	.6812	.6821	.6830	.6839	.6848	.6857	.6866	.6875	.6884	.6893
4.9	.6902	.6911	.6920	.6928	.6937	.6946	.6955	.6964	.6972	.6981
5.0	.6990	.6998	.7007	.7016	.7024	.7033	.7042	.7050	.7059	.7067
5.1	.7076	.7084	.7093	.7101	.7110	.7118	.7126	.7135	.7143	.7152
5.2	.7160	.7168	.7177	.7185	.7193	.7202	.7210	.7218	.7226	.7235
5.3	.7243	.7251	.7259	.7267	.7275	.7284	.7292	.7300	.7308	.7316
5.4	.7324	.7332	.7340	.7348	.7356	.7364	.7372	.7380	.7388	.7396

TABLE II (*continued*) 551

N	0	1	2	3	4	5	6	7	8	9
5.5	.7404	.7412	.7419	.7427	.7435	.7443	.7451	.7459	.7466	.7474
5.6	.7482	.7490	.7497	.7505	.7513	.7520	.7528	.7536	.7543	.7551
5.7	.7559	.7566	.7574	.7582	.7589	.7597	.7604	.7612	.7619	.7627
5.8	.7634	.7642	.7649	.7657	.7664	.7672	.7679	.7686	.7694	.7701
5.9	.7709	.7716	.7723	.7731	.7738	.7745	.7752	.7760	.7767	.7774
6.0	.7782	.7789	.7796	.7803	.7810	.7818	.7825	.7832	.7839	.7846
6.1	.7853	.7860	.7868	.7875	.7882	.7889	.7896	.7903	.7910	.7917
6.2	.7924	.7931	.7938	.7945	.7952	.7959	.7966	.7973	.7980	.7987
6.3	.7993	.8000	.8007	.8014	.8021	.8028	.8035	.8041	.8048	.8055
6.4	.8062	.8069	.8075	.8082	.8089	.8096	.8102	.8109	.8116	.8122
6.5	.8129	.8136	.8142	.8149	.8156	.8162	.8169	.8176	.8182	.8189
6.6	.8195	.8202	.8209	.8215	.8222	.8228	.8235	.8241	.8248	.8254
6.7	.8261	.8267	.8274	.8280	.8287	.8293	.8299	.8306	.8312	.8319
6.8	.8325	.8331	.8338	.8344	.8351	.8357	.8363	.8370	.8376	.8382
6.9	.8388	.8395	.8401	.8407	.8414	.8420	.8426	.8432	.8439	.8445
7.0	.8451	.8457	.8463	.8470	.8476	.8482	.8488	.8494	.8500	.8506
7.1	.8513	.8519	.8525	.8531	.8537	.8543	.8549	.8555	.8561	.8567
7.2	.8573	.8579	.8585	.8591	.8597	.8603	.8609	.8615	.8621	.8627
7.3	.8633	.8639	.8645	.8651	.8657	.8663	.8669	.8675	.8681	.8686
7.4	.8692	.8698	.8704	.8710	.8716	.8722	.8727	.8733	.8739	.8745
7.5	.8751	.8756	.8762	.8768	.8774	.8779	.8785	.8791	.8797	.8802
7.6	.8808	.8814	.8820	.8825	.8831	.8837	.8842	.8848	.8854	.8859
7.7	.8865	.8871	.8876	.8882	.8887	.8893	.8899	.8904	.8910	.8915
7.8	.8921	.8927	.8932	.8938	.8943	.8949	.8954	.8960	.8965	.8971
7.9	.8976	.8982	.8987	.8993	.8998	.9004	.9009	.9015	.9020	.9025
8.0	.9031	.9036	.9042	.9047	.9053	.9058	.9063	.9069	.9074	.9079
8.1	.9085	.9090	.9096	.9101	.9106	.9112	.9117	.9122	.9128	.9133
8.2	.9138	.9143	.9149	.9154	.9159	.9165	.9170	.9175	.9180	.9186
8.3	.9191	.9196	.9201	.9206	.9212	.9217	.9222	.9227	.9232	.9238
8.4	.9243	.9248	.9253	.9258	.9263	.9269	.9274	.9279	.9284	.9289
8.5	.9294	.9299	.9304	.9309	.9315	.9320	.9325	.9330	.9335	.9340
8.6	.9345	.9350	.9355	.9360	.9365	.9370	.9375	.9380	.9385	.9390
8.7	.9395	.9400	.9405	.9410	.9415	.9420	.9425	.9430	.9435	.9440
8.8	.9445	.9450	.9455	.9460	.9465	.9469	.9474	.9479	.9484	.9489
8.9	.9494	.9499	.9504	.9509	.9513	.9518	.9523	.9528	.9533	.9538
9.0	.9542	.9547	.9552	.9557	.9562	.9566	.9571	.9576	.9581	.9586
9.1	.9590	.9595	.9600	.9605	.9609	.9614	.9619	.9624	.9628	.9633
9.2	.9638	.9643	.9647	.9652	.9657	.9661	.9666	.9671	.9675	.9680
9.3	.9685	.9689	.9694	.9699	.9703	.9708	.9713	.9717	.9722	.9727
9.4	.9731	.9736	.9741	.9745	.9750	.9754	.9759	.9763	.9768	.9773
9.5	.9777	.9782	.9786	.9791	.9795	.9800	.9805	.9809	.9814	.9818
9.6	.9823	.9827	.9832	.9836	.9841	.9845	.9850	.9854	.9859	.9863
9.7	.9868	.9872	.9877	.9881	.9886	.9890	.9894	.9899	.9903	.9908
9.8	.9912	.9917	.9921	.9926	.9930	.9934	.9939	.9943	.9948	.9952
9.9	.9956	.9961	.9965	.9969	.9974	.9978	.9983	.9987	.9991	.9996

TABLE III Natural Logarithms

N	$\ln N$	N	$\ln N$	N	$\ln N$
		4.5	1.5041	9.0	2.1972
0.1	−2.3026	4.6	1.5261	9.1	2.2083
0.2	−1.6094	4.7	1.5476	9.2	2.2192
0.3	−1.2040	4.8	1.5686	9.3	2.2300
0.4	−0.9163	4.9	1.5892	9.4	2.2407
0.5	−0.6931	5.0	1.6094	9.5	2.2513
0.6	−0.5108	5.1	1.6292	9.6	2.2618
0.7	−0.3567	5.2	1.6487	9.7	2.2721
0.8	−0.2231	5.3	1.6677	9.8	2.2824
0.9	−0.1054	5.4	1.6864	9.9	2.2925
1.0	0.0000	5.5	1.7047	10	2.3026
1.1	0.0953	5.6	1.7228	11	2.3979
1.2	0.1823	5.7	1.7405	12	2.4849
1.3	0.2624	5.8	1.7579	13	2.5649
1.4	0.3365	5.9	1.7750	14	2.6391
1.5	0.4055	6.0	1.7918	15	2.7081
1.6	0.4700	6.1	1.8083	16	2.7726
1.7	0.5306	6.2	1.8245	17	2.8332
1.8	0.5878	6.3	1.8405	18	2.8904
1.9	0.6419	6.4	1.8563	19	2.9444
2.0	0.6931	6.5	1.8718	20	2.9957
2.1	0.7419	6.6	1.8871	25	3.2189
2.2	0.7885	6.7	1.9021	30	3.4012
2.3	0.8329	6.8	1.9169	35	3.5553
2.4	0.8755	6.9	1.9315	40	3.6889
2.5	0.9163	7.0	1.9459	45	3.8067
2.6	0.9555	7.1	1.9601	50	3.9120
2.7	0.9933	7.2	1.9741	55	4.0073
2.8	1.0296	7.3	1.9879	60	4.0943
2.9	1.0647	7.4	2.0015	65	4.1744
3.0	1.0986	7.5	2.0149	70	4.2485
3.1	1.1314	7.6	2.0281	75	4.3175
3.2	1.1632	7.7	2.0412	80	4.3820
3.3	1.1939	7.8	2.0541	85	4.4427
3.4	1.2238	7.9	2.0669	90	4.4998
3.5	1.2528	8.0	2.0794	95	4.5539
3.6	1.2809	8.1	2.0919	100	4.6052
3.7	1.3083	8.2	2.1041		
3.8	1.3350	8.3	2.1163		
3.9	1.3610	8.4	2.1282		
4.0	1.3863	8.5	2.1401		
4.1	1.4110	8.6	2.1518		
4.2	1.4351	8.7	2.1633		
4.3	1.4586	8.8	2.1748		
4.4	1.4816	8.9	2.1861		

TABLE IV Interest Rates 553

$i = \frac{1}{2}\%$

n	$(1 + i)^n$	n	$(1 + i)^n$
1	1.0050 0000	51	1.2896 4194
2	1.0100 2500	52	1.2960 9015
3	1.0150 7513	53	1.3025 7060
4	1.0201 5050	54	1.3090 8346
5	1.0252 5125	55	1.3156 2887
6	1.0303 7751	56	1.3222 0702
7	1.0355 2940	57	1.3288 1805
8	1.0407 0704	58	1.3354 6214
9	1.0459 1058	59	1.3421 3946
10	1.0511 4013	60	1.3488 5015
11	1.0563 9583	61	1.3555 9440
12	1.0616 7781	62	1.3623 7238
13	1.0669 8620	63	1.3691 8424
14	1.0723 2113	64	1.3760 3016
15	1.0776 8274	65	1.3829 1031
16	1.0830 7115	66	1.3898 2486
17	1.0884 8651	67	1.3967 7399
18	1.0939 2894	68	1.4037 5785
19	1.0993 9858	69	1.4107 7664
20	1.1048 9558	70	1.4178 3053
21	1.1104 2006	71	1.4249 1968
22	1.1159 7216	72	1.4320 4428
23	1.1215 5202	73	1.4392 0450
24	1.1271 5978	74	1.4464 0052
25	1.1327 9558	75	1.4536 3252
26	1.1384 5955	76	1.4609 0069
27	1.1441 5185	77	1.4682 0519
28	1.1498 7261	78	1.4755 4622
29	1.1556 2197	79	1.4829 2395
30	1.1614 0008	80	1.4903 3857
31	1.1672 0708	81	1.4977 9026
32	1.1730 4312	82	1.5052 7921
33	1.1789 0833	83	1.5128 0561
34	1.1848 0288	84	1.5203 6964
35	1.1907 2689	85	1.5279 7148
36	1.1966 8052	86	1.5356 1134
37	1.2026 6393	87	1.5432 8940
38	1.2086 7725	88	1.5510 0585
39	1.2147 2063	89	1.5587 6087
40	1.2207 9424	90	1.5665 5468
41	1.2268 9821	91	1.5743 8745
42	1.2330 3270	92	1.5822 5939
43	1.2391 9786	93	1.5901 7069
44	1.2453 9385	94	1.5981 2154
45	1.2516 2082	95	1.6061 1215
46	1.2578 7892	96	1.6141 4271
47	1.2641 6832	97	1.6222 1342
48	1.2704 8916	98	1.6303 2449
49	1.2768 4161	99	1.6384 7611
50	1.2832 2581	100	1.6466 6849

$i = 1\%$

n	$(1 + i)^n$	n	$(1 + i)^n$
1	1.0100 0000	51	1.6610 7814
2	1.0201 0000	52	1.6776 8892
3	1.0303 0100	53	1.6944 6581
4	1.0406 0401	54	1.7114 1047
5	1.0510 1005	55	1.7285 2457
6	1.0615 2015	56	1.7458 0982
7	1.0721 3535	57	1.7632 6792
8	1.0828 5671	58	1.7809 0060
9	1.0936 8527	59	1.7987 0960
10	1.1046 2213	60	1.8166 9670
11	1.1156 6835	61	1.8348 6367
12	1.1268 2503	62	1.8532 1230
13	1.1380 9328	63	1.8717 4443
14	1.1494 7421	64	1.8904 6187
15	1.1609 6896	65	1.9093 6649
16	1.1725 7864	66	1.9284 6015
17	1.1843 0443	67	1.9477 4475
18	1.1961 4748	68	1.9672 2220
19	1.2081 0895	69	1.9868 9442
20	1.2201 9004	70	2.0067 6337
21	1.2323 9194	71	2.0268 3100
22	1.2447 1586	72	2.0470 9931
23	1.2571 6302	73	2.0675 7031
24	1.2697 3465	74	2.0882 4601
25	1.2824 3200	75	2.1091 2847
26	1.2952 5631	76	2.1302 1975
27	1.3082 0888	77	2.1515 2195
28	1.3212 9097	78	2.1730 3717
29	1.3345 0388	79	2.1947 6754
30	1.3478 4892	80	2.2167 1522
31	1.3613 2740	81	2.2388 8237
32	1.3749 4068	82	2.2612 7119
33	1.3886 9009	83	2.2838 8390
34	1.4025 7699	84	2.3067 2274
35	1.4166 0276	85	2.3297 8997
36	1.4307 6878	86	2.3530 8787
37	1.4450 7647	87	2.3766 1875
38	1.4595 2724	88	2.4003 8494
39	1.4741 2251	89	2.4243 8879
40	1.4888 6373	90	2.4486 3267
41	1.5037 5237	91	2.4731 1900
42	1.5187 8989	92	2.4978 5019
43	1.5339 7779	93	2.5228 2869
44	1.5493 1757	94	2.5480 5698
45	1.5648 1075	95	2.5735 3755
46	1.5804 5885	96	2.5992 7293
47	1.5962 6344	97	2.6252 6565
48	1.6122 2608	98	2.6515 1831
49	1.6283 4834	99	2.6780 3349
50	1.6446 3182	100	2.7048 1383

$i = 1\frac{1}{2}\%$

n	$(1 + i)^n$	n	$(1 + i)^n$
1	1.0150 0000	51	2.1368 2106
2	1.0302 2500	52	2.1688 7337
3	1.0456 7838	53	2.2014 0647
4	1.0613 6355	54	2.2344 2757
5	1.0772 8400	55	2.2679 4398
6	1.0934 4326	56	2.3019 6314
7	1.1098 4491	57	2.3364 9259
8	1.1264 9259	58	2.3715 3998
9	1.1433 8998	59	2.4071 1308
10	1.1605 4083	60	2.4432 1978
11	1.1779 4894	61	2.4798 6807
12	1.1956 1817	62	2.5170 6609
13	1.2135 5244	63	2.5548 2208
14	1.2317 5573	64	2.5931 4442
15	1.2502 3207	65	2.6320 4158
16	1.2689 8555	66	2.6715 2221
17	1.2880 2033	67	2.7115 9504
18	1.3073 4064	68	2.7522 6896
19	1.3269 5075	69	2.7935 5300
20	1.3468 5501	70	2.8354 5629
21	1.3670 5783	71	2.8779 8814
22	1.3875 6370	72	2.9211 5796
23	1.4083 7715	73	2.9649 7533
24	1.4295 0281	74	3.0094 4996
25	1.4509 4535	75	3.0545 9171
26	1.4727 0953	76	3.1004 1059
27	1.4948 0018	77	3.1469 1674
28	1.5172 2218	78	3.1941 2050
29	1.5399 8051	79	3.2420 3230
30	1.5630 8022	80	3.2906 6279
31	1.5865 2642	81	3.3400 2273
32	1.6103 2432	82	3.3901 2307
33	1.6344 7918	83	3.4409 7492
34	1.6589 9637	84	3.4925 8954
35	1.6838 8132	85	3.5449 7838
36	1.7091 3954	86	3.5981 5306
37	1.7347 7663	87	3.6521 2535
38	1.7607 9828	88	3.7069 0723
39	1.7872 1025	89	3.7625 1084
40	1.8140 1841	90	3.8189 4851
41	1.8412 2868	91	3.8762 3273
42	1.8688 4712	92	3.9343 7622
43	1.8968 7982	93	3.9933 9187
44	1.9253 3302	94	4.0532 9275
45	1.9542 1301	95	4.1140 9214
46	1.9835 2621	96	4.1758 0352
47	2.0132 7910	97	4.2384 4057
48	2.0434 7829	98	4.3020 1718
49	2.0741 3046	99	4.3665 4744
50	2.1052 4242	100	4.4320 4565

	$i = 2\%$				$i = 2\frac{1}{2}\%$				$i = 3\%$
n	$(1 + i)^n$	n	$(1 + i)^n$	n	$(1 + i)^n$	n	$(1 + i)^n$	n	$(1 + i)^n$
1	1.0200 0000	51	2.7454 1979	1	1.0250 0000	51	3.5230 3644	1	1.0300 0000
2	1.0404 0000	52	2.8003 2819	2	1.0506 2500	52	3.6111 1235	2	1.0609 0000
3	1.0612 0800	53	2.8563 3475	3	1.0768 9063	53	3.7013 9016	3	1.0927 2700
4	1.0824 3216	54	2.9134 6144	4	1.1038 1289	54	3.7939 2491	4	1.1255 0881
5	1.1040 8080	55	2.9717 3067	5	1.1314 0821	55	3.8887 7303	5	1.1592 7407
6	1.1261 6242	56	3.0311 6529	6	1.1596 9342	56	3.9859 9236	6	1.1940 5230
7	1.1486 8567	57	3.0917 8859	7	1.1886 8575	57	4.0856 4217	7	1.2298 7387
8	1.1716 5938	58	3.1536 2436	8	1.2184 0290	58	4.1877 8322	8	1.2667 7008
9	1.1950 9257	59	3.2166 9685	9	1.2488 6297	59	4.2924 7780	9	1.3047 7318
10	1.2189 9442	60	3.2810 3079	10	1.2800 8454	60	4.3997 8975	10	1.3439 1638
11	1.2433 7431	61	3.3466 5140	11	1.3120 8666	61	4.5097 8449	11	1.3842 3387
12	1.2682 4179	62	3.4135 8443	12	1.3448 8882	62	4.6225 2910	12	1.4257 6089
13	1.2936 0663	63	3.4818 5612	13	1.3785 1104	63	4.7380 9233	13	1.4685 3371
14	1.3194 7876	64	3.5514 9324	14	1.4129 7382	64	4.8565 4464	14	1.5125 8972
15	1.3458 6834	65	3.6225 2311	15	1.4482 9817	65	4.9779 5826	15	1.5579 6742
16	1.3727 8571	66	3.6949 7357	16	1.4845 0562	66	5.1024 0721	16	1.6047 0644
17	1.4002 4142	67	3.7688 7304	17	1.5216 1826	67	5.2299 6739	17	1.6528 4763
18	1.4282 4625	68	3.8442 5050	18	1.5596 5872	68	5.3607 1658	18	1.7024 3306
19	1.4568 1117	69	3.9211 3551	19	1.5986 5019	69	5.4947 3449	19	1.7535 0605
20	1.4859 4740	70	3.9995 5822	20	1.6386 1644	70	5.6321 0286	20	1.8061 1123
21	1.5156 6634	71	4.0795 4939	21	1.6795 8185	71	5.7729 0543	21	1.8602 9457
22	1.5459 7967	72	4.1611 4038	22	1.7215 7140	72	5.9172 2806	22	1.9161 0341
23	1.5768 9926	73	4.2443 6318	23	1.7646 1068	73	6.0651 5876	23	1.9735 8651
24	1.6084 3725	74	4.3292 5045	24	1.8087 2595	74	6.2167 8773	24	2.0327 9411
25	1.6406 0599	75	4.4158 3546	25	1.8539 4410	75	6.3722 0743	25	2.0937 7793
26	1.6734 1811	76	4.5041 5216	26	1.9002 9270	76	6.5315 1261	26	2.1565 9127
27	1.7068 8648	77	4.5942 3521	27	1.9478 0002	77	6.6948 0043	27	2.2212 8901
28	1.7410 2421	78	4.6861 1991	28	1.9964 9502	78	6.8621 7044	28	2.2879 2768
29	1.7758 4469	79	4.7798 4231	29	2.0464 0739	79	7.0337 2470	29	2.3565 6551
30	1.8113 6158	80	4.8754 3916	30	2.0975 6758	80	7.2095 6782	30	2.4272 6247
31	1.8475 8882	81	4.9729 4794	31	2.1500 0677	81	7.3898 0701	31	2.5000 8035
32	1.8845 4059	82	5.0724 0690	32	2.2037 5694	82	7.5745 5219	32	2.5750 8276
33	1.9222 3140	83	5.1738 5504	33	2.2588 5086	83	7.7639 1599	33	2.6523 3524
34	1.9606 7603	84	5.2773 3214	34	2.3153 2213	84	7.9580 1389	34	2.7319 0530
35	1.9998 8955	85	5.3828 7878	35	2.3732 0519	85	8.1569 6424	35	2.8138 6245
36	2.0398 8734	86	5.4905 3636	36	2.4325 3532	86	8.3608 8834	36	2.8982 7833
37	2.0806 8509	87	5.6003 4708	37	2.4933 4870	87	8.5699 1055	37	2.9852 2668
38	2.1222 9879	88	5.7123 5402	38	2.5556 8242	88	8.7841 5832	38	3.0747 8348
39	2.1647 4477	89	5.8266 0110	39	2.6195 7448	89	9.0037 6228	39	3.1670 2698
40	2.2080 3966	90	5.9431 3313	40	2.6850 6384	90	9.2288 5633	40	3.2620 3779
41	2.2522 0046	91	6.0619 9579	41	2.7521 9043	91	9.4595 7774	41	3.3598 9893
42	2.2972 4447	92	6.1832 3570	42	2.8209 9520	92	9.6960 6718	42	3.4606 9589
43	2.3431 8936	93	6.3069 0042	43	2.8915 2008	93	9.9384 6886	43	3.5645 1677
44	2.3900 5314	94	6.4330 3843	44	2.9638 0808	94	10.1869 3058	44	3.6714 5227
45	2.4378 5421	95	6.5616 9920	45	3.0379 0328	95	10.4416 0385	45	3.7815 9584
46	2.4866 1129	96	6.6929 3318	46	3.1138 5086	96	10.7026 4395	46	3.8950 4372
47	2.5363 4352	97	6.8267 9184	47	3.1916 9713	97	10.9702 1004	47	4.0118 9503
48	2.5870 7039	98	6.9633 2768	48	3.2714 8956	98	11.2444 6530	48	4.1322 5188
49	2.6388 1179	99	7.1025 9423	49	3.3532 7680	99	11.5255 7693	49	4.2562 1944
50	2.6915 8803	100	7.2446 4612	50	3.4371 0872	100	11.8137 1635	50	4.3839 0602

	$i = 4\%$		$i = 5\%$		$i = 6\%$		$i = 7\%$		$i = 8\%$
n	$(1 + i)^n$	n	$(1 + i)^n$	n	$(1 + i)^n$	n	$(1 + i)^n$	n	$(1 + i)^n$
1	1.0400 0000	1	1.0500 0000	1	1.0600 0000	1	1.0700 0000	1	1.0800 0000
2	1.0816 0000	2	1.1025 0000	2	1.1236 0000	2	1.1449 0000	2	1.1664 0000
3	1.1248 6400	3	1.1576 2500	3	1.1910 1600	3	1.2250 4300	3	1.2597 1200
4	1.1698 5856	4	1.2155 0625	4	1.2624 7696	4	1.3107 9601	4	1.3604 8896
5	1.2166 5290	5	1.2762 8156	5	1.3382 2558	5	1.4025 5173	5	1.4693 2808
6	1.2653 1902	6	1.3400 9564	6	1.4185 1911	6	1.5007 3035	6	1.5868 7432
7	1.3159 3178	7	1.4071 0042	7	1.5036 3026	7	1.6057 8148	7	1.7138 2427
8	1.3685 6905	8	1.4774 5544	8	1.5938 4807	8	1.7181 8618	8	1.8509 3021
9	1.4233 1181	9	1.5513 2822	9	1.6894 7896	9	1.8384 5921	9	1.9990 0463
10	1.4802 4428	10	1.6288 9463	10	1.7908 4770	10	1.9671 5136	10	2.1589 2500
11	1.5394 5406	11	1.7103 3936	11	1.8982 9856	11	2.1048 5195	11	2.3316 3900
12	1.6010 3222	12	1.7958 5633	12	2.0121 9647	12	2.2521 9159	12	2.5181 7012
13	1.6650 7351	13	1.8856 4914	13	2.1329 2826	13	2.4098 4500	13	2.7196 2373
14	1.7316 7645	14	1.9799 3160	14	2.2609 0396	14	2.5785 3415	14	2.9371 9362
15	1.8009 4351	15	2.0789 2818	15	2.3965 5819	15	2.7590 3154	15	3.1721 6911
16	1.8729 8125	16	2.1828 7459	16	2.5403 5168	16	2.9521 6375	16	3.4259 4264
17	1.9479 0050	17	2.2920 1832	17	2.6927 7279	17	3.1588 1521	17	3.7000 1805
18	2.0258 1652	18	2.4066 1923	18	2.8543 3915	18	3.3799 3228	18	3.9960 1950
19	2.1068 4918	19	2.5269 5020	19	3.0255 9950	19	3.6165 2754	19	4.3157 0106
20	2.1911 2314	20	2.6532 9771	20	3.2071 3547	20	3.8696 8446	20	4.6609 5714
21	2.2787 6807	21	2.7859 6259	21	3.3995 6360	21	4.1405 6237	21	5.0338 3372
22	2.3699 1879	22	2.9252 6072	22	3.6035 3742	22	4.4304 0174	22	5.4365 4041
23	2.4647 1554	23	3.0715 2376	23	3.8197 4966	23	4.7405 2986	23	5.8714 6365
24	2.5633 0416	24	3.2250 9994	24	4.0489 3464	24	5.0723 6695	24	6.3411 8074
25	2.6658 3633	25	3.3863 5494	25	4.2918 7072	25	5.4274 3264	25	6.8484 7520
26	2.7724 6978	26	3.5556 7269	26	4.5493 8296	26	5.8073 5292	26	7.3963 5321
27	2.8833 6858	27	3.7334 5632	27	4.8223 4594	27	6.2138 6763	27	7.9880 6147
28	2.9987 0332	28	3.9201 2914	28	5.1116 8670	28	6.6488 3836	28	8.6271 0639
29	3.1186 5145	29	4.1161 3560	29	5.4183 8790	29	7.1142 5705	29	9.3172 7490
30	3.2433 9751	30	4.3219 4238	30	5.7434 9117	30	7.6122 5504	30	10.0626 5689
31	3.3731 3341	31	4.5380 3949	31	6.0881 0064	31	8.1451 1290	31	10.8676 6944
32	3.5080 5875	32	4.7649 4147	32	6.4533 8668	32	8.7152 7080	32	11.7370 8300
33	3.6483 8110	33	5.0031 8854	33	6.8405 8988	33	9.3253 3975	33	12.6760 4964
34	3.7943 1634	34	5.2533 4797	34	7.2510 2528	34	9.9781 1354	34	13.6901 3361
35	3.9460 8899	35	5.5160 1537	35	7.6860 8679	35	10.6765 8148	35	14.7853 4429
36	4.1039 3255	36	5.7918 1614	36	8.1472 5200	36	11.4239 4219	36	15.9681 7184
37	4.2680 8986	37	6.0814 0694	37	8.6360 8712	37	12.2236 1814	37	17.2456 2558
38	4.4388 1345	38	6.3854 7729	38	9.1542 5235	38	13.0792 7141	38	18.6252 7563
39	4.6163 6599	39	6.7047 5115	39	9.7035 0749	39	13.9948 2041	39	20.1152 9768
40	4.8010 2063	40	7.0399 8871	40	10.2857 1794	40	14.9744 5784	40	21.7245 2150
41	4.9930 6145	41	7.3919 8815	41	10.9028 6101	41	16.0226 6989	41	23.4624 8322
42	5.1927 8391	42	7.7615 8756	42	11.5570 3267	42	17.1442 5678	42	25.3394 8187
43	5.4004 9527	43	8.1496 6693	43	12.2504 5463	43	18.3443 5475	43	27.3666 4042
44	5.6165 1508	44	8.5571 5028	44	12.9854 8191	44	19.6284 5959	44	29.5559 7166
45	5.8411 7568	45	8.9850 0779	45	13.7646 1083	45	21.0024 5176	45	31.9204 4939
46	6.0748 2271	46	9.4342 5818	46	14.5904 8748	46	22.4726 2338	46	34.4740 8534
47	6.3178 1562	47	9.9059 7109	47	15.4659 1673	47	24.0457 0702	47	37.2320 1217
48	6.5705 2824	48	10.4012 6965	48	16.3938 7173	48	25.7289 0651	48	40.2105 7314
49	6.8333 4937	49	10.9213 3313	49	17.3775 0403	49	27.5299 2997	49	43.4274 1899
50	7.1066 8335	50	11.4673 9979	50	18.4201 5427	50	29.4570 2506	50	46.9016 1251

TABLE V Trigonometric Functions of Radians and Real Numbers[a]

t	$\sin t$	$\cos t$	$\tan t$	$\cot t$	$\sec t$	$\csc t$
.00	.0000	1.0000	.0000	—	1.000	—
.01	.0100	1.0000	.0100	99.997	1.000	100.00
.02	.0200	.9998	.0200	49.993	1.000	50.00
.03	.0300	.9996	.0300	33.323	1.000	33.34
.04	.0400	.9992	.0400	24.987	1.001	25.01
.05	.0500	.9988	.0500	19.983	1.001	20.01
.06	.0600	.9982	.0601	16.647	1.002	16.68
.07	.0699	.9976	.0701	14.262	1.002	14.30
.08	.0799	.9968	.0802	12.473	1.003	12.51
.09	.0899	.9960	.0902	11.081	1.004	11.13
.10	.0998	.9950	.1003	9.967	1.005	10.02
.11	.1098	.9940	.1104	9.054	1.006	9.109
.12	.1197	.9928	.1206	8.293	1.007	8.353
.13	.1296	.9916	.1307	7.649	1.009	7.714
.14	.1395	.9902	.1409	7.096	1.010	7.166
.15	.1494	.9888	.1511	6.617	1.011	6.692
.16	.1593	.9872	.1614	6.197	1.013	6.277
.17	.1692	.9856	.1717	5.826	1.015	5.911
.18	.1790	.9838	.1820	5.495	1.016	5.586
.19	.1889	.9820	.1923	5.200	1.018	5.295
.20	.1987	.9801	.2027	4.933	1.020	5.033
.21	.2085	.9780	.2131	4.692	1.022	4.797
.22	.2182	.9759	.2236	4.472	1.025	4.582
.23	.2280	.9737	.2341	4.271	1.027	4.386
.24	.2377	.9713	.2447	4.086	1.030	4.207
.25	.2474	.9689	.2553	3.916	1.032	4.042
.26	.2571	.9664	.2660	3.759	1.035	3.890
.27	.2667	.9638	.2768	3.613	1.038	3.749
.28	.2764	.9611	.2876	3.478	1.041	3.619
.29	.2860	.9582	.2984	3.351	1.044	3.497
.30	.2955	.9553	.3093	3.233	1.047	3.384
.31	.3051	.9523	.3203	3.122	1.050	3.278
.32	.3146	.9492	.3314	3.018	1.053	3.179
.33	.3240	.9460	.3425	2.920	1.057	3.086
.34	.3335	.9428	.3537	2.827	1.061	2.999
.35	.3429	.9394	.3650	2.740	1.065	2.916
.36	.3523	.9359	.3764	2.657	1.068	2.839
.37	.3616	.9323	.3879	2.578	1.073	2.765
.38	.3709	.9287	.3994	2.504	1.077	2.696
.39	.3802	.9249	.4111	2.433	1.081	2.630

[a] Reprinted with permission of the publisher from *Fundamentals of Algebra and Trigonometry*, Fourth Edition, by Earl W. Swokowski, copyright © 1978, Prindle, Weber & Schmidt.

TABLE V (*continued*)   **557**

t	sin t	cos t	tan t	cot t	sec t	csc t
.40	.3894	.9211	.4228	2.365	1.086	2.568
.41	.3986	.9171	.4346	2.301	1.090	2.509
.42	.4078	.9131	.4466	2.239	1.095	2.452
.43	.4169	.9090	.4586	2.180	1.100	2.399
.44	.4259	.9048	.4708	2.124	1.105	2.348
.45	.4350	.9004	.4831	2.070	1.111	2.299
.46	.4439	.8961	.4954	2.018	1.116	2.253
.47	.4529	.8916	.5080	1.969	1.122	2.208
.48	.4618	.8870	.5206	1.921	1.127	2.166
.49	.4706	.8823	.5334	1.875	1.133	2.125
.50	.4794	.8776	.5463	1.830	1.139	2.086
.51	.4882	.8727	.5594	1.788	1.146	2.048
.52	.4969	.8678	.5726	1.747	1.152	2.013
.53	.5055	.8628	.5859	1.707	1.159	1.987
.54	.5141	.8577	.5994	1.668	1.166	1.945
.55	.5227	.8525	.6131	1.631	1.173	1.913
.56	.5312	.8473	.6269	1.595	1.180	1.883
.57	.5396	.8419	.6410	1.560	1.188	1.853
.58	.5480	.8365	.6552	1.526	1.196	1.825
.59	.5564	.8309	.6696	1.494	1.203	1.797
.60	.5646	.8253	.6841	1.462	1.212	1.771
.61	.5729	.8196	.6989	1.431	1.220	1.746
.62	.5810	.8139	.7139	1.401	1.229	1.721
.63	.5891	.8080	.7291	1.372	1.238	1.697
.64	.5972	.8021	.7445	1.343	1.247	1.674
.65	.6052	.7961	.7602	1.315	1.256	1.652
.66	.6131	.7900	.7761	1.288	1.266	1.631
.67	.6210	.7838	.7923	1.262	1.276	1.610
.68	.6288	.7776	.8087	1.237	1.286	1.590
.69	.6365	.7712	.8253	1.212	1.297	1.571
.70	.6442	.7648	.8423	1.187	1.307	1.552
.71	.6518	.7584	.8595	1.163	1.319	1.534
.72	.6594	.7518	.8771	1.140	1.330	1.517
.73	.6669	.7452	.8949	1.117	1.342	1.500
.74	.6743	.7385	.9131	1.095	1.354	1.483
.75	.6816	.7317	.9316	1.073	1.367	1.467
.76	.6889	.7248	.9505	1.052	1.380	1.452
.77	.6961	.7179	.9697	1.031	1.393	1.437
.78	.7033	.7109	.9893	1.011	1.407	1.422
.79	.7104	.7038	1.009	.9908	1.421	1.408

TABLE V (*continued*)

t	$\sin t$	$\cos t$	$\tan t$	$\cot t$	$\sec t$	$\csc t$
.80	.7174	.6967	1.030	.9712	1.435	1.394
.81	.7243	.6895	1.050	.9520	1.450	1.381
.82	.7311	.6822	1.072	.9331	1.466	1.368
.83	.7379	.6749	1.093	.9146	1.482	1.355
.84	.7446	.6675	1.116	.8964	1.498	1.343
.85	.7513	.6600	1.138	.8785	1.515	1.331
.86	.7578	.6524	1.162	.8609	1.533	1.320
.87	.7643	.6448	1.185	.8437	1.551	1.308
.88	.7707	.6372	1.210	.8267	1.569	1.297
.89	.7771	.6294	1.235	.8100	1.589	1.287
.90	.7833	.6216	1.260	.7936	1.609	1.277
.91	.7895	.6137	1.286	.7774	1.629	1.267
.92	.7956	.6058	1.313	.7615	1.651	1.257
.93	.8016	.5978	1.341	.7458	1.673	1.247
.94	.8076	.5898	1.369	.7303	1.696	1.238
.95	.8134	.5817	1.398	.7151	1.719	1.229
.96	.8192	.5735	1.428	.7001	1.744	1.221
.97	.8249	.5653	1.459	.6853	1.769	1.212
.98	.8305	.5570	1.491	.6707	1.795	1.204
.99	.8360	.5487	1.524	.6563	1.823	1.196
1.00	.8415	.5403	1.557	.6421	1.851	1.188
1.01	.8468	.5319	1.592	.6281	1.880	1.181
1.02	.8521	.5234	1.628	.6142	1.911	1.174
1.03	.8573	.5148	1.665	.6005	1.942	1.166
1.04	.8624	.5062	1.704	.5870	1.975	1.160
1.05	.8674	.4976	1.743	.5736	2.010	1.153
1.06	.8724	.4889	1.784	.5604	2.046	1.146
1.07	.8772	.4801	1.827	.5473	2.083	1.140
1.08	.8820	.4713	1.871	.5344	2.122	1.134
1.09	.8866	.4625	1.917	.5216	2.162	1.128
1.10	.8912	.4536	1.965	.5090	2.205	1.122
1.11	.8957	.4447	2.014	.4964	2.249	1.116
1.12	.9001	.4357	2.066	.4840	2.295	1.111
1.13	.9044	.4267	2.120	.4718	2.344	1.106
1.14	.9086	.4176	2.176	.4596	2.395	1.101
1.15	.9128	.4085	2.234	.4475	2.448	1.096
1.16	.9168	.3993	2.296	.4356	2.504	1.091
1.17	.9208	.3902	2.360	.4237	2.563	1.086
1.18	.9246	.3809	2.427	.4120	2.625	1.082
1.19	.9284	.3717	2.498	.4003	2.691	1.077

TABLE V (*continued*) 559

t	$\sin t$	$\cos t$	$\tan t$	$\cot t$	$\sec t$	$\csc t$
1.20	.9320	.3624	2.572	.3888	2.760	1.073
1.21	.9356	.3530	2.650	.3773	2.833	1.069
1.22	.9391	.3436	2.733	.3659	2.910	1.065
1.23	.9425	.3342	2.820	.3546	2.992	1.061
1.24	.9458	.3248	2.912	.3434	3.079	1.057
1.25	.9490	.3153	3.010	.3323	3.171	1.054
1.26	.9521	.3058	3.113	.3212	3.270	1.050
1.27	.9551	.2963	3.224	.3102	3.375	1.047
1.28	.9580	.2867	3.341	.2993	3.488	1.044
1.29	.9608	.2771	3.467	.2884	3.609	1.041
1.30	.9636	.2675	3.602	.2776	3.738	1.038
1.31	.9662	.2579	3.747	.2669	3.878	1.035
1.32	.9687	.2482	3.903	.2562	4.029	1.032
1.33	.9711	.2385	4.072	.2456	4.193	1.030
1.34	.9735	.2288	4.256	.2350	4.372	1.027
1.35	.9757	.2190	4.455	.2245	4.566	1.025
1.36	.9779	.2092	4.673	.2140	4.779	1.023
1.37	.9799	.1994	4.913	.2035	5.014	1.021
1.38	.9819	.1896	5.177	.1913	5.273	1.018
1.39	.9837	.1798	5.471	.1828	5.561	1.017
1.40	.9854	.1700	5.798	.1725	5.883	1.015
1.41	.9871	.1601	6.165	.1622	6.246	1.013
1.42	.9887	.1502	6.581	.1519	6.657	1.011
1.43	.9901	.1403	7.055	.1417	7.126	1.010
1.44	.9915	.1304	7.602	.1315	7.667	1.009
1.45	.9927	.1205	8.238	.1214	8.299	1.007
1.46	.9939	.1106	8.989	.1113	9.044	1.006
1.47	.9949	.1006	9.887	.1011	9.938	1.005
1.48	.9959	.0907	10.983	.0910	11.029	1.004
1.49	.9967	.0807	12.350	.0810	12.390	1.003
1.50	.9975	.0707	14.101	.0709	14.137	1.003
1.51	.9982	.0608	16.428	.0609	16.458	1.002
1.52	.9987	.0508	19.670	.0508	19.695	1.001
1.53	.9992	.0408	24.498	.0408	24.519	1.001
1.54	.9995	.0308	32.461	.0308	32.476	1.000
1.55	.9998	.0208	48.078	.0208	48.089	1.000
1.56	.9999	.0108	92.620	.0108	92.626	1.000
1.57	1.0000	.0008	1255.8	.0008	1255.8	1.000

TABLE VI Trigonometric Functions of Angles in Degrees[a]

t degrees	sin t	cos t	tan t	cot t	sec t	csc t	
0°00′	.0000	1.0000	.0000	—	1.000	—	**90°00′**
10	.0029	1.0000	.0029	343.8	1.000	343.8	50
20	.0058	1.0000	.0058	171.9	1.000	171.9	40
30	.0087	1.0000	.0087	114.6	1.000	114.6	30
40	.0116	.9999	.0116	85.94	1.000	85.95	20
50	.0145	.9999	.0145	68.75	1.000	68.76	10
1°00′	.0175	.9998	.0175	57.29	1.000	57.30	**89°00′**
10	.0204	.9998	.0204	49.10	1.000	49.11	50
20	.0233	.9997	.0233	42.96	1.000	42.98	40
30	.0262	.9997	.0262	38.19	1.000	38.20	30
40	.0291	.9996	.0291	34.37	1.000	34.38	20
50	.0320	.9995	.0320	31.24	1.001	31.26	10
2°00′	.0349	.9994	.0349	28.64	1.001	28.65	**88°00′**
10	.0378	.9993	.0378	26.43	1.001	26.45	50
20	.0407	.9992	.0407	24.54	1.001	24.56	40
30	.0436	.9990	.0437	22.90	1.001	22.93	30
40	.0465	.9989	.0466	21.47	1.001	21.49	20
50	.0494	.9988	.0495	20.21	1.001	20.23	10
3°00′	.0523	.9986	.0524	19.08	1.001	19.11	**87°00′**
10	.0552	.9985	.0553	18.07	1.002	18.10	50
20	.0581	.9983	.0582	17.17	1.002	17.20	40
30	.0610	.9981	.0612	16.35	1.002	16.38	30
40	.0640	.9980	.0641	15.60	1.002	15.64	20
50	.0669	.9978	.0670	14.92	1.002	14.96	10
4°00′	.0698	.9976	.0699	14.30	1.002	14.34	**86°00′**
10	.0727	.9974	.0729	13.73	1.003	13.76	50
20	.0756	.9971	.0758	13.20	1.003	13.23	40
30	.0785	.9969	.0787	12.71	1.003	12.75	30
40	.0814	.9967	.0816	12.25	1.003	12.29	20
50	.0843	.9964	.0846	11.83	1.004	11.87	10
5°00′	.0872	.9962	.0875	11.43	1.004	11.47	**85°00′**
10	.0901	.9959	.0904	11.06	1.004	11.10	50
20	.0929	.9957	.0934	10.71	1.004	10.76	40
30	.0958	.9954	.0963	10.39	1.005	10.43	30
40	.0987	.9951	.0992	10.08	1.005	10.13	20
50	.1016	.9948	.1022	9.788	1.005	9.839	10
6°00′	.1045	.9945	.1051	9.514	1.006	9.567	**84°00′**
10	.1074	.9942	.1080	9.255	1.006	9.309	50
20	.1103	.9939	.1110	9.010	1.006	9.065	40
30	.1132	.9936	.1139	8.777	1.006	8.834	30
40	.1161	.9932	.1169	8.556	1.007	8.614	20
50	.1190	.9929	.1198	8.345	1.007	8.405	10
7°00′	.1219	.9925	.1228	8.144	1.008	8.206	**83°00′**
	cos t	sin t	cot t	tan t	csc t	sec t	t degrees

[a] Reprinted with permission of the publisher from *Fundamentals of Algebra and Trigonometry*, Fourth Edition, by Earl W. Swokowski, copyright © 1978, Prindle, Weber & Schmidt.

TABLE VI (*continued*) 561

t degrees	sin t	cos t	tan t	cot t	sec t	csc t	
7°00′	.1219	.9925	.1228	8.144	1.008	8.206	**83°00′**
10	.1248	.9922	.1257	7.953	1.008	8.016	50
20	.1276	.9918	.1287	7.770	1.008	7.834	40
30	.1305	.9914	.1317	7.596	1.009	7.661	30
40	.1334	.9911	.1346	7.429	1.009	7.496	20
50	.1363	.9907	.1376	7.269	1.009	7.337	10
8°00′	.1392	.9903	.1405	7.115	1.010	7.185	**82°00′**
10	.1421	.9899	.1435	6.968	1.010	7.040	50
20	.1449	.9894	.1465	6.827	1.011	6.900	40
30	.1478	.9890	.1495	6.691	1.011	6.765	30
40	.1507	.9886	.1524	6.561	1.012	6.636	20
50	.1536	.9881	.1554	6.435	1.012	6.512	10
9°00′	.1564	.9877	.1584	6.314	1.012	6.392	**81°00′**
10	.1593	.9872	.1614	6.197	1.013	6.277	50
20	.1622	.9868	.1644	6.084	1.013	6.166	40
30	.1650	.9863	.1673	5.976	1.014	6.059	30
40	.1679	.9858	.1703	5.871	1.014	5.955	20
50	.1708	.9853	.1733	5.769	1.015	5.855	10
10°00′	.1736	.9848	.1763	5.671	1.015	5.759	**80°00′**
10	.1765	.9843	.1793	5.576	1.016	5.665	50
20	.1794	.9838	.1823	5.485	1.016	5.575	40
30	.1822	.9833	.1853	5.396	1.017	5.487	30
40	.1851	.9827	.1883	5.309	1.018	5.403	20
50	.1880	.9822	.1914	5.226	1.018	5.320	10
11°00′	.1908	.9816	.1944	5.145	1.019	5.241	**79°00′**
10	.1937	.9811	.1974	5.066	1.019	5.164	50
20	.1965	.9805	.2004	4.989	1.020	5.089	40
30	.1994	.9799	.2035	4.915	1.020	5.016	30
40	.2022	.9793	.2065	4.843	1.021	4.945	20
50	.2051	.9787	.2095	4.773	1.022	4.876	10
12°00′	.2079	.9781	.2126	4.705	1.022	4.810	**78°00′**
10	.2108	.9775	.2156	4.638	1.023	4.745	50
20	.2136	.9769	.2186	4.574	1.024	4.682	40
30	.2164	.9763	.2217	4.511	1.024	4.620	30
40	.2193	.9757	.2247	4.449	1.025	4.560	20
50	.2221	.9750	.2278	4.390	1.026	4.502	10
13°00′	.2250	.9744	.2309	4.331	1.026	4.445	**77°00′**
10	.2278	.9737	.2339	4.275	1.027	4.390	50
20	.2306	.9730	.2370	4.219	1.028	4.336	40
30	.2334	.9724	.2401	4.165	1.028	4.284	30
40	.2363	.9717	.2432	4.113	1.029	4.232	20
50	.2391	.9710	.2462	4.061	1.030	4.182	10
14°00′	.2419	.9703	.2493	4.011	1.031	4.134	**76°00′**
	cos t	sin t	cot t	tan t	csc t	sec t	t degrees

TABLE VI (*continued*)

t degrees	sin t	cos t	tan t	cot t	sec t	csc t	
14°00′	.2419	.9703	.2493	4.011	1.031	4.134	**76°00′**
10	.2447	.9696	.2524	3.962	1.031	4.086	50
20	.2476	.9689	.2555	3.914	1.032	4.039	40
30	.2504	.9681	.2586	3.867	1.033	3.994	30
40	.2532	.9674	.2617	3.821	1.034	3.950	20
50	.2560	.9667	.2648	3.776	1.034	3.906	10
15°00′	.2588	.9659	.2679	3.732	1.035	3.864	**75°00′**
10	.2616	.9652	.2711	3.689	1.036	3.822	50
20	.2644	.9644	.2742	3.647	1.037	3.782	40
30	.2672	.9636	.2773	3.606	1.038	3.742	30
40	.2700	.9628	.2805	3.566	1.039	3.703	20
50	.2728	.9621	.2836	3.526	1.039	3.665	10
16°00′	.2756	.9613	.2867	3.487	1.040	3.628	**74°00′**
10	.2784	.9605	.2899	3.450	1.041	3.592	50
20	.2812	.9596	.2931	3.412	1.042	3.556	40
30	.2840	.9588	.2962	3.376	1.043	3.521	30
40	.2868	.9580	.2994	3.340	1.044	3.487	20
50	.2896	.9572	.3026	3.305	1.045	3.453	10
17°00′	.2924	.9563	.3057	3.271	1.046	3.420	**73°00′**
10	.2952	.9555	.3089	3.237	1.047	3.388	50
20	.2979	.9546	.3121	3.204	1.048	3.356	40
30	.3007	.9537	.3153	3.172	1.049	3.326	30
40	.3035	.9528	.3185	3.140	1.049	3.295	20
50	.3062	.9520	.3217	3.108	1.050	3.265	10
18°00′	.3090	.9511	.3249	3.078	1.051	3.236	**72°00′**
10	.3118	.9502	.3281	3.047	1.052	3.207	50
20	.3145	.9492	.3314	3.018	1.053	3.179	40
30	.3173	.9483	.3346	2.989	1.054	3.152	30
40	.3201	.9474	.3378	2.960	1.056	3.124	20
50	.3228	.9465	.3411	2.932	1.057	3.098	10
19°00′	.3256	.9455	.3443	2.904	1.058	3.072	**71°00′**
10	.3283	.9446	.3476	2.877	1.059	3.046	50
20	.3311	.9436	.3508	2.850	1.060	3.021	40
30	.3338	.9426	.3541	2.824	1.061	2.996	30
40	.3365	.9417	.3574	2.798	1.062	2.971	20
50	.3393	.9407	.3607	2.773	1.063	2.947	10
20°00′	.3420	.9397	.3640	2.747	1.064	2.924	**70°00′**
10	.3448	.9387	.3673	2.723	1.065	2.901	50
20	.3475	.9377	.3706	2.699	1.066	2.878	40
30	.3502	.9367	.3739	2.675	1.068	2.855	30
40	.3529	.9356	.3772	2.651	1.069	2.833	20
50	.3557	.9346	.3805	2.628	1.070	2.812	10
21°00′	.3584	.9336	.3839	2.605	1.071	2.790	**69°00′**
	cos t	sin t	cot t	tan t	csc t	sec t	t degrees

TABLE VI (*continued*) **563**

t degrees	sin t	cos t	tan t	cot t	sec t	csc t	
21°00′	.3584	.9336	.3839	2.605	1.071	2.790	**69°00′**
10	.3611	.9325	.3872	2.583	1.072	2.769	50
20	.3638	.9315	.3906	2.560	1.074	2.749	40
30	.3665	.9304	.3939	2.539	1.075	2.729	30
40	.3692	.9293	.3973	2.517	1.076	2.709	20
50	.3719	.9283	.4006	2.496	1.077	2.689	10
22°00′	.3746	.9272	.4040	2.475	1.079	2.669	**68°00′**
10	.3773	.9261	.4074	2.455	1.080	2.650	50
20	.3800	.9250	.4108	2.434	1.081	2.632	40
30	.3827	.9239	.4142	2.414	1.082	2.613	30
40	.3854	.9228	.4176	2.394	1.084	2.596	20
50	.3881	.9216	.4210	2.375	1.085	2.577	10
23°00′	.3907	.9205	.4245	2.356	1.086	2.559	**67°00′**
10	.3934	.9194	.4279	2.337	1.088	2.542	50
20	.3961	.9182	.4314	2.318	1.089	2.525	40
30	.3987	.9171	.4348	2.300	1.090	2.508	30
40	.4014	.9159	.4383	2.282	1.092	2.491	20
50	.4041	.9147	.4417	2.264	1.093	2.475	10
24°00′	.4067	.9135	.4452	2.246	1.095	2.459	**66°00′**
10	.4094	.9124	.4487	2.229	1.096	2.443	50
20	.4120	.9112	.4522	2.211	1.097	2.427	40
30	.4147	.9100	.4557	2.194	1.099	2.411	30
40	.4173	.9088	.4592	2.177	1.100	2.396	20
50	.4200	.9075	.4628	2.161	1.102	2.381	10
25°00′	.4226	.9063	.4663	2.145	1.103	2.366	**65°00′**
10	.4253	.9051	.4699	2.128	1.105	2.352	50
20	.4279	.9038	.4734	2.112	1.106	2.337	40
30	.4305	.9026	.4770	2.097	1.108	2.323	30
40	.4331	.9013	.4806	2.081	1.109	2.309	20
50	.4358	.9001	.4841	2.066	1.111	2.295	10
26°00′	.4384	.8988	.4877	2.050	1.113	2.281	**64°00′**
10	.4410	.8975	.4913	2.035	1.114	2.268	50
20	.4436	.8962	.4950	2.020	1.116	2.254	40
30	.4462	.8949	.4986	2.006	1.117	2.241	30
40	.4488	.8936	.5022	1.991	1.119	2.228	20
50	.4514	.8923	.5059	1.977	1.121	2.215	10
27°00′	.4540	.8910	.5095	1.963	1.122	2.203	**63°00′**
10	.4566	.8897	.5132	1.949	1.124	2.190	50
20	.4592	.8884	.5169	1.935	1.126	2.178	40
30	.4617	.8870	.5206	1.921	1.127	2.166	30
40	.4643	.8857	.5243	1.907	1.129	2.154	20
50	.4669	.8843	.5280	1.894	1.131	2.142	10
28°00′	.4695	.8829	.5317	1.881	1.133	2.130	**62°00′**
	cos t	sin t	cot t	tan t	csc t	sec t	t degrees

TABLE VI (*continued*)

t degrees	sin t	cos t	tan t	cot t	sec t	csc t	
28°00′	.4695	.8829	.5317	1.881	1.133	2.130	**62°00′**
10	.4720	.8816	.5354	1.868	1.134	2.118	50
20	.4746	.8802	.5392	1.855	1.136	2.107	40
30	.4772	.8788	.5430	1.842	1.138	2.096	30
40	.4797	.8774	.5467	1.829	1.140	2.085	20
50	.4823	.8760	.5505	1.816	1.142	2.074	10
29°00′	.4848	.8746	.5543	1.804	1.143	2.063	**61°00′**
10	.4874	.8732	.5581	1.792	1.145	2.052	50
20	.4899	.8718	.5619	1.780	1.147	2.041	40
30	.4924	.8704	.5658	1.767	1.149	2.031	30
40	.4950	.8689	.5696	1.756	1.151	2.020	20
50	.4975	.8675	.5735	1.744	1.153	2.010	10
30°00′	.5000	.8660	.5774	1.732	1.155	2.000	**60°00′**
10	.5025	.8646	.5812	1.720	1.157	1.990	50
20	.5050	.8631	.5851	1.709	1.159	1.980	40
30	.5075	.8616	.5890	1.698	1.161	1.970	30
40	.5100	.8601	.5930	1.686	1.163	1.961	20
50	.5125	.8587	.5969	1.675	1.165	1.951	10
31°00′	.5150	.8572	.6009	1.664	1.167	1.942	**59°00′**
10	.5175	.8557	.6048	1.653	1.169	1.932	50
20	.5200	.8542	.6088	1.643	1.171	1.923	40
30	.5225	.8526	.6128	1.632	1.173	1.914	30
40	.5250	.8511	.6168	1.621	1.175	1.905	20
50	.5275	.8496	.6208	1.611	1.177	1.896	10
32°00′	.5299	.8480	.6249	1.600	1.179	1.887	**58°00′**
10	.5324	.8465	.6289	1.590	1.181	1.878	50
20	.5348	.8450	.6330	1.580	1.184	1.870	40
30	.5373	.8434	.6371	1.570	1.186	1.861	30
40	.5398	.8418	.6412	1.560	1.188	1.853	20
50	.5422	.8403	.6453	1.550	1.190	1.844	10
33°00′	.5446	.8387	.6494	1.540	1.192	1.836	**57°00′**
10	.5471	.8371	.6536	1.530	1.195	1.828	50
20	.5495	.8355	.6577	1.520	1.197	1.820	40
30	.5519	.8339	.6619	1.511	1.199	1.812	30
40	.5544	.8323	.6661	1.501	1.202	1.804	20
50	.5568	.8307	.6703	1.492	1.204	1.796	10
34°00′	.5592	.8290	.6745	1.483	1.206	1.788	**56°00′**
10	.5616	.8274	.6787	1.473	1.209	1.781	50
20	.5640	.8258	.6830	1.464	1.211	1.773	40
30	.5664	.8241	.6873	1.455	1.213	1.766	30
40	.5688	.8225	.6916	1.446	1.216	1.758	20
50	.5712	.8208	.6959	1.437	1.218	1.751	10
35°00′	.5736	.8192	.7002	1.428	1.221	1.743	**55°00′**
	cos t	sin t	cot t	tan t	csc t	sec t	t degrees

t degrees	$\sin t$	$\cos t$	$\tan t$	$\cot t$	$\sec t$	$\csc t$	
35°00′	.5736	.8192	.7002	1.428	1.221	1.743	**55°00′**
10	.5760	.8175	.7046	1.419	1.223	1.736	50
20	.5783	.8158	.7089	1.411	1.226	1.729	40
30	.5807	.8141	.7133	1.402	1.228	1.722	30
40	.5831	.8124	.7177	1.393	1.231	1.715	20
50	.5854	.8107	.7221	1.385	1.233	1.708	10
36°00′	.5878	.8090	.7265	1.376	1.236	1.701	**54°00′**
10	.5901	.8073	.7310	1.368	1.239	1.695	50
20	.5925	.8056	.7355	1.360	1.241	1.688	40
30	.5948	.8039	.7400	1.351	1.244	1.681	30
40	.5972	.8021	.7445	1.343	1.247	1.675	20
50	.5995	.8004	.7490	1.335	1.249	1.668	10
37°00′	.6018	.7986	.7536	1.327	1.252	1.662	**53°00′**
10	.6041	.7969	.7581	1.319	1.255	1.655	50
20	.6065	.7951	.7627	1.311	1.258	1.649	40
30	.6088	.7934	.7673	1.303	1.260	1.643	30
40	.6111	.7916	.7720	1.295	1.263	1.636	20
50	.6134	.7898	.7766	1.288	1.266	1.630	10
38°00′	.6157	.7880	.7813	1.280	1.269	1.624	**52°00′**
10	.6180	.7862	.7860	1.272	1.272	1.618	50
20	.6202	.7844	.7907	1.265	1.275	1.612	40
30	.6225	.7826	.7954	1.257	1.278	1.606	30
40	.6248	.7808	.8002	1.250	1.281	1.601	20
50	.6271	.7790	.8050	1.242	1.284	1.595	10
39°00′	.6293	.7771	.8098	1.235	1.287	1.589	**51°00′**
10	.6316	.7753	.8146	1.228	1.290	1.583	50
20	.6338	.7735	.8195	1.220	1.293	1.578	40
30	.6361	.7716	.8243	1.213	1.296	1.572	30
40	.6383	.7698	.8292	1.206	1.299	1.567	20
50	.6406	.7679	.8342	1.199	1.302	1.561	10
40°00′	.6428	.7660	.8391	1.192	1.305	1.556	**50°00′**
10	.6450	.7642	.8441	1.185	1.309	1.550	50
20	.6472	.7623	.8491	1.178	1.312	1.545	40
30	.6494	.7604	.8541	1.171	1.315	1.540	30
40	.6517	.7585	.8591	1.164	1.318	1.535	20
50	.6539	.7566	.8642	1.157	1.322	1.529	10
41°00′	.6561	.7547	.8693	1.150	1.325	1.524	**49°00′**
10	.6583	.7528	.8744	1.144	1.328	1.519	50
20	.6604	.7509	.8796	1.137	1.332	1.514	40
30	.6626	.7490	.8847	1.130	1.335	1.509	30
40	.6648	.7470	.8899	1.124	1.339	1.504	20
50	.6670	.7451	.8952	1.117	1.342	1.499	10
42°00′	.6691	.7431	.9004	1.111	1.346	1.494	**48°00′**
	$\cos t$	$\sin t$	$\cot t$	$\tan t$	$\csc t$	$\sec t$	t degrees

TABLE VI (*continued*)

t degrees	sin t	cos t	tan t	cot t	sec t	csc t	
42°00′	.6691	.7431	.9004	1.111	1.346	1.494	**48°00′**
10	.6713	.7412	.9057	1.104	1.349	1.490	50
20	.6734	.7392	.9110	1.098	1.353	1.485	40
30	.6756	.7373	.9163	1.091	1.356	1.480	30
40	.6777	.7353	.9217	1.085	1.360	1.476	20
50	.6799	.7333	.9271	1.079	1.364	1.471	10
43°00′	.6820	.7314	.9325	1.072	1.367	1.466	**47°00′**
10	.6841	.7294	.9380	1.066	1.371	1.462	50
20	.6862	.7274	.9435	1.060	1.375	1.457	40
30	.6884	.7254	.9490	1.054	1.379	1.453	30
40	.6905	.7234	.9545	1.048	1.382	1.448	20
50	.6926	.7214	.9601	1.042	1.386	1.444	10
44°00′	.6947	.7193	.9657	1.036	1.390	1.440	**46°00′**
10	.6967	.7173	.9713	1.030	1.394	1.435	50
20	.6988	.7153	.9770	1.024	1.398	1.431	40
30	.7009	.7133	.9827	1.018	1.402	1.427	30
40	.7030	.7112	.9884	1.012	1.406	1.423	20
50	.7050	.7092	.9942	1.006	1.410	1.418	10
45°00′	.7071	.7071	1.0000	1.0000	1.414	1.414	**45°00′**
	cos t	sin t	cot t	tan t	csc t	sec t	t degrees

ANSWERS TO ODD-NUMBERED EXERCISES AND TO REVIEW EXERCISES AND PROGRESS TESTS

CHAPTER 1
EXERCISE SET 1.1, page 11

1. F 3. F 5. T 7. T
9. T 11. F 13. F

15. *number line with points marked at 1 through 9 on scale 0–9*

17. *number line with points marked at 3,4,5,6 on scale 2–7*

19. $a \geq 0$ 21. $b \geq 5$ 23. $b \leq -4$
25. multiplication by negative number 27. multiplication by negative number
29. multiplication by positive number
31. 1 33. 4 35. 2 37. 1/5
39. 3 41. 2 43. 8/5
45. commutative (addition) 47. distributive
49. associative (addition) 51. closure (multiplication)
53. commutative (multiplication) 55. commutative, associative (multiplication)
57. multiplicative inverse 63. symmetric 65. substitution

EXERCISE SET 1.2, page 22

1. 13 3. (a) $2160 (b) $2080 (c) $2106.67
5. 9.37 7. 3/2 9. b^7 11. $-20y^9$
13. $-3x^4$ 15. c, d 17. 2; 3 19. 3/5; 4
21. cost of all purchases 23. $5x + 3$
25. $xy^2z - 5x^2yz + xy + yz - x + 5$
27. $-4y^5 - 2y^4 + 2y^3 - 5y^2 - 3y$
29. $4a^5 - 16a^4 + 14a^3 - 3a^2 - 14a + 15$
31. $3b^4 + 3ab^3 + 2b^3 - 7ab^2 + 2b^2 - 4ab - 6a$
33. $-260x + 13y + 17z$ 35. $5b(c + 5)$ 37. $-y^2(3 + 4y^3)$
39. $3x^2(1 + 2y - 3z)$ 41. $(x + 1)(x + 3)$ 43. $(x - 7)(x + 2)$
45. $(2x + 1)(x - 2)$ 47. $(5x + 1)(2x - 3)$ 49. $(3x + 2)(2x + 3)$
51. $4(y + 2)(x - 1)$ 53. $-(x + 2)^2(5x + 31)$ 55. $(y - 1/3)(y + 1/3)$
57. $(4 + 3xy)(4 - 3xy)$ 59. $(x^2 + y^2)(x + y)(x - y)$ 61. $(x + 3y)(x^2 - 3xy + 9y^2)$
63. $(3x - y)(9x^2 + 3xy + y^2)$ 65. $(a + 2)(a^2 - 2a + 4)$
67. $(\frac{1}{2}m - 2n)(\frac{1}{4}m^2 + mn + 4n^2)$ 69. $(x + y - 2)(x^2 + 2xy + y^2 + 2x + 2y + 4)$
71. $(2x^2 - 5y^2)(4x^4 + 10x^2y^2 + 25y^4)$

EXERCISE SET 1.3, page 35

1. y^{8n}
3. $-x^3/y^3$
5. 1
7. $(3/2)^n x^{2n} y^{3n}$
9. $4/3$
11. 3
13. 81
15. $a^9/3b^4$
17. $4a^{10}c^6/b^8$
19. $1/(a - 2b^2)$
21. $(b + a)/(b - a)$
25. 0.074
27. 8
29. $x^{5/36}$
31. $x^2 y^{12}$
33. $\sqrt[5]{1/16}$
35. $\sqrt[4]{a^3}$
37. $8^{3/4}$
39. $(-8)^{-2/5}$
41. not real
43. 5
45. $5/4$
47. $3, 4$
49. $4\sqrt{3}$
51. $3\sqrt[3]{2}$
53. $y^2\sqrt[3]{y}$
55. $2x^2\sqrt[4]{6}\sqrt{x}$
57. $\sqrt{5}/5$
59. $\sqrt{3y}/3y$
61. $3(3 - \sqrt{2})/7$
63. $-3(3\sqrt{a} - 1)/(9a - 1)$
65. $3 + 2\sqrt{2}$
67. $2 + \sqrt{6} + 3\sqrt{2} + 2\sqrt{3}$
69. 1
71. 1
73. $-1/2 + 0i$
75. $0 + 5i$
77. $3 - 7i$
79. $0.3 - 7\sqrt{2}i$
81. $x = 2/3, y = -8$
83. $x = -1, y = -9/2$
85. $3 + i$
87. $5 + i$
89. $-1 - i/2$
91. $5 + 0i$
93. 0
95. 3

EXERCISE SET 1.4, page 43

1. $-10/3$
3. -2
5. 5
7. $-7/2$
9. 1
11. $8/(5 - k), k \neq 5$
13. $(6 + k)/5$
15. I
17. C
19. $x < 4$

21. $x < -6$

23. $x \geq 5$

25. $a > -1$

27. $y < -1/2$

29. $x \geq 0$

31. $r < 2$

33. $x \geq 1$

35. $x > 5/3$

37. $(-\infty, 2]$
39. $[2, \infty)$
41. $(-\infty, 3/2)$
43. $(-12, \infty)$
45. $(-\infty, 9/2)$
47. $(-\infty, -6]$
49. $(-\infty, 3/2)$
51. $(-\infty, 7]$
53. $(-1/2, 5/4]$
55. $[-3, -2]$
57. $(-3, -1]$
59. $[-1, 2)$
61. 98
63. 5
65. 2924
67. $L \leq 20$

EXERCISE SET 1.5, page 47

1. 1, −5
3. 3, 1
5. 2, −4/3
7. 3/2, −3
9. 4, −2
11. $x < -4$ or $x > 2$
13. $x < -1/2$ or $x > 1$
15. $-1/3 < x < 1$

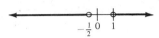
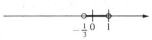

17. $(-\infty, -1], [7, \infty)$
19. $(-7/2, 9/2)$
21. no solution
23. $(-\infty, -7), (17, \infty)$
27. $|x - 100| \le 2;\ 98 \le x \le 102$

EXERCISE SET 1.6, page 51

1. $x < -3, x > -2$
3. $-1/2 < x < 1$
5. $x < 0, x > 2$
7. $-5 \le x < -3$
9. $-1/2 \le r < 3$
11. $s \le -2/3, s > 1/2$
13. $-2 < x < 2/3, x > 1$
15. $x < -3, x > 2$
17. $-1 \le x < 5/2$

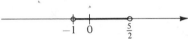

19. $-3/2 < r < 1/2$
21. $x < -1, x \ge 1$

23. $x \le -1, x \ge -1/3$
25. $y \le -2, 2 \le y \le 3$

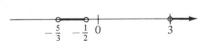

27. $-5/3 < x < -1/2, x > 3$

29. $x \le -1, x \ge 2$
31. $x \le -2, x \ge -3/2$
33. $0 \le x < 100$

REVIEW EXERCISES, page 53

1. T
2. F
3. F
4. F
5. $\pi + (-\pi) = 0$
6. $\pi \cdot 1/\pi = 1$
7. $\sqrt{4} = 2$
8. $a = 5, b = -5$

9.
10.

11.

12. -1

13. $3/2$

14. $51

15. c

16. $-0.5, 7$

17. $-7, 5$

18. $a^2b^2 - 3a^2b + 4b$

19. $2x^3 + 3x^2 - 2x$

20. $12x^3 + 12x^2 + 3x$

21. $2(x + 1)(x - 1)$

22. $(x + 5y)(x - 5y)$

23. $(2a + 3b)(a + 3)$

24. $(4x - 1)(x + 5)$

25. $(x + 1)(x - 1)(x^2 + 1)(x^4 + 1)$

26. $(3r^2 + 2s^2)(9r^4 - 6r^2s^2 + 4s^4)$

27. $b^9/8a^6$

28. 2

29. $1/x^4y^8$

30. x^3

31. $4\sqrt{5}$

32. $\sqrt{3}/3$

33. $(x - \sqrt{xy})/(x - y)$

34. $x = -2, y = 4$

35. $-i$

36. $8 - i$

37. $3 + 4i$

38. $17 + 6i$

39. $8/3$

40. 0

41. $10/3$

42. $k/2(2k + 1)$

43. F

44. F

45. $x \geq 1$

46. $-9/2 \leq x < 5/2$

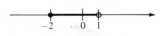

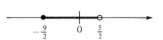

47. $(-\infty, 8)$

48. $(5/2, \infty)$

49. $[-9, \infty)$

50. $5/3, -3$

51. $x = -1, x = 3/2$

52. $x > 3$ or $x < -4$

53. $(1/5, 3/5)$

54. $(-\infty, -4/3], [8/3, \infty)$

55. $(-\infty, -5], [1, \infty)$

56. $(-\infty, -5), [-1/2, \infty)$

57. $(-2, -3/2), (3, \infty)$

58. additive inverse

59. distributive

60. commutative (addition)

61. multiplicative identity

PROGRESS TEST 1A, page 55

1. F

2. F

3. T

4. F

5.

6.

7. -1

8. 2

9. 25

10. $-7/3$

11. b

12. $-2.2, 5$

13. $14, 6$

14. $2xy + 3x + 4y + 1$

15. $3a^3 + 5a^2 + 3a + 10$

16. $4a^2b(2ab^4 - 3a^3b + 4)$

17. $(2 - 3x)(2 + 3x)$

18. $1/x^{17}$

19. y^{n+1}

20. -1

21. $4a^4/b^2$

22. $x > 2$

23. $-1 + 0i$

24. $16 - 11i$

25. $3/4$

26. $8/13$

27. T

28. $-2 \leq x < 1$

29. $(-\infty, \frac{15}{2}]$

30. [−4, 4]   31. 5/2, −2
32. −2 ≤ x ≤ 3   33. (−4/3, 2)

$$\xleftarrow{\hspace{1cm}} \bullet\!\!\!\rule[0.5ex]{2cm}{1pt}\!\!\!\bullet \xrightarrow{\hspace{1cm}}$$
$$\quad -2 \quad\; 0 \quad\;\; 3$$

34. (−∞, 1/2], [1, ∞)   35. [−2, 2/3], [1, ∞)

PROGRESS TEST 1B, page 56

1. T   2. F   3. F   4. T
5.    6.

7. 1   8. 7   9. 14   10. 2
11. a, d   12. 4, 5   13. 1.5, 10
14. $2s^2t^3 − 3s^2t^2 + 2s^2t + 3st^2 + st − s + 2t − 3$
15. $−3b^3 − 7b^2 + 10b + 12$ 16. $5r^3s^3(s − 8rt)$ 17. $(2x − 1)(x + 4)$
18. $4/x$ 19. b^{28} 20. x^6/y^9 21. −1
22. $x ≤ 2$ 23. $2 − \frac{3}{2}i$ 24. $4 − 7i$ 25. −1
26. $k^2/(k + 3)$ 27. F
28. $1 ≤ x ≤ 2$ 29. (−∞, 4/5]

$$\xrightarrow{\hspace{1cm}} \bullet\!\!\!\rule[0.5ex]{1cm}{1pt}\!\!\!\bullet \xrightarrow{\hspace{1cm}}$$
$$\quad\; 0 \quad\; 1 \quad\; 2$$

30. (3, ∞)   31. 8/3, −2
32. $x ≤ 2, x ≥ 6$ 33. (−∞, −1), (1/5, ∞)

$$\xleftarrow{\hspace{1cm}} \bullet\!\!\!\rule[0.5ex]{1cm}{1pt}\!\!\!\bullet\ \bullet \xrightarrow{\hspace{1cm}}$$
$$\quad\; 0 \quad\; 2 \quad\quad\; 6$$

34. (−∞, −5)   35. (−∞, −4), (2/3, 1)

CHAPTER 2
EXERCISE SET 2.1, page 67

1. 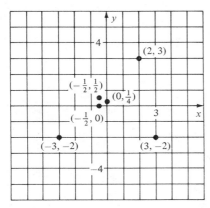   3. $3\sqrt{2}$

5. $4\sqrt{2}$

7. $\sqrt{1345}/6$

9. $\overline{BC} = \sqrt{37}$

11. $\overline{RS} = \sqrt{2}/2$

13. no

15. yes

21. $2\sqrt{10} + 7 + 5\sqrt{2} + \sqrt{37}$

25. any point satisfying $x^2 + y^2 - 10y - 6x + 29 = 0$

27. x-int.: -2
    y-int.: 4

29. x-int.: 0
    y-int.: 0

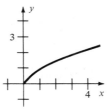

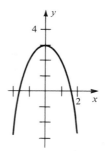

31. x-int.: -3
    y-int.: 3

33. x-int.: $\pm\sqrt{3}$
    y-int.: 3

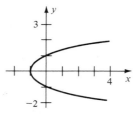

35. x-int.: -1
    y-int.: 1

37. x-int.: -1
    y-int.: ± 1

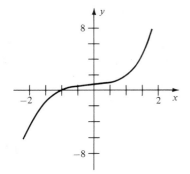

39. none

41. x-axis

43. x-axis

45. x-axis

47. none

49. y-axis

51. all

53. origin

EXERCISE SET 2.2, page 73

1. domain: all reals
    range: all reals

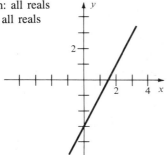

3. domain: all reals
    range: all reals

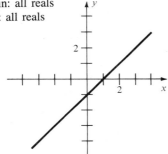

5. domain: $x \geq 1$
    range: $y \geq 0$

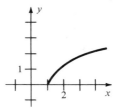

7. $x \geq 3/2$
13. $7/2$
19. $2a^2 + 5$
25. $1/(x^2 + 2x)$
31. $(3x - 1)/(x^2 + 1)$
37. $2(a - 1)/(4a^2 + 4a - 3)$
43. $d(C) = C/\pi$

9. $x > 2$
15. $3/2$
21. $6x^2 + 15$
27. $a^2 + h^2 + 2ah + 2a + 2h$
33. $2(4x^2 + 1)/(6x - 1)$
39. $(a - 1)/a(a + 4)$

11. $x \geq 1, x \neq 2$
17. 5
23. 3
29. -0.92
35. -0.21
41. $I(x) = 0.28x$

EXERCISE SET 2.3, page 82

1. increasing: $(-\infty, \infty)$

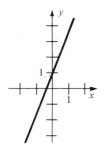

3. increasing: $x \geq 0$
    decreasing: $x \leq 0$

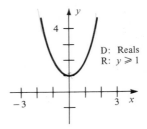

D: Reals
R: $y \geq 1$

5. increasing: $(-\infty, 2]$
    decreasing: $[2, \infty)$

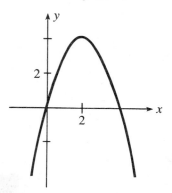

7. increasing: $x \geq -1/2$
    decreasing: $x \leq -1/2$

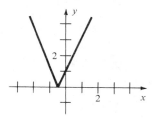

9. increasing: $x \geq 0$

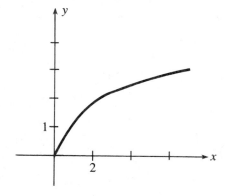

11. constant: $(-\infty, \infty)$

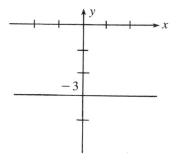

13. increasing: $x > -1$
    decreasing: $x \leq -1$

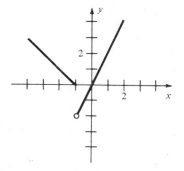

15. increasing: $x \leq 2$
    constant: $x \geq 2$

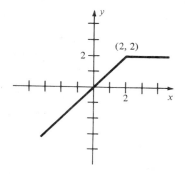

17. increasing: $x \leq 0$
decreasing: $0 \leq x < 1$, $x > 2$
constant: $1 \leq x \leq 2$

19. constant: $x < -2$, $-2 \leq x \leq -1$, $x > -1$

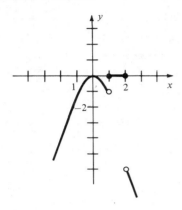

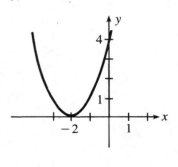

21.
$$C(u) = \begin{cases} 6.50, & 0 \leq u \leq 100 \\ 6.50 + 0.06(u - 100), & 100 < u \leq 200 \\ 12.50 + 0.05(u - 200), & u > 200 \end{cases}$$

23. $R(x) = \begin{cases} 30,000, & 0 \leq x \leq 100 \\ 400x - x^2, & 100 < x < 150 \end{cases}$

25. (a) $C(m) = 14 + 0.08m$
(b) $m \geq 0$
(c) \$22

EXERCISE SET 2.4, page 86

1.

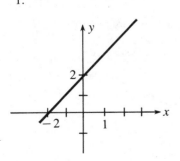

3.

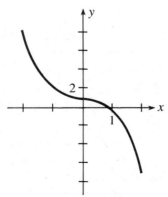

5.

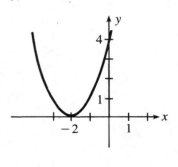

7.

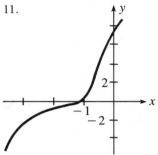

9.

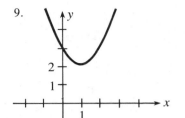

11.

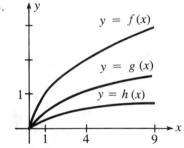

13.

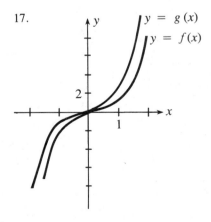

15.

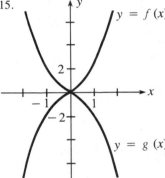

17.

19.

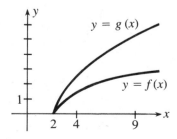

EXERCISE SET 2.5, page 96

1. 2; increasing
3. $-3/2$, decreasing
5. -1; decreasing
9. $2x - y + 5 = 0$
11. $3x - y = 0$
13. $2x - y = 0$
15. $2x - 3y = 0$
17. $2x - y = 0$
19. $3x - y + 2 = 0$
21. $y - 2 = 0$
23. $x - 3y - 15 = 0$
25. $m = -3/4$, $b = 5/4$
27. $m = 0$, $b = 4$
29. $m = -3/4$, $b = -1/2$
31. (a) $y = 3$ (b) $x = -6$
33. (a) $y = 0$ (b) $x = -7$
35. (a) $y = -9$ (b) $x = 9$
37. (a) -3 (b) $1/3$
39. (a) $4/3$ (b) $-3/4$
41. (a) $3x + y - 6 = 0$ (b) $x - 3y + 8 = 0$
43. (a) $3x + 5y - 1 = 0$ (b) $5x - 3y + 21 = 0$
45. (a) $F = \dfrac{9}{5}C + 32$ (b) $68°F$
47. $\$1,000,000$
49. 5
51. $f(x) = 8x + 13$

EXERCISE SET 2.6, page 108

1. ± 3
3. $\pm 8i/3$
5. $-5/2 \pm \sqrt{2}$
7. $5/3 \pm 2\sqrt{2}/3$
9. 1, 2
11. $1, -2$
13. $-2, -4$
15. 0, 4
17. 1/2, 2
19. $4, -2$
21. $-1/2, 4$
23. $1/3, -3$
25. $-1/2 \pm i/2$
27. $1, -3/4$
29. $-1/3 \pm i\sqrt{2}/3$
43. $-1/2 \pm \sqrt{11}/2$
45. $-2, 2/3$
47. $\pm i\sqrt{2}$
49. $\pm \sqrt{c^2 - a^2}$
51. $\pm \sqrt{3V/\pi h}$
53. $(-v \pm \sqrt{v^2 + 2gs})/g$
55. two complex roots
57. double real root
59. two real roots
61. two real roots
63. 4
65. $0, -8$
67. 4
69. 3
71. 0, 4
73. 5
75. $u = x^2$; $\pm i\sqrt{2}$, $\pm \sqrt{3}/3$
77. $u = 1/x$; 2, $-3/2$
79. $u = x^{1/5}$; $-32, -1/32$
81. $u = 1 + 1/x$; $1/3, -2/7$
83. 10 feet
85. $L = 12$ feet, $W = 4$ feet
87. $L = 8$ cm, $W = 6$ cm
89. 3, 7
91. 150 shares
93. 8 days
97. $-1/2$
99. -6
101. ± 9
103. 6

EXERCISE SET 2.7, page 114

1. $(x - 3)^2 + 1$
3. $-2(x - 1)^2 - 3$
5. $2(x + 3/2)^2 + 1/2$
7. $-(x + 1/2)^2 + 1/4$
9. $-2(x - 0)^2 + 5$

11. vertex: $(1, -2)$
$x = 0, 2$
$y = 0$

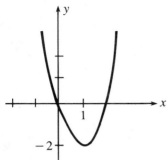

13. vertex: $(1/2, 0)$
$x = 1/2$
$y = -1$

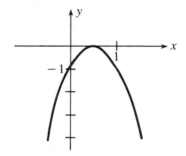

15. vertex: $(-2, 2)$
$y = 4$

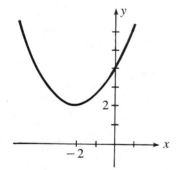

17. vertex: $(3, 1/2)$
$x = 2, 4$
$y = -4$

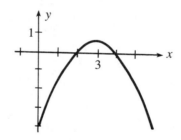

EXERCISE SET 2.8, page 123
1. $x^2 + x - 1$
3. $x^2 - x + 3$
5. $x^3 - 2x^2 + x - 2$
7. $(x^2 + 1)/(x - 2)$
9. domain of f and of g: all reals
11. $4x^2 + 2x + 1$
13. 21
15. $4x^2 + 10x + 7$
17. $8x^2 - 6x + 1$
19. $x + 6, x \geq -2$
21. 29
23. all reals
25. $(f \circ g)(x) = x + 1; (g \circ f)(x) = x + 1$
27. $(f \circ g)(x) = (x - 1)x, x \neq 1$
$(g \circ f)(x) = -(x + 1)/x, x \neq -1$
29. $f(x) = x + 3; g(x) = x^2$
31. $f(x) = x^8; g(x) = 3x + 2$
33. $f(x) = x^{1/3}; g(x) = x^3 - 2x^2$
35. $f(x) = |x|; g(x) = x^2 - 4$
37. $f(x) = \sqrt{x}; g(x) = 4 - x$

45. $f^{-1}(x) = (x - 3)/2$

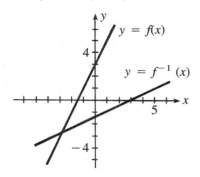

47. $f^{-1}(x) = (3 - x)/2$

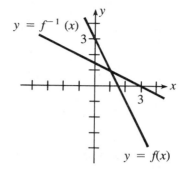

49. $f^{-1}(x) = 3x + 15$

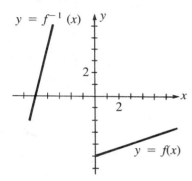

51. $f^{-1}(x) = (x - 1)^{1/3}$

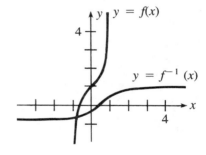

53. yes

55. no

57. yes

59. no

63. $(x - b)/a$

REVIEW EXERCISES, page 126

1. $\sqrt{61}$

2. $\sqrt{65}$

3.

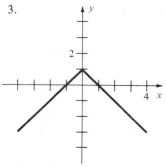

4.

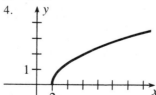

5. x-axis
6. all
7. yes
8. yes
9. $x \geq 5/3$
10. $x \neq -1$
11. 226
12. ± 3
13. 12
14. $y^2 - 3y + 2$
15. $3 + h$
16.

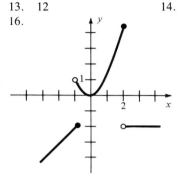

17. increasing: $x \leq -1,\ 0 \leq x \leq 2$
    decreasing: $-1 < x \leq 0$
    constant: $x > 2$
18. -5
19. -2
20.
21.

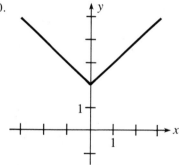

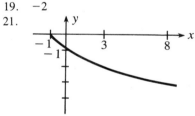

22. 3
23. $y - 3x - 6 = 0$
24. $x = -4$
25. $y = 3$
26. $y - 2x - 2 = 0$
27. $y - 2x - 5 = 0$
28. 5, -4
29. 1/2, 4/3
30. $1 \pm i\sqrt{5}$
31. $(2 \pm i\sqrt{2})/2$
32. -1, 1/3
33. $\pm 3/7$
34. $\pm\sqrt{3\pi k}/k$
35. -4, 3
36. two real, rational roots
37. double root
38. two complex roots
39. 4
40. $\pm\left(\dfrac{-5 \pm \sqrt{85}}{6}\right)^{1/2}$
41. $-1/2, -1$
42. 60
43. vertex: $(-2, 4)$; $x = 0, -4$; $y = 0$
44. vertex: $(5/2, 3/4)$; no x-intercepts; $y = 7$
45. $x^2 + x$
46. 0
47. $(x + 1)/(x^2 - 1)$
48. $x \neq \pm 1$
49. $x^2 + 2x$
50. 4
51. $|x| - 2$
52. $x + 4 - 4\sqrt{x}$
53. 0
54. not defined

56.

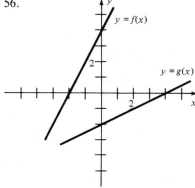

PROGRESS TEST 2A, page 127

1. $3 + \sqrt{26} + \sqrt{41}$

2.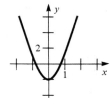

3. origin
4. $x \geq 0, x \neq 1$
5. 17
6. $8t^2 + 3$
7. increasing: $x \geq 0$
    decreasing: $-2 \leq x \leq 0$
    constant: $x < -2$
8. 0
9. 2

10.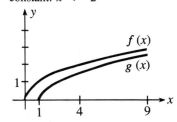

11. $2y - 3x - 19 = 0$
12. $x = -3$
13. $m = 1/2; b = 2$
14. $y = -1$
15. $3y + x - 7 = 0$
16. $-2, 7$
17. $(1 \pm i\sqrt{79})/10$
18. $-3/4, 1/3$
19. two real roots
20. two complex roots
21. -4
22. $\pm\sqrt{2}i, \pm\sqrt{3}/3$

23.

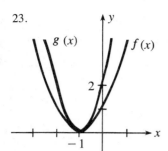

24. -3

25. $x^2(x - 1)$

26. $1/4$

PROGRESS TEST 2B, page 128

1. $6\sqrt{2}$

2.

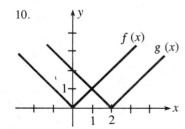

3. origin

4. $x \neq \pm 4$

5. 1

6. 1

7. increasing: $x > 3$
    decreasing: $x \leq -3$
    constant: $-3 < x \leq 3$

8. 10

9. 24

10.

11. $-9/2$

12. $y = -5$

13. 1

14. -3

15. $x + 3y + 3 = 0$

16. $-5/2, 1/3$

17. $1/2, 2$

18. $(1 \pm i\sqrt{83})/6$

19. two complex roots

20. double real root

21. -3

22. $\pm i, \pm(-8)^{3/2}$

23.

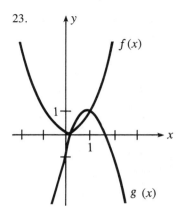

24. 1

25. $\sqrt{2}/2$

26. $\sqrt{x}/x$

CHAPTER 3
EXERCISE SET 3.1, page 135

1. $Q(x) = x - 2, R(x) = 2$
3. $Q(x) = 2x - 4, R(x) = 8x - 4$
5. $Q(x) = 3x^3 - 9x^2 + 25x - 75, R(x) = 226$
7. $Q(x) = 2x - 3, R(x) = -4x + 6$
9. $Q(x) = x^2 - x + 1, R(x) = 0$
11. $Q(x) = x^2 - 3x, R = 5$
13. $Q(x) = x^3 + 3x^2 + 9x + 27, R = 0$
15. $Q(x) = 3x^2 - 4x + 4, R = 4$
17. $Q(x) = x^4 - 2x^3 + 4x^2 - 8x + 16, R = 0$
19. $Q(x) = 6x^3 + 18x^2 + 53x + 159, R = 481$
21. -7
23. -34
25. 0
27. -1
29. 0
31. -62
33. yes
35. no
37. yes
39. yes
41. $r = 3, -1$
43. 5/2

EXERCISE SET 3.2, page 142

	Positive x	Negative x
1.	U	D
3.	D	U
5.	D	D
7.	U	U

9.

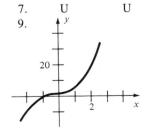

11.

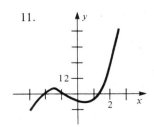

13.

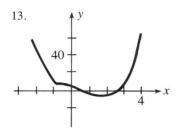

15. *x*-intercepts: -2, $1/2$, 3
$P(x) > 0$: $(-2, 1/2)$, $(3, \infty)$
$P(x) < 0$: $(-\infty, -2)$, $(1/2, 3)$

17. *x*-intercepts: $-5/2$, 0, 1
$P(x) > 0$: $(-5/2, 0)$, $(1, \infty)$
$P(x) < 0$: $(-\infty, -5/2)$, $(0, 1)$

19. *x*-intercepts: -2, 0, 3
$P(x) > 0$: $(-\infty, -2)$, $(3, \infty)$
$P(x) < 0$: $(-2, 0)$, $(0, 3)$

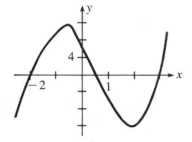
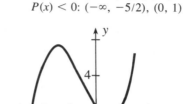
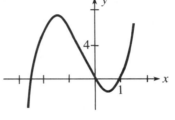

EXERCISE SET 3.3, page 148

1. 1, -2, 3
3. 2, -1, $-1/2$, $2/3$
5. 1, -1, -1, $1/5$

7. 1, $-3/4$
9. 3, 3, $1/2$
11. 2, -1

13. $(3 \pm i\sqrt{3})/2$
15. -1, -2, 4
17. -1, $3/4$, $\pm i$

19. $3/5$, ± 2, $\pm i\sqrt{2}$
21. 0, $1/2$, $2/3$, -1
23. $1/2$, -4, $2 \pm \sqrt{2}$

25. $k = 3$, $r = -2$
27. $k = 7$, $r = 1$; $k = -7$, $r = -1$

EXERCISE SET 3.4, page 158

1. $x \neq 1$
3. $x \neq 0$, 2
5. all real numbers

7. $x = 4$, $y = 0$

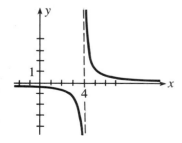

9. $x = -2, y = 0$

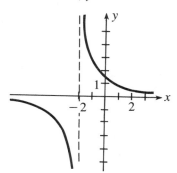

11. $x = -1, y = 0$

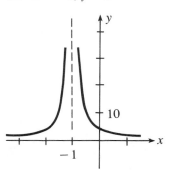

13. $x = 2, y = 1$

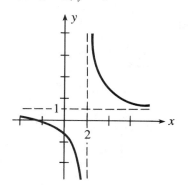

15. $x = 2, x = -2, y = 2$

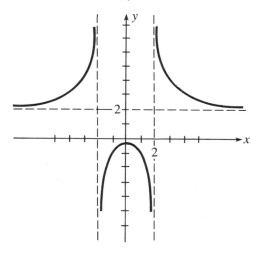

17. $x = 2, x = -3/2, y = 1/2$

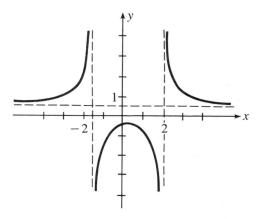

19. $x = 1$

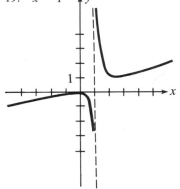

21. $x = 5, x = -5$

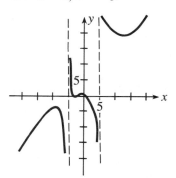

23. $x \neq -2$

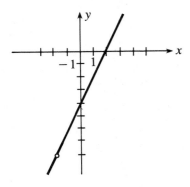

25. $x \neq 2$

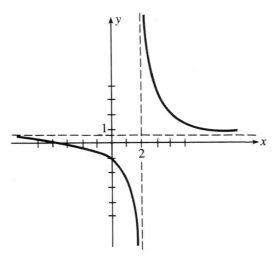

27. $x \neq 0, x \neq -1$

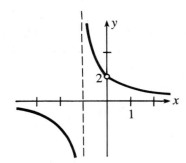

EXERCISE SET 3.5, page 165

1. $\dfrac{3}{x+2} - \dfrac{1}{x-3}$

3. $\dfrac{3}{3x-1} - \dfrac{1}{2x-1}$

5. $\dfrac{-2}{x} + \dfrac{2}{x-1} + \dfrac{1}{x+1}$

7. $\dfrac{2}{x} - \dfrac{1}{x^2} - \dfrac{2}{x+2}$

9. $\dfrac{\frac{1}{2}}{x-1} + \dfrac{\frac{1}{2}}{x+1} - \dfrac{2}{(x+1)^2}$

11. $\dfrac{\frac{1}{4}}{x} - \dfrac{\frac{1}{4}x+2}{x^2+4}$

13. $\dfrac{2x-1}{x^2+3} + \dfrac{3-5x}{(x^2+3)^2}$

15. $\dfrac{\frac{1}{3}}{x+1} + \dfrac{\frac{1}{3} - \frac{1}{3}x}{x^2 - 3x - 1}$

17. $\dfrac{-1}{x+1} + \dfrac{2x-2}{x^2+2} + \dfrac{x-1}{(x^2+2)^2}$

19. $x + \dfrac{2x}{x^2+1} - \dfrac{2}{x+1}$

REVIEW EXERCISES, page 166

1. $Q(x) = 2x^2 + 2x + 8$, $R = 4$
3. $46, -8$
7. large positive x: down
   large negative x: up
9. $3, -2/3, -3/2$
12. $-1/2, 3$
15.

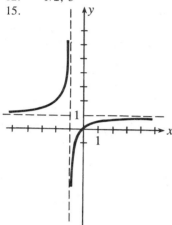

2. $Q(x) = x^3 - 5x^2 + 10x - 18$, $R = 31$
4. $4, 1$
8. large positive x: up
   large negative x: down
10. $1, -2, 2/3, 3/2$ 11. none
13. $-1 \pm \sqrt{2}$
14. $4, 2 + i, 2 - i$
16.

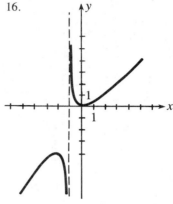

17. $\dfrac{3}{2x - 1} - \dfrac{2}{x + 2}$

18. $\dfrac{3x}{x^2 + 1} + \dfrac{2x - 1}{(x^2 + 1)^2}$

19. $2x + 1 + \dfrac{2}{x - 1} - \dfrac{1 - 2x}{(x - 1)^2}$

PROGRESS TEST 3A, page 167

1. $Q(x) = 2x^2 - 5$, $R(x) = 11$
3. -25
6. extends indefinitely downward
8. 8 9. none
12. $1, -1 \pm i$

2. $Q(x) = 3x^3 - 7x^2 + 14x - 28$, $R(x) = 54$
4. -165
7. extends indefinitely upward
10. $1, 1, -1, -1, 1/2$ 11. $1/2, 1/2$
13.

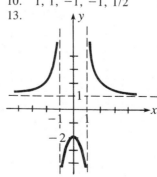

14. $\dfrac{3}{x+3} - \dfrac{2}{x-2}$

PROGRESS TEST 3B, page 168

1. $Q(x) = 3x^3 + 2x^2 + 2, R(x) = x + 3$
2. $Q(x) = -2x^2 + x + 1, R(x) = 0$
3. -1
4. 24
6. extends indefinitely downward
7. extends indefinitely downward
8. 3
9. $-1, 2/3, -2$
10. $-1/2, 3/2, \pm i$
11. $(3 \pm \sqrt{17})/2$
12. $-2 \pm 2\sqrt{2}$
13.

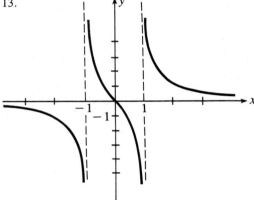

14. $\dfrac{1}{x} - \dfrac{2}{x^2} + \dfrac{4}{x+1}$

CHAPTER 4
EXERCISE SET 4.1, page 180

1.

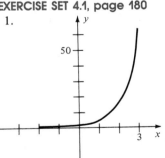

3.

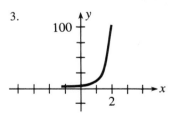

5.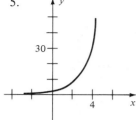

7.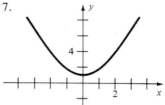

9.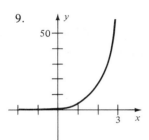

11.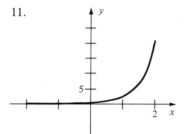

13. 3

15. 4

17. 2

19. 4

21. 2

23. 1

25. (a) 200 (b) 29,682 (c) 256.8, 543.7, 1478, 2436

27. 6.59 billion

29. 670.3 grams

31. $41,611

33. $45,417.50

35. $173.33

37. $2489

39. π^2

EXERCISE SET 4.2, page 188

1. $2^2 = 4$

3. $9^{-2} = 1/81$

5. $e^3 = 20.09$

7. $10^3 = 1000$

9. $e^0 = 1$

11. $3^{-3} = 1/27$

13. $\log_5 25 = 2$

15. $\log_{10} 10,000 = 4$

17. $\log_2 1/8 = -3$

19. $\log_2 1 = 0$

21. $\log_{36} 6 = 1/2$

23. $\log_{16} 64 = 3/2$

25. $\log_{27} 1/3 = -1/3$

27. 25

29. 1/5

31. $e^2 \approx 7.39$

33. $e^{-1/2} \approx 0.61$

35. -2

37. 512

39. 124

41. 2

43. 3

45. 6

47. 2

49. 3

51. 1/2

53. 2

55. 1

57. 0

59. -2

61. 4

63. 2

65.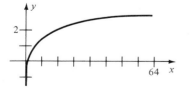

67.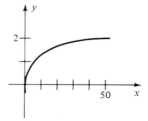

69.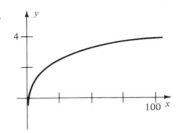

71.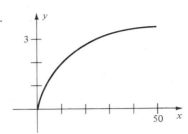

EXERCISE SET 4.3, page 195

1. $\log_{10} 120 + \log_{10} 36$

3. 4

5. $\log_a 2 + \log_a x + \log_a y$

7. $\log_a x - \log_a y - \log_a z$

9. $5 \ln x$

11. $2 \log_a x + 3 \log_a y$

13. $\frac{1}{2}(\log_a x + \log_a y)$

15. $2 \ln x + 3 \ln y + 4 \ln z$

17. $\frac{1}{2} \ln x + \frac{1}{3} \ln y$

19. $2 \log_a x + 3 \log_a y - 4 \log_a z$

21. 0.77

23. 0.94

25. 1.07

27. 0.87

29. 0.435

31. $\log x^2 \sqrt{y}$

33. $\ln \sqrt[3]{xy}$

35. $\log_a \dfrac{x^{1/3} y^2}{z^{3/2}}$

37. $\log_a \sqrt{xy}$

39. $\ln \dfrac{\sqrt[3]{x^2 y^4}}{z^3}$

41. $\log_a \dfrac{\sqrt{x-1}}{(x+1)^2}$

43. $\log_a \dfrac{x^3(x+1)^{1/6}}{(x-1)^2}$

45. 1.2304

47. 4.5046

49. 2.3892

EXERCISE SET 4.4, page 201

1. 2.725×10^3

3. 8.4×10^{-3}

5. 7.16×10^5

7. 2.962×10^2

9. 0.5514

11. 1.5740

13. 1.5476

15. 1.8692

17. 4.6830

19. -0.4660

21. 2.520

23. 2.9

25. 7.9

27. 0.257

29. 0.000607

31. 0.0219

33. 1.028

35. 2.115

37. 103.55

39. 0.002875

41. 1.93×10^{-5}

43. 2.59×10^{-8}

45. $10,453

47. $14,660.72

49. 8.75% compounded quarterly

EXERCISE SET 4.5, page 205

1. $\log 18/\log 5$

3. $1 + \log 7/\log 2$

5. $\log 46/2 \log 3$

7. $5/2 + \log 564/2 \log 5$

9. $(\log 2 + \log 3)/(\log 3 - 2 \log 2)$

11. $-\log 15/\log 2$

13. $1/2 - \log 12/2 \log 4$

15. $\ln 18$

17. $(-3 + \ln 20)/2$

19. 500

21. $1/2$

23. 5

25. 3

27. 8

29. $-1 + \sqrt{17}$

31. $\ln (y + \sqrt{y^2 + 1})$

33. 36.62 years

35. 12.6 hours

37. 8.8 years

39. 27.47 days

41. 1.386 days

REVIEW EXERCISES, page 207

1.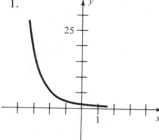

2. 3

3. 2

4. $12,750.40

5. $\log_9 27 = 3/2$

6. $8 = 64^{1/2}$

7. $1/8 = 2^{-3}$

8. $\log_6 1 = 0$

9. 2

10. -2

11. e^{-4}

12. 26

13. 5

14. $-1/3$

15. -1

16. 3

17.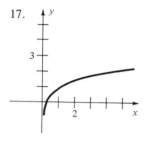

18. $\frac{1}{2} \log_a(x - 1) - \log_a 2 - \log_a x$

19. $\log_a x + 2 \log_a(2 - x) - \frac{1}{2} \log_a(y + 1)$

20. $4 \ln(x + 1) + 2 \ln(y - 1)$

21. $\frac{2}{5} \log y + \frac{1}{5} \log z - \frac{1}{5} \log(z + 3)$

22. 1.15

23. 0.55

24. 0.4

25. -0.15

26. $\log_a \dfrac{\sqrt[3]{x}}{\sqrt{y}}$

27. $\log(x^2 - x)^{4/3}$

28. $\ln \dfrac{3xy^2}{z}$

29. $\log_a \dfrac{(x + 2)^2}{(x + 1)^{3/2}}$

30. 5/3

31. 15/7

32. 4.765×10^2

33. 9.8×10^{-2}

34. 2.6475×10^4

35. 7.767×10^1

36. 803

37. 7.9

38. 3.49×10^{-4}

39. 11.5 hours

40. $\dfrac{1}{3} + \dfrac{\log 14}{3 \log 2}$

41. $\sqrt{5000}$

42. $\dfrac{199}{98}$

PROGRESS TEST 4A, page 209

1.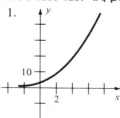

2. $-2/3$

3. $1/9 = 3^{-2}$

4. $\log_{16} 64 = 3/2$

5. 3

6. -1

7. $5/2$

8. $1/2$

9. $3 \log_a x - 2 \log_a y - \log_a z$

10. $2 \log x + \frac{1}{2} \log(2y - 1) - 3 \log y$

11. 0.7

12. 0.45

13. $\log \dfrac{x^2}{(y + 1)^3}$

14. $\log \left(\dfrac{x + 3}{x - 3} \right)^{2/3}$

15. 2.73×10^{-4}

16. 5.972×10^0

17. 4.7×10^{-2}

18. 0.26

19. 34.6 hours

20. 200

21. 4

PROGRESS TEST 4B, page 209

1.

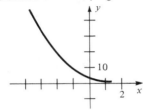

2. 8

3. $\log \dfrac{1}{1000} = -3$

4. $1 = 3^0$

5. $3/2$

6. $3/2$

7. 10

8. $4\sim$

9. $\log_a(x - 1) + \frac{5}{4} \log_a(y + 3)$

10. $\frac{1}{2} \ln x + \frac{1}{2} \ln y + \frac{1}{4} \ln 2z$

11. 0.85

12. 1.5

13. $\dfrac{1}{5} \ln \dfrac{(x - 1)^3 y^2}{z}$

14. $\log \dfrac{x^2}{y^2}$

15. 2.2684321×10^7

16. 2.97×10^{-1}

17. 9397

18. 6.2×10^{-4}

19. \$530.76

20. 3

21. 10

CHAPTER 5

EXERCISE SET 5.1, page 223

1. 0

3. $\pi/7$

5. $3\pi/2$

7. $5\pi/6$

9. π

11. $7\pi/5$

13.

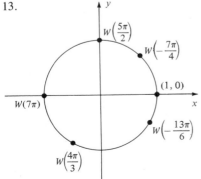

15.

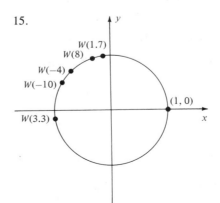

17. $(-1, 0)$
19. $(\sqrt{2}/2, -\sqrt{2}/2)$
21. $(-\sqrt{2}/2, -\sqrt{2}/2)$
23. $(-1/2, -\sqrt{3}/2)$
25. $(-1/2, -\sqrt{3}/2)$
27. $(-\sqrt{3}/2, -1/2)$
29. $(-\sqrt{3}/2, -1/2)$
31. $(1/2, \sqrt{3}/2)$
33. $\pi, -\pi$
35. $3\pi/4, -5\pi/4$
37. $5\pi/6, -7\pi/6$
39. $5\pi/3, -\pi/3$
41. $2\pi/3, -4\pi/3$
43. (a) $(4/5, 3/5)$ (b) $(3/5, -4/5)$ (c) $(-4/5, 3/5)$ (d) $(4/5, -3/5)$

EXERCISE SET 5.2, page 230

1. $\sin t = -\sqrt{3}/2$, $\cos t = 1/2$, $\tan t = -\sqrt{3}$
3. $\sin t = 0$, $\cos t = -1$, $\tan t = 0$
5. $\sin t = -\sqrt{2}/2$, $\cos t = \sqrt{2}/2$, $\tan t = -1$
7. $\sin t = \sqrt{3}/2$, $\cos t = -1/2$, $\tan t = -\sqrt{3}$
9. $\sin t = -\sqrt{3}/2$, $\cos t = 1/2$, $\tan t = -\sqrt{3}$
11. $\sin t = -\sqrt{2}/2$, $\cos t = -\sqrt{2}/2$, $\tan t = 1$
13. II
15. III
17. III
19. $\pi/2, 3\pi/2$
21. $\pi/4, 5\pi/4$
23. $\pi/4, 3\pi/4$
25. $5\pi/6, 7\pi/6$
27. $\pi/3, 4\pi/3$
29. $\pi/3, 2\pi/3$
31. $7\pi/6, 11\pi/6$
33. none
35. $\pi/6, 5\pi/6$
37. $-3/4$
39. $-12/13$
41. $-3/5$
43. $4/5$

EXERCISE SET 5.3, page 236

1.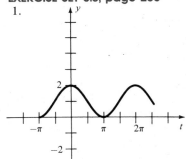

3.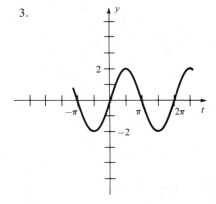

5.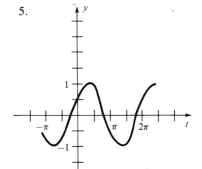

7.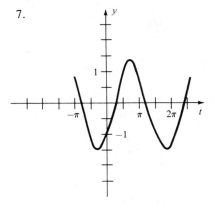

9.

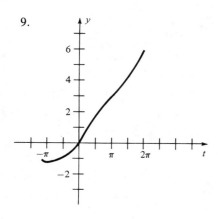

EXERCISE SET 5.4, page 239

1.	0.4357	3.	−5.471	5.	0.3093	7.	0.8415
9.	−0.9737	11.	−0.1307	13.	0.7174	15.	−0.1987
17.	0.1003	19.	1.4592	21.	0.9888	23.	−0.3897
25.	0.7818	27.	−0.4660	29.	0.4983		

EXERCISE SET 5.5, page 245

1. amplitude: 3
   period: 2π

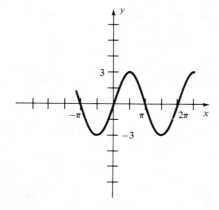

3. amplitude: 1
   period: $\pi/2$

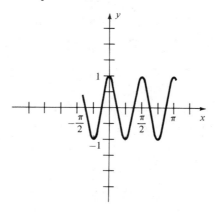

5. amplitude: 2
    period: $\pi/2$

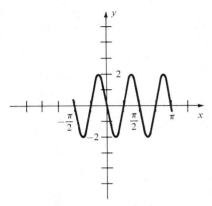

7. amplitude: 2
    period: 6π

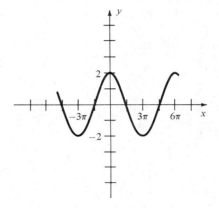

9. amplitude: 1/4
    period: 8π

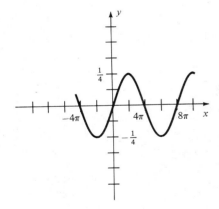

11. amplitude: 3
    period: $2\pi/3$

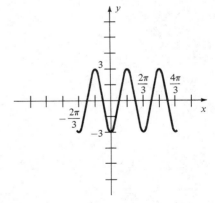

13.   amplitude: 2
      period: 2π
      phase shift: π

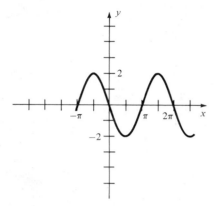

15.   amplitude: 3
      period: π
      phase shift: $\pi/2$

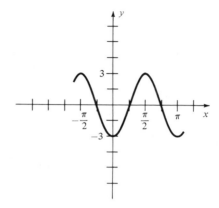

17.   amplitude: 1/3
      period: $2\pi/3$
      phase shift: $-\pi/4$

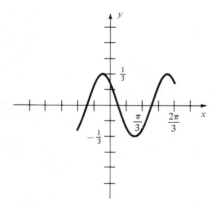

19.   amplitude: 2
      period: 8π
      phase shift: 4π

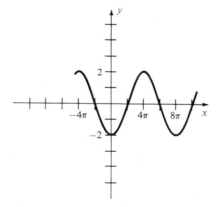

21. $y = 2 \sin(2x - \pi)$

23. $y = 3 \cos(x/3 - 2\pi/3)$

EXERCISE SET 5.6, page 249

1. $\sec t = 2$, $\csc t = 2\sqrt{3}/3$, $\cot t = \sqrt{3}/3$

3. $\sec t = \sqrt{2}$, $\csc t = \sqrt{2}$, $\cot t = 1$

5. $\sec t = -2\sqrt{3}/3$, $\csc t = 2$, $\cot t = -\sqrt{3}$

7. $\sec t$ not defined, $\csc t = -1$, $\cot t = 0$

9. $\sec t = -\sqrt{2}$, $\csc t = \sqrt{2}$, $\cot t = -1$

11. $\sec t = -\sqrt{2}$, $\csc t = -\sqrt{2}$, $\cot t = 1$

13. $0, 2\pi$ 15. $7\pi/6, 11\pi/6$

17. $\pi/4, 5\pi/4$ 19. $3\pi/4, 7\pi/4$

21. $\pi/4, 7\pi/4$ 23. $5\pi/6, 11\pi/6$

25.   III   27.   II

29. III	31. III	33. $5\pi/6$	35. $2\pi/3$
37. $5\pi/4$	39. $3\pi/4$	41. 4.271	43. 1.000
45. 3.270			

EXERCISE SET 5.7, page 257

1. $-\pi/6$	3. $\pi/3$	5. $-\pi/4$	7. $5\pi/6$
9. $-\pi/2$	11. $\pi/2$	13. 0	15. $-\pi/4$
17. $2\pi/3$	19. 0.38	21. 2.44	23. 1.30
25. $\sqrt{2}/2$	27. 0	29. $\pi/4$	31. $2\pi/3$
33. $\arcsin(\pm\sqrt{7}/7)$		35. $\cos^{-1}(1/3), \cos^{-1}(-1/4)$	
37. $\sin^{-1}(2/3)$		39. 0	

REVIEW EXERCISES, page 258

1. $\pi/2$	2. $\pi/2$	3. 0	4. $5\pi/3$
5. $6\pi/5$	6. $(-\sqrt{3}/2, -1/2)$	7. $(-1/2, -\sqrt{3}/2)$	8. $(-\sqrt{3}/2, 1/2)$
9. $(\sqrt{2}/2, \sqrt{2}/2)$	10. $(\sqrt{3}/2, -1/2)$	11. $(-4/5, 3/5)$	12. $(3/5, 4/5)$
13. $(4/5, 3/5)$	14. $(-3/5, -4/5)$	15. $(-4/5, -3/5)$	16. $\sqrt{3}/2$
17. $-\sqrt{2}$	18. $-\sqrt{3}/3$	19. -2	20. $5\pi/4$
21. $11\pi/6$	22. $\pi/3$	23. $2\pi/3$	24. $-3/4$
25. $-5/3$	26. $-12/5$	27. $13/12$	30. -1.8334
31. 0.4228			

32.

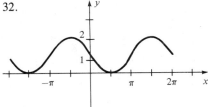

33.

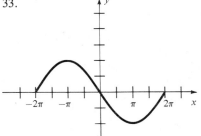

34. amplitude: 1; period: π; phase shift: $\pi/2$
35. amplitude: 4; period: 2π; phase shift: $\pi/2$
36. amplitude: 2; period: 6π; phase shift: $-\pi$
37. $-\pi/6$
38. 0   39. 5
40. $\cos^{-1}(\pm 2\sqrt{5}/5)$

PROGRESS TEST 5A, page 259

1. $\pi/3$	2. 0	3. $\pi/4$	4. $(-\sqrt{3}/2, 1/2)$
5. $(1/2, -\sqrt{3}/2)$	6. $(5/13, -12/13)$	7. $(12/13, 5/13)$	8. $(-5/13, -12/13)$
9. $1/2$	10. $-2\sqrt{3}/3$	11. $5\pi/4$	12. $7\pi/4$
13. $-5/13$	14. $-5/4$	16. -0.5994	17. -0.2509

18.

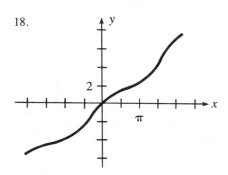

19. amplitude: 2; period: 2π; phase shift: π
21. $-\pi/3$ 22. 1/2

20. amplitude: 2; period: 4π; phase shift: π
                                    23. arctan(2/3), arctan(3/2)

PROGRESS TEST 5B, page 260

1. 0
5. $(-\sqrt{2}/2, -\sqrt{2}/2)$
9. -1
13. $-5/12$
17. -3.1106

2. $\pi/5$
6. $(-4/5, 3/5)$
10. 1
14. $-3/4$

3. $5\pi/6$
7. $(3/5, -4/5)$
11. $\pi/3$
16. 0.6378
18.

4. $(\sqrt{3}/2, -1/2)$
8. $(4/5, -3/5)$
12. $2\pi/3$

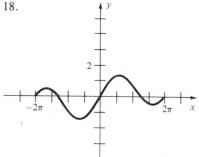

19. amplitude: 4; period: $2\pi/3$; phase shift: $\pi/3$
20. amplitude: 1/2; period: π; phase shift: $-\pi/4$

21. $\pi/6$ 22. 1   23. $\arcsin\left(\dfrac{1 \pm 2\sqrt{19}}{25}\right)$

CHAPTER 6
EXERCISE SET 6.1, page 270

1. IV
9. III
17. I
25. $-5\pi/2$
33. 45°

3. I
11. II
19. $\pi/6$
27. $3\pi/4$
35. 270°

5. II
13. II
21. $-5\pi/6$
29. $2\pi/3$
37. $-90°$

7. I
15. III
23. $5\pi/12$
31. 0.251π
39. 240°

41.	450°	43.	−300°	45.	98°50′	47.	T
49.	F	51.	F	53.	50°	55.	20°
57.	85°	59.	$2\pi/5$	61.	72°	63.	$\pi/4$
65.	4/7; 32°50′			67.	$6/\pi$		
69.	6.8 ft; ≈776.5 rotations			71.	10		

EXERCISE SET 6.2, page 274

		sin	cos	tan	csc	sec	cot
1.	135°	$\sqrt{2}/2$	$-\sqrt{2}/2$	-1	$\sqrt{2}$	$-\sqrt{2}$	-1
3.	−30°	$-1/2$	$\sqrt{3}/2$	$-\sqrt{3}/3$	-2	$2\sqrt{3}/3$	$-\sqrt{3}$
5.	315°	$-\sqrt{2}/2$	$\sqrt{2}/2$	-1	$-\sqrt{2}$	$\sqrt{2}$	-1
7.	270°	-1	0		-1		0

9.	0.7353	11.	0.2504	13.	2.128	15.	−24.54
17.	−0.9147	19.	0.2700	21.	−0.2419		

EXERCISE SET 6.3, page 284

	$\sin\theta$	$\cos\theta$	$\tan\theta$	$\csc\theta$	$\sec\theta$	$\cot\theta$
1.	3/5	4/5	3/4	5/3	5/4	4/3
3.	4/5	3/5	4/3	5/4	5/3	3/4
5.	$2\sqrt{5}/5$	$\sqrt{5}/5$	2	$\sqrt{5}/2$	$\sqrt{5}$	$1/2$
7.	$\dfrac{\sqrt{x^2+1}}{x^2+1}$	$\dfrac{x\sqrt{x^2+1}}{x^2+1}$	$\dfrac{1}{x}$	$\sqrt{x^2+1}$	$\dfrac{\sqrt{x^2+1}}{x}$	x
9.	12/13	−5/13	−12/5	13/12	−13/5	−5/12
11.	$-\sqrt{2}/2$	$-\sqrt{2}/2$	1	$-\sqrt{2}$	$-\sqrt{2}$	1
13.	3/5	−4/5	−3/4	5/3	−5/4	−4/3
15.	−5/13	12/13	−5/12	−13/5	13/12	−12/5
17.	−5/13	−12/13	5/12	−13/5	−13/12	12/5
19.	$\sqrt{5}/5$	$-2\sqrt{5}/5$	$-1/2$	$\sqrt{5}$	$-\sqrt{5}/2$	-2

21.	$5\sin\theta$	23.	$6.5\cot\theta$	25.	$3.7\csc\theta$	27.	36°50′
29.	62.2	31.	30.3	33.	33.9	35.	−5/12
37.	3/5	39.	−4/3	41.	53°10′	43.	61°
45.	7767 feet	47.	970 meters	49.	39°10′, 50°50′	51.	18.7 cm
53.	53 feet	55.	24.19 miles				

EXERCISE SET 6.4, page 289

1.	41°10′	3.	14.4	5.	17.8	7.	62°30′
9.	90°	11.	82°10′	13.	52.5 miles, S29°W	15.	32.8 miles
17.	68.5						

EXERCISE SET 6.5, page 295

1. 29.1
3. 7.2
5. 15.7
7. none
9. 10.9
11. 8.3, 1.6
13. 98.9 meters
15. 682 meters
17. 40.5 miles
19. 18.8 meters
21. 115 cm

REVIEW EXERCISES, page 297

1. $-\pi/3$
2. 270°
3. $-75°$
4. $\pi/4$
5. yes
6. no
7. yes
8. 50°
9. 5°
10. 45°
11. 1.4
12. 15/4 cm
13. 5/13
14. 4/3
15. $\sqrt{65}/7$
16. 2
17. -1
18. $-1/2$
19. $\sqrt{2}/2$
20. 39°50′
21. 14.6
22. 25.4
23. 16.6
24. 5.4 meters
25. 68°10′
26. 36°
27. 51°50′
28. 37°50′

PROGRESS TEST 6A, page 298

1. 300°
2. $-10\pi/9$
3. $5\pi/12$
4. 335°
5. 45°
6. 20°
7. $\pi/4$
8. 4/5
9. 7/5
10. 3
11. -1
12. -1
13. 2
14. 56°30′
15. 23.6
16. 53°10′
17. 138 meters

PROGRESS TEST 6B, page 298

1. $-3\pi/4$
2. 135°
3. $-150°$
4. 70°
5. 240°
6. $7\pi/16$
7. 15°
8. 21/5
9. 5/4
10. 6/7
11. undefined
12. $2\sqrt{3}/3$
13. 1
14. 10.3
15. 66°30′
16. 12.2
17. 67.5 feet

CHAPTER 7
EXERCISE SET 7.1, page 304

47. $3\pi/2$
49. $\pi/4$
51. π

EXERCISE SET 7.2, page 312

1. $s = t = 0$
3. $s = \pi, t = \pi/2$
5. $s = t = \pi/4$
7. $(\sqrt{6} - \sqrt{2})/4$
9. $(\sqrt{2} + \sqrt{6})/4$
11. $-\sqrt{3}/2$
13. $\sqrt{3}$
15. $(\sqrt{6} - \sqrt{2})/4$
17. $(\sqrt{2} - \sqrt{6})/4$
19. $-1/2$
21. $2 - \sqrt{3}$
23. $\cos 43°$
25. $\cot \pi/3$
27. $\sin \pi/6$
29. $-4/5$
31. -7
33. -2.29
35. $-16/65$
37. $2/29$

EXERCISE SET 7.3, page 320

1. $7/25$

3. $-\sqrt{3}/2$

5. $-240/161$

7. $-24/25$

9. $-161/289$

11. 0.1022

13. $\sqrt{2} - \sqrt{3}/2$

15. $\sqrt{(2 - \sqrt{2})}/\sqrt{(2 + \sqrt{2})}$

17. $2/\sqrt{2} - \sqrt{3}$

19. $\sqrt{20}/5$

21. $\sqrt{6}/3$

23. -2

25. $-\sqrt{6}/3$

EXERCISE SET 7.4, page 324

1. $\sin 6\alpha + \sin 4\alpha$

3. $(\cos 5x - \cos x)/2$

5. $-(\cos 7\theta + \cos 3\theta)$

7. $(\cos 2\alpha + \cos 2\beta)/2$

9. $-(2 + \sqrt{2})/4$

11. $\sqrt{3}/4$

13. $2 \sin 3x \cos 2x$

15. $2 \cos 4\theta \cos 2\theta$

17. $2 \sin \alpha \cos \beta$

19. $2 \cos 5x \cos 2x$

21. $\sqrt{6}/2$

23. $-\sqrt{2}$

35. $\dfrac{\sin(a + b)x + \sin(a - b)x}{2}$

EXERCISE SET 7.5, page 329

1. $\pi/6, 5\pi/6; 30°, 150°$

3. $\pi; 180°$

5. $\pi/6, 5\pi/6, 7\pi/6, 11\pi/6; 30°, 150°, 210°, 330°$

7. $\pi/6, 5\pi/6, 7\pi/6, 11\pi/6; 30°, 150°, 210°, 330°$

9. $0°, 30°, 150°, 180°; 0, \pi/6, 5\pi/6, \pi$

11. $0, \pi/3, 5\pi/3; 0°, 60°, 300°$

13. $\pi/10, \pi/2, 9\pi/10, 13\pi/10, 17\pi/10; 18°, 90°, 162°, 234°, 306°$

15. $\pi/3, 5\pi/3; 60°, 300°$

17. $\pi/6, 5\pi/6, 3\pi/2; 30°, 150°, 270°$

19. $0; 0°$

21. $\pi/6 + \pi n; 5\pi/6 + \pi n$

23. $\pi/3 + \pi n, 2\pi/3 + \pi n$

25. $\pi/6 + \pi n; 5\pi/6 + \pi n$

27. $\pi n/4$

29. $\pi/12 + \pi n/2, 5\pi/12 + \pi n/2$

31. $\pi/2 + \pi n$

33. $\pi/2 + 2\pi n/3$

35. $\pi n, \pi/4 + \pi n$

37. $\pi/6 + 2\pi n, 5\pi/6 + 2\pi n$

39. $0.83, 2.31, 3.71, 5.71$ radians

41. 6.05 radians, 5.32 radians

REVIEW EXERCISES, page 331

4. $(\sqrt{2} + \sqrt{6})/4$

5. $-\sqrt{2}/2$

6. $-2 - \sqrt{3}$

7. $(\sqrt{2} + \sqrt{6})/4$

8. $\sec 75°$

9. $\sin 67°$

10. $\cos 3\pi/8$

11. $\cot 3\pi/14$

12. $5/13$

13. $10(3 + 4\sqrt{3})/39$

14. $3/4$

15. 70

16. $-16/65$

17. $-7/25$

18. $-24/25$

19. $24/25$

20. $-\sqrt{3}/2$

21. $120/169$

22. $-\sqrt{10}/10$

23. $-1/3$

24. $-\sqrt{30}/6$

25. $\sqrt{2 + \sqrt{3}}/2$

26. $\sqrt{2 - \sqrt{2}}/2$

27. $-\sqrt{2 + \sqrt{2}}/\sqrt{2 - \sqrt{2}}$

31. $(\cos \alpha - \cos 2\alpha)/2$

32. $-2 \sin 2x \sin x$

33. $1/4$

34. $2(\cos \pi/2)(\cos \pi/4)$

35. $\pi/4, 3\pi/4, 5\pi/4, 7\pi/4$

36. $0, \pi/2, \pi, 3\pi/2$

37. $0, \pi/3, \pi, 5\pi/3$

38. $90° + 360°n, 270° + 360°n$

39. $45° + 60°n$

40. $30° + 90°n, 60° + 90°n$

PROGRESS TEST 7A, page 332

2. $1/2$

3. $\sqrt{3} - 2$

4. $\cos 43°$

5. $3/5$

6. $-81/76$

7. $-119/169$

8. $7/25$

9. $-\sqrt{2} - \sqrt{3}/2$

10. $\sqrt{2} - \sqrt{3}/\sqrt{2} + \sqrt{3}$

12. $2(\sin 5x/2)(\cos x/2)$

13. $2(\cos 90°)(\cos 60°) = 0$

14. $\pi/3,\ 2\pi/3,\ 4\pi/3,\ 5\pi/3$

15. $45° + 90°n$

PROGRESS TEST 7B, page 332

2. 2

3. $(\sqrt{6} + \sqrt{2})/4$

4. $\cot 19°$

5. $-13\sqrt{2}/7$

6. 0

7. $24/25$

8. $336/625$

9. $-3\sqrt{13}/13$

10. $\sqrt{2} - \sqrt{2}/2$

12. $(\sin 7\pi/12 - \sin \pi/12)/2$

13. $(\cos 90° + \cos 60°)/2$

14. $0,\ 3\pi/4,\ \pi,\ 7\pi/4$

15. $90° + 120°n$

CHAPTER 8
EXERCISE SET 8.1, page 338

1. 5

3. 25

5. 20

7. $-13/10 + 11i/10$

9. $-7/25 - 24i/25$

11. $8/5 - i/5$

13. $5/3 - 2i/3$

15. $4/5 + 8i/5$

17. $4/25 - 3i/25$

19. $9/10 + 3i/10$

21. $0 + i/5$

25. $\sqrt{13}$

27. $\sqrt{2}$

29. $2\sqrt{10}$

EXERCISE SET 8.2, page 345

1. $3\sqrt{2}\,(\cos 7\pi/4 + i \sin 7\pi/4)$

3. $2(\cos 11\pi/6 + i \sin 11\pi/6)$

5. $\sqrt{2}\,(\cos 3\pi/4 + i \sin 3\pi/4)$

7. $4(\cos \pi + i \sin \pi)$

9. -4

11. $-1 + i$

13. $-5i$

15. 6

17. $\sqrt{3} - i$

19. $\sqrt{2}\,(\cos 7\pi/4 + i \sin 7\pi/4),\ 2(\cos \pi/2 + i \sin \pi/2);\ 2 + 2i$

21. $4(\cos 2\pi/3 + i \sin 2\pi/3),\ 3\sqrt{2}(\cos \pi/4 + i \sin \pi/4),\ (-6 - 6\sqrt{3}) + (6\sqrt{3} - 6)i$

23. $5(\cos 0 + i \sin 0),\ 2\sqrt{2}\,(\cos 5\pi/4 + i \sin 5\pi/4),\ -10 - 10i$

25. $0 + 512i$

27. $16 - 16i$

29. $-8 + 8i$

31. $\pm\sqrt{2} \pm \sqrt{2}\,i$

33. $\sqrt{2}\,(\cos 150° + i \sin 150°),\ \sqrt{2}\,(\cos 330° + i \sin 330°)$

35. $-2,\ 1 \pm \sqrt{3}\,i$

37. $\pm2,\ \pm2i$

EXERCISE SET 8.3, page 352

1. $x^3 - 2x^2 - 16x + 32$

3. $x^3 + 6x^2 + 11x + 6$

5. $x^3 - 6x^2 + 6x + 8$

7. $x^3/3 + x^2/3 - 7x/12 + 1/6$

9. $x^3 - 4x^2 - 2x + 8$

11. $3, -1, 2$

13. $-2, 4, -4$

15. $-2, -1, 0, -1/2$

17. $5, 5, 5, -5, -5$

19. $x^3 + 6x^2 + 12x + 8$

21. $4x^4 + 4x^3 - 3x^2 - 2x + 1$

23. $x^2 + (1 - 3i)x - (2 + 6i)$

25. $x^2 - 3x + (3 + i)$

27. $x^3 + (1 + 2i)x^2 + (-8 + 8i)x + (-12 + 8i)$

29. $(x^2 - 6x + 10)(x - 1)$

31. $(x^2 + 2x + 5)(x^2 + 2x + 4)$

33. $(x - 2)(x + 2)(x - 3)(x^2 + 6x + 10)$

35. $x - (a + bi)$

EXERCISE SET 8.4, page 356

	Positive roots	Negative roots	Complex roots
1.	3	1	0
	1	1	2
3.	0	0	6
5.	3	2	0
	1	2	2
	3	0	2
	1	0	4
7.	1	2	0
	1	0	2
9.	2	0	2
	0	0	4
11.	1	1	6

13. $3, 3, \pm i$ 15. $2, -2, \pm\sqrt{2}$ 17. $-1/2, -1/2, \pm 1$

REVIEW EXERCISES, page 358

1. $6/25 - 17i/25$
2. $-1/5 + 2i/5$
3. $-5/2 + 5i/2$
4. $1/10 - 3i/10$
5. $i/4$
6. $2/29 + 5i/29$
7. $\sqrt{5}$
8. $\sqrt{13}$
9. $\sqrt{41}$
10. $3\sqrt{2} (\cos 3\pi/4 + i \sin 3\pi/4)$
11. $2i$
12. $1 - i$
13. $2(\cos \pi + i \sin \pi)$
14. $24(\cos 37° + i \sin 37°)$
15. $5(\cos 21° + i \sin 21°)/3$
16. $2(\cos 90° + i \sin 90°)$
17. $-972 + 972i$
18. $0 - 8i$
19. $3(\cos 90° + i \sin 90°), 3(\cos 270° + i \sin 270°)$ 20. $1, -1/2 \pm \sqrt{3}i/2$
21. $x^3 + 6x^2 + 11x + 6$
22. $x^3 - 3x^2 + 3x - 9$
23. $x^4 + x^3 - 5x^2 - 3x + 6$
24. $4x^4 + 4x^3 - 3x^2 - 2x + 1$
25. $x^4 + 2x^2 + 1$
26. $x^4 - 6x^2 - 8x - 3$
27. 1 positive, 1 negative
28. 5 positive, 0 negative
29. 1 positive, 0 negative
30. 2 positive, 2 negative
31. $-1, (-9 \pm \sqrt{321})/12$
32. $2, 3/2, -1 \pm \sqrt{2}$

PROGRESS TEST 8A, page 358

1. $1/13 - 5i/13$
2. $-2/5 - i/5$
3. 5
4. $5(\cos 86° + i \sin 86°)$
5. $(\cos 77° + i \sin 77°)/2$
6. $-\dfrac{1}{2 \cdot 5^4} + \dfrac{\sqrt{3}}{2 \cdot 5^4} i$
7. $3(\cos 60° + i \sin 60°); 3(\cos 180° + i \sin 180°); 3(\cos 300° + i \sin 300°)$
8. $x^3 - 2x^2 - 5x + 6$
9. $x^4 - 6x^3 + 6x^2 + 6x - 7$
10. $2, \pm i$
11. $-1, -1, (3 \pm \sqrt{17})/2$
12. $x^5 + 3x^4 - 6x^3 - 10x^2 + 21x - 9$
13. $16x^5 - 8x^4 + 9x^3 - 9x^2 - 7x - 1$
14. $x^2 - (1 + 2i)x + (-1 + i)$
15. $(x^2 - 4x + 5)(x - 2)$
16. 2
17. 1
18. $2/3, -3, \pm i$

PROGRESS TEST 8B, page 359

1. $-1/4 + i/4$
2. $3/25 + 4i/25$
3. $2\sqrt{2}$
4. $5(\cos 250° + i \sin 250°)$
5. $2(\cos 55° + i \sin 55°)$
6. $-64 + 0i$
7. $-1, 1/2 \pm \sqrt{3}i/2$
8. $2x^4 - x^3 - 3x^2 + x + 1$
9. $x^3 - 4x^2 + 2x + 4$
10. $1, 2, 2, 2$
11. $(-3 \pm \sqrt{13})/2, -3, -3$
12. $8x^4 + 4x^3 - 18x^2 + 11x - 2$
13. $x^4 + 4x^3 - x^2 - 6x + 18$
14. $x^4 - 4x^3 - x^2 + 14x + 10$
15. $(x^2 - 2x + 2)(2x - 1)(x + 2)$
16. 1
17. 2
18. $0, 1/2, \pm\sqrt{2}$

CHAPTER 9

EXERCISE SET 9.1, page 365

1. $(5/2, 5)$
3. $(1, 5/2)$
5. $(-7/2, -1)$
7. $(0, -1/2)$
9. $(-1, 9/2)$
11. $(0, 0)$

EXERCISE SET 9.2, page 369

1. $(x - 2)^2 + (y - 3)^2 = 4$
3. $(x + 2)^2 + (y + 3)^2 = 5$
5. $x^2 + y^2 = 9$
7. $(x + 1)^2 + (y - 4)^2 = 8$
9. $(h, k) = (2, 3); r = 4$
11. $(h, k) = (2, -2); r = 2$
13. $(h, k) = (-4, -3/2); r = 3\sqrt{2}$
15. no graph
17. $(x + 2)^2 + (y - 4)^2 = 16; (h, k) = (-2, 4); r = 4$
19. $(x - 3/2)^2 + (y - 5/2)^2 = 11/2; (h, k) = (3/2, 5/2); r = \sqrt{22}/2$
21. $(x - 1)^2 + y^2 = 7/2; (h, k) = (1, 0); r = \sqrt{14}/2$
23. $(x - 2)^2 + (y + 3)^2 = 8; (h, k) = (2, -3); r = 2\sqrt{2}$
25. $(x - 3)^2 + (y + 4)^2 = 18; (h, k) = (3, -4); r = 3\sqrt{2}$
27. $(x + 3/2)^2 + (y - 5/2)^2 = 3/2; (h, k) = (-3/2, 5/2); r = \sqrt{6}/2$
29. $(x - 3)^2 + y^2 = 11; (h, k) = (3, 0); r = \sqrt{11}$
31. $(x - 3/2)^2 + (y - 1)^2 = 17/4; (h, k) = (3/2, 1); r = \sqrt{17}/2$
33. $(x + 2)^2 + (y - 2/3)^2 = 100/9; (h, k) = (-2, 2/3); r = 10/3$
35. neither

EXERCISE SET 9.3, page 374

1. focus: $(0, 1)$; directrix: $y = -1$

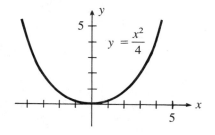

3. focus: $(1/2, 0)$; directrix: $x = -1/2$

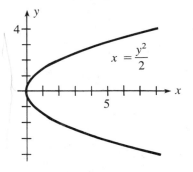

5.   focus: $(0, -5/4)$; directrix: $y = 5/4$

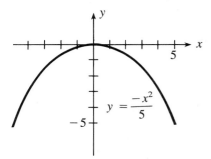

$$y = \frac{-x^2}{5}$$

7.   focus: $(3, 0)$; directrix: $x = -3$

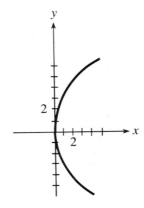

9. $y^2 = 4x$ 11. $y^2 = 6x$

17. $y^2 = -4x$ 19. $y^2 = x$

13. $y^2 = \frac{1}{2}x$ 15. $y^2 = -5x$

21.   downward   23.   to the left

EXERCISE SET 9.4, page 381

1. $\dfrac{x^2}{16} + \dfrac{y^2}{25} = 1$; foci: $(0, \pm 3)$; vertices:

$(0, \pm 5), (\pm 4, 0)$

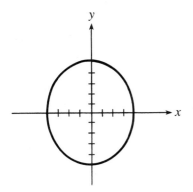

3. $\dfrac{x^2}{4} + \dfrac{y^2}{36} = 1$; foci: $(0, \pm 4\sqrt{2})$; vertices:

$(0, \pm 6), (\pm 2, 0)$

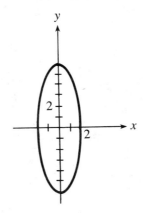

5. $\dfrac{x^2}{64} + \dfrac{y^2}{25} = 1$; foci: $(\pm\sqrt{39},\ 0)$; vertices: $(\pm 8,\ 0),\ (0,\ \pm 5)$

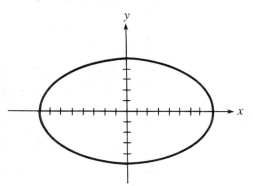

7. $\dfrac{x^2}{9/4} + \dfrac{y^2}{9} = 1$; foci: $(0,\ \pm 3\sqrt{3}/2)$; vertices: $(0,\ \pm 3),\ (\pm 3/2,\ 0)$

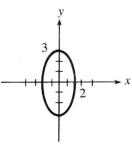

9. $\dfrac{x^2}{20} + \dfrac{y^2}{4} = 1$; foci: $(\pm 4,\ 0)$; vertices: $(\pm 2\sqrt{5},\ 0),\ (0,\ \pm 2)$

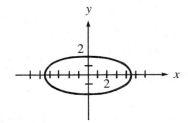

11. $\dfrac{x^2}{40} + \dfrac{y^2}{49} = 1$

13. $\dfrac{x^2}{4} + \dfrac{y^2}{9} = 1$

15. $\dfrac{x^2}{16} + \dfrac{y^2}{1} = 1$

17. $\dfrac{x^2}{1} + \dfrac{y^2}{9} = 1$

19. $\dfrac{x^2}{25} + \dfrac{y^2}{9} = 1$

21. $\dfrac{x^2}{81} + \dfrac{y^2}{1} = 1$

23. $\dfrac{x^2}{49} + \dfrac{y^2}{9/4} = 1$

25. *y*-axis

27. *x*-axis

EXERCISE SET 9.5, page 389

1. foci: $(\pm\sqrt{10}, 0)$; vertices: $(\pm 3, 0)$

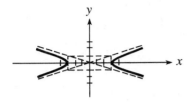

3. foci: $(0, \pm\sqrt{13})$; vertices: $(0, \pm 2)$

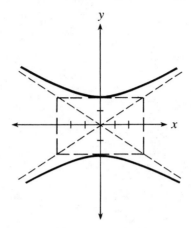

5. foci: $(\pm 3\sqrt{2}, 0)$; vertices: $(\pm 3, 0)$

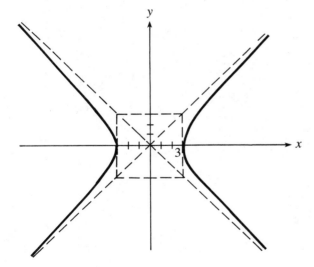

7. foci: $(0, \pm 5\sqrt{5})$; vertices: $(0, \pm 5)$

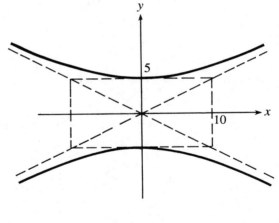

9. x-axis

11. y-axis

13. $\dfrac{y^2}{16} - \dfrac{x^2}{9} = 1$

15. $\dfrac{x^2}{4} - \dfrac{y^5}{5} = 1$

17. $\dfrac{y^2}{1} - \dfrac{x^2}{80} = 1$

19. $\dfrac{y^2}{9} - \dfrac{x^2}{9} = 1$

21. $\dfrac{y^2}{4} - \dfrac{x^2}{16} = 1$

23. $\dfrac{y^2}{81/10} - \dfrac{x^2}{9/10} = 1$

25. $\dfrac{y^2}{16} - \dfrac{x^2}{400/9} = 1$

27. $\dfrac{y^2}{9/2} - \dfrac{x^2}{9/2} = 1$

EXERCISE SET 9.6, page 396

1. $(1, -4)$

3. $(5, -1)$

5. $(-3, 4)$

7. $(1, 7)$

9.

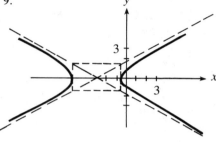

11.

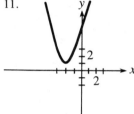

13.

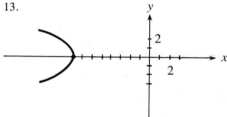

15.

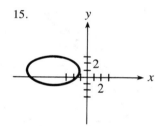

17.

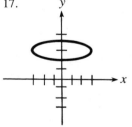

19. parabola

21. ellipse

23. hyperbola

25. hyperbola

EXERCISE SET 9.7, page 404

1. $2x'^2 + 8y'^2 = 36$

3. $y' = -2$

5. $9x'^2 - y'^2 - 18x' - 6y' - 9 = 0$

7. $y' = 6x'$

9. $-3x'^2 - 7y'^2 - 3x' - 3\sqrt{3}y' = 0$

11. $\dfrac{x'^2}{1/2} + y'^2 = 1$

13. $(x' - 1)^2 = -4(y' + 2)$

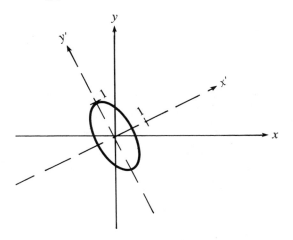

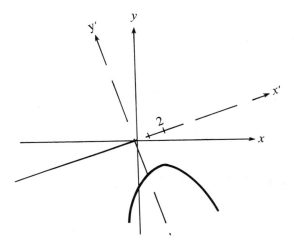

15. $\dfrac{(x' - 3)^2}{4} + y'^2 = 1$

17. $\dfrac{(y' - 3)^2}{4} - \dfrac{(x' - 2)^2}{1} = 1$

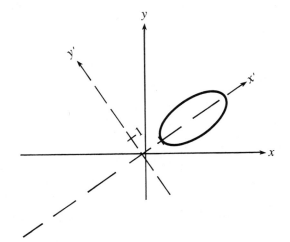

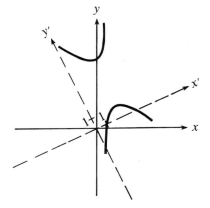

19. $(x' + 3)^2 + \dfrac{(y' - 1)^2}{4} = 1$

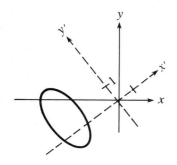

EXERCISE SET 9.8, page 411

1. (a)

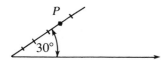

(b)

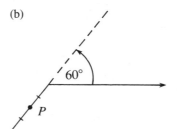

(c)

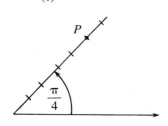

(d)

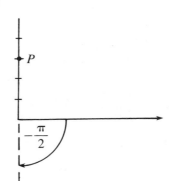

3. (a) $(6, 495°)$, $(-6, 315°)$
    (c) $(4, 17\pi/6)$, $(-4, 11\pi/6)$

(b) $(-2, 480°)$, $(2, 300°)$
(d) $(-4, \pi/4)$, $(4, 5\pi/4)$

5. (a) $(2, 570°)$
    (c) $(3, 11\pi/3)$

(b) $(4, 480°)$
(d) $(1, 11\pi/6)$

7. (a) yes
    (c) yes

(b) no
(d) no

9.   (a) $(5\sqrt{3}/2, -5/2)$ (b) $(0, -2)$

     (c) $(2\sqrt{3}, 2)$ (d) $(3/2, 3\sqrt{3}/2)$

11.   (a) $(2\sqrt{2}, 3\pi/4)$ (b) $(2, -\pi/3)$

     (c) $(2, \pi/6)$ (d) $(4\sqrt{2}, -3\pi/4)$

13.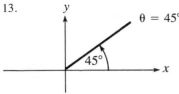

15.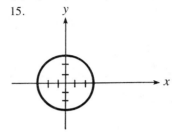

17.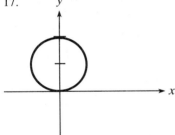

19.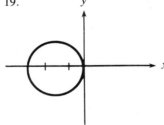

21.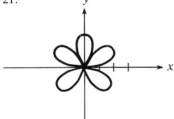

23.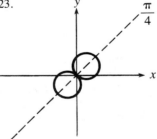

25.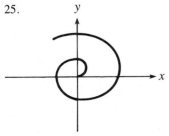

27. $x^2 + y^2 = 16$ 29. $x^2 + y^2 + 2y = 0$ 31. $x = 2$
33. $r = 5$ 35. $r = -5 \sec \theta$ 37. $\tan \theta = 3$

EXERCISE SET 9.9, page 415

1. $y = \frac{1}{2}x + \frac{1}{2}$

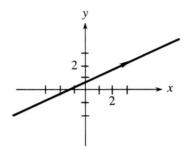

3. $y = x^{3/2}$

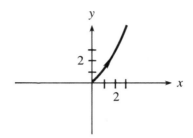

5. $y = 8x^3$

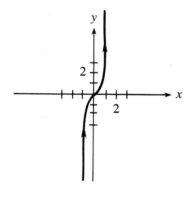

7. $x^3 + y^3 = 3xy$

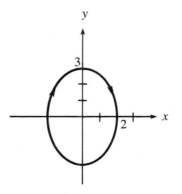

9. $\dfrac{x^2}{16} - \dfrac{y^2}{9} = 1$

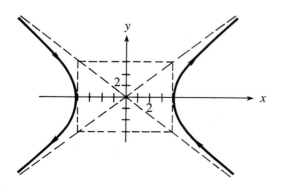

11. $\dfrac{x^2}{4} + \dfrac{y^2}{9} = 1$

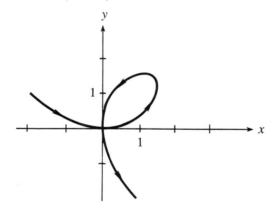

REVIEW EXERCISES, page 416

1. $(-1, -1)$
2. $(-5/2, 5/2)$
3. $(10, 7)$
6. $y = -\frac{5}{8}x - \frac{5}{4}$
7. $(x + 5)^2 + (y - 2)^2 = 16$
8. $(x + 3)^2 + (y - 3)^2 = 4$
9. center: $(2, -3)$; radius: 3
10. center: $(-2, 3)$; radius: $\sqrt{3}$
11. focus: $(0, -1/6)$; directrix: $y = 1/6$

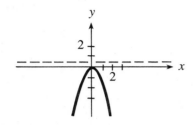

12. focus: $(-1/6, 0)$; directrix: $x = 1/6$

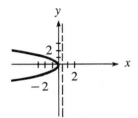

13. $x^2 = -7y$
14. $y = \frac{5}{2}x^2$
15. vertices: $(\pm 1/3, 0)$, $(0, \pm 3/2)$; foci: $(0, \pm\sqrt{77}/6)$

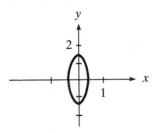

16. vertices: $(-1, 4)$, $(-1, -2)$, $(-5/2, 1)$, and $(1/2, 1)$; foci: $(-1, 1 \pm 3\sqrt{3}/2)$

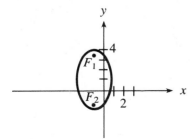

17. $\dfrac{x^2}{81} + \dfrac{y^2}{72} = 1$ 18. $\dfrac{x^2}{1} + \dfrac{y^2}{5} = 1$

19. vertices: $(\pm\sqrt{3}/2, -1)$; foci: $(\pm\sqrt{15}/2, -1)$; asymptotes: $y + 1 = \pm 2x$.

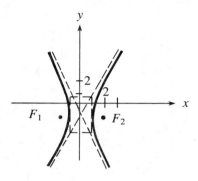

20. vertices: $(0, \pm 2/\sqrt{7})$; foci: $(0, \pm 4\sqrt{2/7}$; asymptotes: $y = \pm\sqrt{7}x/7$

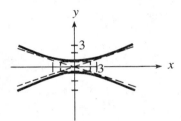

21. $\dfrac{y^2}{4} - \dfrac{x^2}{45} = 1$ 22. $\dfrac{x^2}{4} - \dfrac{y^2}{4/5} = 1$ 23. ellipse

24. parabola   25. no graph   26. hyperbola
27. two intersecting lines 28. hyperbola   29. hyperbola
30. ellipse   31. hyperbola   32. parabola
33. ellipse   34. ellipse

35.

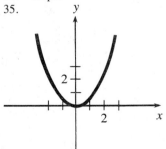

36.

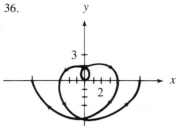

37. $y = 3$

38. $r^2 = \dfrac{1}{1 - (8 \cos^2 \theta)/9}$

39.

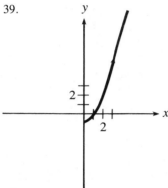

PROGRESS TEST 9A, page 418

1. $(0, 4)$
4. $(x - 2)^2 + (y + 3)^2 = 36$
6. vertex: $(-3, 1)$; axis: $x = -3$

2. $(-1, 2)$
5. center: $(1, -2)$; radius: $\sqrt{6}$

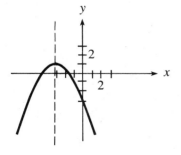

7. $\dfrac{x^2}{4} + \dfrac{y^2}{1} = 1$; intercepts: $(\pm2, 0), (0, \pm1)$

8. $\dfrac{y^2}{9} - \dfrac{x^2}{4} = 1$; intercept: $(0, \pm3)$

9. $\dfrac{x^2}{1/4} - \dfrac{y^2}{1/4} = 1$; intercept: $(\pm1/2, 0)$

10.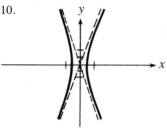

11. ellipse

12. parabola

13. parabola

14. two intersecting lines

15. ellipse

16.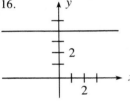

17. $4x + y^2 - 4 = 0$

18. $r^2 = 2/(2 + \sin 2\theta)$

19.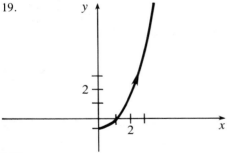

20. $\dfrac{(x - 1)^2}{16} + \dfrac{(y + 1)^2}{9} = 1$; ellipse

PROGRESS TEST 9B, page 418

1. $(-1/2, -1)$

2. $(-7, 1)$

4. $(x + 2)^2 + (y + 5)^2 = 25$

5. center: $(1/2, 1)$; radius: $\sqrt{10}$

6. vertex: $(1/3, -1/3)$; axis: $x = 1/3$

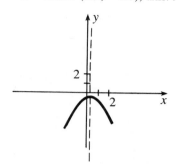

7. $\dfrac{x^2}{5} + \dfrac{y^2}{25/9} = 1$; intercepts: $(\pm\sqrt{5},\ 0),\ (0,\ \pm5/3)$

8. $\dfrac{x^2}{3} + \dfrac{y^2}{7/2}$; intercepts: $(\pm\sqrt{3},\ 0),\ (0,\ \pm\sqrt{7/2})$

9. $\dfrac{y^2}{9} + \dfrac{x^2}{3} = 1$; intercept: $(0,\ \pm3)$

10.

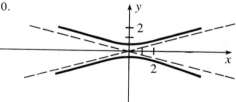

11. hyperbola
14. two intersecting lines
16.

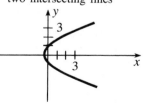

12. point $(-2, 3)$
15. hyperbola

13. hyperbola

17. $x^2 + y^2 - 4x + y = 0$

18. $r^2 = 2/(2\cos 2\theta - \sin 2\theta)$

19.

20. $x + 1 = (y - 2)^2$; parabola

CHAPTER 10
EXERCISE SET 10.1, page 426
1. $x = 2, y = -1$
3. $x = 3, y = 2; x = 1/5, y = -18/5$
5. $x = 1, y = 1; x = 9/16, y = -3/4$
7. $x = 1, y = 2; x = 13/5, y = -6/5$
9. $x = (-1 + \sqrt{5})/2, y = (1 + \sqrt{5})/2; x = (-1 - \sqrt{5})/2, y = (1 - \sqrt{5})/2$
11. $x = 3, y = -1$
13. no solution
15. $x = 3, y = 2; x = 3, y = -2$
17. none
19. $x = 3, y = 2; x = -3, y = 2; x = 3, y = -2; x = -3, y = -2$
21. I
23. I
25. I
27. C; all points on the line $3x - y = 18$
29. C; $x = 1, y = -1; x = 5/2, y = 1/2$
31. I
33. 22 nickels, 12 quarters
35. 6/5 pounds nuts, 4/5 pounds raisins
37. 125 type-A, 75 type-B
39. 6 and 8
41. 4 and 5

EXERCISE SET 10.2, page 434

1.
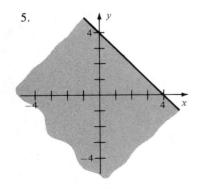

3.
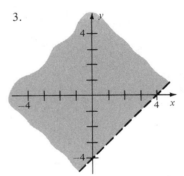

5.
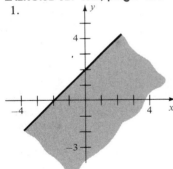

7.
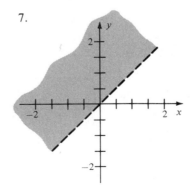

9.

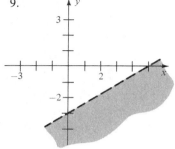

11.

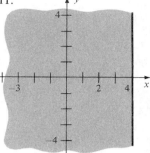

13.

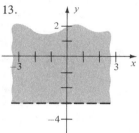

15.

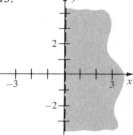

17.

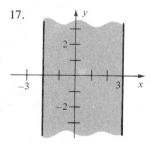

19. $2x + 5y \leq 15;\ x \geq 0;\ y \geq 0$

21.

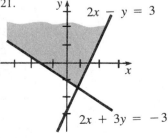

23.

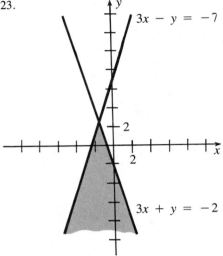

$3x - y = -7$

$3x + y = -2$

25.

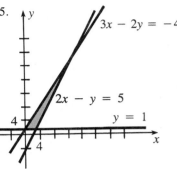

$3x - 2y = -4$

$2x - y = 5$

$y = 1$

27.

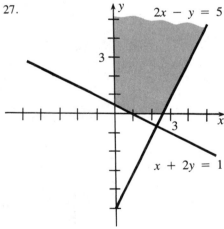

$2x - y = 5$

$x + 2y = 1$

29.   no solution

31.

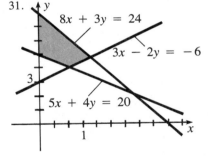

$8x + 3y = 24$

$3x - 2y = -6$

$5x + 4y = 20$

33.

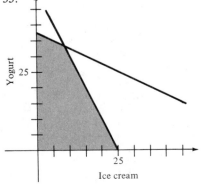

Yogurt

Ice cream

35.

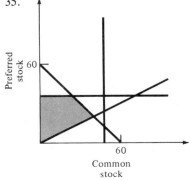

Preferred stock

60

60

Common stock

EXERCISE SET 10.3, page 439

Minimum	Maximum
1. $-2; (5, 4)$	$5; (5, 0)$
3. $-3; (2, 2)$	$3; (6, 0)$
5. $\frac{6}{7}; \left(\frac{6}{7}, \frac{6}{7}\right)$	$5; (4, 3)$
7. $\frac{1}{2}; \left(3, \frac{11}{2}\right)$	$14; (8, 2)$

9. preferred: 190/3 square feet
regular: 30 square feet
13. Java: 2000/11 pounds
Colombian: 4000/11 pounds
17. pack A: 6 pounds
pack B: 12 pounds

11. large: 120
small: 260
15. crop A: 30 acres
crop B: 70 acres

EXERCISE SET 10.4, page 445

1. $x = 2, y = -1, z = -2$
5. no solution
9. $x = 1, y = 1, z = 0$
13. no solution
17. $x = 5, y = -5, z = -20$
21. three 12″-sets, eight 16″-sets, five 19″-sets

3. $x = 1, y = 2/3, z = -2/3$
7. $x = 1, y = 2, z = 2$
11. $x = 1, y = 27/2, z = -5/2$
15. no solution
19. A: 2; B: 3; C: 3

EXERCISE SET 10.5, page 454

1. 2×2
3. 4×3
5. 3×3
7. (a) -4 (b) 7 (c) 6 (d) -3

9. $\begin{bmatrix} 3 & -2 \\ 5 & 1 \end{bmatrix}, \begin{bmatrix} 3 & -2 & | & 12 \\ 5 & 1 & | & -8 \end{bmatrix}$

11. $\begin{bmatrix} 1/2 & 1 & 1 \\ 2 & -1 & -4 \\ 4 & 2 & -3 \end{bmatrix}, \begin{bmatrix} 1/2 & 1 & 1 & | & 4 \\ 2 & -1 & -4 & | & 6 \\ 4 & 2 & -3 & | & 8 \end{bmatrix}$

13. $\frac{3}{2}x + 6y = -1$
$4x + 5y = 3$

15. $x + y + 3z = -4$
$-3x + 4y = 8$
$2x + 7z = 6$

17. $x = -13, y = 8, z = 2$
19. $x = 35, y = 14, z = -4$
21. $x = 2, y = 3$
23. $x = 2, y = -1, z = 3$
25. $x = 3, y = 2, z = -1$
27. $x = -5, y = 2, z = 3$
29. $x = -5/7, y = -2/7, z = -3/7, w = 2/7$

EXERCISE SET 10.6, page 463

1. $a = 3, b = -4, c = 6, d = -2$

3. $\begin{bmatrix} 2 & -1 & 5 \\ 7 & 1 & 6 \\ 4 & 3 & 7 \end{bmatrix}$

5. $\begin{bmatrix} -2 & 18 & 8 \\ -3 & 8 & 11 \end{bmatrix}$

7. not possible

9. $\begin{bmatrix} 17 & 5 \\ 10 & 12 \end{bmatrix}$

11. not possible

13. $\begin{bmatrix} 10 & 4 \\ 12 & 28 \end{bmatrix}$

15. not possible

17. $\begin{bmatrix} 18 & 23 & 29 \\ 17 & -12 & 13 \end{bmatrix}$

19. $AB = \begin{bmatrix} 8 & -6 \\ -8 & 6 \end{bmatrix} \quad AC = \begin{bmatrix} 8 & -6 \\ -8 & 6 \end{bmatrix}$

25. The amount of pesticide 2 eaten by herbivore 3.

27. $A = \begin{bmatrix} 3 & 4 \\ 3 & -1 \end{bmatrix}, \quad X = \begin{bmatrix} x \\ y \end{bmatrix}, \quad B = \begin{bmatrix} -3 \\ 5 \end{bmatrix}$

29. $A = \begin{bmatrix} 3 & -1 & 4 \\ 2 & 2 & 3/4 \\ 1 & -1/4 & 1 \end{bmatrix}, \quad X = \begin{bmatrix} x \\ y \\ z \end{bmatrix}, \quad B = \begin{bmatrix} 5 \\ -1 \\ 1/2 \end{bmatrix}$

31. $\begin{aligned} x_1 - 5x_2 &= 0 \\ 4x_1 + 3x_2 &= 2 \end{aligned}$

33. $\begin{aligned} 4x_1 + 5x_2 - 2x_3 &= 2 \\ 3x_2 - x_3 &= -5 \\ 2x_3 &= 4 \end{aligned}$

EXERCISE SET 10.7, page 474

1. no

3. yes

5. $\begin{bmatrix} 2/3 & 5/6 \\ 1/3 & 1/6 \end{bmatrix}$

7. $\begin{bmatrix} 1 & -1 \\ 2 & -1 \end{bmatrix}$

9. $\begin{bmatrix} 8 & 7 & -1 \\ -4 & -4 & 1 \\ -5 & -5 & 1 \end{bmatrix}$

11. $\begin{bmatrix} 4/7 & -3/7 \\ 1/7 & 1/7 \end{bmatrix}.$

13. none

15. $\begin{bmatrix} 1/2 & 0 \\ 0 & -1/3 \end{bmatrix}$

17. $\begin{bmatrix} -1 & 1 & -1 \\ 2 & -1 & 2 \\ -2 & 1 & -1 \end{bmatrix}$

19. $x = 3, y = -1$

21. $x = 2, y = -3, z = 1$

23. $x = 0, y = 2, z = -3$

43. $x = \begin{bmatrix} 25 \\ 3 \\ 1 \end{bmatrix}, \begin{bmatrix} -2 \\ 5 \\ 10 \end{bmatrix}$

EXERCISE SET 10.8, page 484

1. 22
3. -8
5. 0
7. (a) -6 (b) -1 (c) 1 (d) 7
9. (a) -6 (b) 1 (c) 1 (d) 7
11. 52
13. -3
15. 0
17. -12
19. 0
21. $x = 1, y = -2, z = -1$
23. $x = 3, y = 2, z = -1$
25. $x = -3, y = 0, z = 2$
27. $x = -5/7, y = -2/7, z = -3/7, w = 2/7$

REVIEW EXERCISES, page 487

1. $x = 0, y = -2$
2. $x = -5, y = 1$
3. $x = 5, y = 0; x = -4, y = 3$
4. $x = 1, y = -1; x = 5, y = 3$
5. $x = -7/20, y = -7/10$
6. $x = 2, y = -2$
7. $x = 1/4, y = -1/2$
8. $x = 0, y = 3$
9. $x = 4, y = 4; x = 36/25, y = -12/5$
10. $x = 3, y = -4$
11. 3, 11

12.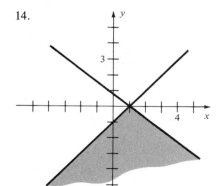

13.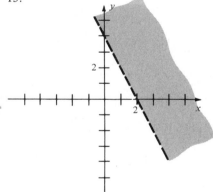

14.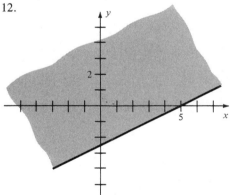

15.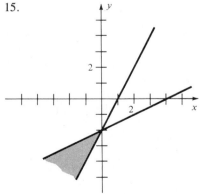

16.

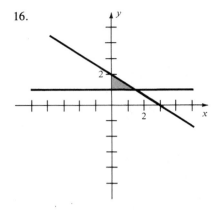

17.

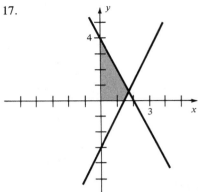

18. $x = 4$, $y = 6$, $z = 26$ 19. $x = 11/2$, $y = 2$, $z = 27/2$

20. $x = -3$, $y = 1$, $z = 4$ 21. $x = -2$, $y = 1/2$, $z = 3$ 22. $x = 1$, $y = -1$, $z = 2$

23. 3×5 24. -1 25. 4   26. 8

27. $\begin{bmatrix} 3 & -7 \\ 1 & 4 \end{bmatrix}$ 28. $\begin{bmatrix} 3 & -7 & | & 14 \\ 1 & 4 & | & 6 \end{bmatrix}$

29. $4x - y = 3$
     $2x + 5y = 0$ 30. $-2x + 4y + 5z = 0$
                                          $6x - 9y + 4z = 0$
                                          $3x + 2y - z = 0$

31. $x = -1$, $y = -4$ 32. $x = -1$, $y = 1$, $z = -3$

33. $x = 1/2$, $y = 3/2$ 34. $x = 3 + 5t/4$, $y = 3 + t/2$, $z = t$

35. -3 36. -3 37. $\begin{bmatrix} 1 & 4 \\ 7 & -1 \end{bmatrix}$

38. $\begin{bmatrix} -3 & 6 \\ 1 & -5 \end{bmatrix}$ 39. not possible   40. $\begin{bmatrix} 5 & 15 & 20 \\ -5 & 0 & -30 \end{bmatrix}$

41. $\begin{bmatrix} -1 & -3 & -4 \\ -4 & 0 & -24 \\ 4 & 6 & 20 \end{bmatrix}$ 42. $\begin{bmatrix} 7 & 4 \\ -11 & 12 \end{bmatrix}$ 43. $\begin{bmatrix} 6 & -13 \\ -5 & -9 \end{bmatrix}$

44. $\begin{bmatrix} -4/11 & 3/11 \\ 1/11 & 2/11 \end{bmatrix}$ 45. $\begin{bmatrix} 2/5 & -7/5 & -8/5 \\ -1 & 3 & 4 \\ -2/5 & 2/5 & 3/5 \end{bmatrix}$ 46. $x = 2$, $y = 3$

47. $x = -1$, $y = -1$, $z = 1/2$ 48. 10   49. -6

50. 0   51. -3 52. $x = 1/2$, $y = 4$

53. $x = 1$, $y = -4$ 54. $x = 1/3$, $y = 2/3$, $z = -1$ 55. $x = 1/4$, $y = -2$, $z = 1/2$

PROGRESS TEST 10A, page 489

1. $x = 2, y = \pm\sqrt{10}; x = 3, y = \pm\sqrt{15}$

2. $x = 2, y = -1$

3. $x = 3, y = \pm 4; x = -3, y = \pm 4$

4. shirts: \$15; ties: \$10

5.

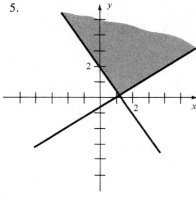

6.

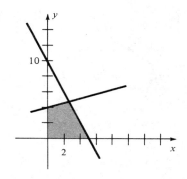

7. $x = -2, y = 4, z = 6$

8. 3×2

9. 0

10. $\begin{bmatrix} -7 & 0 & 6 & | & 3 \\ 0 & 2 & -1 & | & 10 \\ 1 & -1 & 1 & | & 5 \end{bmatrix}$

11. $-5x + 2y = 4$
    $3x - 4y = 4$

12. $x = -1/2, y = 1/2$

13. $x = -6, y = -2$

14. $x = 1/2, y = 1/2, z = 1/2$

15. 3

16. $\begin{bmatrix} 2 & 14 \\ -2 & -4 \\ -5 & 1 \end{bmatrix}$

17. $\begin{bmatrix} -7 & -11 \\ 11 & 15 \end{bmatrix}$

18. $\begin{bmatrix} -10 \\ 2 \\ 0 \end{bmatrix}$

19. not possible

20. $\begin{bmatrix} 1/27 & 12/27 & 4/27 \\ 5/27 & 6/27 & -7/27 \\ 7/27 & 3/27 & 1/27 \end{bmatrix}$

21. $x = -2, y = 1$

22. -2

23. 27

24. $x = 4, y = -3$

PROGRESS TEST 10B, page 490

1. $x = 5, y = \pm 4; x = -5, y = \pm 4$

2. $x = 1, y = -3$

3. $x = 0, y = 2; x = 3, y = 1$

4. 300 of each

5.

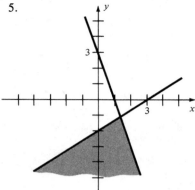

6.

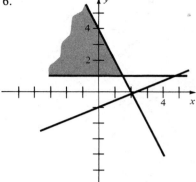

7. $x = -3, y = -10, z = 5$
8. 2×4
9. 1

10. $\begin{bmatrix} 2 & -6 & | & 5 \\ 1 & 3 & | & -2 \end{bmatrix}$

11. $\begin{aligned} 16x \quad\quad + 6z &= 10 \\ -4x - 2y + 5z &= 8 \\ 2x + 3y - z &= -6 \end{aligned}$

12. $x = 0, y = -1, z = -3$
13. $x = -1, y = -3$
14. $x = 5, y = -2, z = 1/2$
15. -5

16. $[10, 0]$

17. $\begin{bmatrix} 6 & -17 \\ 3 & 2 \\ 2 & 13 \end{bmatrix}$

18. not possible
19. not possible

20. $\begin{bmatrix} -3/2 & -5/2 & -3/4 \\ -1/2 & -1/2 & -1/4 \\ -3/4 & -5/4 & -1/8 \end{bmatrix}$

21. $x = -1, y = 1/2$

22. -22
23. -1
24. $x = -3/2, y = 1/2$

CHAPTER 11
EXERCISE SET 11.1, page 501

1. 2, 4, 6, 8; 40
3. 1, 5, 9, 13; 77
5. 5, 5, 5, 5; 5
7. 1/2, 2/3, 3/4, 4/5; 20/21
9. 2.1, 2.01, 2.001, 2.0001; $2 + 0.1^{20}$
11. 1/3, 4/5, 9/7, 16/9; 400/41
13. 9
15. 4/7
17. 256
19. 40
21. 97

23. 49/12
25. 80
27. $\displaystyle\sum_{k=1}^{5} (2k - 1)$

29. $\displaystyle\sum_{k=1}^{5} k^2$

31. $\displaystyle\sum_{k=1}^{4} \frac{(-1)^k}{\sqrt{k}}$

33. $\displaystyle\sum_{k=1}^{4} \frac{(-1)^{k+1}k}{k^2+1}$

35. $\displaystyle\sum_{k=0}^{n} \frac{1}{x^k}$

EXERCISE SET 11.2, page 507

1. 15, 18
3. 1, 5/4
5. 4, 5
7. $\sqrt{5}+6$, $\sqrt{5}+8$
9. 2, 6, 10, 14
11. 3, 5/2, 2, 3/2
13. 1/3, 0, $-1/3$, $-2/3$
15. 25
17. -8
19. -2
21. 19/3
23. 821/160
25. 440
27. -126
29. 1720
31. 30
33. $n = 30$, $d = 3$
35. $n = 6$, $d = 1/4$
37. -2

EXERCISE SET 11.3, page 516

1. 48
3. $-81/64$
5. 0.0096
7. 3, 9, 27, 81
9. 4, 2, 1, 1/2
11. $-3, -6, -12, -24$
13. -384
15. 1/4
17. 1/243
19. 27/8
21. 2
23. 7
25. 1, 3
27. 1/4, 1/16
29. 1093/243
31. $-1353/625$
33. 1020
35. 55/8
37. \$10,235
39. 58,594
41. 2
43. 3/4
45. 8/3
47. 1
49. 1/5
51. 364/990
53. 325/999

EXERCISE SET 11.5, page 527

1. $243x^5 + 810x^4y + 1080x^3y^2 + 720x^2y^3 + 240xy^4 + 32y^5$
3. $256x^4 - 256x^3y + 96x^2y^2 - 16xy^3 + y^4$
5. $32 - 80xy + 80x^2y^2 - 40x^3y^3 + 10x^4y^4 - x^5y^5$
7. $a^8b^4 + 12a^6b^3 + 54a^4b^2 + 108a^2b + 81$
9. $a^8 - 16a^7b + 112a^6b^2 - 448a^5b^3 + 1120a^4b^4 - 1792a^3b^5 + 1792a^2b^6 - 1024ab^7 + 256b^8$
11. $\frac{1}{27}x^3 + \frac{2}{3}x^2 + 4x + 8$
13. $1024 + 5120x + 11,520x^2 + 15,360x^3$
15. $19,683 - 118,098a + 314,928a^2 - 489,888a^3$
17. $16,384x^{14} - 344,064x^{13}y + 3,354,624x^{12}y^2 - 20,127,744x^{11}y^3$
19. $8192x^{13} - 53,248x^{12}yz + 159,744x^{11}y^2z^2 - 292,864x^{10}y^3z^3$
21. 120
23. 12
25. 990
27. 5040
29. 120
31. 210
33. $-35,840x^4$
35. $\frac{495}{256}x^8y^4$
37. $2016x^{-5}$
39. $-540x^3y^3$
41. $181,440x^4y^3$
43. $-144x^6$
45. $\frac{35}{8}x^{12}$
47. 4.8268

REVIEW EXERCISES, page 529

1. 3, 7, 13; 111
2. 0, 7/3, 13/2; 999/11
3. 2
4. 120
5. -16
6. 62
7. 50

8. $\displaystyle\sum_{k=1}^{4} \frac{k}{k+2}$

9. $\displaystyle\sum_{k=0}^{4} (-1)^k x^k$

10. $\displaystyle\sum_{k=1}^{n} \sin kx$

11. 38   12. -9 13. 8   14. $-33/2$
15. 275/3   16. -450 17. -3 18. $-3/2$
19. 5, 1, 1/5, 1/25   20. $-2, 2, -2, 2$ 21. 243/8   22. ±256
23. 1/2, 1/12   24. 21/32   25. -728 26. 10
27. 9/5   29. $16x^4 - 32x^3y + 24x^2y^2 - 8xy^3 + y^4$
30. $x^4/16 - x^3 + 6x^2 - 16x + 16$ 31. $x^6 + 3x^4 + 3x^2 + 1$
32. 720   33. 78   34. $(n + 1)/n$
35. 15   36. 1   37. 90

PROGRESS TEST 11A, page 530

1. 1/4, 2/9, 3/16, 4/25   2. 29/6   3. $-1, 1/2, 2, 7/2$
4. 8   5. 41   6. 55/2
7. 1/3   8. $-2/3, -4/3, -8/3, -16/3$ 9. -2
10. 8, -16 11. $-43/8$ 12. -6
14. $a^{10} + 10a^9/b + 45a^8/b^2 + 120a^7/b^3$
                               15. 22

PROGRESS TEST 11B, page 530

1. 5/3, 5, 51/5, 52/3

2. $\displaystyle\sum_{k=2}^{n} k!$

3. 6, 16/3, 14/3, 4   4. 10   5. -55 6. 13, 1/4
7. 0.2   8. $-1, 1/4, -1/16, 1/64$ 9. $-1/81$ 10. $-4, -8/3$
11. $-31/2$ 12. 25/7   14. $11520x^8$ 15. $n/(n + 1)$

APPENDIX A, page 536

1. \$11,636.36 on 10-speeds, \$4363.64 on 3-speeds   3. \$7000
5. 20 hours   7. 50 and 54 mph   9. 1/12, 1/4; $-1/4, -1/12$
11. 4 meters and 8 meters   13. 40 kph, 80 kph   15. 140 mph
17. $(5F - 160)/9$ 19. $(2A/h) - b'$ 21. $f_1f/(f_1 - f)$
23. $(a + rS)/(r + S)$ 25. $s > 60,000$ sq. ft.
27. roofer: 6 hours; assistant: 12 hours   29. 4 students

APPENDIX B, page 546

1. (a)

x	4.4	4.2	4.1	4.01	4.001
$2x - 3$	5.8	5.4	5.2	5.02	5.002

$$\lim_{x \to 4} (2x - 3) = 5$$

x	3.6	3.8	3.9	3.99	3.999
$2x - 3$	4.2	4.6	4.8	4.98	4.998

(b)

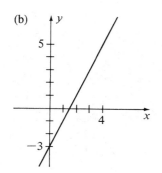

3. 6
9. 12
15. (a) 5 (b)

5. 11
11. −2

7. −4
13. −12

17.

$$\lim_{x \to 4} f(x) = 8$$

19.

$$\lim_{x \to -2} f(x) = -3$$

21.

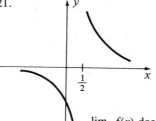

$$\lim_{x \to 1/2} f(x) \text{ does not exist}$$

23.

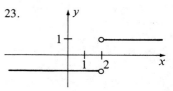

$$\lim_{x \to 2} f(x) \text{ does not exist}$$

25. does not exist

27. does not exist

SOLUTIONS TO SELECTED REVIEW EXERCISES

CHAPTER 1

1. T. (Irrational numbers are a subset of the real numbers.)
3. F. (The negative integers and zero are a subset of the integers.)
12. $|-3| - |1 - 5|$
$$= |-3| - |-4| \qquad \text{Performing operations within bars first}$$
$$= 3 - 4 \qquad \text{Definition of absolute value}$$
$$= -1$$
13. $\overline{PQ} = |9/2 - 6| = |-3/2| = 3/2$
15. c. (Every exponent of a polynomial must be a nonnegative integer.)
19. $x(2x - 1)(x + 2)$
$$= (2x^2 - x)(x + 2)$$
$$= 2x^3 + 3x^2 - 2x$$
23. $2a^2 + 3ab + 6a + 9b$
$$= (2a^2 + 6a) + (3ab + 9b) \qquad \text{Grouping}$$
$$= 2a(a + 3) + 3b(a + 3) \qquad \text{Common factors } 2a, \ 3b$$
$$= (a + 3)(2a + 3b) \qquad \text{Common factor } a + 3$$
25. $x^8 - 1 = (x^4)^2 - (1)^2 \qquad\qquad\quad$ Difference of squares: $a = x^4, \ b = 1$
$$= (x^4 + 1)(x^4 - 1) \qquad\qquad \text{Factoring } a^2 - b^2 = (a + b)(a - b)$$
$$= (x^4 + 1)(x^2 + 1)(x^2 - 1) \qquad \text{Another difference of squares: } x^4 - 1$$
$$= (x^4 + 1)(x^2 + 1)(x + 1)(x - 1) \quad \text{Another difference of squares: } x^2 - 1$$
27. $(2a^2b^{-3})^{-3} = (2)^{-3}(a^2)^{-3}(b^{-3})^{-3} = \frac{1}{8}a^{-6}b^9 = \dfrac{b^9}{8a^6}$

31. $\sqrt{80} = \sqrt{16 \cdot 5} = \sqrt{16} \cdot \sqrt{5} = 4\sqrt{5}$
33. $\dfrac{\sqrt{x}}{\sqrt{x} + \sqrt{y}} = \dfrac{\sqrt{x}}{\sqrt{x} + \sqrt{y}} \cdot \dfrac{\sqrt{x} - \sqrt{y}}{\sqrt{x} - \sqrt{y}} = \dfrac{x - \sqrt{xy}}{x - y}$

34. Equate the real and the imaginary parts.
$$x - 2 = -4 \qquad 2y - 1 = 7$$
$$x = -2 \qquad\qquad y = 4$$

35. $i^{47} = i^{44} \cdot i^3 = i^3 = -i$
37. $(2 + i)(2 + i) = 4 + 2i + 2i + i^2$
$$= 4 + 4i - 1$$
$$= 3 + 4i$$
42. $k - 2x = 4kx$
$$k = 4kx + 2x$$
$$k = x(4k + 2)$$
$$x = \dfrac{k}{4k + 2} = \dfrac{k}{2(2k + 1)}$$

43. F. (The equation does not hold for $x = 0$, and therefore it does not hold for all real values of x.)

46. $-4 < -2x + 1 \le 10$

    $-5 < -2x \le 9$

    $\dfrac{-5}{-2} > \dfrac{-2x}{-2} \ge \dfrac{9}{-2}$

    $\dfrac{5}{2} > x \ge -\dfrac{9}{2}$

    or

    $-\dfrac{9}{2} \le x < \dfrac{5}{2}$

48. Since the numerator is negative, the denominator must be positive if the quotient is to be negative. Note also that the denominator cannot equal 0.

$$2x - 5 > 0$$
$$x > 5/2 \quad \text{or} \quad (5/2, \infty)$$

50. $|3x + 2| = 7$

    $3x + 2 = 7 \qquad -(3x + 2) = 7$

         $3x = 5 \qquad\qquad -3x = 9$

           $x = 5/3 \qquad\qquad x = -3$

53. $|2 - 5x| < 1$

    $-1 < 2 - 5x < 1$

    $-3 < -5x < -1$

    $\dfrac{-3}{-5} > \dfrac{-5x}{-5} > \dfrac{-1}{-5}$

    $\dfrac{3}{5} > x > \dfrac{1}{5}$

    $\dfrac{1}{5} < x < \dfrac{3}{5} \quad \text{or} \quad \left(\dfrac{1}{5}, \dfrac{3}{5} \right)$

56. $2x + 1 \qquad - - - - - 0 + +$

    $x + 5 \qquad - - 0 + + + + +$

    $\dfrac{2x + 1}{x + 5} \qquad + + \quad - - 0 + +$

    $\{x | x < -5 \quad \text{or} \quad x \ge -1/2\} \quad \text{or} \quad (-\infty, -5), [-1/2, \infty)$

    Exclude $x = -5$ since the denominator cannot equal 0.

CHAPTER 2

1. $d = \sqrt{(x_2 - x_1)^2 + (y_2 - y_1)^2}$
   $= \sqrt{(2 + 4)^2 + (-1 + 6)^2}$
   $= \sqrt{36 + 25} = \sqrt{61}$

5.

y-axis test	x-axis test	origin test
Replace x with $-x$:	Replace y with $-y$:	Replace both:
$y^2 = 1 - (-x)^3$	$(-y)^2 = 1 - x^3$	$(-y)^2 = 1 - (-x)^3$
$y^2 = 1 + x^3$	$y^2 = 1 - x^3$	$y^2 = 1 + x^3$
no	yes	no

7. Yes. No vertical line meets the graph in more than one point.

9. The quantity under the radical cannot be negative.

$$3x - 5 \geq 0$$
$$x \geq \frac{5}{3}$$

11. Solve the equation.

$$f(x) = 15 = \sqrt{x - 1}$$
$$225 = x - 1$$
$$x = 226$$

14. Replace x with $y - 1$.

$$f(x) = x^2 - x = (y - 1)^2 - (y - 1)$$
$$= y^2 - 2y + 1 - y + 1$$
$$= y^2 - 3y + 2$$

18. $f(x) = x - 1$ when $x \leq -1$
    $f(-4) = -4 - 1 = -5$

19. $f(x) = -2$ when $x > 2$
    $f(4) = -2$

22. $m = \dfrac{y_2 - y_1}{x_2 - x_1} = \dfrac{3 - (-6)}{-1 - (-4)} = \dfrac{9}{3} = 3$

23. $y - y_1 = m(x - x_1)$
    $y - (-6) = 3[x - (-4)]$
    $y + 6 = 3x + 12$
    $y = 3x + 6$

27. $2y + x - 5 = 0$

$$y = -\frac{1}{2}x + \frac{5}{2}$$

The slope of the given line is $m_1 = -1/2$. The slope m of any line perpendicular to the given line is

$$m = -\frac{1}{m_1} = 2$$
$$y - y_1 = m(x - x_1)$$
$$y - 3 = 2(x + 1)$$
$$y = 2x + 5$$

29. $6x^2 - 11x + 4 = (2x - 1)(3x - 4)$

$2x - 1 = 0$ $3x - 4 = 0$

$$x = \frac{1}{2} \qquad x = \frac{4}{3}$$

30.
$$x^2 - 2x = -6$$
$$x^2 - 2x + 1 = -6 + 1$$
$$(x - 1)^2 = -5$$
$$x - 1 = \pm\sqrt{-5}$$
$$x = 1 \pm i\sqrt{5}$$

34. $kx^2 - 3\pi = 0$

$$kx^2 = 3\pi$$

$$x^2 = \frac{3\pi}{k}$$

$$x = \pm\sqrt{\frac{3\pi}{k}} = \pm\frac{\sqrt{3\pi k}}{k}$$

36. $3r^2 - 2r - 5 = 0$

$a = 3, b = -2, c = -5$

$b^2 - 4ac = 64$

Since $b^2 - 4ac$ is positive and a square, the roots are real and rational.

39. $\sqrt{x} + 2 = x$

$\sqrt{x} = x - 2$

$x = x^2 - 4x + 4$

$0 = x^2 - 5x + 4$

$0 = (x - 1)(x - 4)$

$x = 1 \qquad x = 4$

Checking:

$x = 1$	$x = 4$
$\sqrt{1} + 2 \overset{?}{=} 1$	$\sqrt{4} + 2 \overset{?}{=} 4$
$1 + 2 \overset{?}{=} 1$	$2 + 2 \overset{?}{=} 4$
$3 \neq 1$	$4 \overset{\checkmark}{=} 4$

The solution is 4.

41. Let $u = 1 - \dfrac{2}{x}$.

$u^2 - 8u + 15 = 0$

$(u - 3)(u - 5) = 0$

$u = 3 \qquad u = 5$

Substituting:

$$3 = 1 - \frac{2}{x} \qquad 5 = 1 - \frac{2}{x}$$

$$2 = -\frac{2}{x} \qquad 4 = -\frac{2}{x}$$

$$x = -1 \qquad x = -\frac{1}{2}$$

43. $f(x) = -x^2 - 4x$

$ = -(x^2 + 4x \qquad)$

$ = -(x^2 + 4x + 4) + 4$

$ = -(x + 2)^2 + 4$

vertex: $(-2, 4)$

_x_-intercepts

Let $y = f(x) = 0$.

$0 = -x^2 - 4x$

$0 = -x(x + 4)$

$x = 0 \qquad x = -4$

_y_-intercept

Let $x = 0$.

Then $y = f(0) = 0$.

46. $(f \cdot g)(x) = (x + 1)(x^2 - 1)$
$= x^3 + x^2 - x - 1$
$(f \cdot g)(-1) = (-1)^3 + (-1)^2 - (-1) - 1 = 0$

49. $(g \circ f)(x) = g(x + 1) = (x + 1)^2 - 1 = x^2 + 2x$

50. $g(x) = x^2 - 1$
$g(2) = 2^2 - 1 = 3$
$(f \circ g)(2) = f(3) = 3 + 1 = 4$

51. $(f \circ g)(x) = f(x^2) = \sqrt{x^2} - 2 = |x| - 2$

53. $(f \circ g)(-2) = |-2| - 2 = 0$

55. $(f \circ g)(x) = f\left(\dfrac{x}{2} - 2\right) = 2\left(\dfrac{x}{2} - 2\right) + 4 = x$

$(g \circ f)(x) = g(2x + 4) = \dfrac{2x + 4}{2} - 2 = x$

CHAPTER 3

1.
$$
\begin{array}{r|rrrr}
1] & 2 & 0 & 6 & -4 \\
& & 2 & 2 & 8 \\
\hline
& 2 & 2 & 8 & 4 \\
& \underbrace{} & & | \\
& Q(x) & & R
\end{array}
$$

$Q(x) = 2x^2 + 2x + 8; \; R = 4$

3.
$$
\begin{array}{r|rrrr}
-1] & 7 & -3 & 0 & 2 \\
& & -7 & 10 & -10 \\
\hline
& 7 & -10 & 10 & -8
\end{array}
$$
$P(-1) = -8$

$$
\begin{array}{r|rrrr}
2] & 7 & -3 & 0 & 2 \\
& & 14 & 22 & 44 \\
\hline
& 7 & 11 & 22 & 46
\end{array}
$$
$P(2) = 46$

5.
$$
\begin{array}{r|rrrr}
-2] & 2 & 4 & 3 & 5 & -2 \\
& & -4 & 0 & -6 & 2 \\
\hline
& 2 & 0 & 3 & -1 & 0
\end{array}
$$

Since $P(-2) = 0$, $x + 2$ is a factor.

11. The only possible rational roots are ± 1. Using condensed synthetic division, we find

$$
\begin{array}{r|rrrrr}
& 1 & 3 & 2 & 1 & -1 \\
1] & 1 & 4 & 6 & 7 & \boxed{6} \\
-1] & 1 & 2 & 0 & 1 & \boxed{-2}
\end{array}
$$

Since neither remainder is zero, there are no rational roots.

12. Divide by $x + 2$ to find the depressed equation.

$$
\begin{array}{r|rrrr}
-2] & 2 & -1 & -13 & -6 \\
& & -4 & 10 & 6 \\
\hline
& 2 & -5 & -3 & 0 \\
& \underbrace{} \\
& \text{depressed} \\
& \text{equation}
\end{array}
$$

Solving $2x^2 - 5x - 3 = 0$, we have

$$(2x + 1)(x - 3) = 0$$

$$x = -\frac{1}{2} \qquad x = 3$$

17. $\dfrac{8 - x}{2x^2 + 3x - 2} = \dfrac{8 - x}{(2x - 1)(x + 2)} = \dfrac{A}{2x - 1} + \dfrac{B}{x + 2}$

$8 - x = A(x + 2) + B(2x - 1)$

$x = -2$: $\quad 10 = -5B \quad$ or $\quad B = -2$

$x = \dfrac{1}{2}$: $\quad \dfrac{15}{2} = \dfrac{5}{2}A \quad$ or $\quad A = 3$

$\dfrac{8 - x}{2x^2 + 3x - 2} = \dfrac{3}{2x - 1} - \dfrac{2}{x + 2}$

18. $\dfrac{3x^3 + 5x - 1}{(x^2 + 1)^2} = \dfrac{Ax + B}{x^2 + 1} + \dfrac{Cx + D}{(x^2 + 1)^2}$

$3x^3 + 5x - 1 = (Ax + B)(x^2 + 1) + Cx + D$

$3x^3 + 5x - 1 = Ax^3 + Bx^2 + (A + C)x + (B + D)$

coeff. of x^3: $A = 3$

coeff. of x^2: $B = 0$

coeff. of x: $A + C = 5$

$\qquad\qquad 3 + C = 5$

$\qquad\qquad\quad C = 2$

coeff. of x^0: $B + D = -1$

$\qquad\qquad\quad\; D = -1$

$\dfrac{3x^3 + 5x - 1}{(x^2 + 1)^2} = \dfrac{3x}{x^2 + 1} + \dfrac{2x - 1}{(x^2 + 1)^2}$

CHAPTER 4

2. $2^{2x} = 8^{x-1} = (2^3)^{x-1}$ Write in terms of same base.

$2^{2x} = 2^{3x-3}$ $(a^m)^n = a^{mn}$.

$2x = 3x - 3$ If $a^u = a^v$, then $u = v$.

$x = 3$ Solve for x.

4. $S = P(1 + i)^n$ Compound interest formula.

$i = \dfrac{r}{k} = \dfrac{0.12}{2} = 0.6$ Interest rate i per conversion period.

$n = 4 \times 2 = 8$ Number of conversion periods.

$S = 8000(1 + 0.6)^8$ Substitute for P, i, and n.

$\quad = 8000(1.5938)$ Table IV in Tables Appendix or a calculator.

$\quad = \$12{,}750.40$

10. $\log_5 \dfrac{1}{125} = x - 1$

$5^{x-1} = \dfrac{1}{125}$ Equivalent exponential form.

$5^{x-1} = 5^{-3}$ Write in terms of same base.

$x - 1 = -3$ If $a^u = a^v$, then $u = v$.

$x = -2$ Solve for x.

12. $\log_3(x + 1) = \log_3 27$

    $\quad\quad x + 1 = 27$ If $\log_a u = \log_a v$, then $u = v$.

    $\quad\quad\quad\quad x = 26$ Solve for x.

13. $\log_3 3^5 = 5$ Since $\log_a a^x = x$.   16. $e^{\ln 3} = 3$ Since $a^{\log_a x} = x$.

    or   or

        $\log_3 3^5 = x$ Introduce unknown x. $e^{\ln 3} = x$ Introduce logarithmic form.

    $\quad\quad\quad 3^x = 3^5$ Equivalent exponential form. $\ln x = \ln 3$ Equivalent logarithmic form.

    $\quad\quad\quad\quad x = 5$ If $a^u = a^v$, then $u = v$. $x = 3$ If $\log_a u = \log_a v$, then $u = v$.

18. $\log_a \dfrac{\sqrt{x - 1}}{2x} = \log_a \dfrac{(x - 1)^{1/2}}{2x}$ Exponent form of radical.

    $\quad\quad = \log_a (x - 1)^{1/2} - \log_a 2x$ Property 2.

    $\quad\quad = \log_a (x - 1)^{1/2} - [\log_a 2 + \log_a x]$ Property 1.

    $\quad\quad = \dfrac{1}{2} \log_a (x - 1) - \log_a 2 - \log_a x$ Property 3.

22. $\log 14 = \log (2 \cdot 7)$

    $\quad\quad = \log 2 + \log 7$ Property 1.

    $\quad\quad = 0.30 + 0.85 = 1.15$ Substitute given data.

25. $\log 0.7 = \log \dfrac{7}{10}$

    $\quad\quad = \log 7 - \log 10$ Property 2.

    $\quad\quad = 0.85 - 1$ $\log_a a = 1$.

    $\quad\quad = -0.15$

26. $\dfrac{1}{3} \log_a x - \dfrac{1}{2} \log_a y = \log_a x^{1/3} - \log_a y^{1/2}$ Property 3.

    $\quad\quad\quad\quad = \log_a \dfrac{x^{1/3}}{y^{1/2}}$ Property 2.

    $\quad\quad\quad\quad = \log_a \dfrac{\sqrt[3]{x}}{\sqrt{y}}$ Radical form.

27. $\dfrac{4}{3} [\log x + \log (x - 1)]$

    $\quad\quad = \dfrac{4}{3} \log (x)(x - 1)$ Property 1.

    $\quad\quad = \log (x^2 - x)^{4/3}$ Property 3.

30. $\log_b x = \dfrac{\log_a x}{\log_a b}$ Change of base formula.

    $\log_8 32 = \dfrac{\log 32}{\log 8}$ $x = 32, b = 8, a = 10$.

    $\log_8 32 = \dfrac{1.5}{0.9} = \dfrac{5}{3}$ Substitute given data.

    Checking: $8^{5/3} = 32$ Write in equivalent exponent form.

    $\quad\quad\quad\quad\quad 32 = 32$

38. $x = \dfrac{2.1}{(32.5)^{5/2}}$ Introduce unknown x.

$\log x = \log 2.1 - \dfrac{5}{2} \log 32.5$ Properties of logarithms.

$= 0.3222 - \dfrac{5}{2}(1.5119)$ Table II in Tables Appendix.

$= -3.4575$
$= (4 - 3.4575) - 4$ Mantissa must be positive.
$= 0.5425 - 4$
$x \approx 3.49 \times 10^{-4}$ Table II in Tables Appendix.

39. $Q(t) = q_0 e^{-kt}$ Exponential decay model.

$\dfrac{q_0}{2} = q_0 e^{-0.06t}$ Substitute $k = 0.06$ and $Q(t) = \dfrac{1}{2} q_0$.

$\dfrac{1}{2} = e^{-0.06t}$

$\ln \dfrac{1}{2} = \ln e^{-0.06t}$ Take logs of both sides.

$\ln 0.5 = -0.06t(\ln e) = -0.06t$ Property 3, $\ln e = 1$.

$t = \dfrac{\ln 0.5}{-0.06} = \dfrac{-0.6931}{-0.06} = 11.5$ hours   Table III in Tables Appendix.

40. $2^{3x-1} = 14$

$(3x - 1) \log 2 = \log 14$ Take logs of both sides.

$x = \dfrac{1}{3} + \dfrac{\log 14}{3 \log 2}$ Solve for x.

41. $2 \log x - \log 5 = 3$

$\log x^2 - \log 5 = 3$ Property 3.

$\log \dfrac{x^2}{5} = 3$ Property 2.

$\dfrac{x^2}{5} = 10^3 = 1000$ Equivalent exponent form.

$x = \sqrt{5000}$ Solve for x.

CHAPTER 5

1. $W\left(\dfrac{9\pi}{2}\right) = W\left(\dfrac{9\pi}{2} - 4\pi\right) = W\left(\dfrac{\pi}{2}\right)$

$t' = \dfrac{\pi}{2}$

2. $W\left(-\dfrac{15\pi}{2}\right) = W\left(-\dfrac{15\pi}{2} + 8\pi\right) = W\left(\dfrac{\pi}{2}\right)$

$t' = \dfrac{\pi}{2}$

6. $W(7\pi/6)$ is obtained by reflecting $W(\pi/6) = (\sqrt{3}/2, 1/2)$ about the origin. From the accompanying figure, we see that

$$W\left(\frac{7\pi}{6}\right) = \left(-\frac{\sqrt{3}}{2}, -\frac{1}{2}\right)$$

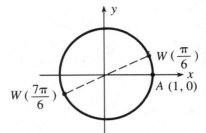

7.
$$W\left(-\frac{8\pi}{3}\right) = W\left(-\frac{8\pi}{3} + 4\pi\right) = W\left(\frac{4\pi}{3}\right)$$

$W(4\pi/3)$ is obtained by reflecting $W(\pi/3) = (1/2, \sqrt{3}/2)$ about the origin. From the accompanying figure, we see that

$$W\left(-\frac{8\pi}{3}\right) = W\left(\frac{4\pi}{3}\right) = \left(-\frac{1}{2}, -\frac{\sqrt{3}}{2}\right)$$

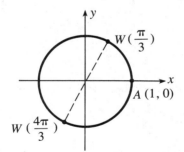

12. If $W(t) = (a, b)$, then $W(t + \pi/2) = (-b, a)$ or $W(t + \pi/2) = (b, -a)$. In the accompanying figure, $W(t + \pi/2)$ is in the first quadrant, so $W(t + \pi/2) = (3/5, 4/5)$.

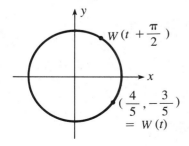

15. $W(-t)$ is the reflection of $W(t)$ about the x-axis. From the accompanying figure, we see that $W(-t) = (4/5, 3/5)$. Then $W(-t - \pi)$ is the reflection of $W(-t)$ about the origin, so $W(-t - \pi) = (-4/5, -3/5)$.

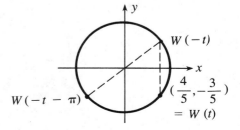

16. $W(2\pi/3)$ is the reflection about the y-axis of $W(\pi/3) = (1/2, \sqrt{3}/2)$. Therefore,

$$W\left(\frac{2\pi}{3}\right) = \left(-\frac{1}{2}, \frac{\sqrt{3}}{2}\right) = \left(\cos\frac{2\pi}{3}, \sin\frac{2\pi}{3}\right)$$

so $\sin 2\pi/3 = \sqrt{3}/2$.

17. $W(-5\pi/4)$ is the reflection of $W(\pi/4) = (\sqrt{2}/2, \sqrt{2}/2)$ about the y-axis. Then

$$W\left(-\frac{5\pi}{4}\right) = \left(-\frac{\sqrt{2}}{2}, \frac{\sqrt{2}}{2}\right) = \left(\cos-\frac{5\pi}{4}, \sin-\frac{5\pi}{4}\right)$$

or

$$\cos\left(-\frac{5\pi}{4}\right) = -\frac{\sqrt{2}}{2}$$

$$\sec\left(-\frac{5\pi}{4}\right) = \frac{1}{\cos\left(-\frac{5\pi}{4}\right)} = -\frac{2}{\sqrt{2}} = -\sqrt{2}$$

20. $W(\pi/4) = (\sqrt{2}/2, \sqrt{2}/2) = (\cos\pi/4, \sin\pi/4)$. Reflecting $W(\pi/4)$ about the origin brings us to the point $W(\pi/4 + \pi) = W(5\pi/4)$ in quadrant III. Then $W(5\pi/4) = (-\sqrt{2}/2, -\sqrt{2}/2) = (\cos 5\pi/4, \sin 5\pi/4)$, so $\sin 5\pi/4 = -\sqrt{2}/2$ and $t = 5\pi/4$.

21. $W(\pi/6) = (\sqrt{3}/2, 1/2) = (\cos\pi/6, \sin\pi/6)$. Reflecting $W(\pi/6)$ about the x-axis brings us to the point $W(-\pi/6) = W(2\pi - \pi/6) = W(11\pi/6)$ in quadrant IV. Then $W(11\pi/6) = (\sqrt{3}/2, -1/2) = (\cos 11\pi/6, \sin 11\pi/6)$, so $\cos 11\pi/6 = \sqrt{3}/2$ and $t = 11\pi/6$.

24. $\sin^2 t + \cos^2 t = 1$

$$\sin^2 t = 1 - \left(\frac{3}{5}\right)^2 = \frac{16}{25}$$

$$\sin t = -\frac{4}{5} \qquad \text{(Since } W(t) \text{ is in quadrant IV, sine is negative.)}$$

$$\tan t = \frac{\sin t}{\cos t} = \frac{-4/5}{3/5} = -\frac{4}{3}$$

$$\cot t = \frac{1}{\tan t} = -\frac{3}{4}$$

25. $\sin^2 t + \cos^2 t = 1$

$$\cos^2 t = 1 - \left(-\frac{4}{5}\right)^2 = \frac{9}{25}$$

$$\cos t = -\frac{3}{5} \qquad \text{(Since } \sin t < 0, \tan t > 0, W(t) \text{ must be in quadrant III.)}$$

$$\sec t = \frac{1}{\cos t} = -\frac{5}{3}$$

29. $\dfrac{\sin t}{\cos^2 t} = \dfrac{\sin t}{\cos t} \cdot \dfrac{1}{\cos t} = (\tan t)(\sec t)$

30. Since $\pi < 3.71 < 3\pi/2$, $W(3.71)$ is in quadrant III. The reference number is
$3.71 - \pi = 3.71 - 3.14 = 0.57$, and

$$\left.\begin{array}{l} \cos 3.71 = -\cos 0.57 = -0.8419 \\ \sin 1.44 = 0.9915 \end{array}\right\} \quad \text{Table V in Tables Appendix.}$$

$$\cos 3.71 - \sin 1.44 = -1.8334$$

34.
$$f(x) = -\cos(2x - \pi) = A\cos(Bx + C)$$

Then $A = -1$, $B = 2$, $C = -\pi$.

$$\text{amplitude} = |A| = 1$$
$$\text{period} = \frac{2\pi}{B} = \frac{2\pi}{2} = \pi$$
$$\text{phase shift} = -\frac{C}{B} = \frac{\pi}{2}$$

Or set $2x - \pi = 0$ to find that $x = \pi/2$ is the phase shift.

35.
$$f(x) = 4\sin\left(-x + \frac{\pi}{2}\right) = A\sin(Bx + C)$$

Since we require $B > 0$, we use the identity $\sin(-t) = -\sin t$ to obtain

$$f(x) = -4\sin\left(x - \frac{\pi}{2}\right) = A\sin(Bx + C)$$

Then $A = -4$, $B = 1$, $C = -\pi/2$.

$$\text{amplitude} = |A| = 4$$
$$\text{period} = \frac{2\pi}{B} = 2\pi$$
$$\text{phase shift} = -\frac{C}{B} = \pi/2$$

Or set $x - \pi/2 = 0$ to find that $x = \pi/2$ is the phase shift.

38. Let $x = \cos^{-1} 1$. Then $\cos x = 1$, $0 \le x \le \pi$, so $x = 0$. Evaluating, we find

$$\tan(\cos^{-1} 1) = \tan 0 = 0$$

39. Let $x = \tan^{-1} 5$. Then $\tan x = 5$, $-\pi/2 < x < \pi/2$, and $\tan(\tan^{-1} 5) = \tan x = 5$. Alternatively, since $f \circ f^{-1}(x) = x$, $\tan(\tan^{-1} 5) = 5$.

40. $5\cos^2 x = 4$

$$\cos^2 x = \frac{4}{5}$$
$$\cos x = \pm\sqrt{\frac{4}{5}} = \pm\frac{2\sqrt{5}}{5}$$
$$x = \arccos\left(\pm\frac{2\sqrt{5}}{5}\right)$$

CHAPTER 6

1. $\dfrac{-60°}{180°} = \dfrac{\theta}{\pi}$

$\theta = -\dfrac{\pi}{3}$ radians

3. $\dfrac{-\dfrac{5\pi}{12}}{\pi} = \dfrac{\theta}{180°}$

$\theta = \left(-\dfrac{5}{12}\right)(180°) = -75°$

5. Convert $\theta = 100°$ to radians:

$\dfrac{100°}{180°} = \dfrac{\theta}{\pi}$

$\theta = \dfrac{5\pi}{9}$ radians

The answer is yes.

7. Convert $5\pi/4$ to degrees:

$\dfrac{\dfrac{5\pi}{4}}{\pi} = \dfrac{\theta}{180°}$

$\theta = 225°$

But $-135°$ is coterminal with

$-135° + 360° = 225°$

The answer is yes.

9. $-185°$ is coterminal with

$-185° + 360° = 175°$

The reference angle for this second-quadrant angle is

$\theta' = 180° - 175° = 5°$

11. $\theta = \dfrac{s}{r} = \dfrac{14}{10} = 1.4$ radians

14. $b^2 = c^2 - a^2 = 25 - 9 = 16$
$b = 4$

$\tan \beta = \dfrac{\text{opposite}}{\text{adjacent}} = \dfrac{4}{3}$

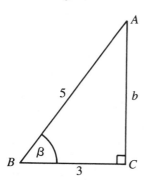

16. hypotenuse $\overline{OP} = \sqrt{3 + 1} = 2$

$\csc \theta' = \dfrac{\text{hypotenuse}}{\text{opposite}} = \dfrac{2}{1} = 2$

Since θ is in quadrant II and cosecant is positive in quadrant II,

$\csc \theta = \csc \theta' = 2$

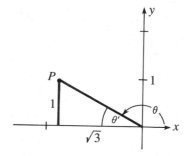

17. $\cot \theta' = \dfrac{\text{adjacent}}{\text{opposite}} = \dfrac{\sqrt{2}}{\sqrt{2}} = 1$

Since θ is in quadrant IV and cotangent is negative in quadrant IV,

$\cot \theta = -\cot \theta' = -1$

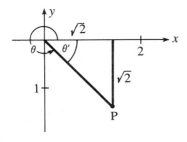

22. $\sin 52° = \dfrac{20}{c}$

$c = \dfrac{20}{\sin 52°} = 25.4$

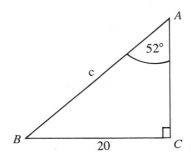

20. $\tan \alpha = \dfrac{50}{60} = 0.8333$

$\alpha = 39°50'$

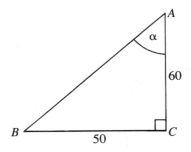

25. $\tan \theta = \dfrac{\text{tree height}}{\text{shadow length}} = \dfrac{25}{10} = 2.5$

$\theta = 68°10'$

27. $a^2 = b^2 + c^2 - 2bc \cos \alpha$

$144 = 49 + 225 - 2(7)(15) \cos \alpha$

$\dfrac{-130}{-210} = 0.6190 = \cos \alpha$

$\alpha = 51°50'$

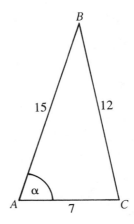

CHAPTER 7

1. $\sin \sigma \sec \sigma + \tan \sigma$

$$= \sin \sigma \, \frac{1}{\cos \sigma} + \tan \sigma$$

$$= \tan \sigma + \tan \sigma$$
$$= 2 \tan \sigma$$

3. $\sin \alpha + \sin \alpha \cot^2 \alpha$
$$= \sin \alpha (1 + \cot^2 \alpha)$$
$$= \sin \alpha \csc^2 \alpha$$

$$= \frac{\csc^2 \alpha}{\csc \alpha} = \csc \alpha$$

5. $\cos(45° + 90°)$
$$= \cos 45° \cos 90° - \sin 45° \sin 90°$$

$$= \left(\frac{\sqrt{2}}{2}\right)(0) - \left(\frac{\sqrt{2}}{2}\right)(1) = -\frac{\sqrt{2}}{2}$$

7. $\sin \dfrac{7\pi}{12} = \sin\left(\dfrac{3\pi}{12} + \dfrac{4\pi}{12}\right) = \sin\left(\dfrac{\pi}{4} + \dfrac{\pi}{3}\right)$

$$= \sin\left(\frac{\pi}{4}\right) \cos\left(\frac{\pi}{3}\right) + \cos\left(\frac{\pi}{4}\right) \sin\left(\frac{\pi}{3}\right)$$

$$= \left(\frac{\sqrt{2}}{2}\right)\left(\frac{1}{2}\right) + \left(\frac{\sqrt{2}}{2}\right)\left(\frac{\sqrt{3}}{2}\right)$$

$$= \frac{\sqrt{2} + \sqrt{6}}{4}$$

8. $\csc 15° = \sec(90° - 15°) = \sec 75°$

10. $\sin \dfrac{\pi}{8} = \cos\left(\dfrac{\pi}{2} - \dfrac{\pi}{8}\right) = \cos \dfrac{3\pi}{8}$

13.

$$\csc\left(\sigma + \frac{\pi}{3}\right) = \frac{1}{\sin\left(\sigma + \pi/3\right)}$$

$$\sin\left(\sigma + \frac{\pi}{3}\right) = \sin \sigma \cos \frac{\pi}{3} + \cos \sigma \sin \frac{\pi}{3}$$

$$\cos \sigma = \frac{1}{\sec \sigma} = \frac{8}{10}$$

$$\sin^2 \sigma = 1 - \cos^2 \sigma = \frac{36}{100}$$

$$\sin \sigma = -\frac{6}{10} \qquad \text{Since } \sigma \text{ is in quadrant IV.}$$

Substituting, we have

$$\sin\left(\sigma + \frac{\pi}{3}\right) = \left(-\frac{6}{10}\right)\left(\frac{1}{2}\right) + \left(\frac{8}{10}\right)\left(\frac{\sqrt{3}}{2}\right) = \frac{4\sqrt{3} - 3}{10}$$

$$\csc\left(\sigma + \frac{\pi}{3}\right) = \frac{10}{4\sqrt{3} - 3} = \frac{10(4\sqrt{3} + 3)}{39}$$

15.

$$\tan(\alpha + \beta) = \frac{\tan \alpha + \tan \beta}{1 - \tan \alpha \tan \beta}, \quad \tan \beta = -\frac{5}{2}$$

$$\tan^2 \alpha = \sec^2 \alpha - 1 = \frac{1}{\cos^2 \alpha} - 1$$

$$= \left(-\frac{13}{12}\right)^2 - 1 = \frac{25}{144}$$

$$\tan \alpha = -\frac{5}{12} \qquad \text{Since } \alpha \text{ is in quadrant II.}$$

Substituting, we have

$$\tan(\alpha + \beta) = 70$$

17.

$$\cos 2u = \cos^2 u - \sin^2 u$$

$$\sin u = \frac{1}{\csc u} = -\frac{4}{5}$$

$$\cos^2 u = 1 - \sin^2 u = 1 - \left(-\frac{4}{5}\right)^2 = \frac{9}{25}$$

Substituting, we have

$$\cos 2u = \frac{9}{25} - \left(-\frac{4}{5}\right)^2 = -\frac{7}{25}$$

19.

$$\sin 4t = \sin 2u, \text{ where } u = 2t$$

$$\sin 2u = 2 \sin u \cos u = 2 \sin 2t \cos 2t$$

$$\sin 2t = \frac{3}{5}$$

$$\cos^2 2t = 1 - \sin^2 2t = \frac{16}{25}$$

$$\cos 2t = \frac{4}{5} \qquad \text{Since } W(2t) \text{ is in quadrant I.}$$

Substituting, we have

$$\sin 4t = 2\left(\frac{3}{5}\right)\left(\frac{4}{5}\right) = \frac{24}{25}$$

21.

$$\cos \frac{\sigma}{2} = \frac{12}{13} = \pm\sqrt{\frac{1 + \cos \sigma}{2}}$$

$$\cos \sigma = 2\left(\frac{12}{13}\right)^2 - 1 = \frac{119}{169}$$

$$\sin^2 \sigma = 1 - \cos^2 \sigma = \frac{14400}{28561}$$

$$\sin \sigma = \frac{120}{169} \qquad \text{Since } \sigma \text{ is acute (quadrant I).}$$

23.

$$\tan \frac{t}{2} = \pm \sqrt{\frac{1 - \cos t}{1 + \cos t}}$$

$$\csc^2 t = \cot^2 t + 1 = \left(-\frac{4}{3}\right)^2 + 1 = \frac{25}{9}$$

$$\sin^2 t = \frac{1}{\csc^2 t} = \frac{9}{25}$$

$$\cos^2 t = 1 - \sin^2 t = \frac{16}{25}$$

$$\cos t = \frac{4}{5} \qquad \text{Since } W(t) \text{ is in quadrant IV.}$$

$$\tan \frac{t}{2} = \pm \sqrt{\frac{1 - \frac{4}{5}}{1 + \frac{4}{5}}} = \pm \sqrt{\frac{1}{9}} = \pm \frac{1}{3}$$

Since $270° < t < 360°$, $135° < t/2 < 180°$, so $t/2$ is in quadrant II. Then $\tan t/2 = -1/3$.

25. $\cos 15° = \cos \dfrac{30°}{2} = \sqrt{\dfrac{1 + \cos 30°}{2}}$

$\cos 15° = \dfrac{1}{2} \sqrt{2 + \sqrt{3}}$

27.

$$\tan 112.5° = \tan \frac{225°}{2} = -\sqrt{\frac{1 - \cos 225°}{1 + \cos 225°}}$$

Since $\cos 225° = -\cos 45° = -\dfrac{\sqrt{2}}{2}$,

$$\tan 112.5° = -\sqrt{\frac{2 + \sqrt{2}}{2 - \sqrt{2}}}$$

28. Let $u = 15x$.
Then $2u = 30x$.

$$\cos 2u = \cos^2 u - \sin^2 u$$
$$= 1 - 2 \sin^2 u$$

Since $u = 15x$,

$$\cos 30x = 1 - 2 \sin^2 15x$$

30.

$$\tan \frac{\alpha}{2} = \pm \sqrt{\frac{1 - \cos \alpha}{1 + \cos \alpha}} \cdot \sqrt{\frac{1 - \cos \alpha}{1 - \cos \alpha}}$$

$$= \pm \frac{1 - \cos \alpha}{\sqrt{1 - \cos^2 \alpha}} = \pm \frac{1 - \cos \alpha}{\sqrt{\sin^2 \alpha}}$$

$$= \pm \frac{1 - \cos \alpha}{\sin \alpha}$$

Since $1 - \cos \alpha \geq 0$ for all α, the sign of $\tan \alpha/2$ is determined by the sign of $\sin \alpha$. Therefore,

$$\tan \frac{\alpha}{2} = \frac{1 - \cos \alpha}{\sin \alpha}$$

31. $\sin \dfrac{3\alpha}{2} \sin \dfrac{\alpha}{2} = \dfrac{1}{2} \left[\cos \left(\dfrac{3\alpha}{2} - \dfrac{\alpha}{2} \right) - \cos \left(\dfrac{3\alpha}{2} + \dfrac{\alpha}{2} \right) \right]$

$\qquad = \dfrac{1}{2} [\cos \alpha - \cos 2\alpha]$

32. $\cos 3x - \cos x = -2 \sin \dfrac{3x + x}{2} \sin \dfrac{3x - x}{2}$

$\qquad\qquad\qquad = -2 \sin 2x \sin x$

36. $\qquad\qquad\qquad 2 \sin \sigma \cos \sigma = 0$

$\qquad\qquad \sin \sigma = 0 \qquad \text{or} \qquad \cos \sigma = 0$

$\qquad\qquad \sigma = 0, \pi \qquad\qquad \sigma = \dfrac{\pi}{2}, \dfrac{3\pi}{2}$

37. $\qquad\qquad\qquad\qquad \sin 2t - \sin t = 0$

$\qquad\qquad\qquad 2 \sin t \cos t - \sin t = 0$

$\qquad\qquad\qquad \sin t(2 \cos t - 1) = 0$

$\qquad \sin t = 0 \qquad \text{or} \qquad 2 \cos t - 1 = 0$

$\qquad t = 0, \pi \qquad\qquad\qquad \cos t = \dfrac{1}{2}$

$\qquad\qquad\qquad\qquad\qquad t = \dfrac{\pi}{3}, \dfrac{5\pi}{3}$

CHAPTER 8

2. $\dfrac{2 + i}{0 - 5i} \left(\dfrac{0 + 5i}{0 + 5i} \right) = \dfrac{10i + 5i^2}{-25i^2} = \dfrac{-5 + 10i}{25} = -\dfrac{1}{5} + \dfrac{2}{5} i$

7. $|2 - i| = \sqrt{2^2 + (-1)^2} = \sqrt{5}$

10. $r = \sqrt{a^2 + b^2} = \sqrt{(-3)^2 + (3)^2} = \sqrt{18} = 3\sqrt{2}$

$\qquad \tan \theta = \dfrac{3}{-3} = -1$

$\qquad \theta = 135°$

$\qquad -3 + 3i = r(\cos \theta + i \sin \theta)$

$\qquad\qquad = 3\sqrt{2}(\cos 135° + i \sin 135°)$

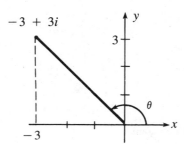

11. $2(\cos 90° + i \sin 90°) = 2(0 + i) = 0 + 2i$

15. $\dfrac{5(\cos 71° + i \sin 71°)}{3(\cos 50° + i \sin 50°)} = \dfrac{5}{3}[\cos(71° - 50°) + i \sin(71° - 50°)]$

$$= \dfrac{5}{3}(\cos 21° + i \sin 21°)$$

17. $r = \sqrt{a^2 + b^2} = \sqrt{3^2 + (-3)^2} = \sqrt{18} = 3\sqrt{2}$

$\tan \theta = \dfrac{-3}{3} = -1$

$\theta = 315°$

$(3 - 3i)^5 = [3\sqrt{2}(\cos 315° + i \sin 315°)]^5$
$= (3\sqrt{2})^5(\cos 1575° + i \sin 1575°)$
$= 972\sqrt{2}(\cos 135° + i \sin 135°)$

$= 972\sqrt{2}\left(-\dfrac{\sqrt{2}}{2} + i\dfrac{\sqrt{2}}{2}\right)$

$= -972 + 972i$

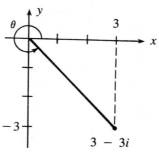

19. In trigonometric form,

$$-9 = 9(\cos 180° + i \sin 180°)$$

so $r = 9$, $\theta = 180°$, and $n = 2$.
The square roots are

$$\sqrt{9}\left[\cos\left(\dfrac{180° + 360°k}{2}\right) + i \sin\left(\dfrac{180° + 360°k}{2}\right)\right]$$

for $k = 0, 1$. Substituting for k, we have

$$3(\cos 90° + i \sin 90°) = 3i$$
$$3(\cos 270° + i \sin 270°) = -3i$$

22. With $\sqrt{-3} = \sqrt{3}i$, form the product:

$$(x - 3)(x - \sqrt{3}i)(x + \sqrt{3}i) = (x - 3)(x^2 + 3)$$
$$= x^3 - 3x^2 + 3x - 9$$

24. The number $1/2$ is a root of the linear factor $(2x - 1)$, and -1 is a root of the linear factor $(x + 1)$. Form the product:

$$(2x - 1)^2(x + 1)^2 = 4x^4 + 4x^3 - 3x^2 - 2x + 1$$

28. The polynomial

$$P(x) = x^5 - x^4 + 3x^3 - 4x^2 + x - 5$$

has 5 variations in sign and therefore has a maximum of 5 positive real roots. The polynomial

$$P(-x) = -x^5 - x^4 - 3x^3 - 4x^2 - x - 5$$

has no variations in sign, and therefore there are no negative real roots.

30. The polynomial

$$P(x) = 3x^4 - 2x^2 + 1$$

has 2 variations in sign, so there can be at most 2 positive real roots. $P(-x) = P(x)$, so there can be at most 2 negative real roots.

31. Since the coefficients are all integers, the Rational Root Theorem restricts the possible rational roots to

$$\pm 1, \quad \pm\frac{1}{2}, \quad \pm\frac{1}{3}, \quad \pm\frac{1}{6}, \quad \pm 2, \quad \pm\frac{2}{3}, \quad \pm 5, \quad \pm\frac{5}{2}, \quad \pm\frac{5}{3}, \quad \pm\frac{5}{6}, \quad \pm 10, \quad \pm\frac{10}{3}$$

Testing by synthetic division,

$$
\begin{array}{r|rrrr}
-1 & 6 & 15 & -1 & -10 \\
   &   & -6 & -9 & 10 \\
\hline
   & 6 & 9 & -10 & 0
\end{array}
$$

we show that -1 is a root. The remaining roots are those of the depressed equation

$$6x^2 + 9x - 10 = 0$$

and are found by the quadratic formula:

$$x = \frac{-9 \pm \sqrt{81 + 240}}{12} = \frac{-9 \pm \sqrt{321}}{12}$$

CHAPTER 9

1. $x = \dfrac{x_1 + x_2}{2} = \dfrac{-5 + 3}{2} = -1$

    $y = \dfrac{y_1 + y_2}{2} = \dfrac{4 + (-6)}{2} = -1$

    The midpoint is $(-1, -1)$.

3. By the midpoint formula,

    $$2 = \frac{x - 6}{2} \quad \text{and} \quad 2 = \frac{y - 3}{2}$$

    so $x = 10$ and $y = 7$.

6. We first find the coordinates of the midpoint of AB:

    $$x = \frac{x_1 + x_2}{2} = \frac{-4 + 1}{2} = -\frac{3}{2}$$

    $$y = \frac{y_1 + y_2}{2} = \frac{-3 + 3}{2} = 0$$

    The slope of AB is

    $$m_{AB} = \frac{y_2 - y_1}{x_2 - x_1} = \frac{3 - (-3)}{1 - (-4)} = \frac{6}{5}$$

    Then the slope of the perpendicular bisector of AB is $-5/6$. Using the point-slope form for the equation of a line

    $$y - y_1 = m(x - x_1)$$

we now obtain an equation of the perpendicular bisector:

$$y - 0 = -\frac{5}{6}\left[x - \left(-\frac{3}{2}\right)\right]$$

$$y = -\frac{5}{6}\left(x + \frac{3}{2}\right)$$

$$y = -\frac{5}{6}x - \frac{5}{4}$$

10. Completing the square in both x and y, we have

$$x^2 + 4x + y^2 - 6y = -10$$
$$(x + 2)^2 + (y - 3)^2 = -10 + 4 + 9 = 3$$

The center is at $(-2, 3)$, and the radius is $\sqrt{3}$.

11. We have $4p = -2/3$, so $p = -1/6$. Thus, the focus is at $(0, -1/6)$, and the directrix is $y = 1/6$.

13. Since the directrix is $y = 7/4$, $p = -7/4$ and $4p = -7$. The equation of the parabola is then

$$x^2 = 4py = -7y$$

16. Completing the squares in both x and y, we have

$$4(x^2 + 2x) + y^2 - 2y = 4$$
$$4(x + 1)^2 + (y - 1)^2 = 4 + 4 + 1 = 9$$
$$\frac{(x + 1)^2}{\frac{9}{4}} + \frac{(y - 1)^2}{9} = 1$$

The center of the ellipse is at $(-1, 1)$. Since $a > b$, $a^2 = 9$ and $b^2 = 9/4$, and the major axis is on the line $x = -1$. The vertices are at $(-1, 1 \pm a)$ and $(-1 \pm b, 1)$, or $(-1, 4)$, $(-1, -2)$, $(1/2, 1)$, and $(-5/2, 1)$. Also,

$$c^2 = a^2 - b^2 = 9 - \frac{9}{4} = \frac{27}{4}$$

The foci are on the major axis at $(-1, 1 \pm c)$, or $(-1, 1 \pm 3\sqrt{3}/2)$.

17. Since the foci are at $(\pm 3, 0)$, the center must be at the origin. We are given $a^2 = 81$ and $c^2 = 9$. Then $b^2 = a^2 - c^2 = 81 - 9 = 72$. An equation of the ellipse is then

$$\frac{x^2}{81} + \frac{y^2}{72} = 1$$

19. Completing the squares in both x and y, we have

$$4x^2 - (y^2 + 2y) = 4$$
$$4x^2 - (y + 1)^2 = 4 - 1 = 3$$
$$\frac{x^2}{\frac{3}{4}} - \frac{(y + 1)^2}{3} = 1$$

The center of the hyperbola is at $(0, -1)$; $a^2 = 3/4$ and $b^2 = 3$. The vertices are at $(\pm\sqrt{3}/2, -1)$. Since $c^2 = a^2 + b^2$, we have $c^2 = 3/4 + 3 = 15/4$. Thus, the foci are at $(\pm\sqrt{15}/2, -1)$. The asymptotes are $y + 1 = \pm 2x$.

21. We are given $a^2 = 4$ and $c^2 = 49$, so $b^2 = c^2 - a^2 = 49 - 4 = 45$. An equation of the hyperbola is then

$$\frac{y^2}{4} - \frac{x^2}{45} = 1$$

29. We eliminate the xy term by performing a rotation of axes. Since $\cot 2\theta = 0$, $2\theta = \dfrac{\pi}{2}$ and $\theta = \dfrac{\pi}{4}$. Then

$$x = x' \cos\theta - y' \sin\theta = \frac{\sqrt{2}}{2}x' - \frac{\sqrt{2}}{2}y'$$

$$y = x' \sin\theta + y' \cos\theta = \frac{\sqrt{2}}{2}x' + \frac{\sqrt{2}}{2}y'$$

Then

$$\left(\frac{\sqrt{2}}{2}x' - \frac{\sqrt{2}}{2}y'\right)\left(\frac{\sqrt{2}}{2}x' + \frac{\sqrt{2}}{2}y'\right) + 2\left(\frac{\sqrt{2}}{2}x' - \frac{\sqrt{2}}{2}y'\right) - 2\left(\frac{\sqrt{2}}{2}x' + \frac{\sqrt{2}}{2}y'\right) = 6$$

$$= x'^2 - y'^2 - 4\sqrt{2}y' = 12$$

$$x'^2 - (y' + 2\sqrt{2})^2 = 4$$

$$\frac{x'^2}{4} - \frac{(y' + 2\sqrt{2})^2}{4} = 1$$

The conic section is a hyperbola.

31. We eliminate the xy term by performing a rotation of axes. Since

$$\cot 2\theta = \frac{A - C}{B} = \frac{3}{4}$$

from the accompanying figure, $\cos 2\theta = 3/5$. Then

$$\cos\theta = \sqrt{\frac{1 + \cos 2\theta}{2}} = \sqrt{\frac{1 + 3/5}{2}} = \frac{2}{\sqrt{5}}$$

$$\sin\theta = \sqrt{\frac{1 - \cos 2\theta}{2}} = \sqrt{\frac{1 - 3/5}{2}} = \frac{1}{\sqrt{5}}$$

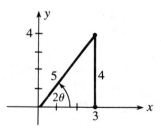

and

$$x = x' \cos\theta - y' \sin\theta = \frac{2}{\sqrt{5}}x' - \frac{1}{\sqrt{5}}y'$$

$$y = x' \sin\theta + y' \cos\theta = \frac{1}{\sqrt{5}}x' + \frac{2}{\sqrt{5}}y'$$

Substituting these equations in the given equation of the conic, we have

$$2\left(\frac{2}{\sqrt{5}}x' - \frac{1}{\sqrt{5}}y'\right)^2 - \left(\frac{1}{\sqrt{5}}x' + \frac{2}{\sqrt{5}}y'\right)^2 + 4\left(\frac{2}{\sqrt{5}}x' - \frac{1}{\sqrt{5}}y'\right)\left(\frac{1}{\sqrt{5}}x' + \frac{2}{\sqrt{5}}y'\right)$$

$$-2\left(\frac{2}{\sqrt{5}}x' - \frac{1}{\sqrt{5}}y'\right) + 3\left(\frac{1}{\sqrt{5}}x' + \frac{2}{\sqrt{5}}y'\right) = 6$$

$$3x'^2 - 2y'^2 - \frac{1}{\sqrt{5}}x' + \frac{8}{\sqrt{5}}y' = 6$$

$$3\left(x'^2 - \frac{1}{3\sqrt{5}}x'\right) - 2\left(y'^2 + \frac{4y'}{\sqrt{5}}\right) = 6$$

Completing the squares in both x' and y', we obtain

$$3\left(x' - \frac{1}{6\sqrt{5}}\right)^2 - 2\left(y' + \frac{2}{\sqrt{5}}\right)^2 = 6 + \frac{1}{60} - \frac{8}{5} = \frac{265}{60}$$

The conic section is a hyperbola.

37. Since $r = 3 \csc \theta$, $r = \dfrac{3}{\sin \theta}$, or $r \sin \theta = 3$, so $y = 3$.

38. We have

$$r^2 \cos^2 \theta + 9r^2 \sin^2 \theta = 9$$

$$r^2 = \frac{9}{\cos^2 \theta + 9 \sin^2 \theta}$$

40. We have

$$\frac{x}{y} = \sqrt{t - 1} \quad \text{or} \quad t = 1 + \frac{x^2}{y^2}$$

Then

$$y^2(t + 2) = 1$$

Substituting the expression for t in this last equation, we have

$$y^2(t + 2) = 1$$

$$y^2\left(3 + \frac{x^2}{y^2}\right) = 1$$

$$x^2 + 3y^2 = 1$$

$$x^2 + \frac{y^2}{\dfrac{1}{3}} = 1$$

which is the standard form of the equation of an ellipse with its center at the origin.

CHAPTER 10

3. Substituting $x = 5 - 3y$, we have

$$(5 - 3y)^2 + y^2 = 25$$

$$25 - 30y + 9y^2 + y^2 = 25$$

$$10y^2 - 30y = 0$$
$$10y(y - 3) = 0$$

$$y = 0 \qquad \text{or} \qquad y = 3$$
$$x = 5 - 3y = 5 \qquad x = 5 - 3y = -4$$

7. To eliminate x, multiply the first equation by 3 and the second equation by 2, and add:

$$6x + 9y = -3$$
$$-6x + 8y = -\frac{11}{2}$$
$$\overline{ \quad 17y = -\frac{17}{2}}$$

$$y = -\frac{1}{2}$$

$$2x + 3\left(-\frac{1}{2}\right) = -1$$

$$x = \frac{1}{4}$$

8. Rewriting the equations and adding,

$$x^2 + y^2 - 9 = 0$$
$$-x^2 + y - 3 = 0$$
$$\overline{ y^2 + y - 12 = 0}$$
$$(y - 3)(y + 4) = 0$$

$$y = 3 \qquad \text{or} \qquad y = -4$$
$$x^2 = y - 3 = 0 \qquad x^2 = y - 3 = -7$$
$$x = 0 \qquad\qquad \text{no real solutions}$$

The circle and parabola are tangent at $(0, 3)$.

18. The figure shows the set of feasible solutions and the coordinates of the vertices. Evaluating the objective function at these points gives us the following information:

x	y	$z = 5y - x$
0	1	5
0	$\frac{9}{2}$	$\frac{45}{2}$
4	6	26
$\frac{29}{6}$	1	$\frac{1}{6}$

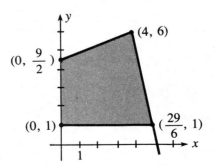

The maximum value, $z = 26$, occurs at $x = 4$, $y = 6$.

20. Interchange equations 1 and 3:

$$-x + 4y + 2z = 15$$
$$2x + 5y - 2z = -9$$
$$-3x - y + z = 12$$

Multiply equation 1 by 2 and add it to equation 2; multiply equation 1 by -3 and add it to equation 3:

$$-x + 4y + 2z = 15$$
$$13y + 2z = 21$$
$$-13y - 5z = -33$$

Add equation 2 to equation 3:

$$-x + 4y + 2z = 15$$
$$13y + 2z = 21$$
$$-3z = -12$$

Use back-substitution:

$$-3z = -12 \quad \text{or} \quad z = 4$$
$$13y + 2(4) = 21 \quad \text{or} \quad y = 1$$
$$-x + 4(1) + 2(4) = 15 \quad \text{or} \quad x = -3$$
$$x = -3, \quad y = 1, \quad z = 4$$

32. From the third row, $x_3 = -3$. Then, from row 2,

$$x_2 + 3x_3 = -8$$
$$x_2 + 3(-3) = -8$$
$$x_2 = 1$$

From row 1,

$$x_1 - 2x_2 + 2x_3 = -9$$
$$x_1 - 2(1) + 2(-3) = -9$$
$$x_1 = -1, \quad x_2 = 1, \quad x_3 = -3$$

33. In matrix form,

$$\begin{bmatrix} 1 & 1 & | & 2 \\ 2 & -4 & | & -5 \end{bmatrix}$$

Add -2 times row 1 to row 2:

$$\begin{bmatrix} 1 & 1 & | & 2 \\ 0 & -6 & | & -9 \end{bmatrix}$$

Multiply row 2 by $-1/6$:

$$\begin{bmatrix} 1 & 1 & | & 2 \\ 0 & 1 & | & 3/2 \end{bmatrix}$$

Add -1 times row 2 to row 1:

$$\begin{bmatrix} 1 & 0 & | & 1/2 \\ 0 & 1 & | & 3/2 \end{bmatrix}$$

The solution is $x = 1/2$, $y = 3/2$.

36. Two matrices of the same dimension are equal if corresponding entries are equal. This requires that

$$x^2 = 9 \quad \text{and} \quad 4x = -12$$

The only value satisfying both equations is $x = -3$.

38. $B - A = \begin{bmatrix} -1 & 5 \\ 4 & -3 \end{bmatrix} - \begin{bmatrix} 2 & -1 \\ 3 & 2 \end{bmatrix} = \begin{bmatrix} -1-2 & 5-(-1) \\ 4-3 & -3-2 \end{bmatrix} = \begin{bmatrix} -3 & 6 \\ 1 & -5 \end{bmatrix}$

39. Addition of matrices is defined only when the matrices are of the same dimension.

44.

$$\begin{bmatrix} -2 & 3 & | & 1 & 0 \\ 1 & 4 & | & 0 & 1 \end{bmatrix}$$

$$\begin{bmatrix} 1 & 4 & | & 0 & 1 \\ -2 & 3 & | & 1 & 0 \end{bmatrix} \qquad \text{Interchanged rows.}$$

$$\begin{bmatrix} 1 & 4 & | & 0 & 1 \\ 0 & 11 & | & 1 & 2 \end{bmatrix} \qquad \begin{array}{l}\text{Added 2 times} \\ \text{row 1 to row 2.}\end{array}$$

$$\begin{bmatrix} 1 & 4 & | & 0 & 1 \\ 0 & 1 & | & 1/11 & 2/11 \end{bmatrix} \qquad \begin{array}{l}\text{Multiplied row 2} \\ \text{by } 1/11.\end{array}$$

$$\begin{bmatrix} 1 & 0 & | & -4/11 & 3/11 \\ 0 & 1 & | & 1/11 & 2/11 \end{bmatrix} \qquad \begin{array}{l}\text{Added } -4 \text{ times} \\ \text{row 2 to row 1.}\end{array}$$

$$\begin{bmatrix} -4/11 & 3/11 \\ 1/11 & 2/11 \end{bmatrix} \qquad \text{Inverse.}$$

46. Appending the identity matrix I_2 to the coefficient matrix, we have

$$\begin{bmatrix} 2 & -1 & | & 1 & 0 \\ 1 & 1 & | & 0 & 1 \end{bmatrix}$$

$$\begin{bmatrix} 1 & 1 & | & 0 & 1 \\ 2 & -1 & | & 1 & 0 \end{bmatrix} \qquad \text{Interchanged rows.}$$

$$\begin{bmatrix} 1 & 1 & | & 0 & 1 \\ 0 & -3 & | & 1 & -2 \end{bmatrix} \qquad \begin{array}{l}\text{Added } -2 \text{ times} \\ \text{row 1 to row 2.}\end{array}$$

$$\begin{bmatrix} 1 & 1 & | & 0 & 1 \\ 0 & 1 & | & -1/3 & 2/3 \end{bmatrix} \qquad \text{Multiplied row 2 by } -1/3.$$

$$\begin{bmatrix} 1 & 0 & | & 1/3 & 1/3 \\ 0 & 1 & | & -1/3 & 2/3 \end{bmatrix} \qquad \begin{array}{l}\text{Added } -1 \text{ times} \\ \text{row 2 to row 1.}\end{array}$$

Multiplying the inverse by the coefficients of the right-hand side, we have

$$\begin{bmatrix} 1/3 & 1/3 \\ -1/3 & 2/3 \end{bmatrix} \begin{bmatrix} 1 \\ 5 \end{bmatrix} = \begin{bmatrix} 2 \\ 3 \end{bmatrix}$$

so $x = 2$, $y = 3$ is the unique solution.

50. Expanding by the cofactors of the first column, we have

$$1 \begin{vmatrix} 5 & 4 \\ 3 & 8 \end{vmatrix} + 2 \begin{vmatrix} -1 & 2 \\ 5 & 4 \end{vmatrix} = (40 - 12) + 2(-4 - 10) = 0$$

52.
$$D = \begin{vmatrix} 2 & -1 \\ -2 & 3 \end{vmatrix} = 6 - 2 = 4$$

$$x = \frac{\begin{vmatrix} -3 & -1 \\ 11 & 3 \end{vmatrix}}{4} = \frac{2}{4} = \frac{1}{2}$$

$$y = \frac{\begin{vmatrix} 2 & -3 \\ -2 & 11 \end{vmatrix}}{4} = \frac{16}{4} = 4$$

CHAPTER 11

3. $a_n = n - a_{n-1}$
$a_1 = 0$
$a_2 = 2 - 0 = 2$
$a_3 = 3 - 2 = 1$
$a_4 = 4 - 1 = 3$
$a_5 = 5 - 3 = 2$

5. $\displaystyle\sum_{k=1}^{4} (1 - 2k) = (1 - 2) + (1 - 4) + (1 - 6) + (1 - 8) = -16$

11. $a_n = a_1 + (n - 1)d$
$a_{21} = -2 + (21 - 1)(2) = 38$

15. $S_n = \dfrac{n}{2}[2a_1 + (n - 1)d]$

$$= \frac{25}{2}\left[-\frac{2}{3} + 24\left(\frac{1}{3}\right)\right] = \frac{275}{3}$$

17. $r = \dfrac{a_2}{a_1} = \dfrac{-6}{2} = -3$

19. $a_2 = a_1 r = 5\left(\dfrac{1}{5}\right) = 1$

$a_3 = a_2 r = 1\left(\dfrac{1}{5}\right) = \dfrac{1}{5}$

$a_4 = a_3 r = \left(\dfrac{1}{5}\right)\left(\dfrac{1}{5}\right) = \dfrac{1}{25}$

21. $r = \dfrac{a_2}{a_1} = \dfrac{6}{-4} = -\dfrac{3}{2}$

$a_n = a_1 r^{n-1}$

$a_6 = (-4)\left(-\dfrac{3}{2}\right)^5 = \dfrac{243}{8}$

23. The sequence is

$$3, a_2, a_3, 1/72$$

With $a_1 = 3$, $a_4 = 1/72$, and $n = 4$,

$$a_n = a_1 r^{n-1}$$
$$a_4 = a_1 r^3$$
$$r^3 = \frac{1}{216}$$
$$r = \frac{1}{6}$$

Then

$$a_2 = a_1 r = 3\left(\frac{1}{6}\right) = \frac{1}{2}$$

$$a_3 = a_2 r = \left(\frac{1}{2}\right)\left(\frac{1}{6}\right) = \frac{1}{12}$$

24. $r = \dfrac{a_2}{a_1} = \dfrac{1}{2}$

$$S_n = \frac{a_1(1 - r^n)}{1 - r}$$

$$S_6 = \frac{\dfrac{1}{3}\left[1 - \left(\dfrac{1}{2}\right)^6\right]}{1 - \dfrac{1}{2}} = \frac{21}{32}$$

27. $r = \dfrac{a_2}{a_1} = -\dfrac{2}{3}$

$$S = \frac{a_1}{1 - r} = \frac{3}{1 - \left(-\dfrac{2}{3}\right)} = \frac{9}{5}$$

30. By the binomial formula,

$$\left(\frac{x}{2} - 2\right)^4 = \left(\frac{x}{2}\right)^4 + \frac{4}{1}\left(\frac{x}{2}\right)^3(-2) + \frac{4 \cdot 3}{1 \cdot 2}\left(\frac{x}{2}\right)^2(-2)^2 + \frac{4 \cdot 3 \cdot 2}{1 \cdot 2 \cdot 3}\left(\frac{x}{2}\right)(-2)^3 + (-2)^4$$

$$= \frac{x^4}{16} - x^3 + 6x^2 - 16x + 16$$

33. $\dfrac{13!}{11!2!} = \dfrac{13 \cdot 12 \cdot 11!}{11!2!} = \dfrac{13 \cdot 12}{2} = 78$

34. $\dfrac{(n - 1)!(n + 1)!}{n!n!} = \dfrac{(n - 1)!(n + 1)n!}{n!n(n - 1)!} = \dfrac{n + 1}{n}$

35. $\dbinom{6}{4} = \dfrac{6!}{4!2!} = \dfrac{6 \cdot 5}{2} = 15$

INDEX

  4
B 5
C 6
D 7
E 8
F 9
G 0

TRIGONOMETRIC FUNCTIONS OF REAL NUMBERS

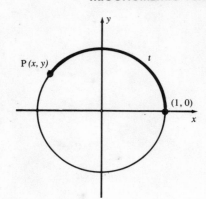

$$\sin t = y \qquad\qquad \csc t = \frac{1}{y}$$

$$\cos t = x \qquad\qquad \sec t = \frac{1}{x}$$

$$\tan t = \frac{y}{x} \qquad\qquad \cot t = \frac{x}{y}$$

TRIGONOMETRIC FUNCTIONS OF ANGLES

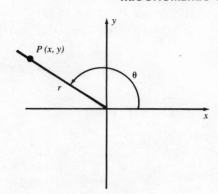

$$\sin \theta = \frac{y}{r} \qquad\qquad \csc \theta = \frac{r}{y}$$

$$\cos \theta = \frac{x}{r} \qquad\qquad \sec \theta = \frac{r}{x}$$

$$\tan \theta = \frac{y}{x} \qquad\qquad \cot \theta = \frac{x}{y}$$

RIGHT TRIANGLE TRIGONOMETRY

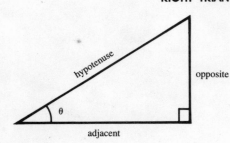

$$\sin \theta = \frac{\text{opposite}}{\text{hypotenuse}} \qquad \csc \theta = \frac{\text{hypotenuse}}{\text{opposite}}$$

$$\cos \theta = \frac{\text{adjacent}}{\text{hypotenuse}} \qquad \sec \theta = \frac{\text{hypotenuse}}{\text{adjacent}}$$

$$\tan \theta = \frac{\text{opposite}}{\text{adjacent}} \qquad \cot \theta = \frac{\text{adjacent}}{\text{opposite}}$$

LAW OF COSINES

$$a^2 = b^2 + c^2 - 2bc \cos \alpha$$

LAW OF SINES

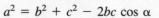

$$\frac{a}{\sin \alpha} = \frac{b}{\sin \beta} = \frac{c}{\sin \gamma}$$